ALGEBRAIC TOPOLOGY

C. R. F. MAUNDER
Christ's College, University of Cambridge
Cambridge, England

DOVER PUBLICATIONS, INC.
Mineola, New York

Copyright

Published in Canada by General Publishing Company, Ltd., 30 Lesmill Road, Don Mills, Toronto, Ontario.
Published in the United Kingdom by Constable and Company, Ltd., 3 The Lanchesters, 162–164 Fulham Palace Road, London W6 9ER.

Bibliographical Note

This Dover edition, first published in 1996, is an unabridged, unaltered republication of the corrected printing, published by the Cambridge University Press, Cambridge, England, 1980, of the work first published by the Van Nostrand Reinhold Company, London, 1970.

Library of Congress Cataloging-In-Publication Data

Maunder, C. R. F.
Algebraic topology / C.R.F. Maunder.
p. cm.
Originally published: London : Van Nostrand Reinhold Co., 1970.
Includes bibliographical references and index.
ISBN 0-486-69131-4 (pbk.)
1. Algebraic topology. I. Title.
QA612.M38 1996
514′.2—dc20 95-51359
CIP

Manufactured in the United States of America
Dover Publications, Inc., 31 East 2nd Street, Mineola, N.Y. 11501

INTRODUCTION

Most of this book is based on lectures to third-year undergraduate and first-year postgraduate students. It aims to provide a thorough grounding in the more elementary parts of algebraic topology, although these are treated wherever possible in an up-to-date way. The reader interested in pursuing the subject further will find suggestions for further reading in the notes at the end of each chapter.

Chapter 1 is a survey of results in algebra and analytic topology that will be assumed known in the rest of the book. The knowledgeable reader is advised to read it, however, since in it a good deal of standard notation is set up. Chapter 2 deals with the topology of simplicial complexes, and Chapter 3 with the fundamental group. The subject of Chapters 4 and 5 is homology and cohomology theory (particularly of simplicial complexes), with applications including the Lefschetz Fixed-Point Theorem and the Poincaré and Alexander duality theorems for triangulable manifolds. Chapters 6 and 7 are concerned with homotopy theory, homotopy groups and CW-complexes, and finally in Chapter 8 we shall consider the homology and cohomology of CW-complexes, giving a proof of the Hurewicz theorem and a treatment of products in cohomology.

A feature of this book is that we have included in Chapter 2 a proof of Zeeman's version of the *relative* Simplicial Approximation Theorem. We believe that the small extra effort needed to prove the relative rather than the absolute version of this theorem is more than repaid by the easy deduction of the equivalence of singular and simplicial homology theory for polyhedra.

Each chapter except the first contains a number of exercises, most of which are concerned with further applications and extensions of the theory. There are also notes at the end of each chapter, which are partly historical and partly suggestions for further reading.

Each chapter is divided into numbered sections, and Definitions, Propositions, Theorems, etc., are numbered consecutively within each section: thus for example Definition 1.2.6 follows Theorem 1.2.5 in the second section (Section 1.2) of Chapter 1. A reference to Exercise n denotes Exercise n at the end of the chapter in which the reference is made; if reference is made to an exercise in a different chapter, then the number of that chapter will also be specified. The symbol ▮ denotes the end (or absence) of a proof, and is also used to indicate the end of an example in the text. References are listed and numbered at the end of the book, and are referred to in the text by numbers in brackets: thus for example [73] denotes the book *Homotopy Theory* by S.-T. Hu.

Finally, it is a pleasure to acknowledge the help I have received in writing this book. My indebtedness to the books of Seifert and Threlfall [124] and Hu [73], and papers by Puppe [119], G. W. Whitehead [155], J. H. C. Whitehead [160] and Zeeman [169] will be obvious to anyone who has read them, but I should also like to thank D. Barden, R. Brown, W. B. R. Lickorish, N. Martin, R. Sibson, A. G. Tristram and the referee for many valuable conversations and suggestions.

CONTENTS

Introduction

CHAPTER 1 ALGEBRAIC AND TOPOLOGICAL PRELIMINARIES

1.1 Introduction 1
1.2 Set theory 1
1.3 Algebra 3
1.4 Analytic topology 15

CHAPTER 2 HOMOTOPY AND SIMPLICIAL COMPLEXES

2.1 Introduction 23
2.2 The classification problem; homotopy 23
2.3 Simplicial complexes 31
2.4 Homotopy and homeomorphism of polyhedra 40
2.5 Subdivision and the Simplicial Approximation Theorem 45
Exercises 59
Notes on Chapter 2 61

CHAPTER 3 THE FUNDAMENTAL GROUP

3.1 Introduction 63
3.2 Definition and elementary properties of the fundamental group 63
3.3 Methods of calculation 71
3.4 Classification of triangulable 2-manifolds 87
Exercises 99
Notes on Chapter 3 103

CHAPTER 4 HOMOLOGY THEORY

4.1 Introduction 104
4.2 Homology groups 104
4.3 Methods of calculation: simplicial homology 113
4.4 Methods of calculation: exact sequences 124

4.5 Homology groups with arbitrary coefficients, and the Lefschetz Fixed-Point Theorem 141
Exercises 151
Notes on Chapter 4 156

CHAPTER 5 COHOMOLOGY AND DUALITY THEOREMS

5.1 Introduction 158
5.2 Definitions and calculation theorems 159
5.3 The Alexander–Poincaré Duality Theorem 170
5.4 Manifolds with boundary and the Lefschetz Duality Theorem 184
Exercises 195
Notes on Chapter 5 198

CHAPTER 6 GENERAL HOMOTOPY THEORY

6.1 Introduction 200
6.2 Some geometric constructions 201
6.3 Homotopy classes of maps 222
6.4 Exact sequences 233
6.5 Fibre and cofibre maps 243
Exercises 250
Notes on Chapter 6 254

CHAPTER 7 HOMOTOPY GROUPS AND CW-COMPLEXES

7.1 Introduction 257
7.2 Homotopy groups 258
7.3 CW-complexes 273
7.4 Homotopy groups of CW-complexes 287
7.5 The theorem of J. H. C. Whitehead and the Cellular Approximation Theorem 298
Exercises 304
Notes on Chapter 7 309

CHAPTER 8
HOMOLOGY AND COHOMOLOGY OF CW-COMPLEXES

8.1 Introduction 311
8.2 The Excision Theorem and cellular homology 311
8.3 The Hurewicz theorem 322

8.4 Cohomology and Eilenberg–MacLane spaces 330
8.5 Products 339
Exercises 352
Notes on Chapter 8 359

References 361

Index 369

CHAPTER 1

ALGEBRAIC AND TOPOLOGICAL PRELIMINARIES

1.1 Introduction

In this chapter we collect together some elementary results in set theory, algebra and analytic topology that will be assumed known in the rest of the book. Since the reader will probably be familiar with most of these results, we shall usually omit proofs and give only definitions and statements of theorems. Proofs of results in set theory and analytic topology will be found in Kelley [85], and in algebra in Jacobson [77]; or indeed in almost any other standard textbook. It will be implicitly assumed that the reader is familiar with the concepts of sets (and subsets), integers, and rational, real and complex numbers.

1.2 Set theory

The notation $a \in A$ means that a is an element of the set A; $A \subset B$ that A is a subset of B. $\{a \in A \mid \ldots\}$ means the subset of A such that $\ldots$ is true, and if A, B are subsets of some set C, then $A \cup B$, $A \cap B$ denote the *union* and *intersection* of A and B respectively: thus $A \cup B = \{c \in C \mid c \in A \text{ or } c \in B\}$ and $A \cap B = \{c \in C \mid c \in A \text{ and } c \in B\}$. Unions and intersections of arbitrary collections of sets are similarly defined.

Definition 1.2.1 Given sets A and B, the *product set* $A \times B$ is the set of all ordered pairs (a, b), for all $a \in A$, $b \in B$. A *relation* between the sets A and B is a subset R of $A \times B$; we usually write aRb for the statement '$(a, b) \in R$'.

Definition 1.2.2 A *partial ordering* on a set A is a relation $<$ between A and itself such that, whenever $a < b$ and $b < c$, then $a < c$. A *total ordering* on A is a partial ordering $<$ such that
(a) if $a < b$ and $b < a$, then $a = b$;
(b) given $a, b \in A$, either $a < b$ or $b < a$.

Proposition 1.2.3 *Given a finite set A containing n distinct elements, there exist $n!$ distinct total orderings on A.* ▮

Definition 1.2.4 A relation R between a set A and itself is called an *equivalence relation on A* if

(a) for all $a \in A$, aRa;
(b) if aRb, then bRa;
(c) if aRb and bRc, then aRc.

The *equivalence class* $[a]$ of an element $a \in A$ is defined by $[a] = \{b \in A \mid aRb\}$.

Theorem 1.2.5 *If R is an equivalence relation on A, then each element of A is in one and only one equivalence class.* ∎

Definition 1.2.6 Given sets A and B, a *function f* from A to B is a relation between A and B such that, for each $a \in A$, there exists a unique $b \in B$ such that afb. We write $b = f(a)$, or $f(a) = b$, for the statement 'afb', and $f: A \to B$ for 'f is a function from A to B'.

Example 1.2.7 Given any set A, the *identity function* $1_A: A \to A$ is defined by $1_A(a) = a$ for all $a \in A$ (we shall often abbreviate 1_A to 1, if no ambiguity arises). ∎

Definition 1.2.8 If $f: A \to B$ is a function and C is a subset of A, the *restriction* $(f|C): C \to B$ is defined by $(f|C)(c) = f(c)$ for all $c \in C$. Given two functions $f: A \to B$, $g: B \to C$, the *composite function* $gf: A \to C$ is defined by $gf(a) = g(f(a))$. The *image* $f(A)$ of $f: A \to B$ is the subset of B of elements of the form $f(a)$, for some $a \in A$; f is *onto* if $f(A) = B$; f is *one-to-one* (written (1-1) if, whenever $f(a_1) = f(a_2)$, then $a_1 = a_2$; f is a (1-1)-*correspondence* if it is both onto and (1-1). Two sets A and B are said to be *in* (1-1)-*correspondence* if there exists a (1-1)-correspondence $f: A \to B$.

Proposition 1.2.9 *Let $f: A \to B$ be a function.*

(a) *$f: A \to B$ is onto if and only if there exists a function $g: B \to A$ such that $fg = 1_B$.*

(b) *$f: A \to B$ is* (1-1) *if and only if there exists a function $g: B \to A$ such that $gf = 1_A$ (provided A is non-empty).*

(c) *$f: A \to B$ is a* (1-1)-*correspondence if and only if there exists a function $g: B \to A$ such that $fg = 1_B$* and *$gf = 1_A$. In this case g is unique and is called the 'inverse function' to f.* ∎

Definition 1.2.10 A set A is *countable* (or *enumerable*) if it is in (1-1)-correspondence with a subset of the set of positive integers.

Proposition 1.2.11 *If the sets A and B are countable, so is $A \times B$.* ∎

Definition 1.2.12 A *permutation* of a set A is a (1-1)-correspondence from A to itself; a *transposition* is a permutation that leaves fixed all but two elements of A, which are interchanged. If A is a finite set, a permutation is *even* if it is a composite of an even number of transpositions and *odd* if it is a composite of an odd number of transpositions.

1.3 Algebra

Definition 1.3.1 A *group* G is a set, together with a function $m: G \times G \to G$, called a *multiplication*, satisfying the following rules.

(a) $m(m(g_1, g_2), g_3) = m(g_1, m(g_2, g_3))$ for all $g_1, g_2, g_3 \in G$.

(b) There exists an element $e \in G$, called the *unit element*, such that $m(g, e) = g = m(e, g)$ for all $g \in G$.

(c) For each $g \in G$, there exists $g' \in G$ such that $m(g, g') = e = m(g', g)$.

The element $m(g_1, g_2)$ is regarded as the 'product' of g_1 and g_2, and is normally written $g_1 g_2$, so that rule (a), for example, becomes $(g_1 g_2)g_3 = g_1(g_2 g_3)$ (this is usually expressed by saying that the product is *associative*; we may unambiguously write $g_1 g_2 g_3$ for either $(g_1 g_2)g_3$ or $g_1(g_2 g_3)$). We shall often write 1 instead of e in rule (b), and g^{-1} instead of g' in rule (c) (g^{-1} is the *inverse* of g).

The *order* of G is the number of elements in it, if this is finite; the *order* of the element $g \in G$ is the smallest positive integer n such that $g^n = e$ (where g^n means the product of g with itself n times).

A group with just one element is called a *trivial group*, often written 0.

A subset H of a group G is called a *subgroup* if $m(H \times H) \subset H$ and H satisfies rules (a)–(c) with respect to m.

Proposition 1.3.2 *A non-empty subset H of G is a subgroup if and only if $g_1 g_2^{-1} \in H$ for all $g_1, g_2 \in H$.* ▮

Theorem 1.3.3 *If H is a subgroup of a finite group G, the order of H divides the order of G.* ▮

Definition 1.3.4 Given groups G and H, a *homomorphism* $\theta: G \to H$ is a function such that $\theta(g_1 g_2) = \theta(g_1)\theta(g_2)$ for all $g_1, g_2 \in G$. θ is an *isomorphism* (or is *isomorphic*) if it is also a (1-1)-correspondence; in this case G and H are said to be *isomorphic*, written $G \cong H$. We write Im θ for $\theta(G)$, and the *kernel* of θ, Ker θ, is the subset $\{g \in G \mid \theta(g) = e\}$, where e is the unit element of H.

Example 1.3.5 The identity function $1_G: G \to G$ is an isomorphism, usually called the *identity isomorphism.* ▮

Proposition 1.3.6

(a) *The composite of two homomorphisms is a homomorphism.*

(b) *If θ is an isomorphism, the inverse function is also an isomorphism.*

(c) *If $\theta: G \to G$ is a homomorphism,* Im θ *is a subgroup of H and* Ker θ *is a subgroup of G. θ is* (1-1) *if and only if* Ker θ *contains only the unit element of G.* ▮

Definition 1.3.7 Two elements $g_1, g_2 \in G$ are *conjugate* if there exists $h \in G$ such that $g_2 = h^{-1}g_1h$. A subgroup H of G is *normal* (*self-conjugate*) if $g^{-1}hg \in H$ for all $h \in H$ and $g \in G$.

Given a normal subgroup H of a group G, define an equivalence relation R on G by the rule g_1Rg_2 if and only if $g_1g_2^{-1} \in H$; then R is an equivalence relation and the equivalence class $[g]$ is called the *coset* of g.

Theorem 1.3.8 *The set of distinct cosets can be made into a group by setting* $[g_1][g_2] = [g_1g_2]$. ▮

Definition 1.3.9 The group of Theorem 1.3.8 is called the *quotient group* of G by H, and is written G/H.

Proposition 1.3.10 *The function $p: G \to G/H$, defined by $p(g) = [g]$, is a homomorphism, and is onto.* Ker $p = H$. ▮

Theorem 1.3.11 *Given groups G, G', normal subgroups H, H' of G, G' respectively, and a homomorphism $\theta: G \to G'$ such that $\theta(H) \subset H'$, there exists a unique homomorphism $\bar{\theta}: G/H \to G'/H'$ such that* $\bar{\theta}[g] = [\theta(g)]$. ▮

Proposition 1.3.12 *Given a homomorphism $\theta: G \to H$,* Ker p *is a normal subgroup of G, and $\bar{\theta}: G/$*Ker $\theta \to$ Im θ *is an isomorphism.* ▮

Definition 1.3.13 Given a collection of groups G_a, one for each element a of a set A (not necessarily finite), the *direct sum* $\bigoplus_{a \in A} G_a$ is the set of collections of elements (g_a), one element g_a in each G_a, where all but a finite number of the g_a's are unit elements. The multiplication in $\bigoplus_{a \in A} G_a$ is defined by $(g_a)(g_a') = (g_ag_a')$, that is, corresponding elements in each G_a are multiplied together.

We shall sometimes write $\oplus G_a$ instead of $\bigoplus_{a\in A} G_a$, if no ambiguity can arise; and if A is the set of positive integers we write $\bigoplus_{n=1}^{\infty} G_n$ (similarly $\bigoplus_{r=1}^{n} G_r$ or even $G_1 \oplus G_2 \oplus \cdots \oplus G_n$ if A is the set of the first n positive integers). In the latter case, we prefer the notation $g_1 \oplus g_2 \oplus \cdots \oplus g_n$ rather than (g_r) for a typical element.

Proposition 1.3.14 *Given homomorphisms* $\theta_a\colon G_a \to H_a$ $(a \in A)$, *the function* $\oplus\, \theta_a\colon \bigoplus_{a\in A} G_a \to \bigoplus_{a\in A} H_a$, *defined by* $\oplus\, \theta_a(g_a) = (\theta_a(g_a))$, *is a homomorphism, which is isomorphic if each* θ_a *is.* ∎

Once again, we prefer the notation $\theta_1 \oplus \theta_2 \oplus \cdots \oplus \theta_n$ if A is the set of the first n integers.

Definition 1.3.15 Given a set A, the *free group generated by* A, Gp $\{A\}$, is defined as follows. A *word* w in A is a formal expression

$$w = a_1^{\epsilon_1} \cdots a_n^{\epsilon_n},$$

where $a_1, \ldots, a_n$ are (not necessarily distinct) elements of A, $\epsilon_i = \pm 1$, and $n \geqslant 0$ (if $n = 0$, w is the 'empty word', and is denoted by 1). Define an equivalence relation R on the set of words in A by the rule: $w_1 R w_2$ if and only if w_2 can be obtained from w_1 by a finite sequence of operations of the form 'replace $a_1^{\epsilon_1} \cdots a_n^{\epsilon_n}$ by $a_1^{\epsilon_1} \cdots a_r^{\epsilon_r} a^1 a^{-1} a_{r+1}^{\epsilon_{r+1}} \cdots a_n^{\epsilon_n}$ or $a_1^{\epsilon_1} \cdots a_r^{\epsilon_r} a^{-1} a^1 a_{r+1}^{\epsilon_{r+1}} \cdots a_n^{\epsilon_n}$ $(0 \leqslant r \leqslant n)$, or vice versa'. The elements of Gp $\{A\}$ are the equivalence classes $[w]$ of words in A, and the multiplication is defined by

$$[a_1^{\epsilon_1} \cdots a_n^{\epsilon_n}][a_{n+1}^{\epsilon_{n+1}} \cdots a_m^{\epsilon_m}] = [a_1^{\epsilon_1} \cdots a_n^{\epsilon_n} a_{n+1}^{\epsilon_{n+1}} \cdots a_m^{\epsilon_m}].$$

Normally the elements of Gp $\{A\}$ are written without square brackets, and by convention we write a for a^1, a^2 for $a^1 a^1$, a^{-2} for $a^{-1}a^{-1}$, and so on. The omission of square brackets has the effect of introducing equalities such as $a^2 a^{-1} = a$, $aa^{-1} = 1$ (note that 1 is the unit element of Gp $\{A\}$).

Example 1.3.16 The group of integers under addition (usually denoted by Z) is isomorphic to Gp $\{a\}$, where a denotes a set consisting of just one element a. ∎

Proposition 1.3.17 *Given a set* A, *a group* G *and a function* $\theta\colon A \to G$, *there exists a unique homomorphism* $\bar{\theta}\colon$ Gp $\{A\} \to G$ *such that* $\bar{\theta}(a) = \theta(a)$ *for each* $a \in A$. ∎

Definition 1.3.18 Given a set B of elements of Gp $\{A\}$, let $\bar{B}$ be the intersection of all the normal subgroups of Gp $\{A\}$ that contain B. $\bar{B}$ is itself a normal subgroup (called the *subgroup generated by* B), and the quotient group Gp $\{A\}/\bar{B}$ is called the group *generated by* A, *subject to the relations* B, and is written Gp $\{A; B\}$. The elements of Gp $\{A; B\}$ are still written in the form of words in A, and the effect of the relations B is to introduce new equalities of the form $b = 1$, for each element $b \in B$.

A group G is *finitely generated* if $G \cong$ Gp $\{A; B\}$ for some finite set A; in particular, if A has only one element, G is said to be *cyclic*.

Example 1.3.19 For each integer $n \geqslant 2$, the group Z_n of integers modulo n, under addition mod n, is a cyclic group, since $Z_n \cong$ Gp $\{a; a^n\}$.

In fact every group G is isomorphic to a group of the form Gp $\{A; B\}$, since we could take A to be the set of all the elements of G. Of course, this representation is not in general unique: for example, Gp $\{a; a^2\} \cong$ Gp $\{a, b; a^2, b\}$.

Proposition 1.3.20 *A function* $\theta\colon A \to G$, *such that* $\theta(b) = e$ *(the unit element of* G*) for all* $b \in B$, *defines a unique homomorphism* $\bar{\theta}\colon$ Gp $\{A; B\} \to G$, *such that* $\bar{\theta}(a) = \theta(a)$ *for all* $a \in A$. ∎

Definition 1.3.21 A group G is said to be *abelian* (*commutative*) if $g_1g_2 = g_2g_1$ for all $g_1, g_2 \in G$. In an abelian group, the notation $g_1 + g_2$ is normally used instead of g_1g_2 (and the unit element is usually written 0). Similarly, one writes $-g$ instead of g^{-1}.

Observe that every subgroup of an abelian group is normal, and that every quotient group of an abelian group is abelian, as also is every direct sum of a collection of abelian groups.

Definition 1.3.22 Given a group G (not necessarily abelian), the *commutator subgroup* $[G, G]$ is the set of all (finite) products of elements of the form $g_1g_2g_1^{-1}g_2^{-1}$.

Proposition 1.3.23 $[G, G]$ *is a normal subgroup of* G, *and* $G/[G, G]$ *is abelian. Given any homomorphism* $\theta\colon G \to H$ *into an abelian group,* $[G, G] \subset \operatorname{Ker} \theta$. ∎

Proposition 1.3.24 *If* $G \cong H$, *then* $G/[G, G] \cong H/[H, H]$. ∎

Definition 1.3.25 Given a set A, the *free abelian group generated by* A, Ab $\{A\}$, is the group Gp $\{A\}/[$Gp $\{A\}$, Gp $\{A\}]$.

Proposition 1.3.26 Ab $\{A\} \cong$ Gp $\{A; B\}$, *where B is the set of all elements of* Gp $\{A\}$ *of the form* $a_1a_2a_1^{-1}a_2^{-1}$. ∎

The elements of Ab $\{A\}$ will normally be written in the form $\epsilon_1 a_1 + \cdots + \epsilon_n a_n$ ($\epsilon_i = \pm 1$), and the coset of 1 will be denoted by 0.

Definition 1.3.27 If B is a set of elements of Ab $\{A\}$, let $\bar{B}$ be the intersection of all the subgroups of Ab $\{A\}$ that contain B: thus $\bar{B}$ is a subgroup and consists of all finite sums of elements of B (or their negatives), together with 0. The quotient group Ab $\{A\}/\bar{B}$ is called the *abelian group generated by A, subject to the relations B*, and is written Ab $\{A; B\}$.

As in Definition 1.3.18, the elements of Ab $\{A; B\}$ are still written in the form of 'additive' words in A.

Proposition 1.3.28 *If $G =$ Gp $\{A; B\}$, and $p: G \to G/[G, G]$ is the homomorphism of Proposition 1.3.10, then $G/[G, G] \cong$ Ab$\{A; p(B)\}$.* ∎

Examples 1.3.29 Particular examples of abelian groups include Z and Z_n: observe that $Z \cong$ Ab $\{a\}$ and $Z_n \cong$ Ab $\{a; na\}$. We shall also make frequent use of the groups of rational, real and complex numbers, under addition: these are denoted by R, Q and C respectively. ∎

There is a very useful theorem giving a standard form for the finitely generated abelian groups.

Theorem 1.3.30 *Let G be a finitely generated abelian group. There exists an integer $n \geqslant 0$, primes $p_1, \ldots, p_m$ and integers $r_1, \ldots, r_m$ ($m \geqslant 0$, $r_i \geqslant 1$), such that*

$$G \cong nZ \oplus Z_{p_1^{r_1}} \oplus \cdots \oplus Z_{p_m^{r_m}}.$$

(Here, nZ denotes the direct sum of n copies of Z.) Moreover, if

$$H \cong lZ \oplus Z_{q_1^{s_1}} \oplus \cdots \oplus Z_{q_k^{s_k}},$$

then $G \cong H$ if and only if $n = l$, $m = k$, and the numbers $p_1^{r_1}, \ldots, p_m^{r_m}$ and $q_1^{s_1}, \ldots, q_k^{s_k}$ are equal in pairs. ∎

Definition 1.3.31 A sequence of groups and homomorphisms

$$\cdots \longrightarrow G_i \xrightarrow{\theta_i} G_{i+1} \xrightarrow{\theta_{i+1}} G_{i+2} \longrightarrow \cdots$$

is called an *exact sequence* if, for each i, Ker $\theta_i =$ Im θ_{i-1} (if the sequence terminates in either direction, for example $G_0 \xrightarrow{\theta_0} G_1 \to \cdots$

or $\cdots \longrightarrow G_{n-1} \xrightarrow{\theta_{n-1}} G_n$, then no restriction is placed on Ker θ_0 or Im θ_{n-1}).

Example 1.3.32 The sequence $0 \to G \xrightarrow{\theta} H \to 0$ is exact if and only if θ is an isomorphism. (Here, 0 denotes the trivial group, and $0 \to G$, $H \to 0$ the only possible homomorphisms.) This follows immediately from the definitions.

Similarly, if H is a normal subgroup of G and $i: H \to G$ is defined by $i(h) = h$ for all $h \in H$, then

$$0 \to H \xrightarrow{i} G \xrightarrow{p} G/H \to 0$$

is an exact sequence. ∎

Proposition 1.3.33 *Given exact sequences*

$$0 \to G_a \xrightarrow{\theta_a} H_a \xrightarrow{\phi_a} K_a \to 0,$$

one for each element a of a set A, the sequence

$$0 \longrightarrow \bigoplus_{a \in A} G_a \xrightarrow{\oplus \theta_a} \bigoplus_{a \in A} H_a \xrightarrow{\oplus \phi_a} \bigoplus_{a \in A} K_a \longrightarrow 0$$

is also exact. ∎

Definition 1.3.34 A square of groups and homomorphisms

$$\begin{array}{ccc} G_1 & \xrightarrow{\theta_1} & G_2 \\ {\scriptstyle \phi_1}\downarrow & & \downarrow{\scriptstyle \phi_2} \\ H_1 & \xrightarrow[\theta_2]{} & H_2 \end{array}$$

is said to be *commutative* if $\phi_2\theta_1 = \theta_2\phi_1$. Commutative triangles, etc., are similarly defined, and in general any diagram of groups and homomorphisms is *commutative* if each triangle, square, ... in it is commutative.

Proposition 1.3.35 *Given a commutative diagram of groups and homomorphisms*

$$\begin{array}{ccccccccc} G_1 & \xrightarrow{\theta_1} & G_2 & \xrightarrow{\theta_2} & G_3 & \xrightarrow{\theta_3} & G_4 & \xrightarrow{\theta_4} & G_5 \\ {\scriptstyle \psi_1}\downarrow & & {\scriptstyle \psi_2}\downarrow & & {\scriptstyle \psi_3}\downarrow & & {\scriptstyle \psi_4}\downarrow & & {\scriptstyle \psi_5}\downarrow \\ H_1 & \xrightarrow[\phi_1]{} & H_2 & \xrightarrow[\phi_2]{} & H_3 & \xrightarrow[\phi_3]{} & H_4 & \xrightarrow[\phi_4]{} & H_5, \end{array}$$

in which the rows are exact sequences, and ψ_2, ψ_4 are isomorphisms, ψ_1 is onto and ψ_5 is (1-1), *then ψ_3 is an isomorphism.*

Proof. To show that ψ_3 is (1-1), consider an element $x \in G_3$ such that $\psi_3(x) = 1$ (we shall write 1 indiscriminately for the unit element of each group). Then $\psi_4\theta_3(x) = \phi_3\psi_3(x) = 1$, so that $\theta_3(x) = 1$ since ψ_4 is isomorphic. By exactness, therefore, $x = \theta_2(y)$ for some $y \in G_2$; and then $\phi_2\psi_2(y) = \psi_3\theta_2(y) = 1$. By exactness again, $\psi_2(y) = \phi_1(z)$ for some $z \in H_1$; and $z = \psi_1(w)$ for some $w \in G_1$ since ψ_1 is onto. Thus $\psi_2\theta_1(w) = \phi_1\psi_1(w) = \psi_2(y)$, so that $\theta_1(w) = y$; but then $x = \theta_2(y) = \theta_2\theta_1(w) = 1$.

The proof that ψ_3 is onto is rather similar. This time, choose an element $x \in H_3$; then $\phi_3(x) = \psi_4(y)$ for some $y \in G_4$, since ψ_4 is isomorphic. Thus $\psi_5\theta_4(y) = \phi_4\psi_4(y) = \phi_4\phi_3(x) = 1$, so that $\theta_4(y) = 1$ since ψ_5 is (1-1). Hence by exactness $y = \theta_3(z)$ for some $z \in G_3$. Unfortunately there is no reason why $\psi_3(z)$ should be x, but it is at least true that $\phi_3((\psi_3(z))^{-1}x) = (\psi_4\theta_3(z))^{-1}(\phi_3(x)) = 1$, so that $(\psi_3(z))^{-1}x = \phi_2\psi_2(w)$ for some $w \in G_2$, since ψ_2 is isomorphic. Thus $\psi_3(z . \theta_2(w)) = (\psi_3(z)).\phi_2\psi_2(w) = (\psi_3(z))(\psi_3(z))^{-1}x = x$, and hence ψ_3 is onto. ∎

Proposition 1.3.36 *Given an exact sequence of abelian groups and homomorphisms*

$$0 \to G \xrightarrow{\theta} H \xrightarrow{\phi} K \to 0,$$

and a homomorphism $\psi: K \to H$ such that $\phi\psi = 1_K$, then $H \cong G \oplus K$.

Proof. Define $\alpha: G \oplus K \to H$ by $\alpha(g \oplus k) = \theta(g) + \psi(k)$: it is easy to see that α is a homomorphism. Also α is (1-1), for if $\alpha(g \oplus k) = 0$, we have

$$0 = \phi(\theta(g) + \psi(k)) = \phi\psi(k) = k;$$

but then $\theta(g) = 0$, so that $g = 0$ since θ is (1-1).

Moreover α is onto, since given $h \in H$ we have

$$\phi(h - \psi\phi(h)) = \phi(h) - \phi\psi\phi(h) = 0.$$

Thus there exists $g \in G$ such that $h - \psi\phi(h) = \theta(g)$, that is,

$$h = \theta(g) + \psi\phi(h) = \alpha(g \oplus \phi(h)). \ ∎$$

An exact sequence as in the statement of Proposition 1.3.36 is called a *split exact sequence.*

Of course, it is not true that all exact sequences $0 \to G \to H \to K \to 0$ split. However, this is true if K is a *free* abelian group.

Proposition 1.3.37 *Given abelian groups and homomorphisms $G \xrightarrow{\theta} H \xleftarrow{\phi} K$, where θ is onto and K is free abelian, there exists a homomorphism $\psi: K \to G$ such that $\theta\psi = \phi$.*

Proof. Suppose $K = \text{Ab}\{A\}$. For each $a \in A$, choose $g_a \in G$ such that $\theta(g_a) = \phi(a)$. By Proposition 1.3.20, there is a unique homomorphism $\psi: K \to G$ such that $\psi(a) = g_a$; and then clearly $\theta\psi = \phi$. ∎

Corollary 1.3.38 *Given an exact sequence of abelian groups*

$$0 \to G \xrightarrow{\theta} H \xrightarrow{\phi} K \to 0,$$

if K is free abelian, the sequence splits and $H \cong G \oplus K$.

Proof. By Proposition 1.3.37 there exists a homomorphism $\psi: K \to H$ such that $\phi\psi = 1_K$. ∎

Definition 1.3.39 A *ring* R is an abelian group, together with a function $m: R \times R \to R$, such that the following rules are satisfied for all r_1, r_2 and r_3 in R.

(a) $m(m(r_1, r_2), r_3) = m(r_1, m(r_2, r_3))$.
(b) $m(r_1, r_2 + r_3) = m(r_1, r_2) + m(r_1, r_3)$.
(c) $m(r_1 + r_2, r_3) = m(r_1, r_2) + m(r_2, r_3)$.

Since R, considered as a group, is abelian, we use the notation $+$ for the addition, and refer to m as the *multiplication*; and following the convention for groups we shall write r_1r_2 for $m(r_1, r_2)$.

A ring R is *commutative* if $r_1r_2 = r_2r_1$ for all $r_1, r_2 \in R$, and R has an *identity element* (or *has a* 1) if there exists an element $1 \in R$ such that $1r = r = r1$ for all $r \in R$.

Examples 1.3.40 Z and Z_n are commutative rings with 1, as also are Q, R and C. If R is any ring with a 1, we can form a new ring $R[x]$, the *polynomial ring*, whose elements are formal polynomials

$$r_0 + r_1x + r_2x^2 + \cdots + r_nx^n \qquad (r_1, \ldots, r_n \in R, n \geqslant 0),$$

with the obvious addition and multiplication. ∎

Definition 1.3.41 A subgroup S of a ring R is called a *subring* if $s_1s_2 \in S$ for all $s_1, s_2 \in S$, and an *ideal* if $rs, sr \in S$ for all $s \in S$, $r \in R$.

Given two rings R and S, a homomorphism $\theta: R \to S$ is a *ring homomorphism* if $\theta(r_1r_2) = \theta(r_1)\theta(r_2)$ for all $r_1, r_2 \in R$. θ is a *ring isomorphism* if it is a ring homomorphism and a (1-1)-correspondence. In any case, $\text{Im}\,\theta$ is a subring of S and $\text{Ker}\,\theta$ is an ideal of R.

Given rings R and S, the *direct sum* $R \oplus S$ can be made into a ring by defining $(r_1 \oplus s_1)(r_2 \oplus s_2) = (r_1s_1) \oplus (r_2s_2)$.

Definition 1.3.42 A *field* F is a commutative ring with 1, in which the non-zero elements form a group under multiplication.

Examples 1.3.43 Q, R and C are fields, as also is Z_p if p is a prime. However, Z and Z_n (n not prime) are not fields. ∎

Definition 1.3.44 A *vector space* V over a field F is an abelian group V, together with a function $F \times V \to V$, in which the image of (λ, v) is written λv. The following rules are also satisfied.

(a) $1v = v$, and $\lambda_1(\lambda_2 v) = (\lambda_1\lambda_2)v$ for all $\lambda_1, \lambda_2 \in F$, $v \in V$.

(b) $\lambda(v_1 + v_2) = \lambda v_1 + \lambda v_2$, $(\lambda_1 + \lambda_2)v = \lambda_1 v + \lambda_2 v$, for all λ, λ_1, $\lambda_2 \in F$, $v_1, v_2, v \in V$.

A subgroup W of V is called a *subspace* if $\lambda w \in W$ for all $\lambda \in F$, $w \in W$; the quotient group V/W is also a vector space over F, called the *quotient space*. If V and W are vector spaces over F, the *direct sum* $V \oplus W$ is the direct sum of the groups, with $\lambda(v \oplus w)$ defined to be $(\lambda v) \oplus (\lambda w)$.

Examples 1.3.45 Any field F is a vector space over itself, using the multiplication in F. More generally, so is F^n, the direct sum of n copies of F. Rather perversely, it is more usual to revert to the notation $(\lambda_1, \ldots, \lambda_n)$ instead of $\lambda_1 \oplus \cdots \oplus \lambda_n$, for elements of F^n Often $(\lambda_1, \ldots, \lambda_n)$ is abbreviated to a single letter, x say, so that λx means $(\lambda\lambda_1, \ldots, \lambda\lambda_n)$. ∎

Definition 1.3.46 If x, y are two points (elements) in a vector space V over F, the *straight-line segment* joining x and y is the subset of points of the form $\lambda x + (1 - \lambda)y$ $(0 \leqslant \lambda \leqslant 1)$. A subset A of V is *convex* if, for all $x, y \in A$, the straight-line segment joining x and y is contained in A.

Definition 1.3.47 Given vector spaces V and W over F, a homomorphism $\theta: V \to W$ is called a *linear map* if $\theta(\lambda v) = \lambda\theta(v)$, for all $\lambda \in F$, $v \in V$. If θ is also a (1-1)-correspondence, it is called a *linear* (or *vector space*) *isomorphism*.

Definition 1.3.48 A set of elements $v_1, \ldots, v_n$ in a vector space V over a field F is *linearly dependent* if there exist elements $\lambda_1, \ldots, \lambda_n \in F$, not all zero, such that $\lambda_1 v_1 + \cdots + \lambda_n v_n = 0$; otherwise $v_1, \ldots, v_n$ are *linearly independent*. A set of elements $v_1, \ldots, v_n$ forms a *base* of V if it is linearly independent, and given any element $v \in V$ there exist elements $\lambda_1, \ldots, \lambda_n \in F$ such that $v = \lambda_1 v_1 + \cdots + \lambda_n v_n$. If V possesses a (finite) base, V is *finite-dimensional*.

Proposition 1.3.49

(a) *If V is a finite-dimensional vector space over F, any two bases have the same number of elements.*

(b) *If W is a subspace of a finite-dimensional vector space V, then W is finite-dimensional and any base of W can be extended to a base of V.* ▌

The number of elements in a base is called the *dimension* of V. It is easy to see that two finite-dimensional vector spaces over F are isomorphic if and only if they have the same dimension; in particular, if V has dimension n, then $V \cong F^n$.

Proposition 1.3.50 *Given an exact sequence of vector spaces over F and linear maps:*

$$0 \to V_1 \xrightarrow{\theta} V_2 \xrightarrow{\phi} V_3 \to 0,$$

in which V_3 is finite-dimensional, the sequence splits and

$$V_2 \cong V_1 \oplus V_3. \;▌$$

Definition 1.3.51 Finite-dimensional spaces V and W over F are said to be *dual spaces* if there exists a function $V \times W \to F$, the image of (v, w) being written $\langle v, w\rangle$, with the following properties.

(a) $\langle v_1 + v_2, w\rangle = \langle v_1, w\rangle + \langle v_2, w\rangle$, $\langle v, w_1 + w_2\rangle = \langle v, w_1\rangle + \langle v, w_2\rangle$, $\langle \lambda v, w\rangle = \lambda\langle v, w\rangle = \langle v, \lambda w\rangle$, for all $v_1, v_2, v \in V$, $w, w_1, w_2 \in W$ and $\lambda \in F$.

(b) $\langle v, w\rangle = 0$ for all $w \in W$ implies $v = 0$; $\langle v, w\rangle = 0$ for all $v \in V$ implies $w = 0$.

Proposition 1.3.52 *Given V, of dimension n, there exists W such that V, W are dual spaces. Moreover, any such W has dimension n.* ▌

Proposition 1.3.53 *Given pairs of dual spaces V_1, W_1 and V_2, W_2, and a linear map $\theta: V_1 \to V_2$, there exists a unique linear map $\theta': W_2 \to W_1$ such that*

$$\langle \theta(v_1), w_2\rangle = \langle v_1, \theta'(w_2)\rangle,$$

for all $v_1 \in V_1$, $w_2 \in W_2$. ▌

θ' is called the *dual linear map* to θ.

Definition 1.3.54 Given a pair of dual spaces V, W, and a subspace U of V, the *annihilator* of U is the subspace $\mathscr{A}(U)$ of W of elements w such that $\langle u, w\rangle = 0$ for all $u \in U$.

Proposition 1.3.55 *For any subspace U of V, $\mathscr{A}(\mathscr{A}(U)) = U$. If $U_1 \subset U_2$ are subspaces of V, then $\mathscr{A}(U_2) \subset \mathscr{A}(U_1)$, and U_2/U_1, $\mathscr{A}(U_1)/\mathscr{A}(U_2)$ are dual spaces.* ▌

Definition 1.3.56 An $(m \times n)$ *matrix* A over a field F is a set (A_{ij}) of elements of F $(1 \leqslant i \leqslant m, 1 \leqslant j \leqslant n)$. Given two $(m \times n)$ matrices A and B, the *sum* $A + B$ is the $(m \times n)$ matrix defined by $(A + B)_{ij} = A_{ij} + B_{ij}$, and given an $(n \times p)$ matrix C, the *product* AC is the $(m \times p)$ matrix defined by $(AC)_{ij} = \sum_{k=1}^{n} A_{ik}C_{kj}$. The *identity* $(n \times n)$ matrix I is defined by

$$(I)_{ij} = \begin{cases} 1, & i = j \\ 0, & \text{otherwise,} \end{cases}$$

and the $(n \times n)$ matrix A has an *inverse* A^{-1} if $AA^{-1} = I = A^{-1}A$ if A has an inverse it is said to be *non-singular*.

Proposition 1.3.57 *Given finite-dimensional vector spaces V, W over F, and bases $v_1, \ldots, v_n$ of V, $w_1, \ldots, w_m$ of W, there is a* (1-1)-*correspondence between the linear maps $\theta\colon V \to W$ and the* $(m \times n)$ *matrices A over F, defined by $\theta(v_j) = \sum_{i=1}^{m} A_{ij}w_i$. Moreover the product of two matrices corresponds to the composite of the corresponding linear maps.* ∎

Definition 1.3.58 The *trace* of an $(n \times n)$ matrix $A = (A_{ij})$, written tr (A), is $\sum_i A_{ii}$, the sum of the diagonal elements.

Proposition 1.3.59 *Let $\theta\colon V \to V$ be a linear map of the n-dimensional vector space V. Let A, B be the matrices representing θ with respect to two bases $(v_1, \ldots, v_n)$ and $(w_1, \ldots, w_n)$ of V. Then* tr $(A) =$ tr (B).

Proof. By Proposition 1.3.57, there exists a non-singular matrix P such that $A = P^{-1}BP$. Let $P = (P_{ij})$ and $P^{-1} = (Q_{ij})$; then

$$\begin{aligned} \operatorname{tr}(A) &= \sum_i A_{ii} \\ &= \sum_{i,j,k} Q_{ij}B_{jk}P_{ki} \\ &= \sum_{j,k} \left(B_{jk} \sum P_{ki}Q_{ij}\right) \\ &= \sum_{j,k} B_{jk}\delta_{kj}, \qquad \text{where } \delta_{kj} = \begin{cases} 1, & \text{if } k = j \\ 0, & \text{otherwise} \end{cases} \\ &= \sum_j B_{jj} \\ &= \operatorname{tr}(B). \end{aligned}$$ ∎

Thus we can unambiguously write tr (θ) for this common value.

Proposition 1.3.60 *Let $\theta: V \to V$ be a linear map, and let W be a subspace of V such that $\theta(W) \subset W$. Let $\phi: W \to W$ be the restriction of θ to W, and let $\psi: V/W \to V/W$ be the linear map induced by θ. Then*

$$\operatorname{tr}(\theta) = \operatorname{tr}(\phi) + \operatorname{tr}(\psi).$$

Proof. Let $w_1, \ldots, w_r$ be a base of W; extend to a base $w_1, \ldots, w_r, v_{r+1}, \ldots, v_n$ of V. If A is the matrix of θ with respect to this base,

$$\operatorname{tr}(\theta) = \sum_{i=1}^{r} A_{ii} + \sum_{i=r+1}^{n} A_{ii}.$$

But $\sum_{i=1}^{r} A_{ii} = \operatorname{tr}(\phi)$, and $\sum_{i=r+1}^{n} A_{ii} = \operatorname{tr}(\psi)$, since the cosets $[v_{r+1}], \ldots, [v_n]$ obviously form a base for V/W. ∎

Definition 1.3.61 The *determinant* det A of an $(n \times n)$ matrix A over F is the element of F defined by

$$\det A = \sum_{\rho} \epsilon_\rho A_{1,\rho(1)} \cdots A_{n,\rho(n)},$$

where ρ runs over all permutations of $1, \ldots, n$, and ϵ_ρ is $+1$ or -1 according as ρ is even or odd.

Proposition 1.3.62

(a) $\det(AB) = \det(A)\det(B)$.
(b) $\det A \neq 0$ *if and only if* A *is non-singular.* ∎

Corollary 1.3.63 *A set of equations*

$$\sum_{i=1}^{n} A_{ij} x_i = 0 \qquad (j = 1, 2, \ldots, n)$$

has a solution, other than $x_i = 0$ for all i, if and only if $\det A = 0$. ∎

Definition 1.3.64 Let V be a vector space over R. An *inner product* on V is a function $V \times V \to R$, where the image of (v_1, v_2) is written $[v_1, v_2]$, satisfying the following rules.

(a) $[v_1, v_2] = [v_2, v_1]$ for all $v_1, v_2 \in V$.
(b) $[v, v] \geqslant 0$; $[v, v] = 0$ if and only if $v = 0$.
(c) $[v_1 + v_2, v_3] = [v_1, v_2] + [v_1, v_3]$, for all $v_1, v_2, v_3 \in V$.
(d) $[rv_1, v_2] = r[v_1, v_2]$, for all $r \in R$, $v_1, v_2 \in V$.

The *length* (or *norm*) of v, $\|v\|$, is defined to be $[v, v]^{1/2}$.

Example 1.3.65 There is an inner product on R^n, defined by

$$[(r_1, \ldots, r_n), (r'_1, \ldots, r'_n)] = r_1 r'_1 + \cdots + r_n r'_n.$$

R^n, together with this inner product, is sometimes referred to as n-dimensional *Euclidean space*. ▮

1.4 Analytic topology

Definition 1.4.1 A *topological space* X (or just *space* when no ambiguity arises) is a set, together with a set of subsets called *open sets*, such that the following rules are satisfied.

(a) The empty set $\varnothing$, and X itself, are open sets.
(b) The intersection of two open sets is an open set.
(c) The union of any collection of open sets is an open set.

The set of open sets is called a *topology* for X, and the elements of X are usually called *points*.

A subset of X is called *closed* if its complement is open. Both X and $\varnothing$ are closed, as also are the union of two closed sets and the intersection of any collection of closed sets. Given a subset U of X, the *closure* $\bar{U}$ is the intersection of all closed sets that contain U; $\bar{U}$ is itself closed, and $\bar{U} = U$ if and only if U is closed.

If x is a point of X, an (open) *neighbourhood* of x is an open set that contains x.

Example 1.4.2 Any set X can be made into a topological space, by calling *every* subset an open set. This is called the *discrete topology* on X, and X with this topology is called a *discrete space*. ▮

Definition 1.4.3 The *subspace topology* for a subset Y of a space X consists of all subsets of Y of the form $Y \cap U$, where U is an open set of X. A *subspace* Y of X is a subset Y with the subspace topology.

Definition 1.4.4 A space X is *Hausdorff* if, given two distinct points $x_1, x_2 \in X$, there exist neighbourhoods U_1, U_2 of x_1, x_2 respectively, such that $U_1 \cap U_2 = \varnothing$. X is *regular* if, given a point $x \in X$ and a closed set F, not containing x, there exist open sets U_1, U_2 such that $x \in U_1$, $F \subset U_2$ and $U_1 \cap U_2 = \varnothing$. X is *normal* if a similar property holds given two closed sets F_1, F_2 whose intersection is empty.

Definition 1.4.5 A space X is *connected* if, given any two non-empty open sets U_1, U_2 such that $X = U_1 \cup U_2$, we have $U_1 \cap U_2 \neq \varnothing$. If X is not connected, it is said to be *disconnected*.

Proposition 1.4.6 *If U is a connected subspace of a space X, and $U \subset V \subset \bar{U}$, then V is a connected subspace.* ∎

Definition 1.4.7 A space X is *compact* if, given any *open covering* $\{U_a\}$ $(a \in A)$ (that is, a set of open sets, indexed by a set A, whose union is X) there exists a finite subset of A, $a_1, \ldots, a_n$, such that $X = U_{a_1} \cup \cdots \cup U_{a_n}$. A subset Y of X is *compact* if it is compact as a subspace. X is *locally compact* if, given any point $x \in X$, there exists a neighbourhood U and a compact subset C such that $x \in U \subset C$.

Proposition 1.4.8

(a) *A compact space is locally compact.*
(b) *In a Hausdorff space, a compact subset is closed.*
(c) *A compact Hausdorff space is regular.* ∎

Proposition 1.4.9 *Let X be a locally compact Hausdorff space. Given a point $x \in X$ and a neighbourhood U of x, there exists an open set V such that $x \in V \subset \bar{V} \subset U$, and $\bar{V}$ is compact.* ∎

Definition 1.4.10 Given a space X, a *base* of open sets of X is a set of open sets U_a $(a \in A)$ such that every open set of X is a union of sets U_a. A set of open sets U_a $(a \in A)$ is called a *sub-base* if every open set of X is a union of finite intersections of sets U_a.

Proposition 1.4.11 *The set of U_a $(a \in A)$ is a sub-base of open sets of X if and only if, given $x \in X$ and a neighbourhood V of x, there exist $a_1, \ldots, a_n \in A$ such that $x \in U_{a_1} \cap \cdots \cap U_{a_n} \subset V$.* ∎

Definition 1.4.12 Given spaces X and Y, a function $f\colon X \to Y$ is said to be *continuous* (or a *continuous map*, or usually just a *map*) if, for each open set $U \subset Y$, the set $f^{-1}(U) = \{x \in X \mid f(x) \in U\}$ is open in X. Alternatively, f is continuous if $f^{-1}(V)$ is closed for each closed set $V \subset Y$. f is a *homeomorphism* if it is also a (1-1)-correspondence, and the inverse function is continuous; in this case, X and Y are said to be *homeomorphic*.

Example 1.4.13 If Y is a subspace of a space X, the *inclusion map* $i\colon Y \to X$, defined by $i(y) = y$ for all $y \in Y$, is a continuous map. And for any space X, the identity function $1_X\colon X \to X$ is a homeomorphism (usually called the *identity map*). ∎

Proposition 1.4.14 *The relation between spaces of being homeomorphic is an equivalence relation on any set of spaces.* ∎

Proposition 1.4.15

(a) *The composite of continuous maps is again continuous.*

(b) *Given a function $f: X \to Y$, f is continuous if $f^{-1}(U)$ is open for each member U of a base of open sets of Y, or even for each member of a sub-base.*

(c) *If $f: X \to Y$ is continuous, and $C \subset X$ is compact, then $f(C)$ is compact.*

(d) *If A and B are closed subspaces of a space X, where $X = A \cup B$, and if $f: A \to Y$, $g: B \to Y$ are continuous maps such that $f(x) = g(x)$ for all $x \in A \cap B$, then $h: X \to Y$, defined by*

$$h(x) = \begin{cases} f(x), & x \in A \\ g(x), & x \in B, \end{cases}$$

is also continuous. ∎

Proposition 1.4.16 *The properties of being Hausdorff, regular, normal, connected, compact or locally compact are preserved under homeomorphism.* ∎

Definition 1.4.17 A *pair* of spaces (X, Y) is a space X, together with a subspace Y. Given pairs (X, Y) and (A, B), a *map of pairs* $f: (X, Y) \to (A, B)$ is a map $f: X \to A$ such that $f(Y) \subset B$. f is a *homeomorphism of pairs* if f is a homeomorphism and the inverse map to f is a map of pairs $(A, B) \to (X, Y)$ (thus $f|Y: Y \to B$ is also a homeomorphism). *Triples*, etc., of spaces, and maps between them, are similarly defined: a *triple* (X, Y, Z) for example consists of a space X, a subspace Y and a subspace Z of Y.

Definition 1.4.18 Given a collection of (disjoint) spaces X_a $(a \in A)$, the *disjoint union* $\bigcup_{a \in A} X_a$ is the union of the sets, with topology given by open sets of the form $\bigcup_{a \in A} U_a$, where each U_a is an open set in X_a. As usual, we shall use the notation $X \cup Y$, for example, if the set A is finite.

Proposition 1.4.19 *Given a collection of spaces X_a $(a \in A)$, and maps $f_a: X_a \to Y$, for all $a \in A$, the function $f: \bigcup_{a \in A} X_a \to Y$ defined by $f(x) = f_a(x)$, $x \in X_a$, is continuous.* ∎

Definition 1.4.20 Given a collection of spaces X_a $(a \in A)$, the *product* $\times_{a \in A} X_a$ is the set of collections of elements (x_a), one element

x_a in each X_a. The topology is given by a base of open sets of the form

$$U_{a_1,\ldots,a_n} = \{(x_a) \mid x_{a_1} \in U_{a_1}, \ldots, x_{a_n} \in U_{a_n}\},$$

where $a_1, \ldots, a_n$ is any *finite* set of elements of A, and each U_{a_r} is open in X_{a_r}. Once again, we shall write $X \times Y$, for example, if A is finite: note that the open sets of $X \times Y$ are all unions of sets of the form $U \times V$, where U is open in X and V is open in Y.

Proposition 1.4.21

(a) *Given spaces X and Y, and points $x \in X$, $y \in Y$, the subspaces $X \times y$, $x \times Y$ of $X \times Y$ are homeomorphic to X, Y respectively.*

(b) *The product of a collection of Hausdorff spaces is Hausdorff, and the product of a collection of compact spaces is compact.*

(c) *Each projection map $p_a: \bigtimes_{a \in A} X_a \to X_a$, defined by $p_a((x_a)) = x_a$, is continuous, and a function $f: Y \to \bigtimes_{a \in A} X_a$ is continuous if and only if each $p_a f$ is continuous. In particular, given maps $f_a: X_a \to Y_a$, for each $a \in A$, the product map $\times f_a: \bigtimes_{a \in A} X_a \to \bigtimes_{a \in A} Y_a$, defined by $\times f_a((x_a)) = (f(x_a))$, is continuous.* ∎

Definition 1.4.22 Given a space X and an equivalence relation R on X, the *identification space* X/R consists, as a set, of the disjoint equivalence classes $[x]$ of elements of X, and the topology is defined by specifying that a set $U \subset X/R$ is open if and only if $p^{-1}(U)$ is open, where $p: X \to X/R$ is the function defined by $p(x) = [x]$ (thus p is certainly continuous). Alternatively, we can specify that $V \subset X/R$ is closed if and only if $p^{-1}(V)$ is closed.

In particular, given a subspace Y of X, the *quotient space* X/Y is defined to be X/R, where R is the equivalence relation on X defined by $x_1 R x_2 \Leftrightarrow x_1 = x_2$ or $x_1, x_2 \in Y$. Thus the points of X/Y are those of $X - Y$, together with a single point (Y) representing the whole of Y. If Y happens to be the empty set, it is usually convenient to interpret X/Y as the disjoint union of X with another point.

More generally, a map $p: X \to Y$ is called an *identification map* if it is onto, and $U \subset Y$ is open if and only if $p^{-1}(U)$ is open. Clearly such a map defines an equivalence relation R on X, by setting $x_1 R x_2 \Leftrightarrow p(x_1) = p(x_2)$, and X/R is homeomorphic to Y. Conversely, if R is an equivalence relation on X, then $p: X \to X/R$, defined by $p(x) = [x]$, is an identification map.

Proposition 1.4.23

(a) *If $p: X \to Y$ is an identification map, a function $f: Y \to Z$ is continuous if and only if fp is continuous.*

(b) *The composite of two identification maps is an identification map.* ∎

Example 1.4.24 Let (X, Y) and (A, B) be pairs of spaces, and let $f: Y \to B$ be any function. Define an equivalence relation R on the disjoint union $X \cup A$, by setting $pRq \Leftrightarrow p = q$, or $p \in Y$ and $q = f(p)$, or $q \in Y$ and $p = f(q)$, or $p, q \in Y$ and $f(p) = f(q)$. The space $(X \cup A)/R$ is often referred to as the space obtained from X and A by 'identifying together corresponding points of Y and B'. For example, if A and B are closed subspaces of a space X, then on $A \cup B \subset X$, the topologies as a subspace of X, and as the space obtained from the *disjoint* union of disjoint copies of A and B by identifying together corresponding points of $A \cap B$ (using the identity map), are the same. ∎

Example 1.4.25 In particular, given (disjoint) spaces X and Y, a subspace A of X, and a map $f: A \to Y$, the *adjunction space* $Y \cup_f X$ is the space $(X \cup Y)/R$, where R is the equivalence relation defined by $pRq \Leftrightarrow p = q$, or $p \in A$ and $q = f(p)$, or $q \in A$ and $p = f(q)$, or $p, q \in A$ and $f(p) = f(q)$. This is sometimes thought of as the space obtained from Y by 'attaching the space X by the map f'. ∎

Proposition 1.4.26 *Given spaces X and Y, identification maps $p: X \to Z$, $q: Y \to W$, and a map $f: X \to Y$, a function $g: Z \to W$, such that $gp = qf$, is continuous. In particular, a map of pairs $f: (X, Y) \to (A, B)$ gives rise to a (unique) map $\bar{f}: X/Y \to A/B$ in this way, which is a homeomorphism if f is a homeomorphism of pairs.* ∎

Definition 1.4.27 A *metric space* X is a set, together with a function $d: X \times X \to R$ (the real numbers), called a *metric* or *distance*, satisfying the following rules.

(a) $d(x, y) = d(y, x)$.
(b) $d(x, y) = 0$ if and only if $x = y$.
(c) $d(x, y) + d(y, z) \geqslant d(x, z)$. (For all $x, y, z \in X$.)

Definition 1.4.28 A subset U of a metric space X is called *open* if, given any point $x \in U$, there exists $\delta > 0$ such that the set $\{y \in X \mid d(x, y) < \delta\}$ is contained in U. The set $\{y \in X \mid d(x, y) < \delta\}$ is called the δ-*neighbourhood* of x.

Proposition 1.4.29 *This definition of open set makes X into a topological space, which is Hausdorff, regular and normal. Any subset Y of X is a metric space, with the same metric; and the topologies of Y, as a subspace and given by the metric, coincide. Moreover, a function $f: X \to Y$, between metric spaces with metrics d, d' respectively, is continuous if and only if, for each point $x \in X$ and each $\epsilon > 0$, there exists $\delta > 0$ such that $d'(f(x), f(y)) < \epsilon$ whenever $d(x, y) < \delta$.* ▮

Definition 1.4.30 A topological space X is *metrizable* if there exists a metric d on X, such that the topology on X defined by d coincides with the original topology.

Example 1.4.31 The most important example of a metric space is R^n, in which a metric can easily be constructed from the inner product of Example 1.3.65 by setting

$$d(x, y) = \|x - y\| = [x - y, x - y]^{1/2}.$$

With this metric, R^n is a connected locally compact space, and is homeomorphic to the product of n copies of R^1. ▮

Definition 1.4.32 Let X be a metric space, and let x be a point and Y be a subset. The *distance* $d(x, Y)$ is defined to be $\inf_{y \in Y} d(x, y)$. The *diameter* of Y is $\sup_{y_1, y_2 \in Y} d(y_1, y_2)$.

Proposition 1.4.33 *If Y is closed, $d(x, Y) = 0$ if and only if $x \in Y$.* ▮

Proposition 1.4.34 *A subset X of R^n is compact if and only if it is closed, and has finite diameter.* ▮

Theorem 1.4.35 *Let X be a compact metric space. Given an open covering $\{U_a\}$ $(a \in A)$, there exists a real number $\delta > 0$ (called a Lebesgue number of $\{U_a\}$), such that any subset of diameter less than δ is contained in one of the sets U_a.*

Proof. Since X is compact, we may as well assume that A is finite, say $A = \{1, 2, \ldots, n\}$. For each $x \in X$ and $r \in A$, let $f_r(x) = d(x, X - U_r)$; it is easy to see that f_r is continuous, as also is $f(x) = \max f_r(x)$. Now by Proposition 1.4.15(c) $f(X)$ is a compact subspace of R^1, and so by Proposition 1.4.33 there exists $\delta > 0$ such that $f(x) > \delta$ for all $x \in X$. It follows that any set containing x, of diameter less than δ, must be contained in one U_r. ▮

We end this chapter with a description of a few particular spaces and maps, that will be of importance in the rest of this book.

Definition 1.4.36 The *unit interval* I is the subspace of R^1 consisting of points x such that $0 \leqslant x \leqslant 1$; similarly the *double unit interval* J is the subspace of x such that $-1 \leqslant x \leqslant 1$. The *$n$-cell* E^n is the subspace $\{x \in R^n \mid d(x, 0) \leqslant 1\}$, and the *$(n-1)$-sphere* S^{n-1} is $\{x \in R^n \mid d(x, 0) = 1\}$ (E^0 is a single point, and S^{-1} is empty). When necessary, R^n is regarded as the subspace $R^n \times 0$ of $R^{n+m} = R^n \times R^m$, and similarly with E^n and S^{n-1}; thus S^{n-1}, for example, is the subspace of S^n of points $(x_1, \ldots, x_{n+1})$ such that $x_{n+1} = 0$.

Proposition 1.4.37 *I, J, E^n and S^{n-1} $(n > 1)$ are compact connected spaces.* ▌

We shall frequently need to use certain standard maps between these spaces, for example $l \colon I \to J$, defined by $l(x) = 2x - 1$, $0 \leqslant x \leqslant 1$, and a map of pairs $\theta \colon (E^n, S^{n-1}) \to (S^n, (-1, 0, \ldots, 0))$ $(n \geqslant 0)$, defined by $\theta(x_1, \ldots, x_n) = (\cos \pi r, (x_1/r) \sin \pi r, \ldots, (x_n/r) \sin \pi r)$, where $r = (x_1^2 + \cdots + x_n^2)^{1/2}$, and $(\sin \pi r)/r$ is interpreted as π if $r = 0$. It is easy to see that l and θ are continuous; moreover by Proposition 1.4.26 θ gives rise to a map $\bar{\theta} \colon E^n/S^{n-1} \to S^n$ (if $n = 0$ we interpret E^0/S^{-1} as the disjoint union of E^0 with another point, this point being mapped by θ to -1 in S^0).

Proposition 1.4.38 *$\bar{\theta} \colon E^n/S^{n-1} \to S^n$ is a homeomorphism $(n \geqslant 0)$.* ▌

Another useful map is the homeomorphism $\rho \colon E^n \to J^n$ (the product of n copies of J), defined by magnifying straight lines through the origin by suitable amounts: more precisely, for points $x \in E^n$ other than the origin, we define $\rho(x) = \lambda x$, where $x = (x_1, \ldots, x_n)$, and $\lambda = \|x\|/\max |x_i|$. Since $1 \leqslant \lambda \leqslant n^{1/2}$, Proposition 1.4.29 shows that ρ and its inverse are continuous.

Lastly, the standard homeomorphism $h_{m,n} \colon E^{m+n} \to E^m \times E^n$ is defined to be the composite

$$E^{m+n} \xrightarrow{\rho} J^{m+n} = J^m \times J^n \xleftarrow{\rho \times \rho} E^m \times E^n.$$

Definition 1.4.39 *Real projective space RP^n $(n \geqslant 0)$* is defined to be $(R^{n+1} - 0)/S$, where S is the equivalence relation defined by $xSy \Leftrightarrow x = ry$ for some real number r. We write $[x_1, \ldots, x_{n+1}]$ for the equivalence class of $(x_1, \ldots, x_{n+1})$.

Proposition 1.4.40 *RP^n is homeomorphic to*

(a) *S^n/T, where $xTy \Leftrightarrow x = \pm y$;*
(b) *E^n/U, where $xUy \Leftrightarrow x = y$, or $x, y \in S^{n-1}$ and $x = -y$.*

Proof.

(a) The inclusion map $i: S^n \to (R^{n+1} - 0)$ induces $\bar{i}: S^n/T \to RP^n$, by Proposition 1.4.26, and the map $f: (R^{n+1} - 0) \to S^n$ given by $f(x) = x/\|x\|$ induces $\bar{f}: RP^n \to S^n/T$. Clearly $\bar{i}$ and $\bar{f}$ are inverses to each other, and so are homeomorphisms.

(b) There is a homeomorphism $\phi: (E^n, S^{n-1}) \to (S^n_+, S^{n-1})$ (where S^n_+ is the subspace of S^n defined by $x_{n+1} \geqslant 0$), given by

$$\phi(x_1, \ldots, x_n) = ((x_1/r) \sin \pi r/2, \ldots, (x_n/r) \sin \pi r/2, \cos \pi r/2).$$

This induces $\bar{\phi}: E^n/U \to S^n/T$, which can easily be seen to be a homeomorphism. ∎

CHAPTER 2

HOMOTOPY AND SIMPLICIAL COMPLEXES

2.1 Introduction

We have seen in Section 1.4 that the relation between spaces of being homeomorphic is an equivalence relation, and so divides any set of spaces into disjoint equivalence classes. The main problem of topology is thus the classification of topological spaces: given two spaces X and Y, are they homeomorphic? This is usually a very difficult question to answer without employing some fairly sophisticated machinery, and the idea of algebraic topology is that one should transform such topological problems into algebraic problems, which may have a better chance of solution. This transformation process will be explained in a little more detail in Section 2.2. It turns out, however, that the algebraic techniques are usually not delicate enough to classify spaces up to homeomorphism, and so in Section 2.2 we shall also introduce the notion of *homotopy*, in order to define a somewhat coarser classification.

In the rest of this chapter we shall make a start on the general classification problem. Instead of considering all topological spaces, we shall show in Section 2.3 how a large class of spaces, called *polyhedra*, may be built up from certain very simple spaces called *simplexes*. This not only simplifies the geometry, but gives a reasonable hope of constructing algebraic invariants, by examining how the simplexes are fitted together. The general theory will be explained in Section 2.3, and in Section 2.4 we shall establish some geometrical properties of polyhedra that will be useful in later chapters. Finally, Section 2.5 is concerned with the homotopy theory of polyhedra, the vital result being the Simplicial Approximation Theorem. This theorem is the most important tool in the study of polyhedra, and is the fundamental result used in Chapters 3, 4 and 5.

2.2 The classification problem; homotopy

If we are presented with two spaces X and Y, the problem of deciding whether or not they are homeomorphic is formidable: we have either to construct a homeomorphism $f\colon X \to Y$ or, worse still, to prove that no such homeomorphism exists. We therefore wish to reflect the problem algebraically. Suppose there is some means of associating a group with each topological space: say the group $G(X)$ is associated with the space X. Suppose also that, whenever we have a continuous map (not necessarily a homeomorphism) $f\colon X \to Y$, there is associated with f a homomorphism $f_*\colon G(X) \to G(Y)$, in such a way that

(a) the identity isomorphism $1\colon G(X) \to G(X)$ is associated with the identity homeomorphism $1\colon X \to X$; and

(b) given another continuous map $g\colon Y \to Z$, where Z is a third space, then $(gf)_* = g_* f_*$.

Given this machinery, we can readily see that if $f\colon X \to Y$ happens to be a homeomorphism, then $f_*\colon G(X) \to G(Y)$ is an isomorphism. For if $g\colon Y \to X$ is the inverse map to f, we have

$$gf = 1\colon X \to X \quad \text{and} \quad fg = 1\colon Y \to Y.$$

Hence, using properties (a) and (b), we obtain

$$g_* f_* = 1\colon G(X) \to G(X), \qquad f_* g_* = 1\colon G(Y) \to G(Y),$$

whence it follows that f_* is an isomorphism. Thus if X and Y are homeomorphic, $G(X)$ and $G(Y)$ are isomorphic. The converse to this result is not in general true, however, since there is nothing to guarantee that $G(X)$ and $G(Y)$ will be non-isomorphic if X and Y are not homeomorphic. As a general principle, therefore, if we wish to prove that X and Y are homeomorphic, we must construct an explicit homeomorphism, but if we wish to prove that they are not homeomorphic, we design algebraic machinery of the sort outlined above, and try to show that $G(X)$ and $G(Y)$ are not isomorphic. Most of this book will be concerned with ways of constructing such algebraic invariants.

In practice, however, the situation is a little more complicated. Virtually all the algebraic invariants known at present are 'homotopy-type' invariants, that is, all 'homotopy equivalences' give rise to isomorphisms. Since 'homotopy equivalence' is a weaker relation than homeomorphism, this means that the algebraic invariants will never distinguish between spaces that are homotopy equivalent but not homeomorphic. Thus we may as well abandon—temporarily, at least—any attempt to make a homeomorphism classification, and concentrate on homotopy instead.

The first step is obviously to define homotopy precisely. Two continuous maps $f, g: X \to Y$ are said to be *homotopic* if f can be continuously deformed into g, that is to say, if there exists a continuous family of maps $f_t: X \to Y$ $(0 \leqslant t \leqslant 1)$, such that $f_0 = f$ and $f_1 = g$: see Fig. 2.1.

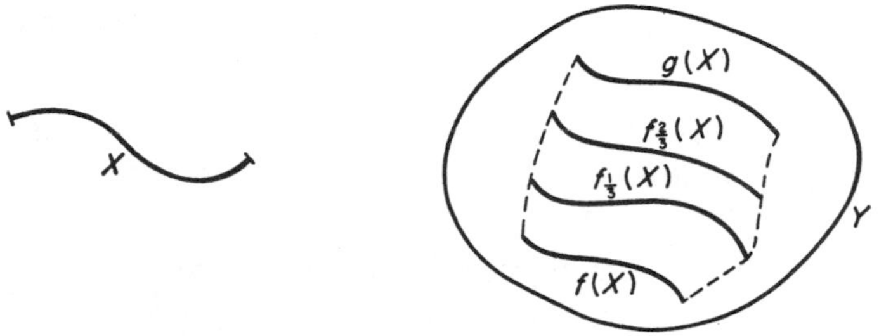

Fig. 2.1

This definition is still not quite precise, since we have not made clear what is meant by a 'continuous family'. However, it will be seen that instead of considering a family of maps $f_t: X \to Y$, we can equally well consider a single map $F: X \times I \to Y$ (where as usual I is the unit interval), defined by the rule

$$F(x, t) = f_t(x) \qquad (x \in X, t \in I).$$

When we say that the maps f_t form a continuous family, we merely mean that F is continuous with respect to t as well as x, that is, F is continuous as a map of the product $X \times I$ to Y. To sum up, the following is the official definition.

Definition 2.2.1 Two continuous maps $f, g: X \to Y$ are *homotopic* (or 'f is *homotopic* to g') if there exists a continuous map $F: X \times I \to Y$, such that

$$F(x, 0) = f(x)$$

and

$$F(x, 1) = g(x),$$

for all $x \in X$. The map F is said to be a *homotopy*, and we write $f \simeq g$ for 'f is homotopic to g' (or $F: f \simeq g$ if we wish to specify the homotopy).

Proposition 2.2.2 *Given any map $f: X \to Y$, $f \simeq f$.*

Proof. Define $F: X \times I \to I$ by $F(x, t) = f(x)$ $(x \in X, t \in I)$. Then F is continuous (why?) and is clearly a homotopy between f and itself. ∎

For a more interesting example of a homotopy, we prove the very useful result that, given $f, g: X \to Y$ with the property that for all x, $f(x)$ and $g(x)$ can be joined by a straight line in Y, then f and g are homotopic. For this to make sense, of course, we must assume that Y is a subspace of some Euclidean space R^n.

Theorem 2.2.3 *Let Y be a subspace of R^n, and let $f, g: X \to Y$ be two maps. If, for each $x \in X$, $f(x)$ and $g(x)$ can be joined by a straight-line segment in Y, then $f \simeq g$.*

Proof. Define a homotopy (called a *linear* homotopy) $F: X \times I \to Y$ by the rule $F(x, t) = (1 - t).f(x) + t.g(x)$ $(x \in X, t \in I)$; in other words 'deform f to g along the straight-line segments'. Certainly $F(x, 0) = f(x)$ and $F(x, 1) = g(x)$, so it remains to prove that F is continuous.

Now if $x' \in X$ and $t' \in I$, we have

$$\begin{aligned} F(x', t') - F(x, t) = {} & (t' - t).(g(x') - f(x')) \\ & + (1 - t).(f(x') - f(x)) \\ & + t.(g(x') - g(x)), \end{aligned}$$

so that if d is the metric in R^n we have

$$\begin{aligned} d(F(x', t'), F(x, t)) \leqslant {} & |t - t'|.d(g(x'), f(x')) \\ & + (1 - t).d(f(x'), f(x)) \\ & + t.d(g(x'), g(x)). \end{aligned}$$

But given $\epsilon > 0$, there exist open neighbourhoods U_1, U_2 of x in X, such that

$$x' \in U_1 \Rightarrow d(f(x'), f(x)) < \epsilon/3,$$

$$x' \in U_2 \Rightarrow d(g(x'), g(x)) < \epsilon/3.$$

Thus if $x' \in U_1 \cap U_2$, then $d(g(x'), f(x')) < K$, where K is the constant $d(g(x), f(x)) + 2\epsilon/3$; and if also $|t' - t| < \epsilon/3K$, then $d(F(x', t'), F(x, t)) < \epsilon$. Since the set

$$(U_1 \cap U_2) \times (t - \epsilon/3K, t + \epsilon/3K)$$

is open in $X \times I$, this proves that F is continuous. ∎

Thus for example any two maps $f, g: X \to R^n$ must be homotopic. Indeed, almost any two maps into S^{n-1} are homotopic.

Corollary 2.2.4 *Let X be any space, and let $f, g\colon X \to S^{n-1}$ be two maps such that $f(x) \neq -g(x)$ for all $x \in X$. Then $f \simeq g$.*

Proof. Considered as maps into $R^n - 0$, f and g are homotopic by Theorem 2.2.3, since the line segment joining $f(x)$ and $g(x)$ does not pass through 0. Compose this homotopy with the map $\phi\colon (R^n - 0) \to S^{n-1}$ defined by $\phi(x) = x/\|x\|$ (this is the identity map on S^{n-1} itself). ∎

Sometimes it is necessary to consider homotopies between maps of pairs, triples, etc., of spaces. Definition 2.2.1 is easily extended.

Definition 2.2.5 Given pairs (X, A) and (Y, B), two maps of pairs $f, g\colon (X, A) \to (Y, B)$ are *homotopic* if there exists a map of pairs $F\colon (X \times I, A \times I) \to (Y, B)$, such that

$$F(x, 0) = f(x)$$

and

$$F(x, 1) = g(x) \quad \text{for all } x \in X.$$

As before, we write $f \simeq g$. Homotopies of triples, etc., are similarly defined. Sometimes it is useful to consider a more restrictive kind of homotopy of pairs: if $f, g\colon (X, A) \to (Y, B)$ are maps of pairs such that $f|A = g|A$, f and g are homotopic *relative to A* if there exists a homotopy $F\colon (X \times I, A \times I) \to (Y, B)$ such that $F(a, t) = f(a) = g(a)$ for all $a \in A$, $t \in I$ (that is, F is 'fixed' on A). In this case we write $f \simeq g \text{ rel } A$.

For example, in Theorem 2.2.3 or Corollary 2.2.4, if A is the subspace of X of those points x such that $f(x) = g(x)$, then $f \simeq g \text{ rel } A$.

The notion of *homotopy equivalence* of topological spaces (pairs, triples, etc.) follows easily from Definitions 2.2.1 and 2.2.5.

Definition 2.2.6 Two spaces X and Y are *homotopy-equivalent* (or *of the same homotopy type*) if there exist maps $f\colon X \to Y$ and $g\colon Y \to X$, such that $gf \simeq 1_X$ and $fg \simeq 1_Y$, where 1_X and 1_Y are the identity maps of X and Y respectively. In this case f is a *homotopy equivalence* and g is a *homotopy inverse* to f. We write $X \simeq Y$ for 'X is homotopy-equivalent to Y' (notice that the symbol $\simeq$ has two distinct meanings, depending on the context).

Similarly two pairs (X, A) and (Y, B) are *homotopy-equivalent* (written $(X, A) \simeq (Y, B)$) if there exist maps (of pairs) $f\colon (X, A) \to (Y, B)$, $g\colon (Y, B) \to (X, A)$ such that $gf \simeq 1_X$ and $fg \simeq 1_Y$, the homotopies being homotopies of pairs.

As the name suggests, homotopy equivalence is an equivalence relation on any set of spaces. In order to prove this, we first prove that homotopy is an equivalence relation on the set of all maps between two given spaces.

Proposition 2.2.7 *Given two spaces X and Y, the relation between maps from X to Y of being homotopic is an equivalence relation. Similarly, given two pairs (X, A) and (Y, B), the relation between maps of pairs from (X, A) to (Y, B) of being homotopic as maps of pairs is an equivalence relation, and the relation between a set of maps coinciding on each point of A, of being homotopic relative to A, is also an equivalence relation.*

Proof. Consider maps $f, g, \ldots$ from X to Y. Certainly $f \simeq f$ for each f, by Proposition 2.2.2. Moreover if $F: f \simeq g$, then $G: g \simeq f$, where $G: X \times I \to Y$ is defined by

$$G(x, t) = F(x, 1 - t).$$

Lastly, if $F: f \simeq g$ and $G: g \simeq h$, then $H: f \simeq h$, where

$$H(x, t) = \begin{cases} F(x, 2t) & (0 \leqslant t \leqslant \frac{1}{2}) \\ G(x, 2t - 1) & (\frac{1}{2} \leqslant t \leqslant 1). \end{cases}$$

Here, H is continuous by Proposition 1.4.15(d).

This proves that homotopy is an equivalence relation on the set of maps from X to Y; the other two statements are proved similarly. ∎

Corollary 2.2.8 *Given spaces X, Y and Z, and maps $f_0, f_1: X \to Y$, $g_0, g_1: Y \to Z$, such that $f_0 \simeq f_1$ and $g_0 \simeq g_1$, then $g_0 f_0 \simeq g_1 f_1$.*

Proof. Let F be the homotopy between f_0 and f_1, and G that between g_0 and g_1. Let $H_1 = g_0 F: X \times I \to Z$: it is clear that H_1 is a homotopy between $g_0 f_0$ and $g_0 f_1$. But $H_2 = G(f_1 \times 1_I)$ is a homotopy between $g_0 f_1$ and $g_1 f_1$; hence $g_0 f_0 \simeq g_1 f_1$ by Proposition 2.2.7. ∎

Of course, similar results hold for homotopies of pairs and for homotopies relative to a subspace. The details are left to the reader.

Proposition 2.2.9 *The relation between spaces (pairs, triples, etc.) of being homotopy-equivalent is an equivalence relation.*

Proof. Clearly every space is homotopy-equivalent to itself (the identity map is a homotopy equivalence). Equally obviously, if $X \simeq Y$, then $Y \simeq X$. It remains only to show that if $X \simeq Y$ and

$Y \simeq Z$, then $X \simeq Z$. But if the relevant homotopy equivalences and homotopy inverses are $f: X \to Y$, $f': Y \to X$, $g: Y \to Z$, $g': Z \to Y$, then

$$\begin{aligned} f'g'gf &\simeq f'f, \quad \text{by Corollary 2.2.8} \\ &\simeq 1_X, \end{aligned}$$

and similarly $gff'g' \simeq 1_Z$.

Again, the proof for pairs, triples, etc. is similar. ▮

It is easy to see that two homeomorphic spaces are homotopy-equivalent (just use Proposition 2.2.2 again). Thus the classification of spaces up to homotopy equivalence is coarser than the homeomorphism classification. Indeed, it is strictly coarser, as the following example shows.

Example 2.2.10 Let X be the unit circle S^1 in R^2, and let Y be S^1, together with the closed line segment joining (1, 0) and (2, 0): see Fig. 2.2.

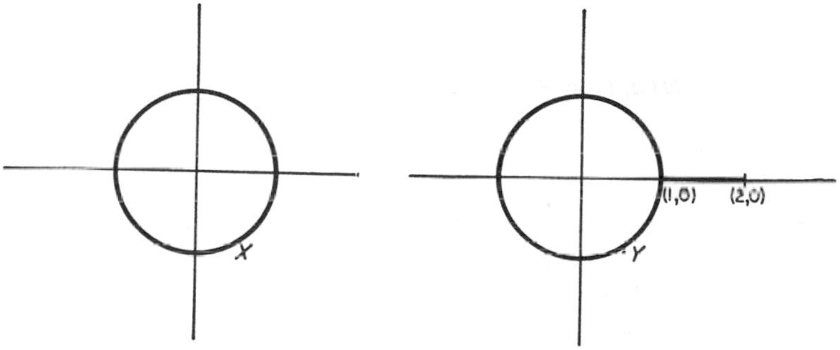

Fig. 2.2

Now X and Y are not homeomorphic, since the removal of the point (1, 0) from Y disconnects Y, whereas the removal of any point from X leaves X connected. On the other hand X and Y are homotopy-equivalent. To prove this, define $f: X \to Y$ by $f(x) = x$, for all $x \in X$, and $g: Y \to X$ by

$$g(y) = \begin{cases} y, & \text{if } y \in S^1 \\ (1, 0), & \text{if } y \text{ lies between } (1, 0) \text{ and } (2, 0). \end{cases}$$

Clearly f and g are continuous, and $gf = 1_X$. Also $fg \simeq 1_Y$ by Theorem 2.2.3, since $fg: Y \to Y$ is given by the same formula as g. Hence f is a homotopy equivalence. ▮

In fact the equivalence between X and Y in Example 2.2.10 is of a special type, known as a (strong) deformation retraction.

Definition 2.2.11 A subspace A of a topological space X is a *retract* of X if there exists a map $r: X \to A$ (called a *retraction*), such that $r(a) = a$ for all $a \in A$. If $i: A \to X$ denotes the inclusion map, then r is a *deformation retraction* (and A is a *deformation retract* of X) if $ir \simeq 1_X$. If also $ir \simeq 1_X$ rel A, then r is a *strong deformation retraction*, and A is a *strong deformation retract* of X.

For example, the map g in Example 2.2.10 is a strong deformation retraction.

Proposition 2.2.12 *If A is a deformation retract of X, then $A \simeq X$.* ▮

Example 2.2.13 If E^2 is the standard 2-cell in R^2, and 0 is the origin, then 0 is a strong deformation retract of E^2. For $r: E^2 \to 0$, defined by $r(x) = 0$ for all $x \in E^2$, is clearly a retraction, and is a strong deformation retraction by Theorem 2.2.3. ▮

Thus E^2 is homotopy-equivalent to the point 0. It is convenient to have a special name for such spaces.

Definition 2.2.14 A space X, homotopy-equivalent to a point, is called *contractible*.

Other examples of contractible spaces are E^n, the letter Y, and an empty bottle. The space S^1 is an example of a space that is not contractible (see Exercise 17).

We end this section with some remarks about the set of all continuous maps from a space X to a space Y. Now by Proposition 2.2.7 this set of maps splits up into disjoint equivalence classes, called *homotopy classes*. Let us write $[X, Y]$ for the set of homotopy classes of maps from X to Y; by keeping X fixed and varying Y, this set is an invariant of the homotopy type of Y, in the sense that there is a (1-1) correspondence between the sets corresponding to homotopy-equivalent spaces: see Exercise 5. Indeed, as we shall see in Chapter 6, the set $[X, Y]$ can often be endowed, in a natural way, with the structure of a group, and we then obtain exactly the sort of algebraic invariant described at the beginning of this section. Alternatively, we can keep Y fixed and vary X: once again a homotopy invariant results, which is in some ways easier to handle.

Of course, given two pairs (X, A) and (Y, B), we can similarly consider the set of homotopy classes of maps of pairs from (X, A) to

(Y, B), written $[(X, A), (Y, B)]$. This arises most frequently in the case where A and B are single points of X and Y respectively, called *base points*: a map of pairs is then called a *base-point-preserving* (or *based*) map, and a homotopy of maps of pairs is called a *based* homotopy. Notice that in this situation 'homotopy of maps of pairs' and 'homotopy relative to A' mean exactly the same.

2.3 Simplicial complexes

This section is concerned with building up spaces called *polyhedra*, from certain elementary spaces called *simplexes*. A simplex is just a generalization to n dimensions of a triangle or tetrahedron, and these are fitted together in such a way that two simplexes meet (if at all) in a common edge or face. In order to give the precise definition of a simplex, we must first explain what is meant by 'independent points' in Euclidean space.

Definition 2.3.1 A set of $(n + 1)$ points $a^0, a^1, \ldots, a^n$ in R^m is said to be *independent* if the vectors $a^1 - a^0, a^2 - a^0, \ldots, a^n - a^0$ are linearly independent. It is easy to see that this is equivalent to the statement that the equations

$$\sum_{i=0}^{n} \lambda_i a^i = 0, \qquad \sum_{i=0}^{n} \lambda_i = 0$$

(where $\lambda_0, \lambda_1, \ldots, \lambda_n$ are real numbers) imply that $\lambda_0 = \lambda_1 = \cdots = \lambda_n = 0$; hence the definition of independence does not depend on the order of the points $a^0, a^1, \ldots, a^n$.

For example, points a^0, a^1, a^2 in R^2 are independent if they are not collinear.

Definition 2.3.2 A *geometric n-simplex* σ_n is the set of points $\sum_{i=0}^{n} \lambda_i a^i$, where $a^0, a^1, \ldots, a^n$ are independent points in some Euclidean space R^m, and the λ_i are real numbers such that $\lambda_i \geqslant 0$ for all i and $\sum_{i=0}^{n} \lambda_i = 1$. This defines σ_n as a subset of R^m; σ_n is given the subspace topology.

The points $a^0, a^1, \ldots, a^n$ are called the *vertices* of σ_n, and are said to *span* σ_n; we write $(a^0, a^1, \ldots, a^n)$ for σ_n if we wish to specify the vertices.

The subspace of σ_n of those points $\sum \lambda_i a^i$ such that $\lambda_i > 0$ for all i is called the *interior* of σ_n (note that this is not the same as the 'interior' as defined in analytic topology: for example a 0-simplex coincides with its interior). One particular point in the interior of σ_n is the *barycentre*

$$\hat{\sigma}_n = \left(\frac{1}{n+1}\right)(a^0 + a^1 + \cdots + a^n).$$

If $a^{i_0}, a^{i_1}, \ldots, a^{i_r}$ is any subset of vertices of σ_n, the subspace of σ_n of those points linearly dependent on $a^{i_0}, a^{i_1}, \ldots, a^{i_r}$ is called a *face* of σ_n. Note that a face could quite well be empty or, at the other extreme, the whole of σ_n; a face is *proper* if it is neither of these.

Finally, the number n is called the *dimension* of σ_n.

Proposition 2.3.3 *A geometric n-simplex σ_n is a closed convex compact connected subspace of R^m, and is the closure of its interior. A face is a closed subspace of σ_n, and is itself a simplex. Moreover, a simplex determines its vertices, so that two simplexes coincide if and only if they have the same set of vertices.*

Proof. We prove only the assertion that a simplex σ_n determines its vertices. And this is almost immediate, since a point of σ_n is a vertex if and only if it is not a point of an open line segment lying within σ_n. ∎

We write $\tau < \sigma$ (or $\sigma > \tau$) for the statement 'the simplex τ is a face of the simplex σ'.

Now suppose that $\sigma_n = (a^0, a^1, \ldots, a^n)$ is a geometric n-simplex in R^m, and that $\tau_n = (b^0, b^1, \ldots, b^n)$ is a geometric n-simplex in R^p. Then σ_n and τ_n are homeomorphic, in a rather special way.

Proposition 2.3.4 *σ_n and τ_n are linearly homeomorphic, that is, there exists a homeomorphism $f: \sigma_n \to \tau_n$, such that*

$$f\left(\sum_{i=0}^{n} \lambda_i a^i\right) = \sum_{i=0}^{n} \lambda_i b^i,$$

for all points of σ_n.

Proof. *Define* $f: \sigma_n \to \tau_n$ by the formula $f(\sum \lambda_i a^i) = \sum \lambda_i b^i$. It is easy to see that f is continuous, and it is then obvious that f is a homeomorphism. ∎

It follows that a geometric n-simplex is completely characterized, up to homeomorphism, by its dimension.

We now wish to consider how simplexes may be fitted together to make more complicated spaces.

Definition 2.3.5 A *geometric simplicial complex* K is a finite set of simplexes, all contained in some Euclidean space R^m. Furthermore

(a) if σ_n is a simplex of K, and τ_p is a face of σ_n, then τ_p is in K;

(b) if σ_n and τ_p are simplexes of K, then $\sigma_n \cap \tau_p$ either is empty, or is a common face of σ_n and τ_p.

The *dimension* of K, dim K, is the maximum of the dimensions of its simplexes.

A *subcomplex* L of K is a subset of simplexes of K, satisfying property (a) (and hence also (b): see Proposition 2.3.6(c)). In particular, for each $r \geqslant 0$ the *r-skeleton* of K, K^r, is the subset of simplexes of dimension at most r.

A *simplicial pair* (K, L) consists of a simplicial complex K and a subcomplex L. Simplicial *triples*, etc., are similarly defined.

It is important to remember that a geometric simplicial complex K is *not a topological space*; it is merely a set whose elements are geometric simplexes. However, the set of points of R^m that lie in at least one of the simplexes of K, topologized as a subspace of R^m, is a topological space, called the *polyhedron* of K, written $|K|$; if L is a subcomplex of K, then $|L|$ is called a *subpolyhedron* of $|K|$. To illustrate this point, consider a single n-simplex σ_n in R^m. It is not itself a simplicial complex, but we can form a simplicial complex $K(\sigma_n)$ by taking as its elements σ_n, together with all faces of σ_n. The reader is invited to prove that $K(\sigma_n)$ is indeed a simplicial complex, and that $|K(\sigma_n)| = \sigma_n$; also that the set of all faces of σ_n other than σ_n itself forms a subcomplex of $K(\sigma_n)$, called the *boundary* of σ_n, written $\dot{\sigma}_n$.

Some elementary but important properties of simplicial complexes and polyhedra are collected together in the next proposition.

Proposition 2.3.6

(a) *If K is a simplicial complex, $|K|$ is a closed compact subspace of R^m.*

(b) *Every point of $|K|$ is in the interior of exactly one simplex of K. Conversely, if K is a set of simplexes in R^m satisfying Definition 2.3.5(a), and such that the interiors of distinct simplexes have empty intersection, then K is a simplicial complex.*

(c) *A subcomplex L of a simplicial complex K is itself a simplicial complex, and $|L|$ is a closed subspace of $|K|$.*

(d) *If L and M are subcomplexes of K, so are $L \cup M$ and $L \cap M$.*

Proof. Parts (a), (c) and (d) are easy, and are left as exercises for the reader. As for part (b), if K is a simplicial complex, every point of $|K|$ is obviously in the interior of at least one simplex; and if the interiors of two simplexes σ and τ meet, the common face μ, in which σ and τ intersect, meets the interiors of σ and τ, so that $\sigma = \mu = \tau$. Conversely, suppose that K is a set of simplexes in R^m, satisfying (a) of Definition 2.3.5, and such that the interiors of distinct simplexes are disjoint. Let

$$\sigma = (a^0, \ldots, a^r, b^{r+1}, \ldots, b^s) \quad \text{and} \quad \tau = (a^0, \ldots, a^r, c^{r+1}, \ldots, c^t)$$

be two simplexes of K, with no b^i equal to any c^j. Obviously the simplex $(a^0, \ldots, a^r)$ is contained in $\sigma \cap \tau$; and if x is any point of $\sigma \cap \tau$ we can write

$$x = \sum_{i=0}^{r} \lambda_i a^i + \sum_{i=r+1}^{s} \lambda_i b^i = \sum_{i=0}^{r} \mu_i a^i + \sum_{i=r+1}^{t} \mu_i c^i,$$

where $\sum \lambda_i = \sum \mu_i = 1$. Then $\lambda_{r+1} = \cdots = \lambda_s = \mu_{r+1} = \cdots = \mu_t = 0$, for otherwise by Proposition 2.3.3 x would be in the interior of two distinct simplexes. Hence also $\sigma \cap \tau$ is contained in $(a^0, \ldots, a^r)$, so that $\sigma \cap \tau$ is exactly the common face $(a^0, \ldots, a^r)$. Thus K satisfies (b) of Definition 2.3.5 and so is a simplicial complex. ∎

It is clear that if L is a subcomplex of K, $K - L$ is not in general a subcomplex, since a face of a simplex in $K - L$ could quite well be in L. However, we do at least have the following result.

Proposition 2.3.7 *There exists a subcomplex M of K, such that $|M| = \overline{|K| - |L|}$ (M is called the closure of $K - L$, written* cl $(K - L)$*).*

Proof. Let M be the set of simplexes of K, that are faces of simplexes of $K - L$. Clearly M is a subcomplex, and since each point of K is in the interior of a unique simplex, $|K| - |L| \subset |M|$, which is closed. But if x is any point in $|M|$, x is in a simplex that is a face of a simplex σ of $K - L$. Hence $x \in \sigma$, and every open neighbourhood of x meets $|K| - |L|$, at a point in the interior of σ. Thus

$$|M| \subset \overline{|K| - |L|}. \; ∎$$

$|K|$ has of course already been topologized, as a subspace of R^m. However, it is often more convenient to have the following alternative description of the topology.

Proposition 2.3.8 *A subset X of $|K|$ is closed if and only if $X \cap \sigma$ is closed in σ, for each simplex σ in K.*

Proof. Since each simplex σ is closed in R^m, it is also closed in $|K|$. Hence if $X \cap \sigma$ is closed in σ, it is also closed in $|K|$. Thus $X =$

$\bigcup_{\sigma \in K} X \cap \sigma$ is closed, since K is a finite set of simplexes. The converse is trivial. ∎

Corollary 2.3.9 *The topology of $|K|$ as a subspace of R^m is the same as the topology of $|K|$, considered as the space obtained from its simplexes by identifying together the various intersections.* ∎

So far we have been concerned exclusively with spaces, in the form of polyhedra and their associated simplicial complexes, and have said nothing about continuous maps. At first sight there is nothing to be said: given polyhedra $|K|$ and $|L|$ there seems to be no reason why a map from $|K|$ to $|L|$ should be anything more than continuous. However, $|K|$ and $|L|$ are more than just topological spaces: the simplicial complexes K and L endow them with further structure, and we ought to concentrate our attention on those maps $f\colon |K| \to |L|$ that in some sense preserve the simplicial structure. (The reader may like to compare the notion of a *ring homomorphism*: although every ring is a group, there is little point in considering *group* homomorphisms between rings that do not also 'preserve the multiplication'.) To this end, we make the following definition.

Definition 2.3.10 Given simplicial complexes K and L, a *simplicial map* $f\colon |K| \to |L|$ is a function from $|K|$ to $|L|$ with the following properties.

(a) If a is a vertex of a simplex of K, then $f(a)$ is a vertex of a simplex of L.

(b) If $(a^0, a^1, \ldots, a^n)$ is a simplex of K, then $f(a^0), f(a^1), \ldots, f(a^n)$ span a simplex of L (possibly with repeats).

(c) If $x = \sum \lambda_i a^i$ is in a simplex $(a^0, a^1, \ldots, a^n)$ of K, then $f(x) = \sum \lambda_i f(a^i)$; in other words, f is 'linear' on each simplex.

A *simplicial map of simplicial pairs* $f\colon (|K|, |L|) \to (|M|, |N|)$ is, of course, just a simplicial map $f\colon |K| \to |M|$ such that $f(|L|) \subset |N|$.

It is clear that the composite of two simplicial maps is another simplicial map.

We did not specify in Definition 2.3.10 that f was continuous, since this follows automatically from properties (a)–(c).

Proposition 2.3.11 *A simplicial map $f\colon |K| \to |L|$ is continuous.*

Proof. If X is a closed subset of $|L|$, $X \cap \tau$ is closed in τ for each simplex τ of L. But the restriction of f to any simplex of K, being linear, is continuous: thus $f^{-1}(X) \cap \sigma$ is closed in σ for each σ in K.

Hence $f^{-1}(X)$ is closed in $|K|$, by Proposition 2.3.8, and so f is continuous. ∎

Simplicial maps, then, are the correct 'structure-preserving' continuous maps between polyhedra. Indeed, as we shall see in the Simplicial Approximation Theorem, every continuous map between polyhedra can be approximated by a simplicial map, so that there is hardly any loss of generality in confining attention to simplicial maps.

There is a slight difficulty in the use of polyhedra, in that not every topological space that is homeomorphic to a polyhedron is itself a polyhedron. This difficulty is evaded by making another definition.

Definition 2.3.12 Given a topological space X, a *triangulation* of X consists of a simplicial complex K and a homeomorphism $h: |K| \to X$. A space with a triangulation is called a *triangulated space*. Similarly, if (X, A) is a pair of spaces, a *triangulation* consists of a simplicial pair (K, L) and a homeomorphism (of pairs) $h: (|K|, |L|) \to (X, A)$; (X, A) is a *triangulated pair*. Usually the particular homeomorphism h involved does not matter, and so we shall often refer—loosely—to K alone as a 'triangulation of X'.

It follows from Proposition 2.3.6 that a triangulated space is compact, normal and metrizable.

Example 2.3.13 In R^n, let $\bar{E}^n$ be the set of points $(x_1, x_2, \ldots, x_n)$ satisfying $\sum_{i=1}^{n} |x_i| \leqslant 1$, and let $\bar{S}^{n-1}$ be the subset where $\sum_{i=1}^{n} |x_i| = 1$. As in Section 1.4, the pair $(\bar{E}^n, \bar{S}^{n-1})$ is homeomorphic to the pair (E^n, S^{n-1}), by a homeomorphism that magnifies lines through the origin by suitable amounts: see Fig. 2.3 in the case $n = 2$.

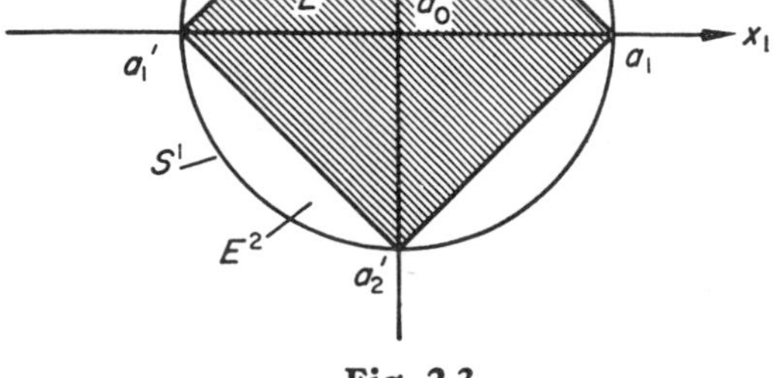

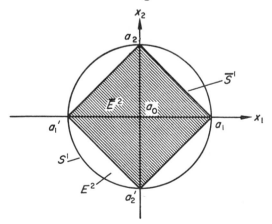

Fig. 2.3

We claim that $\bar{E}^n$ is a polyhedron, and $\bar{S}^{n-1}$ is a subpolyhedron. To prove this, take vertices a_0 at 0, a_i at $x_i = 1$ and a'_i at $x_i = -1$. Let K be the simplicial complex whose simplexes are all those of the form $(b_{i_0}, b_{i_1}, \ldots, b_{i_r})$, where $i_0 < i_1 < \cdots < i_r$ and b_{i_s} denotes a_{i_s} or a'_{i_s}. Certainly all such sets of vertices are independent, and K satisfies (a) and (b) of Definition 2.3.5, so that K is indeed a simplicial complex. Moreover if L denotes the subset of those simplexes not involving a_0, then L is a subcomplex of K, and $(|K|, |L|) = (\bar{E}^n, \bar{S}^{n-1})$. Hence (K, L) is a triangulation of (E^n, S^{n-1}).

Alternatively, another triangulation of (E^n, S^{n-1}) is $(K(\sigma), \dot{\sigma})$, where σ is any n-simplex. For if σ is an n-simplex in R^n, whose barycentre $\hat{\sigma}$ is at the origin, then since σ is convex, suitable magnification of lines through the origin provides a homeomorphism of the pair $(|K(\sigma)|, |\dot{\sigma}|)$ with (E^n, S^{n-1}). And by Proposition 2.3.4 $(|K(\sigma)|, |\dot{\sigma}|)$ is determined up to homeomorphism by the dimension n of σ. ∎

So far simplicial complexes have been sets of simplexes lying in one particular Euclidean space R^m, and we should now like to free ourselves of this restriction, by establishing an analogue for simplicial complexes of Proposition 2.3.4. In order to state this result precisely, it is necessary to introduce the notion of an *abstract simplicial complex.*

Definition 2.3.14 An *abstract simplicial complex* $\mathscr{K}$ is a finite set of elements $a^0, a^1, \ldots$, called (*abstract*) *vertices*, together with a collection of subsets $(a^{i_0}, a^{i_1}, \ldots, a^{i_n}), \ldots$, called (*abstract*) *simplexes*, with the property that any subset of a simplex is itself a simplex. The *dimension* of an abstract simplex is one less than the number of vertices in it, and the *dimension* of $\mathscr{K}$ is the maximum of the dimensions of its simplexes.

Let K be a geometric simplicial complex, and let $\mathscr{K}$ be an abstract simplicial complex whose vertices are in (1-1) correspondence with the vertices of K, a subset of vertices being a simplex of $\mathscr{K}$ if and only if they correspond to the vertices of some simplex of K. $\mathscr{K}$ is called an *abstraction* of K, and any geometric simplicial complex having $\mathscr{K}$ as an abstraction is called a *realization* of $\mathscr{K}$.

The point of this definition is that we can now state the analogue of Proposition 2.3.4 in the form: 'if K_1 and K_2 are any two realizations of an abstract simplicial complex $\mathscr{K}$, then $|K_1|$ and $|K_2|$ are simplicially homeomorphic'.

Theorem 2.3.15 *Let K_1 and K_2 be two realizations of an abstract simplicial complex $\mathscr{K}$. Then there exists a simplicial map $f\colon |K_1| \to |K_2|$,* *such that f is a homeomorphism (that is, f is a 'simplicial homeomorphism').*

Proof. Since K_1 and K_2 are both realizations of $\mathscr{K}$, there is a (1-1) correspondence between their vertices. Denote the vertices of K_1 by $a^0, a^1, \ldots$, and the vertices of K_2 by $b^0, b^1, \ldots$, where b^i corresponds to a^i. Thus $a^{i_0}, a^{i_1}, \ldots, a^{i_n}$ span a simplex of K_1 if and only if $b^{i_0}, b^{i_1}, \ldots, b^{i_n}$ span a simplex of K_2. We can therefore define a simplicial map $f\colon |K_1| \to |K_2|$ by setting $f(a^i) = b^i$ (all i), and requiring that f is linear on each simplex. Since it is obvious that f has a (simplicial) inverse, f is also a homeomorphism. ∎

This theorem allows us to forget about the particular Euclidean space in which a geometric simplicial complex lies, and to specify it by an abstraction. To justify this approach, however, we ought to establish that not only does every geometric simplicial complex have an abstraction, but also every abstract simplicial complex has a realization.

Theorem 2.3.16 *An n-dimensional abstract simplicial complex $\mathscr{K}$ has a realization in R^{2n+1}.*

Proof. Let the vertices of $\mathscr{K}$ be $a^0, a^1, \ldots, a^m$. We first choose $(m + 1)$ points in R^{2n+1}, with the property that any $(2n + 2)$ of them are independent (such points are said to be *in general position*). This can be done by defining

$$A^r = (r, r^2, \ldots, r^{2n+1}) \qquad (0 \leqslant r \leqslant m);$$

if say $A^{r_1}, A^{r_2}, \ldots, A^{r_{2n+2}}$ are not independent, there exist real numbers $\lambda_1, \lambda_2, \ldots, \lambda_{2n+2}$, not all zero, such that

$$\lambda_1 + \lambda_2 + \cdots + \lambda_{2n+2} = 0,$$

$$\lambda_1 r_1 + \lambda_2 r_2 + \cdots + \lambda_{2n+2} r_{2n+2} = 0,$$

$$\cdots$$

$$\lambda_1 r_1^{2n+1} + \lambda_2 r_2^{2n+1} + \cdots + \lambda_{2n+2} r_{2n+2}^{2n+1} = 0.$$

But the determinant of this set of linear equations is $\prod_{i>j} (r_i - r_j)$. This is non-zero, so that no such numbers λ_i can exist, and $A^0, A^1, \ldots, A^m$ are in general position.

Now let the point A^r correspond to a^r $(0 \leqslant r \leqslant m)$, and 'fill in' simplexes in R^{2n+1} corresponding to the simplexes of $\mathscr{K}$: since $\mathscr{K}$ is n-dimensional, the points corresponding to any simplex of $\mathscr{K}$ are independent. It is also clear that property (a) of Definition 2.3.5 is satisfied, so that it remains to check property (b). To do so, let σ_p and τ_q be two of the simplexes (of dimensions p and q respectively) that we have 'filled in', and suppose that σ_p and τ_q have r vertices in common. The number of vertices in either σ_p or τ_q is thus $p + q - r + 2 \leqslant 2n + 2$, so that these vertices are independent, and could be taken to be the vertices of a $(p + q - r + 1)$-simplex having σ_p and τ_q as faces. Thus $\sigma_p \cap \tau_q$ is either empty or a common face. ∎

The result of Theorem 2.3.16 is 'best possible', in the sense that for each $n \geqslant 0$, there exists an n-dimensional abstract simplicial complex that cannot be realized in R^{2n}: see Exercise 9. Of course, a particular complex $\mathscr{K}$ may be realizable in Euclidean space of dimension less than $(2n + 1)$: the determination of this dimension in special cases is one of the most interesting problems of algebraic topology.

We end Section 2.3 with another example of the use of abstract simplicial complexes, in defining the *join* of two simplicial complexes.

Definition 2.3.17 Let K and L be two geometric simplicial complexes, and let $\mathscr{K}$ and $\mathscr{L}$ be abstractions, with vertices $a^0, a^1, \ldots$ and $b^0, b^1, \ldots$ respectively. The *join* $\mathscr{K} * \mathscr{L}$ is defined to be the abstract simplicial complex whose vertices are $a^0, a^1, \ldots, b^0, b^1, \ldots$, and whose simplexes are all subsets $(a^{i_0}, a^{i_1}, \ldots, b^{j_0}, b^{j_1}, \ldots)$ such that $(a^{i_0}, a^{i_1}, \ldots)$ is a simplex of $\mathscr{K}$ and $(b^{j_0}, b^{j_1}, \ldots)$ is a simplex of $\mathscr{L}$ (the special cases $(a^{i_0}, a^{i_1}, \ldots)$ and $(b^{j_0}, b^{j_1}, \ldots)$ are allowed as simplexes of $\mathscr{K} * \mathscr{L}$). Any realization of $\mathscr{K} * \mathscr{L}$ is called the *join* of K and L, written $K * L$; this is defined up to simplicial homeomorphism, by Theorem 2.3.15.

It is clear that the join construction is associative, in the sense that $(K * L) * M = K * (L * M)$. Thus we can write $K * L * M * \cdots$ unambiguously for the join of more than two simplicial complexes.

Example 2.3.18 The triangulation L of S^{n-1}, constructed in Example 2.3.13, can be regarded as $L_1 * L_2 * \cdots * L_n$, where L_r is the simplicial complex consisting only of the two 0-simplexes a_r and a'_r. Similarly the triangulation K of E^n is $a_0 * L$. ∎

If K is a simplicial complex in R^m and L is a simplicial complex in R^n, we can construct a representative for $K * L$ in R^{m+n+1} as follows. Since $R^{m+n+1} = R^m \times R^n \times R^1$, a point of R^{m+n+1} can be specified by three co-ordinates (x, y, z), where $x \in R^m$, $y \in R^n$ and $z \in R^1$; also K and L may be thought of as simplicial complexes in R^{m+n+1}, by regarding R^m as $(R^m, 0, 0)$ and R^n as $(0, R^n, 1)$. Now if $(a^{i_0}, \ldots, a^{i_r})$ and $(b^{j_0}, \ldots, b^{j_s})$ are simplexes of K and L respectively, the points

$(a^{i_0}, 0, 0), \ldots, (a^{i_r}, 0, 0)$, $(0, b^{j_0}, 1), \ldots, (0, b^{j_s}, 1)$ are independent, since the equations

$$\lambda_0 + \cdots + \lambda_r + \mu_0 + \cdots + \mu_s = 0,$$

$$\lambda_0(a^{i_0}, 0, 0) + \cdots + \lambda_r(a^{i_r}, 0, 0) + \mu_0(0, b^{j_0}, 1) + \cdots + \mu_s(0, b^{j_s}, 1) = (0, 0, 0)$$

clearly imply $\lambda_0 = \cdots = \lambda_r = \mu_0 = \cdots = \mu_s = 0$. Thus all the simplexes of $K * L$ can be filled in; and to show that this has constructed a realization of $\mathcal{K} * \mathcal{L}$, it is sufficient, by Proposition 2.3.6(b), to show that the interiors of distinct simplexes are disjoint. This is obvious for simplexes of K or L, and if x is in the interior of $(a^{i_0}, \ldots, a^{i_r}, b^{j_0}, \ldots, b^{j_s})$, it is easy to see that x has the form $((1 - \lambda)y, \lambda z, \lambda)$, where y is in the interior of $(a^{i_0}, \ldots, a^{i_r})$, z is in the interior of $(b^{j_0}, \ldots, b^{j_s})$, and $0 < \lambda < 1$. But the co-ordinates of x fix λ, y and z, so that x cannot be in the interior of any other simplex.

It follows that $|K * L|$ may be regarded as the set of points $((1 - \lambda)y, \lambda z, \lambda)$ in R^{m+n+1}, for all $y \in |K|$, $z \in |L|$ and $0 \leqslant \lambda \leqslant 1$. Consequently, given two more simplicial complexes M and N, and continuous (not necessarily simplicial) maps $f\colon |K| \to |M|$, $g\colon |L| \to |N|$, we obtain a continuous map $f * g\colon |K * L| \to |M * N|$ by setting $(f * g)((1 - \lambda)y, \lambda z, \lambda) = ((1 - \lambda)f(y), \lambda g(z), \lambda)$. In particular, if f and g are homeomorphisms, so is $f * g$, since it has an obvious inverse. This means that we can unambiguously write $|K| * |L|$ for $|K * L|$: for example, since each L_r in Example 2.3.18 is a triangulation of S^0, it makes sense to say that S^{n-1} is homeomorphic to the join of n copies of S^0. (Indeed, one can define the join of any two topological spaces: see Chapter 6, Exercise 3.)

2.4 Homotopy and homeomorphism of polyhedra

This section is concerned with some general results about homotopy and homeomorphism of polyhedra that will be needed later. The reader may care to miss this section at first reading, therefore, and return to it when necessary for the proofs of these results.

The first theorem states that polyhedral pairs possess the *absolute homotopy extension property*: that is, any homotopy of the subpolyhedron can be extended to a homotopy of the large polyhedron, so as to start with any given continuous map.

Theorem 2.4.1 *Let (K, L) be a simplicial pair. Given a space X, a homotopy $F\colon |L| \times I \to X$, and a map $g\colon |K| \to X$ such that the* *restriction of g to* $|L|$ *is the restriction of F to* $|L| \times 0$, *there exists a homotopy* $G: |K| \times I \to X$, *such that the restriction of G to* $|K| \times 0$ *is g and the restriction of G to* $|L| \times I$ *is F.*

Proof. Given a simplex σ of K, let $\rho: \sigma \times I \to (|\dot{\sigma}| \times I) \cup (\sigma \times 0)$ be the projection map from $(\hat{\sigma}, 2)$, where $\hat{\sigma}$ is the barycentre of σ. Clearly ρ is a retraction: see Fig. 2.4.

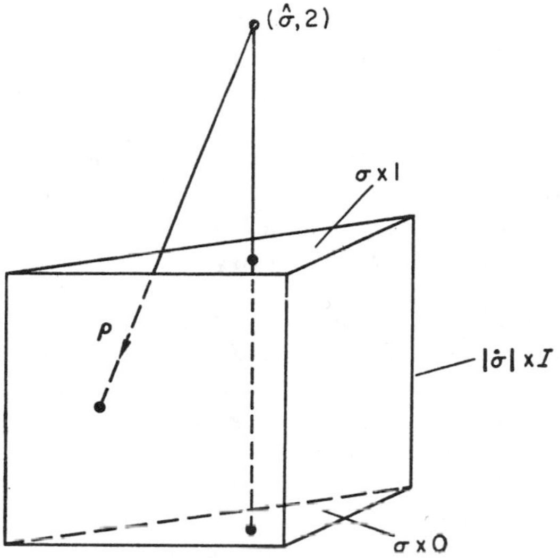

Fig. 2.4

Now if we write M^r for $|K^r| \cup |L|$, these retractions can be fitted together, by Proposition 1.4.15(d), to yield a retraction

$$\rho: (M^r \times I) \cup (|K| \times 0) \to (M^{r-1} \times I) \cup (|K| \times 0)$$

and hence, by induction on r, a retraction

$$\rho: (|K| \times I) \to (|L| \times I) \cup (|K| \times 0).$$

But F and g fit together to give a map F, say, from $(|L| \times I) \cup (|K| \times 0) \to X$; the composite $F\rho: |K| \times I \to X$ is then the homotopy G that we require. ∎

For an example of a pair of spaces that does not possess the absolute homotopy extension property, see Exercise 11.

The other important theorem in this section concerns the problem of deciding when two polyhedra are homeomorphic. The usual practical method is that outlined at the beginning of Section 2.2, which will be developed further in Chapters 3 and 4. As has already been pointed out, however, the algebraic invariants constructed there suffer from the disadvantage that they are homotopy-type invariants, and so at first sight are useless for distinguishing between two polyhedra that are homotopy-equivalent, but not homeomorphic.

There is however a trick that can sometimes be used to overcome this disadvantage. The idea is that, given simplicial complexes K and L, one should construct certain subcomplexes whose polyhedra are homotopy-equivalent if $|K|$ and $|L|$ are homeomorphic, but not necessarily if $|K|$ and $|L|$ are merely homotopy-equivalent. The algebraic machinery can be applied in favourable circumstances to show that the subpolyhedra are not homotopy-equivalent, so that $|K|$ and $|L|$ are not homeomorphic, even though it may happen that $|K| \simeq |L|$.

In order to state and prove the theorem involved, a few preliminary definitions and results are necessary.

Definition 2.4.2 Let K be a simplicial complex. For each point x of $|K|$, the *simplicial neighbourhood* of x, $\mathrm{N}_K(x)$, is the set of simplexes of K that contain x, together with all their faces. The ***link*** of x, $\mathrm{Lk}_K(x)$, is the subset of simplexes of $\mathrm{N}_K(x)$ that do not contain x. Clearly $\mathrm{N}_K(x)$ and $\mathrm{Lk}_K(x)$ are subcomplexes of K.

For each simplex σ of K, the ***star*** of σ, $\mathrm{st}_K(\sigma)$, is the union of the interiors of the simplexes of K that have σ as a face.

The suffix K will often be omitted, if it is clear to which complex we refer.

For examples of $\mathrm{N}(x)$, $\mathrm{Lk}(x)$ and $\mathrm{st}(\sigma)$, see Fig. 2.5.

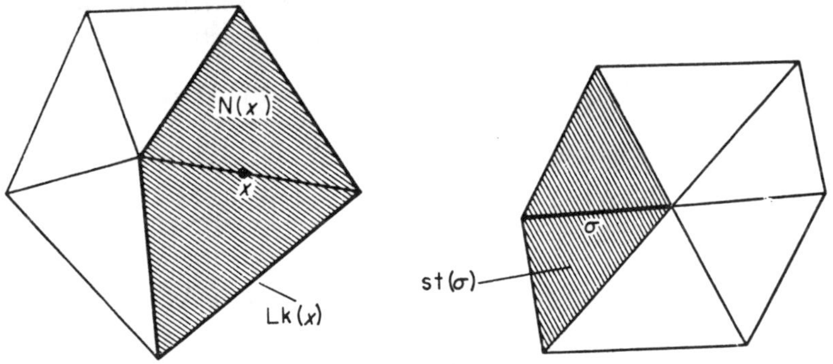

Fig. 2.5

Now Proposition 2.3.6(b) shows that each x in $|K|$ is in the interior of a unique simplex σ of K, so that it is easy to relate $\mathrm{N}(x)$, $\mathrm{Lk}(x)$ and $\mathrm{st}(\sigma)$.

Proposition 2.4.3 *For each simplex σ of K,* st(σ) *is an open set. If x is any point in the interior of σ, then*

$$\mathrm{st}(\sigma) = |\mathrm{N}(x)| - |\mathrm{Lk}(x)|.$$

Proof. Let K_0 be the set of simplexes of K that do not have σ as a face. Clearly K_0 is a subcomplex, so that $|K_0|$ is closed by Proposition 2.3.6(c). But Proposition 2.3.6(b) shows that $\mathrm{st}(\sigma) = |K| - |K_0|$, which is therefore open. Similarly $\mathrm{st}(\sigma) = |\mathrm{N}(x)| - |\mathrm{Lk}(x)|$ for any x in the interior of σ. ∎

N(x) and Lk(x) also have convenient 'convexity' properties.

Proposition 2.4.4 *If $y \in |\mathrm{N}(x)|$, then all points on the straight-line segment xy lie in $|\mathrm{N}(x)|$. Moreover, each straight-line segment starting from x meets* Lk(x) *in exactly one point.*

Proof. If $y \in |\mathrm{N}(x)|$, then y is in a simplex τ that contains x. By Proposition 2.3.3, τ is convex, so that all points of the segment xy lie in τ, and hence are in $|\mathrm{N}(x)|$.

Now consider a straight-line segment l starting from x, and let y be the 'last point' in $l \cap |\mathrm{N}(x)|$; more precisely, let y be the point on l for which $d(x, y) = \sup \{d(x, y') \mid y' \in l \cap |\mathrm{N}(x)|\}$: see Fig. 2.6.

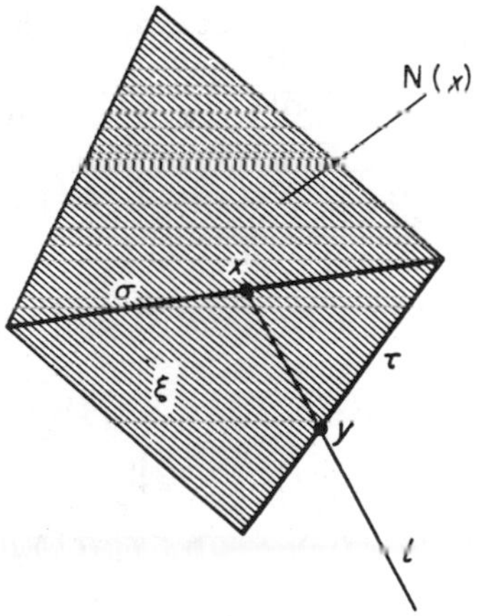

Fig. 2.6

Then $y \neq x$, since $|\mathrm{N}(x)| \supset \mathrm{st}(\sigma)$, which is open ($\sigma$ is the simplex that contains x in its interior). On the other hand $|\mathrm{N}(x)|$ is closed, and so contains y. Thus y is in the interior of τ, say, where $x \notin \tau$, for otherwise we could produce the segment xy further and still remain in $|\mathrm{N}(x)|$. That is, $y \in |\mathrm{Lk}(x)|$.

It remains to prove that no other point of l is in $|\mathrm{Lk}(x)|$. Now points of l further from x than y are not in $|\mathrm{N}(x)|$, and so are certainly not in $|\mathrm{Lk}(x)|$. On the other hand, if ξ is the simplex spanned by the vertices of σ and τ, then all points of xy other than x and y are in the interior of ξ (ξ exists since τ must be a face of a simplex containing x, and so containing σ). Thus all points of xy except y are in $\mathrm{st}(\sigma)$, and so are not in $|\mathrm{Lk}(x)|$. ∎

The main theorem states that, given simplicial complexes K and L, and a homeomorphism $f\colon |K| \to |L|$, then $|\mathrm{Lk}_K(x)| \simeq |\mathrm{Lk}_L(f(x))|$ for each $x \in |K|$. It is convenient, however, to prove a slightly more general result.

Theorem 2.4.5 *Let K and L be simplicial complexes, and let $f\colon |K| \to |L|$ be a homeomorphism onto a subspace of $|L|$. Then for each $x \in |K|$ such that $f(x)$ is contained in an open set U of $|L|$, with U contained in $f(|K|)$, we have $|\mathrm{Lk}_K(x)| \simeq |\mathrm{Lk}_L(f(x))|$.*

Proof. Suppose that $f(x)$ is in the interior of a simplex σ of L. Then $f(x) \in U \cap \mathrm{st}(\sigma) \subset |\mathrm{N}_L(f(x))|$, so that $f^{-1}(U \cap \mathrm{st}(\sigma))$ is an open set containing x, whose image under f is contained in $|\mathrm{N}(f(x))|$. For each real number λ, with $0 < \lambda \leqslant 1$, let $\lambda|\mathrm{N}_K(x)|$ be the set of points of $|\mathrm{N}(x)|$ of the form $(1 - \lambda)x + \lambda y$, where $y \in |\mathrm{N}(x)|$: thus $\lambda|\mathrm{N}(x)|$ is $|\mathrm{N}(x)|$ 'magnified by a factor λ', and $\lambda|\mathrm{N}(x)|$ is homeomorphic to $|\mathrm{N}(x)|$. Since $f^{-1}(U \cap \mathrm{st}(\sigma))$ is open, and $|\mathrm{N}(x)|$ is bounded, there exists such a λ, so that

$$x \in \lambda|\mathrm{N}(x)| \subset f^{-1}(U \cap \mathrm{st}(\sigma)),$$

and hence

$$f(x) \in f(\lambda|\mathrm{N}(x)|) \subset |\mathrm{N}(f(x))|.$$

Similarly, there exist μ and ν, such that

$$f(x) \in f(\nu|\mathrm{N}(x)|) \subset \mu|\mathrm{N}(f(x))| \subset f(\lambda|\mathrm{N}(x)|) \subset |\mathrm{N}(f(x))|:$$

see Fig. 2.7.

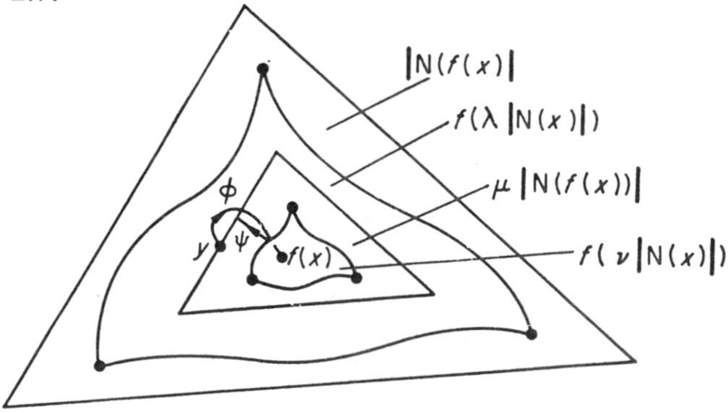

Fig. 2.7

With an obvious notation, let y be a point of $\mu|\text{Lk}(f(x))|$. By Proposition 2.4.4, we can define a map $\phi: \mu|\text{Lk}(f(x))| \to f(\nu|\text{Lk}(x)|)$ by projecting $f^{-1}(y)$ along the straight line through x in $\lambda|\text{N}(x)|$ and then applying f. Similarly, we can define $\psi: f(\nu|\text{Lk}(x)|) \to \mu|\text{Lk}(f(x))|$ by radial projection from $f(x)$ in $\mu|\text{N}(f(x))|$: see Fig. 2.7.

Now let $F: \mu|\text{Lk}(f(x))| \times I \to |\text{N}(f(x))| - f(x)$ be the homotopy formed by 'sticking together', as in the proof of Proposition 2.2.7, the linear homotopy between f^{-1} and $f^{-1}\phi$, composed with f, and the linear homotopy between ϕ and $\psi\phi$. Thus F is a homotopy between 1 and $\psi\phi$. So if $g: |\text{N}(f(x))| - f(x) \to \mu|\text{Lk}(f(x))|$ is the radial projection map from $f(x)$, the composite

$$gF: \mu|\text{Lk}(f(x))| \times I \to \mu|\text{Lk}(f(x))|$$

is a homotopy between 1 and $\psi\phi$. Similarly $\phi\psi \simeq 1$, so that $\mu|\text{Lk}(f(x))| \simeq f(\mu|\text{Lk}(x)|)$. Since $\mu|\text{Lk}(f(x))|$ is homeomorphic to $|\text{Lk}(f(x))|$ and $f(\mu|\text{Lk}(x)|)$ to $|\text{Lk}(x)|$, this proves that $|\text{Lk}(f(x))| \simeq |\text{Lk}(x)|$. ∎

The result of Theorem 2.4.5 need not be true if $f: |K| \to |L|$ is merely a homotopy equivalence. For example, let $K = K(\sigma)$, where σ is a 2-simplex, and let L be a single vertex. Certainly $|K| \simeq |L|$ as in Example 2.2.13; but if x is in the interior of σ, then $|\text{Lk}_K(x)|$ is homeomorphic to S^1, and if a is the vertex of L, $|\text{Lk}_L(a)|$ is empty.

It follows, of course, that $|K|$ is not homeomorphic to $|L|$. Indeed, the same method will show that no two simplicial complexes of different dimensions can have homeomorphic polyhedra, although since the proof involves some homology theory, we must postpone it to Chapter 4.

2.5 Subdivision and the Simplicial Approximation Theorem

When simplicial maps were introduced in Section 2.3, it was remarked that any continuous map between polyhedra may be approximated by a simplicial map. The purpose of this section is to make this statement precise, and to prove it.

A map g is regarded as an 'approximation' to a map f if f and g are homotopic. Thus we seek to construct simplicial maps that are homotopic to a given continuous map, and these simplicial maps will usually be ***simplicial approximations***, in the sense of the following definition.

Definition 2.5.1 Given simplicial complexes K and L, and a continuous map $f: |K| \to |L|$, a simplicial map $g: |K| \to |L|$ is called a *simplicial approximation* to f if, for each vertex a of K, $f(\mathrm{st}_K(a)) \subset \mathrm{st}_L(g(a))$ (see Definition 2.4.2).

Notice that a simplicial map is always a simplicial approximation to itself. For if f is a simplicial map, any simplex of K having a as a vertex is mapped by f to a simplex of L having $f(a)$ as a vertex; hence $f(\mathrm{st}_K(a)) \subset \mathrm{st}_L(f(a))$ for each vertex a of K.

Before justifying the introduction of simplicial approximations, by showing that they are always homotopic to the original maps, it is useful to have a criterion for their existence.

Proposition 2.5.2 *Let K and L be simplicial complexes, and let $f\colon |K| \to |L|$ be a continuous map. If, for each vertex a of K, a vertex b of L can be found, such that $f(\mathrm{st}_K(a)) \subset \mathrm{st}_L(b)$, then there exists a simplicial approximation g to f, such that $g(a) = b$ for each vertex of K.*

Proof. It is necessary only to check that $g(a^0), g(a^1), \ldots, g(a^n)$ span a simplex of L whenever $a^0, a^1, \ldots, a^n$ span a simplex of K, since g can then be extended linearly to the interiors of the simplexes of K.

Let x be a point in the interior of the simplex $(a^0, a^1, \ldots, a^n)$. Then

$$x \in \mathrm{st}(a^0) \cap \mathrm{st}(a^1) \cap \cdots \cap \mathrm{st}(a^n).$$

Thus

$$\begin{aligned} f(x) &\in f(\mathrm{st}(a^0)) \cap f(\mathrm{st}(a^1)) \cap \cdots \cap f(\mathrm{st}(a^n)) \\ &\subset \mathrm{st}(g(a^0)) \cap \mathrm{st}(g(a^1)) \cap \cdots \cap \mathrm{st}(g(a^n)). \end{aligned}$$

So the unique simplex of L that contains $f(x)$ in its interior must have each $g(a^i)$ as a vertex, and so has a face spanned by $g(a^0), g(a^1), \ldots, g(a^n)$. ∎

We show now that a simplicial approximation is homotopic to the original map.

Theorem 2.5.3 *Let K and L be simplicial complexes, and let $f\colon |K| \to |L|$ be a continuous map. Then any simplicial approximation g to f is homotopic to f. Moreover, the homotopy is relative to the subspace of $|K|$ of those points x such that $f(x) = g(x)$.*

Proof. Take a point x of $|K|$, and suppose that x is in the interior of the simplex $(a^0, a^1, \ldots, a^n)$. By the proof of Proposition 2.5.2, $f(x)$ lies in the interior of a simplex of L that has each $g(a^i)$ as a vertex, and so also contains $g(x)$. It follows that the straight-line segment joining $f(x)$ and $g(x)$ is contained in $|L|$, and so f and g are homotopic by Theorem 2.2.3. By construction, this homotopy is relative to the subspace of $|K|$ where f and g coincide. ∎

Corollary 2.5.4 *Let (K, L) and (M, N) be simplicial pairs, and let $f: (|K|, |L|) \to (|M|, |N|)$ be a map of pairs. If g is any simplicial approximation to $f: |K| \to |M|$, then $g(|L|) \subset |N|$, and $f \simeq g$ as maps of pairs.*

Proof. Let x be any point of $|L|$. Then $f(x)$, being in $|N|$, is in the interior of a unique simplex of N, that also contains $g(x)$: that is, $g(x) \in |N|$. Moreover, the line segment joining $f(x)$ and $g(x)$ is also contained in $|N|$. ∎

Not surprisingly, the composite of two simplicial approximations is again a simplicial approximation.

Proposition 2.5.5 *Given simplicial complexes K, L and M, continuous maps $f_1: |K| \to |L|$ and $f_2: |L| \to |M|$, and simplicial approximations g_1, g_2 to f_1, f_2 respectively, then $g_2 g_1$ is a simplicial approximation to $f_2 f_1$.*

Proof. For each vertex a of K,

$$\begin{aligned} f_2 f_1(\mathrm{st}_K(a)) &\subset f_2(\mathrm{st}_L(g_1(a))) \\ &\subset \mathrm{st}_M(g_2 g_1(a)). \end{aligned}$$ ∎

Example 2.5.6 Let K be the simplicial complex consisting of 1-simplexes (a^0, a^1), (a^1, a^2), (a^2, a^3) and all their vertices, and let L be the simplicial complex consisting of 2-simplexes (b^0, b^1, b^2), (b^0, b^2, b^3), (b^1, b^2, b^3), (b^1, b^3, b^4) and all their faces. Let $f: |K| \to |L|$ be the continuous map taking a^i to c^i $(0 \leqslant i \leqslant 3)$, as shown in Fig. 2.8.

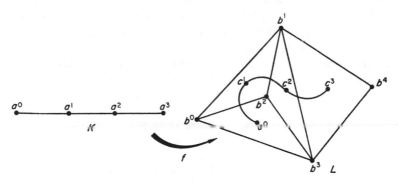

Fig. 2.8

Now $\qquad f(\mathrm{st}(a^0)) \subset \mathrm{st}(b^0) \cap \mathrm{st}(b^2),$

$\qquad f(\mathrm{st}(a^1)) \subset \mathrm{st}(b^2),$

$\qquad f(\mathrm{st}(a^2)) \subset \mathrm{st}(b^1),$

and $\qquad f(\mathrm{st}(a^3)) \subset \mathrm{st}(b^1) \cap \mathrm{st}(b^3).$

Thus one possible simplicial approximation to f is g, the simplicial map sending a^0, a^1, a^2, a^3 to b^0, b^2, b^1, b^3 respectively. ∎

Two points should be noticed about Example 2.5.6.

(a) Another simplicial approximation to f is g', the simplicial map sending a^0, a^1, a^2, a^3 to b^2, b^2, b^1, b^1 respectively. Thus if a simplicial approximation exists, it may not be unique. However, Theorem 2.5.3 assures us that any two simplicial approximations to f are each homotopic to f, and so are themselves homotopic.

(b) If the vertices a^1 and a^2 were removed from K, thus making K into $K(a^0, a^3)$, there would be no simplicial approximation to f, since then $f(\text{st}(a^0))$ would be $f(|K|)$, which is not contained in the star of any vertex of L. Thus not every map has a simplicial approximation.

At first sight the situation revealed in (b) means that our search for simplicial approximations is bound to fail in general. However, the reason for the lack of a simplicial approximation in (b) is that the simplexes of $K(a^0, a^3)$ are too large: if (a^0, a^3) is subdivided by reintroducing the vertices a^1 and a^2, the simplexes are then small enough to make the method of Proposition 2.5.2 work. This is the situation we face in general: there may be no simplicial approximation to a given continuous map $f\colon |K| \to |L|$, but if the simplexes of K are subdivided enough, a simplicial approximation can always be shown to exist, by using Theorem 1.4.35.

We must obviously investigate more closely the idea of subdivision. In general, a subdivision of a simplicial complex K is another simplicial complex K', obtained by 'chopping up' the simplexes of K. A systematic way of doing this is to introduce a new vertex at the barycentre of each simplex of K, and then to join up the vertices. For example, let K be the complex $K(a^0, a^1, a^2)$ formed from a single 2-simplex. The new vertices are $b^0 = \frac{1}{2}(a^1 + a^2)$, $b^1 = \frac{1}{2}(a^2 + a^0)$, $b^2 = \frac{1}{2}(a^0 + a^1)$ and $c = \frac{1}{3}(a^0 + a^1 + a^2)$; these are joined up as in Fig. 2.9.

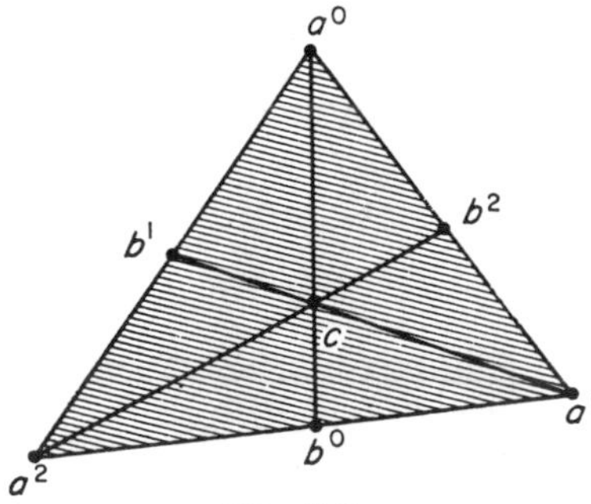

Fig. 2.9

In this way K is replaced by a new simplicial complex K', that has more, but smaller, simplexes than K. Obviously the process of 'barycentric subdivision' could be repeated as often as necessary to make the simplexes as small as we please.

In practice it may be necessary to subdivide only a part of a simplicial complex K, so as to leave alone a given subcomplex. For example, in Fig. 2.9 we might not wish to subdivide the subcomplex $L = K(a^0, a^2)$. This can be done by omitting the vertex b^1 and the simplex (c, b^1) in the subdivision, so as to retain (a^0, a^2, c) as a single simplex. Such a subdivision is called a subdivision *relative to* L, and the subdivided simplicial complex is called the *derived complex* of K, relative to L. The precise definition proceeds by induction on the dimensions of the skeletons of K.

Definition 2.5.7 Let L be a subcomplex of a simplicial complex K. The *derived complex of* K, *relative to* L, written $(K, L)'$, is defined as follows.

Let $M^n = K^n \cup L$, a subcomplex of K that contains L. Define $(M^0, L)' = M^0$, and suppose, inductively, that $(M^r, L)'$ has been defined for all $0 \leqslant r < n$, in such a way that

(a) $(M^r, L)'$ is a simplicial complex, containing L as a subcomplex;
(b) $|(M^r, L)'| = |M^r|$;
(c) each simplex of $(M^r, L)'$ is contained in a simplex of M^r;
(d) if N is a subcomplex of M^r, there exists a subcomplex N' of $(M^r, L)'$ such that $|N| = |N'|$.

Certainly (a)–(d) are satisfied if $r = 0$. Now if σ is an n-simplex of $K - L$, the boundary $\dot{\sigma}$ is a subcomplex of M^{n-1}, so that by (d) there exists a subcomplex $(\dot{\sigma})'$ of $(M^{n-1}, L)'$ such that $|\dot{\sigma}| = |(\dot{\sigma})'|$. If $\tau = (b^0, b^1, \ldots, b^r)$ is a simplex of $(\dot{\sigma})'$, write $\hat{\sigma}\tau$ for the simplex $(\hat{\sigma}, b^0, \ldots, b^r)$, where $\hat{\sigma}$ is the barycentre of σ (it follows from (c) that the vertices of $\hat{\sigma}\tau$ are independent). Define

$$(M^n, L)' = (M^{n-1}, L)' \cup \{\hat{\sigma}\tau\} \cup \{(\hat{\sigma})\},$$

where σ runs through all n-simplexes of $K - L$, and τ through all simplexes in each $(\dot{\sigma})'$.

To justify this definition, we must check that $(M^n, L)'$ also satisfies conditions (a)–(d).

Proposition 2.5.8 *$(M^n, L)'$ satisfies* (a)–(d).

Proof.

(a) We have to check (a) and (b) of Definition 2.3.5. In (b), three cases arise.

(i) Let $\xi \in (M^{n-1}, L)'$, and η be of form $\hat{\sigma}\tau$: then $\xi \cap \eta = \xi \cap \tau$ since $\sigma \notin M^{n-1}$, which by induction either is empty or is a common face.

(ii) If ξ, η are of form $\hat{\sigma}\tau$, $\hat{\sigma}\mu$ respectively, then $\xi \cap \eta = \hat{\sigma}(\tau \cap \mu)$; again $\tau \cap \mu$ is either empty or a common face.

(iii) Lastly, if ξ, η are of form $\hat{\sigma}\tau$, $\hat{\nu}\mu$, where $\sigma \neq \nu$, then $\xi \cap \eta = \tau \cap \mu$.

To prove (a), it is sufficient to consider a simplex of the form $\hat{\sigma}\tau$, where σ is an n-simplex of $K - L$. Its faces are of the form $(\hat{\sigma})$, ξ, or $\hat{\sigma}\xi$, where ξ is a face of τ; clearly each of these is in $(M^n, L)'$.

(b) We have

$$\begin{aligned} |(M^n, L)'| &= |(M^{n-1}, L)'| \cup \bigcup (\hat{\sigma}\tau) \\ &= |M^{n-1}| \cup \bigcup (\hat{\sigma}\tau). \end{aligned}$$

On the other hand $|M^n| = |M^{n-1}| \cup \bigcup (\sigma)$, where σ runs over the n-simplexes of $K - L$. For each such σ, and $\tau \in (\dot{\sigma})'$, we have $\hat{\sigma}\tau \subset \sigma$ since $\tau \subset |\dot{\sigma}|$; conversely $\sigma \subset \bigcup (\hat{\sigma}\tau)$, for $\tau \in (\dot{\sigma})'$, since the union of such τ is $|(\dot{\sigma})'| = |\dot{\sigma}|$. Hence $\bigcup (\hat{\sigma}\tau) = \bigcup (\sigma)$, and $|(M^n, L)'| = |M^n|$.

(c) Obviously if τ is contained in a simplex of $\dot{\sigma}$, then $\hat{\sigma}\tau \subset \sigma$.

(d) $N \cap M^{n-1}$ is a subcomplex of M^{n-1}, by Proposition 2.3.6(d).

Thus there exists a subcomplex P' of $(M^{n-1}, L)'$, such that $|N \cap M^{n-1}| = |P'|$. Define

$$N' = P' \cup \{\hat{\sigma}\tau\} \cup \{(\hat{\sigma})\},$$

for all n-simplexes σ of $(K - L) \cap N$, and all τ in each $(\dot{\sigma})'$. As in the proof of (a) and (b), N' is a subcomplex of $(M^n, L)'$, and $|N| = |N'|$. (Really all we have done is to define $N' = (N, L \cap N)'$.) ∎

Finally, define $(K, L)' = (M^m, L)'$, where K has dimension m.

Thus $(K, L)'$ is a simplicial complex such that $|(K, L)'| = |K|$, every simplex of $(K, L)'$ is contained in a simplex of K, and for any subcomplex N of K, there exists a subcomplex N' of $(K, L)'$ such that $|N| = |N'|$.

If L happens to be empty, K is called just the *derived complex of* K, and is usually written K'.

Example 2.5.9 Let K be the simplicial complex consisting of the 2-simplexes (a^0, a^1, a^2), (a^0, a^2, a^3) and (a^2, a^3, a^4), together with all their faces, and let L be the subcomplex consisting of (a^0, a^1, a^2) and its faces: see Fig. 2.10.

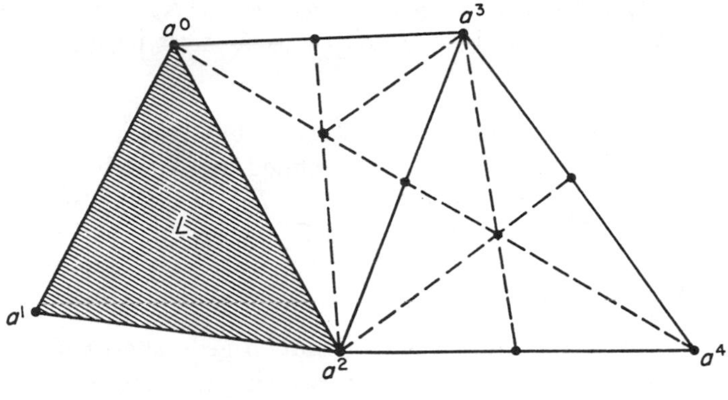

Fig. 2.10

First, $(M^0, L)' = M^0 = L \cup (a^3) \cup (a^4)$. $(M^1, L)'$ is next obtained by filling in the barycentre of each 1-simplex of $K - L$, thus chopping each of these 1-simplexes in half. Finally $(K, L)' = (M^2, L)'$ is constructed by filling in the barycentres of (a^0, a^2, a^3) and (a^2, a^3, a^4), and joining them up to the (chopped-up) boundaries of (a^0, a^2, a^3) and (a^2, a^3, a^4). The resulting simplicial complex has twelve 2-simplexes, as shown in Fig. 2.10. ∎

The following alternative description of $(K, L)'$ may help the reader to familiarize himself with the idea of the derived complex.

Proposition 2.5.10 *The vertices of $(K, L)'$ are the barycentres of the simplexes of $K - L$, together with the vertices of L. Distinct points $\hat{\sigma}_m, \ldots, \hat{\sigma}_0, a^0, \ldots, a^n$ (with $\dim \sigma_r \geqslant \dim \sigma_{r-1}$) span a simplex of $(K, L)'$ if and only if $a^0, \ldots, a^n$ span a simplex σ of L, and $\sigma_m > \cdots > \sigma_0 > \sigma$.*

Proof. That the vertices of $(K, L)'$ are as stated, follows immediately from the definition. If $\sigma = (a^0, \ldots, a^n)$ is a simplex of L, and $\sigma_m > \cdots > \sigma_0 > \sigma$, then $(\hat{\sigma}_m, \ldots, \hat{\sigma}_0, a^0, \ldots, a^n)$ is a simplex of $(K, L)'$, since we may assume inductively that $(\hat{\sigma}_{m-1}, \ldots, \hat{\sigma}_0, a^0, \ldots, a^n) \in (K, L)'$, and then use Definition 2.5.7. On the other hand if $(\hat{\sigma}_m, \ldots, \hat{\sigma}_0, a^0, \ldots, a^n)$ is a simplex of $(K, L)'$ then so is $(\hat{\sigma}_{m-1}, \ldots, \hat{\sigma}_0, a^0, \ldots, a^n)$,and we may assume inductively that this implies that $(a^0, \ldots, a^n)$ is a simplex of L, and $\sigma_{m-1} > \cdots > \sigma_0 > \sigma$. But

$$(\hat{\sigma}_{m-1}, \ldots, \hat{\sigma}_0, a^0, \ldots, a^n) \subset \sigma_{m-1},$$

and we must have $\sigma_{m-1} \in \dot{\sigma}_m$, so that $\sigma_m > \sigma_{m-1}$. ∎

Corollary 2.5.11 *If L is a subcomplex of K, then L is 'full' in $(K, L)'$, that is, each simplex of $(K, L)' - L$ meets $|L|$ (if at all) in a face.* ▮

The process of subdivision can of course be iterated. The *rth derived complex of K, relative to L*, is defined inductively by the rule $(K, L)^{(0)} = K$, $(K, L)^{(r)} = ((K, L)^{(r-1)}, L)'$ $(r > 0)$. Similarly, we write $K^{(0)} = K$, $K^{(r)} = (K^{(r-1)})'$, if L is empty.

Corollary 2.5.12 *L is full in $(K, L)^{(r)}$, for all $r > 0$.* ▮

The result of Corollary 2.5.12 is not in general true if $r = 0$: for example, let $K = K(\sigma)$, where σ is a 2-simplex, and let $L = \dot{\sigma}$. It is clear that $\sigma \cap |L| = |\dot{\sigma}|$, which is more than just a face of σ.

We have seen that $|(K, L)'| = |K|$, although in neither direction is the identity map simplicial. However, by using Proposition 2.5.2 it is easy to construct a simplicial approximation to $1: |(K, L)'| \to |K|$. Now each vertex of $(K, L)'$ is a barycentre of a simplex σ of K (possibly a 0-simplex of L); for each σ, choose any vertex a of σ.

Proposition 2.5.13 *There exists a simplicial approximation h to $1: |(K, L)'| \to |K|$, such that $h(\hat{\sigma}) = a$ for each $\hat{\sigma}$.*

Proof. By Proposition 2.5.2, it is sufficient to show that $\text{st}_{(K,L)'}(\hat{\sigma}) \subset \text{st}_K(a)$, for each $\hat{\sigma}$. If τ is a simplex of $(K, L)'$ having $\hat{\sigma}$ as a vertex, there exists a simplex μ of K such that $\tau \subset \mu$, and the interior of τ is contained in the interior of μ. Since $\hat{\sigma} \in \mu$, σ must be a face of μ; thus a is a vertex of μ, so that the interior of τ is contained in $\text{st}_K(a)$. ▮

Notice that for each n-simplex σ of K, and for each n-simplex τ of $(K, L)'$ that is contained in σ, $h(\tau) \subset \sigma$. It follows that $h(\tau) = \sigma$ for just one such τ. For suppose this is true for simplexes of dimension less than n (it is certainly true for 0-simplexes). If $\sigma \in L$, then $h(\sigma) = \sigma$; if $\sigma \notin L$, then each n-simplex of $(K, L)'$ contained in σ is of the form $(\hat{\sigma}, b^1, \ldots, b^n)$. If $h(\hat{\sigma}) = a$, then $h(\hat{\sigma}, b^1, \ldots, b^n) = \sigma$ if and only if $(b^1, \ldots, b^n)$ is contained in the $(n - 1)$-face μ of σ obtained by omitting a, and $h(b^1, \ldots, b^n) = \mu$; but by induction this is true for just one such $(b^1, \ldots, b^n)$.

The purpose of introducing subdivisions was that their simplexes should be in some sense 'smaller' than those of the original simplicial complex. In order to make this precise, we make the following definition.

Definition 2.5.14 The *star covering* of a simplicial complex K is the set of stars of vertices of K. By Proposition 2.4.3, the star covering is an open covering of $|K|$. The *mesh* of an open covering of a metric space is defined to be the supremum of the diameters of the open sets of the covering, and the *mesh* of a simplicial complex K, written mesh K, is the mesh of its star covering.

If we consider only the 'non-relative' derived complexes, the mesh can be made as small as we please by subdividing enough times.

Proposition 2.5.15 *Given a simplicial complex K, and a number $\epsilon > 0$, there exists an integer r such that* mesh $K^{(r)} < \epsilon$.

Proof. Let λ be the maximum of the lengths of the 1-simplexes of K. It is easy to see that the diameter of each simplex of K cannot exceed λ. Thus if a is a vertex and $x \in \mathrm{st}(a)$, then $d(x, a) \leqslant \lambda$, so that the diameter of $\mathrm{st}(a)$ is at most 2λ, and mesh $K \leqslant 2\lambda$.

Similarly, let λ' be the maximum of the lengths of the 1-simplexes of K': say λ' is the length of τ. Now τ is contained in some n-simplex σ of K, so that $\lambda' \leqslant [n/(n+1)]l$, where l is the length of some line segment in σ. Hence

$$\lambda' \leqslant [N/(N+1)]\lambda,$$

where N is the dimension of K. Hence if $\lambda^{(r)}$ is the maximum of the lengths of the 1-simplexes of $K^{(r)}$, we have

$$\text{mesh } K^{(r)} \leqslant 2\lambda^{(r)} \leqslant 2[N/(N+1)]^r\lambda.$$

Since $[N/(N+1)]^r \to 0$ as $r \to \infty$, the required result follows. ∎

One form of the Simplicial Approximation Theorem can be deduced immediately.

Theorem 2.5.16 *Let K and L be simplicial complexes, and let $f\colon |K| \to |L|$ be a continuous map. Then there exists an integer r such that $f\colon |K^{(r)}| \to |L|$ has a simplicial approximation.*

Proof. Consider the sets $f^{-1}(\mathrm{st}(b))$, for each vertex b of L. These sets form an open covering of $|K|$, and by Theorem 1.4.35 this open covering has a Lebesgue number δ, say. Choose r, so that mesh $K^{(r)} < \delta$: then for each vertex a of $K^{(r)}$ there exists a vertex b of L such that $\mathrm{st}(a) \subset f^{-1}(\mathrm{st}(b))$, or $f(\mathrm{st}(a)) \subset \mathrm{st}(b)$. Hence by Proposition 2.5.2 f has a simplicial approximation. ∎

Corollary 2.5.17 *Given simplicial complexes K and L, the set $[|K|, |L|]$ is countable.*

Proof. We need consider only simplicial maps $f\colon |K^{(r)}| \to |L|$, for various r, since each homotopy class of maps contains such a map.

But for each r there exists only a finite number of simplicial maps $f\colon |K^{(r)}| \to |L|$, since $K^{(r)}$ and L have only a finite number of vertices. ∎

Theorem 2.5.16 is what is usually referred to as the Simplicial Approximation Theorem. However, for many purposes it is useful to have a somewhat more refined version. Suppose that M is a subcomplex of K, and that $f\colon |K| \to |L|$ is a continuous map such that $f| \, |M|$ is already simplicial. We should like to find a simplicial approximation to f that actually coincides with f on $|M|$; and this is clearly not possible unless K is subdivided relative to M. A difficulty then arises, since Proposition 2.5.15 is no longer true, because the simplexes of M are unchanged under subdivision. Indeed, it is not even true that all simplexes *not* in M get smaller, because those that meet M have a face in M that is not subdivided. The most that can be said is the following.

Definition 2.5.18 Given a simplicial complex K and a subcomplex M, the *supplement* of M in K, $\overline{M}$, is the set of simplexes of $(K, M)'$ that have no vertices in M. Clearly $\overline{M}$ is a subcomplex of $(K, M)'$, and is the same as the subcomplex of K' of simplexes having no vertices in M'.

Proposition 2.5.19 *For each* $r \geqslant 0$, *let* α_r *denote the star covering of* $(K, M)^{(r)}$, *and let* α_r' *be the subset of stars of vertices in* $|\overline{M}|$. *Given* $\epsilon > 0$, *there exists* r *such that* mesh $\alpha_r' < \epsilon$.

Proof. Let (a^0, a^1) be a 1-simplex of $(K, M)^{(2)}$, and suppose that $a^1 \in M$. Then either $a^0 \in M$, or $a^0 = \hat{\sigma}$, where σ is a simplex of $(K, M)'$ that has a^1 as a vertex. Thus $\sigma \notin \overline{M}$, and so $\hat{\sigma} \notin |\overline{M}|$. In other words, no 1-simplex, and hence no n-simplex, of $(K, M)^{(2)}$ can have vertices in both M and $|\overline{M}|$.

It follows that each simplex of $(K, M)^{(2)}$ that has a vertex in $|\overline{M}|$ must be in $\hat{M}$, the supplement of M in $(K, M)'$. But for $r \geqslant 2$, the subdivision $(K, M)^{(r)}$ includes the 'non-relative' subdivision $\hat{M}^{(r-2)}$ of $\hat{M}$, and hence α_r' is contained in the star covering of $\hat{M}^{(r-2)}$. Now use Proposition 2.5.15. ∎

Suppose that $f\colon |K| \to |L|$ is a continuous map such that f is simplicial on $|M|$. We would hope to use Proposition 2.5.19 in the same way as Proposition 2.5.15 to obtain a simplicial approximation to f that coincides with f on $|M|$. Unfortunately this is not quite possible, because the simplexes that are in neither M nor $\overline{M}$ do not get smaller under subdivision; on the other hand, f is not itself simplicial on them. These simplexes need special treatment, and the price we must pay is that the simplicial map we finally obtain is not a simplicial approximation to f, although it is homotopic to f.

Theorem 2.5.20 *Let K and L be simplicial complexes, let M be a subcomplex of K, and let $f: |K| \to |L|$ be a continuous map such that $f| \, |M|$ is simplicial. Then there exists an integer r and a simplicial map $g: |(K, M)^{(r)}| \to |L|$ such that $g = f$ on $|M|$, and $g \simeq f$ rel $|M|$.*

Proof. As we have just remarked, special treatment is necessary for the simplexes of $(K, M)'$ that are in neither M nor $\overline{M}$, and we start by pushing all their barycentres into $|M|$.

Let $K^+ = ((K, M)', M \cup \overline{M})'$: this is obtained from $(K, M)'$ by subdividing these exceptional simplexes, and so is a subdivision of K 'between' $(K, M)'$ and $(K, M)^{(2)}$: see Fig. 2.11, in which $K = K(a^0, a^1, a^2) \cup K(a^1, a^2, a^3)$ and $M = K(a^0, a^1, a^2)$.

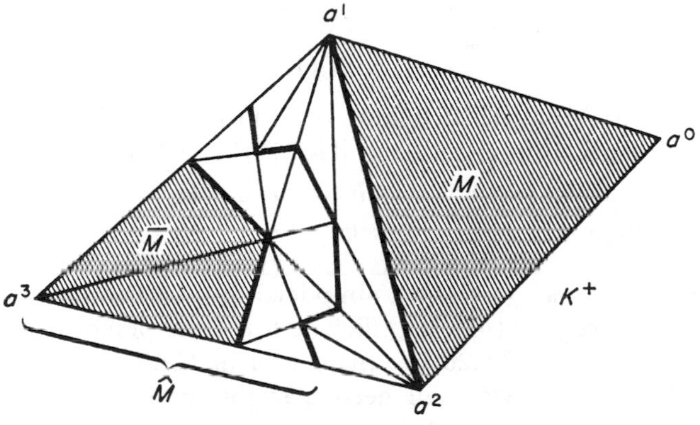

Fig. 2.11

Now a vertex of K^+ is either a vertex of $M \cup \overline{M}$, or the barycentre $\hat{\sigma}$ of a unique simplex σ of $(K, M)'$ meeting both $|M|$ and $|\overline{M}|$ (σ meets $|M|$ since it is not in $\overline{M}$, and $|\overline{M}|$ since by Corollary 2.5.11 it cannot have all its vertices in M). Hence by Proposition 2.5.13 there exists a simplicial approximation h to $1: |K^+| \to |(K, M)'|$ such that

(a) if a is a vertex of $M \cup \overline{M}$, then $h(a) = a$;
(b) otherwise, $h(\hat{\sigma})$ is a vertex of σ lying in $|M|$.

Notice that $h \simeq 1$ rel $|M|$, since h leaves fixed all vertices of M, and that $h(\text{st}_{K^+}(a)) \subset \text{st}_M(a)$ for each vertex a in M, since a simplex τ of K^+ having a as a vertex can have no vertex in $\overline{M}$: thus h maps all vertices of τ into M and so $h(\tau) \in M$ by Corollary 2.5.11.

It is now fairly easy to construct a simplicial approximation to fh. Let β be $(fh)^{-1}$ (star covering of $|L|$), an open covering of $|K|$, let α_r be the star covering of $(K, M)^{(r)}$, and let α'_r be the subset of α_r of stars of vertices in $|\hat{M}|$, where as in Proposition 2.5.19 $\hat{M}$ is the supplement of M in $(K, M)'$. By Proposition 2.5.19 there exists r such that mesh α'_r is less than a Lebesgue number of β. That is, for each vertex a of $(K, M)^{(r)}$ lying in $|\hat{M}|$, there exists a vertex b in L such that $fh(\text{st}(a)) \subset \text{st}(b)$. On the other hand, if a is a vertex of $(K, M)^{(r)}$ $(r \geqslant 2)$ that does not lie in $|\hat{M}|$, then by iteration of Proposition 2.5.13 there exists a vertex b of $(K, M)^{(2)}$, not in $|\hat{M}|$ and so a vertex of M, such that

$$\text{st}_{(K,M)^{(r)}}(a) \subset \text{st}_{(K,M)^{(2)}}(b) = \text{st}_{K^+}(b).$$

Thus

$$\begin{aligned} fh(\text{st}(a)) &\subset fh(\text{st}_{K^+}(b)) \\ &\subset f(\text{st}_M(b)) \\ &\subset \text{st}_L(f(b)), \end{aligned}$$

since $f \mid |M|$ is simplicial. It follows from Proposition 2.5.2 that there exists a simplicial approximation $g\colon |(K, M)^{(r)}| \to |L|$ to fh. Moreover, if a is a vertex of M, which is certainly not in $|\hat{M}|$, we may as well take $b = a$, so that $g(a) = fh(a)$. Thus $g = fh = f$ on $|M|$, and $g \simeq fh \simeq f$ rel $|M|$. ▮

Although the map g is not a simplicial approximation to f itself, the fact that $g \simeq f$ rel $|M|$ is sufficient for most practical purposes, and makes Theorem 2.5.20 the main tool in Chapters 3 and 4. In Chapter 4, however, we shall need a slight modification in which $f| \, |M|$ is not itself simplicial, but we are given a homotopy between $f| \, |M|$ and a simplicial map from $|M|$ to $|L|$: in this case we wish to extend the homotopy and the simplicial map to the whole of $|K|$. This result is an easy deduction from Theorem 2.5.20, provided that the homotopy involved is 'semi-constant'.

Definition 2.5.21 A homotopy $F\colon X \times I \to Y$ is *semi-constant* if there exists s, $0 \leqslant s < 1$, such that $F(x, t) = F(x, 1)$ for all $s \leqslant t \leqslant 1$.

Corollary 2.5.22 *Let (K, M) be a simplicial pair, let L be a simplicial complex, and let $f\colon |K| \to |L|$ be a continuous map. Given a simplicial map $g\colon |M| \to |L|$ and a semi-constant homotopy G between $f| \, |M|$ and g, there exists an integer r and a simplicial map $h\colon |(K, M)^{(r)}| \to |L|$, such that $h = g$ on $|M|$, and $f \simeq h$ by a homotopy that extends G.*

Proof. By Theorem 2.4.1, there exists a homotopy $F: |K| \times I \to |L|$, whose restriction to $|K| \times 0$ is f and whose restriction to $|M| \times I$ is G. By Theorem 2.5.20, applied to the final map of F, there exists an integer r and a simplicial map $h: |(K, M)^{(r)}| \to |L|$ such that $h = g$ on $|M|$, and there is a homotopy H, rel $|M|$, between the final map of F and h. If $G(x, t) = G(x, 1)$ for all $s \leqslant t \leqslant 1$, the required homotopy J between f and h can be constructed by setting

$$J(x, t) = \begin{cases} F(x, t), & 0 \leqslant t \leqslant s \\ F(x, 2t - s), & s \leqslant t \leqslant (1 + s)/2 \\ H(x, (2t - 1 - s)/(1 - s)), & (1 + s)/2 \leqslant t \leqslant 1. \end{cases}$$

That is, we compose F and H as in Proposition 2.2.7, but adjust the t-co-ordinate so that the restriction of J to $|M| \times I$ is G. (J is continuous, by Proposition 1.4.15(d).) ∎

Observe that J can be made semi-constant if necessary, by composing with a constant homotopy and re-adjusting the t-co-ordinate.

Although Theorem 2.5.20 is useful mainly in later chapters, it can also be used directly to obtain some interesting geometrical results. For example, we can prove the following theorem on 'fixed points' of maps of E^n to itself.

Theorem 2.5.23 *Any continuous map $f: E^n \to E^n$ $(n \geqslant 0)$ has a fixed point, that is, there exists a point x in E^n such that $f(x) = x$.*

Proof. Suppose that, on the contrary, $f(x) \neq x$ for each point x of E^n (this is immediately a contradiction if $n = 0$, so we may as well assume that $n \geqslant 1$ from now on). We can construct a retraction $\rho: E^n \to S^{n-1}$ as follows.

For each point $x \in E^n$, join $f(x)$ to x by a straight line, and produce the line beyond x until it meets S^{n-1} at a point x', say: see Fig. 2.12.

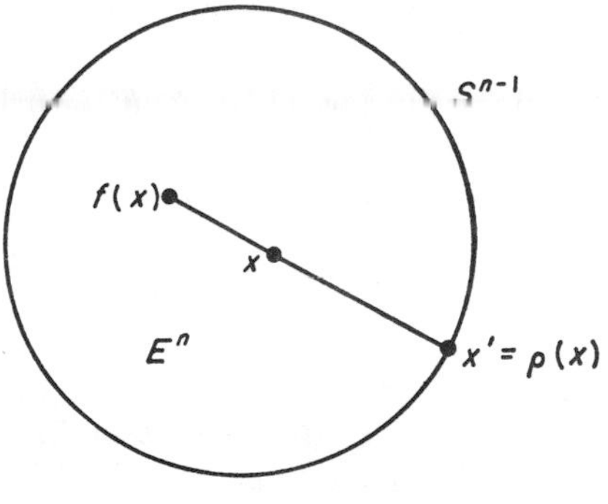

Fig. 2.12

Define $\rho(x) = x'$: clearly $\rho(x) = x$ if $x \in S^{n-1}$, so that ρ is indeed a retraction (the proof that ρ is continuous is left as an exercise for the reader).

Now let $h: (|K|, |L|) \to (E^n, S^{n-1})$ be a triangulation, as in Example 2.3.13. Then $h^{-1}\rho h: |K| \to |L|$ is also a retraction, and is simplicial (being the identity map) on $|L|$. By Theorem 2.5.20, there exists an integer r, and a simplicial map $g: |(K, L)^{(r)}| \to |L|$, such that $g \mid |L| = 1$. That is, g is also a retraction.

Let x be the barycentre of an $(n-1)$-simplex σ of L. The idea is to show that $g^{-1}(x)$ is a 'broken line' starting from x, and ending at another point of $|L|$, thus contradicting the fact that g is a retraction. To prove this, consider $g^{-1}(x) \cap \tau$, for each n-simplex τ of $(K, L)^{(r)}$. We claim that $g^{-1}(x) \cap \tau$, if non-empty, is a straight-line segment joining two points in the interiors of $(n-1)$-faces of τ: see Fig. 2.13.

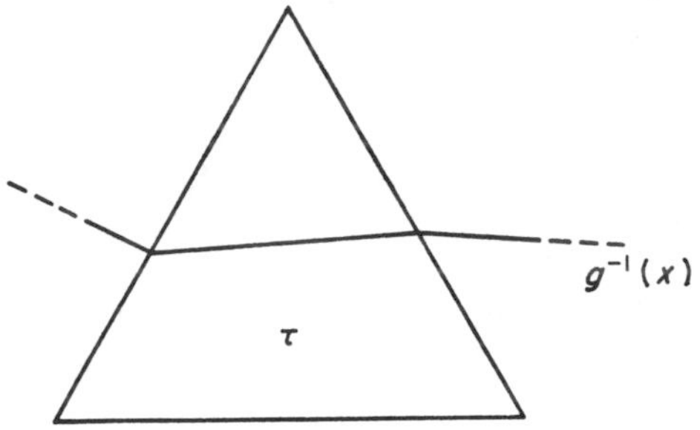

Fig. 2.13

For suppose $x \in g(\tau)$. Then, since $g(\tau)$ is a simplex meeting the interior of σ, we must have $g(\tau) = \sigma$. Let

$$\tau = (a^0, \ldots, a^n) \qquad \text{and} \qquad \sigma = (b^0, \ldots, b^{n-1}),$$

where $g(a^r) = b^r$ $(r < n)$ and $g(a^n) = b^{n-1}$. Then

$$g\left(\sum_{r=0}^{n} \lambda_r a^r\right) = (1/n) \sum_{r=0}^{n} b^r = x$$

if and only if $\lambda_r = 1/n$ $(r < n-1)$ and $\lambda_{n-1} + \lambda_n = 1/n$. Thus $g^{-1}(x) \cap \tau$ is as claimed.

It follows that $g^{-1}(x)$ is a 'string' of line segments, which starts at x, each segment joining on to the next one at a point in the interior of some $(n-1)$-simplex: this is because each $(n-1)$-simplex is a face of exactly two n-simplexes unless it is in L, in which case it is a face of just one n-simplex (see Exercise 15). Since each $g^{-1}(x) \cap \tau$ consists of at most one line segment, the 'string' can never cross itself, and so must continue until it meets $|L|$ again, at y, say. Then $x \neq y$, but $g(y) = x$, which contradicts the fact that g is a retraction. Hence g cannot exist and so f must have a fixed point. ∎

EXERCISES

1. Use Corollary 2.2.4 to show that two maps $f, g\colon X \to S^{n-1}$ that both fail to be onto must be homotopic.
2. Define maps $f, g\colon RP^1 \to RP^2$ by $f[x, y] = [x, y, 0]$, $g[x, y] = [x, -y, 0]$. Construct an explicit homotopy between f and g.
3. Given two maps $f, g\colon X \to Y$, show that $f \simeq g$ if Y is contractible.
4. Let X be the subspace of R^2 consisting of straight-line segments joining $(0, 1)$ to the points $(1/n, 0)$ $(n = 1, 2, 3, \ldots)$, and the segment joining $(0, 1)$ to $(0, 0)$. Show that X is contractible, but that the map $f\colon (X, (0, 0)) \to (X, (0, 0))$, defined by $f(x) = (0, 0)$ for all $x \in X$, is not homotopic to the identity map as a map of pairs (that is, $(X, (0, 0))$ is not 'pairwise contractible').
5. Consider the set $[A, X]$, where A is a fixed space. Show that a continuous map $f\colon X \to Y$ gives rise to a function $f_*\colon [A, X] \to [A, Y]$, with the following properties.
   (a) If $f \simeq g$, then $f_* = g_*$.
   (b) If $1\colon X \to X$ is the identity map, then 1_* is the identity function.
   (c) If $g\colon Y \to Z$ is another continuous map, then $(gf)_* = g_* f_*$.
   Deduce that if $X \simeq Y$ there is a (1-1)-correspondence between the sets $[A, X]$ and $[A, Y]$.
   What are the corresponding results for the sets $[X, A]$, for a fixed space A?
6. Complete the proof of Proposition 2.3.6.
7. Construct a triangulation of RP^2. (*Hint:* use Proposition 1.4.40(b).)
8. The *torus* and the *Klein bottle* are defined as follows. Let $ABCD$ be the unit square in R^2: see Fig. 2.14 overleaf.
   The torus is the space obtained from $ABCD$ by identifying the sides AD and BC, and then AB and DC; more precisely, we identify $(x_1, 0)$ with $(x_1, 1)$ $(0 \leqslant x_1 \leqslant 1)$ and also $(0, x_2)$ with $(1, x_2)$ $(0 \leqslant x_2 \leqslant 1)$. Similarly the Klein bottle is obtained by identifying $(x_1, 0)$ with $(x_1, 1)$ and $(0, x_2)$ with $(1, 1 - x_2)$, that is, AD with BC and AB with CD. Construct triangulations of these two spaces.
9. Let $\mathscr{K}$ be the abstract 1-dimensional simplicial complex with vertices a^0, a^1, a^2, a^3, a^4, each pair of vertices being an abstract 1-simplex. Show that $\mathscr{K}$ has no realization in R^2. (*Hint:* suppose the contrary, and consider the vertices $a^0, \ldots, a^3$. Prove that these must be placed in such a way that three of them span a 2-simplex with the fourth in its

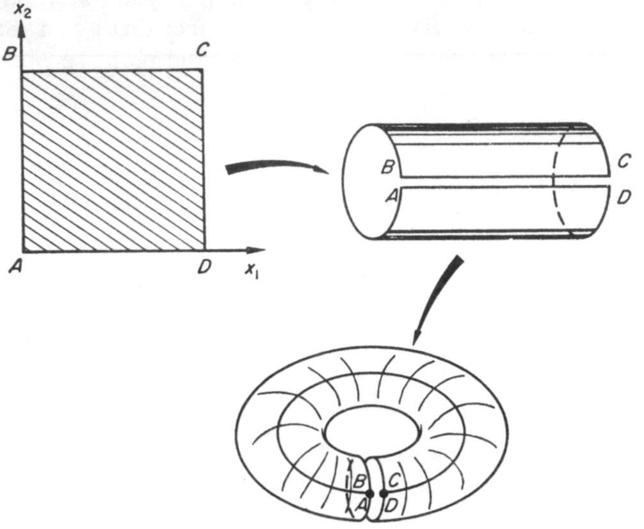

Fig. 2.14

interior, and deduce that the fifth vertex cannot be placed anywhere at all.) This example can be generalized to provide an example of an n-dimensional abstract simplicial complex that has no realization in R^{2n}.

10. Given simplicial complexes K, L, M and N, and simplicial maps $f\colon |K| \to |M|$, $g\colon |L| \to |N|$, show that $f * g\colon |K * L| \to |M * N|$ is also a simplicial map.
11. Show that the pair of spaces $(S^1, S^1 - (1, 0))$ does not have the absolute homotopy extension property. (*Hint:* use Theorem 2.5.23 to show that S^1 is not contractible.)
12. Let $\sigma = (a^0, \ldots, a^n)$ be a simplex in a simplicial complex K. Prove that $\mathrm{st}_K(\sigma) = \mathrm{st}_K(a^0) \cap \cdots \cap \mathrm{st}_K(a^n)$.
13. Let λ be the maximum of the lengths of the 1-faces of a simplex σ. Show that λ is the diameter of σ.
14. Prove that the retraction ρ defined in the proof of Proposition 2.5.23 is continuous.
15. Let (K, L) be a simplicial pair, where $\dim K = n$. The pair (K, L) is said to have the property (M) if each $(n - 1)$-simplex of $K - L$ is a face of an even number of n-simplexes of K, and each $(n - 1)$-simplex of L is a face of an odd number of n-simplexes of K. Prove that the pair $((K, L)', L)$ also has the property (M). (*Hint:* consider the various types of $(n - 1)$-simplexes in $(K, L)'$.) Deduce that $((K, L)^{(r)}, L)$ has the property (M) for each $r \geqslant 0$.

16. Let (K, L) be a simplicial pair, where $\dim K = n$ and $\dim L = n - 1$. Suppose also that (K, L) has the property (M). Prove that $|L|$ is not a retract of $|K|$.
17. Use Theorem 2.5.23 to show that S^n is not contractible, for each $n \geqslant 0$.

NOTES ON CHAPTER 2

Categories and functors. The transformation process from geometry to algebra, outlined at the beginning of Section 2.2, is a particular example of a *functor*, in the sense of Eilenberg and MacLane [53] (see also Eilenberg and Steenrod [56], Chapter 4). One first defines a *category* $\mathscr{C}$ to be a collection of 'objects' $X, Y, \ldots$ and 'maps' $f, g, \ldots$ between objects, such that the following rules are satisfied.

(a) Given maps $f: X \to Y$, $g: Y \to Z$, there exists a unique 'composite map' $gf: X \to Z$.
(b) For each object X in $\mathscr{C}$, there exists an 'identity map' $1_X: X \to X$, such that $1_X f = f$ and $g 1_X = g$ whenever these composites are defined.
(c) If gf and hg are defined, then $h(gf) = (hg)f$.

For example, the class of all topological spaces and continuous maps, and the class of all groups and homomorphisms, are categories.

Given two categories $\mathscr{C}$ and $\mathscr{D}$, a *functor* $\mathscr{F}: \mathscr{C} \to \mathscr{D}$ is a function that assigns an object of $\mathscr{D}$ to each object of $\mathscr{C}$, and a map of $\mathscr{D}$ to each map of $\mathscr{C}$, in such a way that

(a) if $f: X \to Y$ in $\mathscr{C}$, then $\mathscr{F}(f): \mathscr{F}(X) \to \mathscr{F}(Y)$ in $\mathscr{D}$;
(b) $\mathscr{F}(1_X) = 1_{\mathscr{F}(X)}$;
(c) $\mathscr{F}(gf) = \mathscr{F}(g)\mathscr{F}(f)$.

Thus, for example, the process described at the beginning of Section 2.2 is a functor from the category of topological spaces and continuous maps to the category of groups and homomorphisms.

Homotopy. The concept of homotopy, at least for maps of the unit interval I, is due to Jordan [81]. The word 'homotopy' was first introduced by Dehn and Heegaard [43].

Simplicial complexes. The study of 1- and 2-dimensional simplicial complexes is one of the oldest parts of topology, and dates back at least to Euler. The earliest treatment of simplicial complexes of higher dimensions seems to be that of Listing [92] (who was also the first to use the word 'topology', in 1847).

Simplicial complexes can be generalized in various directions. For a description of *infinite* simplicial complexes, which contain more than a finite number of simplexes, see Lefschetz [89], Chapter 7. By relaxing all 'linearity' conditions, one arrives at the notion of a *CW-complex*, for which see Chapter 7 of this book (the original reference is a paper of J. H. C. Whitehead [160]). Lastly, a generalization of the idea of an abstract simplicial complex, known as a *semi-simplicial complex*, has been very fruitful in recent years (see for example Eilenberg and Zilber [57] and Kan [84]).

The *join* of two simplicial complexes was first defined by Poincaré [117] (see also Newman [109]).

Section 2.4. Theorem 2.4.1 is due to Hurewicz [74] and Theorem 2.4.5 to Seifert and Threlfall [124], Chapter 5.

The Simplicial Approximation Theorem. Theorem 2.5.16 is the original version of this theorem, and was first proved by Alexander [7, 9] and Veblen [147]. The more refined version, Theorem 2.5.20, is due to Zeeman [169].

Theorem 2.5.23. This is usually known as the Brouwer Fixed-Point Theorem, for which the original reference is Brouwer [25]. The proof we give is that of Hirsch [65].

CHAPTER 3

THE FUNDAMENTAL GROUP

3.1 Introduction

In this chapter we shall define and study a first example of an algebraic invariant of a topological space X, namely the *fundamental group* $\pi_1(X)$: this is defined to be the set of homotopy classes of maps of the unit interval I to X, that send 0 and 1 to some fixed point. We shall prove that $\pi_1(X)$ can be given the structure of a group in a natural way, and that it is a homotopy-type invariant of X. If X is triangulable, it is not too difficult to give a method of calculating $\pi_1(X)$: as might be expected, this method is based on the Simplicial Approximation Theorem.

The general theory will be presented in Section 3.2, and the calculation theorem for triangulated spaces will be proved in Section 3.3. In Section 3.4 we shall show how the fundamental group can be used to prove the classification theorem for triangulated surfaces; thus the fundamental group is quite a powerful algebraic invariant.

3.2 Definition and elementary properties of the fundamental group

Let X be a topological space, and let x_0 be a fixed point of X, called a *base point*.

Definition 3.2.1 If x and y are points of X, a *path in X from x to y* is a continuous map $u: I \to X$ such that $u(0) = x$ and $u(1) = y$. If $x = y = x_0$, such a path is called a *loop in X, based at x_0*.

By Proposition 2.2.7, the relation between paths and loops of being homotopic relative to 0 and 1 is an equivalence relation. This justifies the following definition.

Definition 3.2.2 The *fundamental group* of X, with base point x_0, written $\pi_1(X, x_0)$, is the set of homotopy classes, relative to 0 and 1, of loops in X based at x_0.

We are a little premature, of course, in calling $\pi_1(X, x_0)$ a *group*, since we have not yet endowed it with any group structure. In order to do so, we first define a 'product' and 'inverse' for paths in X, and then check that this definition extends to homotopy classes.

Definition 3.2.3 Given paths $u, v: I \to X$, such that $u(1) = v(0)$, the *product path* $u.v: I \to X$ is obtained by 'sticking u and v together'. More precisely, $u.v$ is defined by the rule

$$(u.v)(t) = \begin{cases} u(2t) & (0 \leqslant t \leqslant \frac{1}{2}) \\ v(2t-1) & (\frac{1}{2} \leqslant t \leqslant 1). \end{cases}$$

($u.v$ is continuous, by Proposition 1.4.15(d).) Similarly, given n paths $u_1, u_2, \ldots, u_n: I \to X$, such that $u_r(1) = u_{r+1}(0)$ for $1 \leqslant r \leqslant n-1$, the *product path* $u_1.u_2\ldots u_n: I \to X$ is defined by

$$(u_1.u_2\ldots u_n)(t) = u_r(nt - r + 1) \quad ((r-1)/n \leqslant t \leqslant r/n, 1 \leqslant r \leqslant n).$$

The *inverse path* $u^{-1}: I \to X$ is defined by $u^{-1}(t) = u(1-t)$ $(0 \leqslant t \leqslant 1)$; obviously u^{-1} is continuous, and $(u_1.u_2\ldots u_n)^{-1} = u_n^{-1}.u_{n-1}^{-1}\ldots u_1^{-1}$.

The following proposition shows that this definition can be extended to 'homotopy classes of paths'.

Proposition 3.2.4

(a) *Given paths* $u_1, \ldots, u_n$ *and* $v_1, \ldots, v_n$ *in* X, *such that* $u_1(0) = v_1(0)$, $u_r(1) = u_{r+1}(0) = v_r(1) = v_{r+1}(0)$ $(1 \leqslant r \leqslant n-1)$, *and* $u_n(1) = v_n(1)$, *then if* $u_r \simeq v_r$ rel 0, 1 $(1 \leqslant r \leqslant n)$, *we have* $u_1\ldots u_n \simeq v_1\ldots v_n$ rel 0, 1.

(b) *Given paths* u, v *in* X, *such that* $u(0) = v(0)$ *and* $u(1) = v(1)$, *then if* $u \simeq v$ rel 0, 1, *we have* $u^{-1} \simeq v^{-1}$ rel 0, 1.

Proof.

(a) Let the homotopies be $F_r: u_r \simeq v_r$ $(1 \leqslant r \leqslant n)$. A homotopy G between $u_1\ldots u_n$ and $v_1\ldots v_n$ can be constructed by 'sticking together' $F_1, \ldots, F_n$, that is, by defining

$$G(t_1, t_2) = F_r(nt_1 - r + 1, t_2) \quad ((r-1)/n \leqslant t_1 \leqslant r/n, 1 \leqslant r \leqslant n, t_2 \in I).$$

As usual, G is continuous, and it is obviously a homotopy relative to 0 and 1.

(b) If the homotopy is $F: u \simeq v$, then the required homotopy between u^{-1} and v^{-1} is F^{-1}, where $F^{-1}(t_1, t_2) = F(1-t_1, t_2)$. ∎

It follows that the definition of product and inverse can be extended unambiguously to homotopy classes (relative to 0 and 1) of paths. Moreover, when we pass to homotopy classes, the product is associative and the inverse behaves as an inverse should. These results follow from the following trivial corollary of Theorem 2.2.3.

Proposition 3.2.5 *Given paths $u, v: I \to I$, such that $u(0) = v(0)$ and $u(1) = v(1)$, then $u \simeq v$ rel 0, 1.* ∎

Corollary 3.2.6

(a) *If $u_1, \ldots, u_n$ are paths in X as in Definition 3.2.3, then for each r, $1 \leqslant r < n$, $(u_1 \ldots u_r).(u_{r+1} \ldots u_n) \simeq u_1 \ldots u_n$ rel 0, 1.*

(b) *If u is a path in X from x to y, and e_x is the 'constant path at x', defined by $e_x(t) = x$ for all $t \in I$, then*

$$e_x.u \simeq u \simeq u.e_y \text{ rel } 0, 1.$$

(c) *If u is as in* (b), *then $u.u^{-1} \simeq e_x$ rel 0, 1 and $u^{-1}.u \simeq e_y$ rel 0, 1.*

Proof.

(a) $[(u_1 \ldots u_r).(u_{r+1} \ldots u_n)](t) = (u_1 \ldots u_n)(f(t))$, where $f: I \to I$ is the map that sends $0, \frac{1}{2}, 1$ to $0, r/n, 1$ respectively and is linear in between. But $f \simeq 1_I$ rel 0, 1 by Proposition 3.2.5.

(b) Again, $(e_x.u)(t) = u(f(t))$, where $f: I \to I$ is the map that sends $0, \frac{1}{2}, 1$ to $0, 0, 1$ respectively.

(c) This time $u.u^{-1}(t) = u(f(t))$, where f sends $0, \frac{1}{2}, 1$ to $0, 1, 0$. But $f \simeq e_0$ rel 0, 1, and $ue_0 = e_x$. ∎

Corollary 3.2.6 applies in particular to loops in X based at x_0, and the product of such loops is always defined. It follows easily that $\pi_1(X, x_0)$ can be given the structure of a group.

Theorem 3.2.7 *$\pi_1(X, x_0)$ is a group.*

Proof. If u is a loop in X based at x_0, write $[u]$ for the equivalence class of u under the relation of homotopy relative to 0 and 1. By Proposition 3.2.4(a) the product of two equivalence classes can be unambiguously defined by the rule $[u][v] = [u.v]$, and by Corollary 3.2.6(a) this product is associative. There is an identity element $[e_{x_0}]$, since by Corollary 3.2.6(b) $[e_{x_0}][u] = [u] = [u][e_{x_0}]$. Finally, by using Proposition 3.2.4(b) and Corollary 3.2.6(c), the element $[u]$ has an inverse $[u^{-1}]$, since $[u][u^{-1}] = [u^{-1}][u] = [e_{x_0}]$. ∎

Notice also that if $u_1, \ldots, u_n$ are loops in X based at x_0, then by Corollary 3.2.6(a) we have $[u_1][u_2] \ldots [u_n] = [u_1 \ldots u_n]$.

At this stage, then, we have a method for associating a group $\pi_1(X, x_0)$ with each topological space X, and we shall see later that homotopy-equivalent spaces have isomorphic fundamental groups. However, the discussion at the beginning of Section 2.2 shows that, in order to make this sort of method work, it is necessary to deal with continuous maps as well as topological spaces: we ought to show that each continuous map $f\colon X \to Y$ gives rise to a homomorphism $f_*\colon \pi_1(X, x_0) \to \pi_1(Y, y_0)$. This is indeed the case, at least if f is a based map.

Theorem 3.2.8 *Let X and Y be topological spaces with base points x_0 and y_0 respectively, and let $f\colon X \to Y$ be a based map, that is, a map such that $f(x_0) = y_0$. Then f gives rise to a homomorphism*

$$f_*\colon \pi_1(X, x_0) \to \pi_1(Y, y_0),$$

with the following properties.

(a) *If $f'\colon X \to Y$ is another based map, and $f \simeq f'$ rel x_0, then $f_* = f'_*$.*

(b) *If $1\colon X \to X$ is the identity map, then 1_* is the identity isomorphism.*

(c) *If $g\colon Y \to Z$ is another based map, then $(gf)_* = g_* f_*$.*

Proof. Let $u\colon I \to X$ be a loop based at x_0. Define f_* by the rule $f_*[u] = [fu]$. It is clear that $fu\colon I \to Y$ is a loop based at y_0, and that if $u \simeq v$ rel 0, 1 then $fu \simeq fv$ rel 0, 1; thus the definition of f_* is unambiguous. To show that f_* is a homomorphism, consider $u.v$, where $u, v\colon I \to X$ are loops based at x_0. Now

$$(u.v)(t) = \begin{cases} u(2t) & (0 \leqslant t \leqslant \frac{1}{2}) \\ v(2t-1) & (\frac{1}{2} \leqslant t \leqslant 1), \end{cases}$$

from which it is clear that $f(u.v) = (fu).(fv)$, so that

$$f_*([u][v]) = f_*[u]f_*[v].$$

Properties (a)–(c) are now obvious from the definition of f_*. ∎

Corollary 3.2.9 *Let X and Y be spaces with base points x_0 and y_0 respectively, and suppose that X and Y are of the same 'based homotopy type', that is, there exist based maps $f\colon X \to Y$ and $g\colon Y \to X$ such that $gf \simeq 1_X$ rel x_0 and $fg \simeq 1_Y$ rel y_0. Then $\pi_1(X, x_0) \cong \pi_1(Y, y_0)$.*

Proof. By Theorem 3.2.8, $g_* f_* = (gf)_* = (1_X)_* = 1$, the identity isomorphism. Similarly, $f_* g_*$ is the identity isomorphism, so that f_* and g_* are isomorphisms. ∎

The situation still leaves something to be desired, however, since $\pi_1(X, x_0)$ appears to depend on the particular choice of base point x_0.

We should like to prove a theorem to the effect that, if x_1 is another choice of base point, then $\pi_1(X, x_0) \cong \pi_1(X, x_1)$, but unfortunately this is not true without some restriction on the space X: see Exercise 1. In fact X must be *path-connected*, in the sense of the next definition.

Definition 3.2.10 Define a relation on the points of a space X by the rule: x and y are related if there exists a path in X from x to y. By Definition 3.2.3 this is an equivalence relation, and the resulting equivalence classes are called the *path components* of X. If in particular X has only one path component, X is said to be *path-connected.*

The set of path components of a space X is often denoted by $\pi_0(X)$. There is of course no question of giving $\pi_0(X)$ the structure of a group, in general.

Example 3.2.11 E^n is path-connected for all $n \geqslant 0$, and S^n is path-connected if $n \geqslant 1$. For clearly each point of E^n can be connected by a path to the origin, and each point of S^n can be connected to the point $(1, 0, \ldots, 0)$, at least if $n > 0$. ▌

Path-connectedness is a stronger notion than connectedness in the sense of Definition 1.4.5, as the next proposition and example show.

Proposition 3.2.12 *If X is path-connected, it is connected.*

Proof. Suppose, if possible, that X is path-connected, but disconnected in the sense of Definition 1.4.5. Then we may write $X = U_1 \cup U_2$, where U_1 and U_2 are disjoint open sets. Choose points $x \in U_1$, $y \in U_2$, and let $f: I \to X$ be a path from x to y. Now the sets $f^{-1}(U_1), f^{-1}(U_2)$ are open in I, since f is continuous; also $f^{-1}(U_1) \cup f^{-1}(U_2) = f^{-1}(X) = I$, and $f^{-1}(U_1) \cap f^{-1}(U_2) = \varnothing$. Thus I is disconnected, which contradicts Proposition 1.4.37. ▌

On the other hand, a space may well be connected, without being path-connected.

Example 3.2.13 In R^2, let X be the set of points $(0, x_2)$ for $-1 \leqslant x_2 \leqslant 1$, and let Y be the set of points $(x_1, \sin(\pi/x_1))$, for $0 < x_1 \leqslant 1$: see Fig. 3.1 overleaf.

Now Y is path-connected, since $(1, 0)$ can be connected to $(1 - a, \sin(\pi/(1 - a)))$ by the path $u: I \to Y$, where $u(t) = (1 - at, \sin(\pi/(1 - at)))$, $0 \leqslant t \leqslant 1$. Hence Y is also connected; but $X \cup Y \subset \bar{Y}$, so that $X \cup Y$ is connected, by Proposition 1.4.6.

On the other hand $X \cup Y$ is not path-connected. For suppose, if possible, that u is a path in $X \cup Y$ from $(0, 0)$ to $(1, 0)$; write $u(t) =$

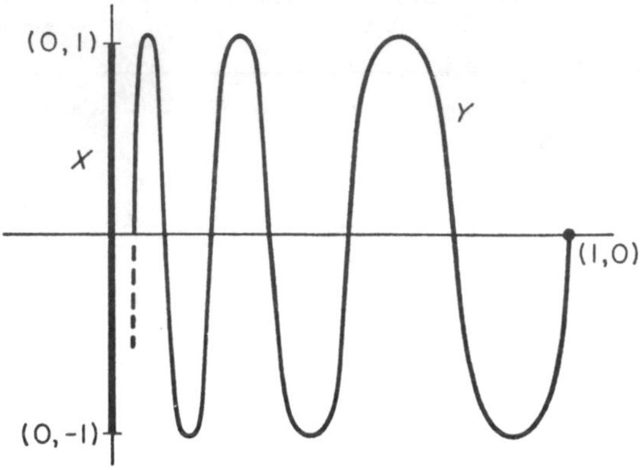

Fig. 3.1

$(u_1(t), u_2(t))$. Now $u^{-1}(X)$ is a closed set in I that contains 0, and so contains its least upper bound b, say, where $0 < b < 1$. We shall show that u_2 cannot be continuous at b.

Suppose that $u_2(b) \leqslant 0$. Then for any $\delta > 0$, with $b + \delta \leqslant 1$, we have $u_1(b + \delta) > 0$, so that there exists an integer n such that $0 = u_1(b) < 2/(4n + 1) < u_1(b + \delta)$, and there exists t such that $b < t < b + \delta$ and $u_1(t) = 2/(4n + 1)$. Thus $u_2(t) = 1$, and $u_2(t) - u_2(b) \geqslant 1$, so that u_2 is discontinuous at b. A similar argument applies if $u_2(b) \geqslant 0$, so that u_2 cannot be continuous. Hence no such path u can exist, and so $X \cup Y$ is not path-connected. ∎

For very well-behaved spaces, however, the notions of connectedness and path-connectedness coincide: see Exercise 2.

The point of Definition 3.2.10 is that $\pi_1(X, x_0)$ will yield information only about the path component of X that contains x_0.

Proposition 3.2.14 *Let X_0 be the path component of X that contains x_0, and let $i: X_0 \to X$ be the inclusion map. Then*

$$i_*: \pi_1(X_0, x_0) \to \pi_1(X, x_0)$$

is an isomorphism.

Proof. Clearly any loop in X based at x_0 must in fact be a loop in X_0, so that it is necessary only to check that two loops that are homotopic rel 0, 1 in X are homotopic rel 0, 1 in X_0. But this is immediate, since if $F: I \times I \to X$ is a homotopy whose image contains x_0, its image must lie entirely in X_0, because $I \times I$ is itself path-connected. ∎

In fact the set $\pi_0(X)$ is a homotopy-type invariant of the space X.

Proposition 3.2.15 *If $X \simeq Y$, there is a (1-1) correspondence between the sets $\pi_0(X)$ and $\pi_0(Y)$.*

Proof. Let $f: X \to Y$ and $g: Y \to X$ be the homotopy equivalence and homotopy inverse. Now f gives rise to a function $f_*: \pi_0(X) \to \pi_0(Y)$, by sending the path component of x in X to the path component of $f(x)$ in Y. Moreover homotopic maps give the same function, since $I \times I$ is path-connected. Thus an argument similar to that of Corollary 3.2.9 shows that f_* is a (1-1) correspondence. ▮

We are now ready to prove the theorem on the behaviour of $\pi_1(X, x_0)$ under a change of base point.

Theorem 3.2.16 *Let x_0 and x_1 be two base points lying in the same path component of X. A path u in X from x_0 to x_1 gives rise to an isomorphism $u_\#: \pi_1(X, x_0) \to \pi_1(X, x_1)$, with the following properties.*

(a) *If $u \simeq v$ rel 0, 1, then $u_\# = v_\#$.*

(b) *$(e_{x_0})_\#$ is the identity isomorphism.*

(c) *If w is a path in X from x_1 to x_2, then $(u.w)_\# = w_\# u_\#$.*

(d) *If $f: X \to Y$ is a map such that $f(x_0) = y_0$ and $f(x_1) = y_1$, then $f_* u_\# = (fu)_\# f_*: \pi_1(X, x_0) \to \pi_1(Y, y_1)$.*

Proof. If v is a loop in X based at x_0, it is clear that $u^{-1}.v.u$ is a loop based at x_1, whose class in $\pi_1(X, x_1)$ depends only on that of v. Moreover if w is another loop based at x_0,

$$u^{-1}.(v.w).u \simeq (u^{-1}.v.u).(u^{-1}.w.u) \text{ rel } 0, 1$$

by Corollary 3.2.6, so that the rule $u_\#[v] = [u^{-1}.v.u]$ defines a homomorphism $u_\#: \pi_1(X, x_0) \to \pi_1(X, x_1)$.

Properties (a)–(d) are immediate from the definition of $u_\#$, so that in particular $u_\#(u^{-1})_\# = (u.u^{-1})_\# = (e_{x_0})_\# = 1$. Similarly $(u^{-1})_\# u_\# = 1$, so that $u_\#$ is indeed an isomorphism. ▮

In particular, if X is path-connected, $\pi_1(X, x_0)$ is determined up to isomorphism by X alone, and does not depend on the choice of base point. It therefore makes sense to write $\pi_1(X)$ instead of $\pi_1(X, x_0)$, if we do not wish to distinguish between isomorphic groups.

An important special case of Theorem 3.2.16 is obtained by taking $x_0 = x_1$: each loop u based at x_0 gives rise to an isomorphism $u_\#: \pi_1(X, x_0) \to \pi_1(X, x_0)$, and this isomorphism depends only on the class of u in $\pi_1(X, x_0)$. Indeed, it is clear from the definition that if $[v]$ is any element of $\pi_1(X, x_0)$ we have $u_\#[v] = [u]^{-1}[v][u]$; such an isomorphism $u_\#$ is called the *inner automorphism* of $\pi_1(X, x_0)$ determined by $[u]$. Notice that the set of all isomorphisms $u_\#$ reduces to the identity isomorphism alone if and only if $\pi_1(X, x_0)$ is abelian.

Theorem 3.2.16 can be used to prove that two path-connected spaces of the same homotopy type have isomorphic fundamental groups. This result should be carefully distinguished from Corollary 3.2.9: two homotopy-equivalent spaces need not be of the same 'based homotopy type': see Exercise 3.

Theorem 3.2.17 *Let $f: X \to Y$ be a homotopy equivalence, let x_0 be a base point for X, and let $y_0 = f(x_0)$. Then*

$$f_*: \pi_1(X, x_0) \to \pi_1(Y, y_0)$$

is an isomorphism.

Proof. Let $g: Y \to X$ be a homotopy inverse to f, and let F be the homotopy between gf and 1_X. Let $g(y_0) = x_1$, $f(x_1) = y_1$, and define a path u in X from x_0 to x_1 by the rule

$$u(t) = F(x_0, 1 - t) \qquad (t \in I).$$

If v is any loop in X based at x_0, we have $gfv \simeq u^{-1}.v.u$ rel 0, 1, by the homotopy $G: I \times I \to X$, defined by

$$G(t_1, t_2) = \begin{cases} u(1 - 3t_1) & (0 \leqslant t_1 \leqslant t_2/3) \\ F(v\{(3t_1 - t_2)/(3 - 2t_2)\}, t_2) & (t_2/3 \leqslant t_1 \leqslant 1 - t_2/3) \\ u(3t_1 - 2) & (1 - t_2/3 \leqslant t_1 \leqslant 1). \end{cases}$$

Since these formulae may appear rather unenlightening, we offer an alternative description of G in Fig. 3.2, in which the square $QRPL$ is $I \times I$.

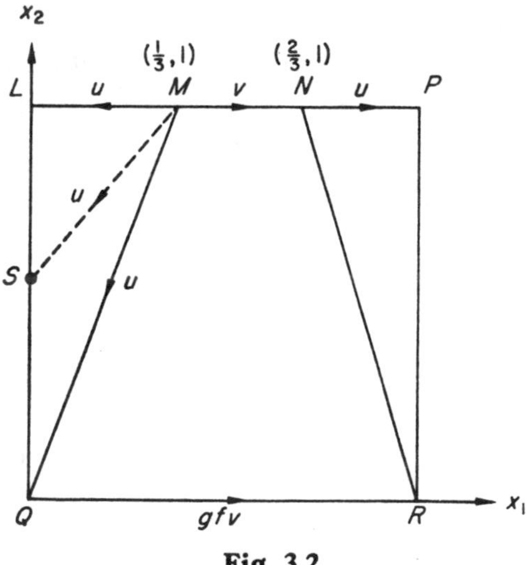

Fig. 3.2

The trapezium $QRNM$ is stretched horizontally until it becomes $I \times I$, and is then mapped by $F(v \times 1)$. The triangle QML is dealt with by mapping straight lines MS by u, after suitable magnification, where S is a general point of LQ; similarly for the triangle RPN. Clearly these definitions coincide on QM and RN, so that G is continuous. Moreover, G is a homotopy between gfv and $u^{-1}.v.u$, and is relative to 0 and 1, since the lines QL and RP are both mapped to x_1.

It follows that $g_*f_*[v] = u_\#[v]$, so that

$$g_*f_*: \pi_1(X, x_0) \to \pi_1(X, x_1)$$

is an isomorphism. A similar argument shows that f_*g_* is an isomorphism, so that finally both f_* and g_* are themselves isomorphisms. ∎

Definition 3.2.18 A space X is said to be *simply-connected* (or 1-*connected*) if it is path-connected, and $\pi_1(X) = 0$, the trivial group with just one element. (By Theorem 3.2.16, the choice of base point is immaterial here.)

Clearly a path-connected space X is simply-connected if and only if each loop in X (based anywhere) is homotopic rel 0, 1 to a 'constant loop'. As we shall see in Section 3.3, S^1 is an example of a space that is path-connected but not simply-connected, whereas S^n is simply-connected for all $n > 1$. By Theorem 3.2.17 a contractible space is simply-connected (it is easy to see that such a space is path-connected), though the converse is not true, as is demonstrated by S^n for $n > 1$ (see Chapter 2, Exercise 17).

3.3 Methods of calculation

So far we have defined the fundamental group and established some of its properties; but it would be useless for proving topological theorems if there were no means of calculating $\pi_1(X)$ for a given space X. In general the problem of calculation is formidable, but if X is a polyhedron the Simplicial Approximation Theorem can be used to reduce the problem considerably. Indeed, it even allows one to write down a finite set of generators and relations for $\pi_1(X)$.

In outline, the method is the following. First note that, by the Simplicial Approximation Theorem, each homotopy class of loops based at x_0 contains a simplicial map of some subdivision of I into X (at least if x_0 is a vertex). Thus in defining $\pi_1(X, x_0)$ it is sufficient to consider only such 'simplicial loops', and divide them into equivalence classes under homotopy rel 0, 1. But such a homotopy is a map of

$I \times I$ into X, which is simplicial on the 'boundary': so we can use the Simplicial Approximation Theorem again to show that the homotopy itself may as well be taken to be a simplicial map. It follows that we can take for generators of $\pi_1(X, x_0)$ all simplicial loops, and use 'simplicial homotopies' to give all the relations; and in fact this method can be refined a little so as to produce only a finite number of generators and relations.

In order to simplify the classification of simplicial homotopies between simplicial loops, we start the detailed work by introducing the idea of *collapsing* a simplicial complex onto a subcomplex.

Definition 3.3.1 Let K be a simplicial complex. An n-simplex σ of K is said to have a *free face* τ, if τ is an $(n-1)$-face of σ but is a face of no other n-simplex of K. If σ has a free face, it is easy to see that σ is not a proper face of any simplex of K, so that $K - \sigma - \tau$ is a subcomplex of K. The process of passing from K to $K - \sigma - \tau$ is called an *elementary collapse*, and if L is a subcomplex of K, K is said to *collapse* to L, written $K \searrow L$, if L can be obtained from K by a sequence of elementary collapses.

Example 3.3.2 Let K be the simplicial complex shown in Fig. 3.3. $K \searrow a^0$, by the sequence of elementary collapses illustrated. ∎

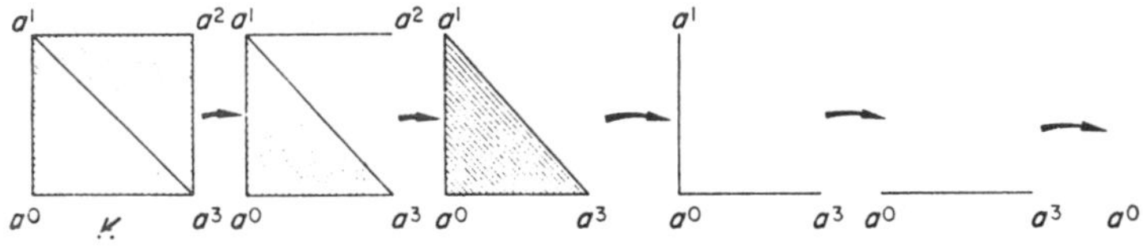

Fig. 3.3

An important property of collapsibility is that it is invariant under subdivision. We shall not prove the general result here (though see Exercise 5), since the following special case will be sufficient for our purposes.

Proposition 3.3.3 *Let K be a 1- or 2-dimensional simplicial complex, that collapses onto a subcomplex L. Then if M is any subcomplex of K, $(K, M)' \searrow L'$.*

Proof. It is clearly sufficient to prove this result in the special case where L is obtained from K by an elementary collapse, so that $L = K - \sigma - \tau$, where τ is a free face of σ. Now the result is obvious if σ is a 1-simplex, and if σ has dimension 2, it is easy to see that $[K(\sigma)]' \searrow [K(\sigma) - \sigma - \tau]'$: Fig. 3.4 illustrates a possible method of collapse, in the case where $K(\sigma) \cap M$ is empty; the other cases are dealt with similarly.

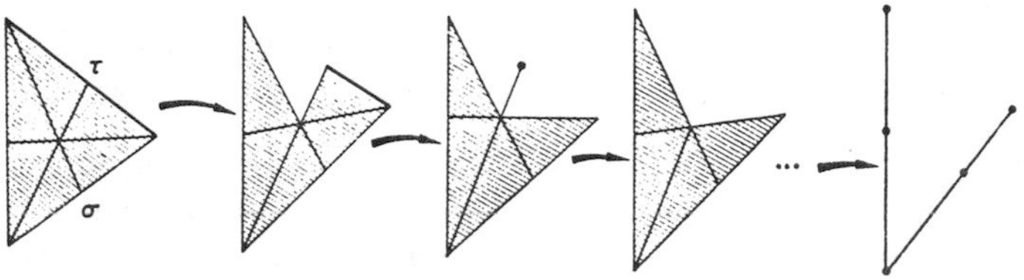

Fig. 3.4

This is sufficient to prove that $(K, M)' \searrow L'$. ∎

Corollary 3.3.4 *$(K, M)^{(r)} \searrow L^{(r)}$, for each $r \geqslant 0$.* ∎

The first step in the programme outlined at the beginning of this section is the construction of simplicial analogues of paths and loops. For these, let K be a simplicial complex, and let $L = K(\sigma)$, where σ is the 1-simplex in R^1 whose vertices are 0 and 1: thus L is a triangulation of I. If the vertices of $L^{(r)}$ are $0 = b^0 < b^1 < \cdots < b^n = 1$, a simplicial map $u: |L^{(r)}| \to |K|$ is completely determined by the sequence of vertices $u(b^0), u(b^1), \ldots, u(b^n)$. This suggests the following definition.

Definition 3.3.5 An *edge-path* in K, from a vertex a^0 to a vertex a^n, is a sequence α of vertices $a^0a^1\ldots a^n$, such that for each $r = 1, 2, \ldots, n$, the vertices a^{r-1}, a^r span a simplex of K (we allow $a^{r-1} = a^r$). If $a^0 = a^n$, α is called an *edge-loop*, based at a^0.

Given another edge-path $\beta = a^na^{n+1}\ldots a^{n+m}$, whose first vertex is the same as the last vertex of α, the *product* edge-path is defined by $\alpha.\beta = a^0a^1\ldots a^na^{n+1}\ldots a^{n+m}$, and the *inverse* of α is $\alpha^{-1} = a^na^{n-1}\ldots a^0$. Clearly $(\alpha.\beta).\gamma = \alpha.(\beta.\gamma)$ (so that we may unambiguously write $\alpha.\beta.\gamma$), and $(\alpha.\beta)^{-1} = \beta^{-1}.\alpha^{-1}$. (Compare Definition 3.2.3.)

We need next a convenient definition of 'equivalence' between edge-paths, analogous to the relation 'homotopic rel 0, 1' for ordinary paths. The reader may not immediately perceive the correspondence between the following definition and that for ordinary paths; however he is assured there is one, which will become apparent in the proof of Theorem 3.3.9 (it is based on the notion of collapsing).

Definition 3.3.6 Two edge-paths α and β are *equivalent* if one can be obtained from the other by a finite sequence of operations of the form

(a) if $a^{r-1} = a^r$, replace $\dots a^{r-1}a^r \dots$ by $\dots a^r \dots$, or conversely replace $\dots a^r \dots$ by $\dots a^r a^r \dots$; or

(b) if a^{r-1}, a^r, a^{r+1} span a simplex of K (not necessarily 2-dimensional), replace $\dots a^{r-1}a^r a^{r+1} \dots$ by $\dots a^{r-1}a^{r+1} \dots$, or conversely.

This clearly sets up an equivalence relation between edge-paths, and we write $\alpha \sim \beta$ for 'α and β are equivalent'. Notice that if α is an edge-path from a^0 to a^n, and $\alpha \sim \beta$, then β also is an edge-path from a^0 to a^n.

Proposition 3.3.7 *Let α_0, β_0 be edge-paths from a^0 to a^n, and let α_1, β_1 be edge-paths from a^n to a^{n+m}, such that $\alpha_0 \sim \beta_0$ and $\alpha_1 \sim \beta_1$. Then*

(a) $\alpha_0\alpha_1 \sim \beta_0\beta_1$;
(b) $\alpha_0^{-1} \sim \beta_0^{-1}$;
(c) $a^0 . \alpha_0 = \alpha_0 = \alpha_0 . a^n$;
(d) $\alpha_0 . \alpha_0^{-1} \sim a^0$ *and* $\alpha_0^{-1} . \alpha_0 \sim a^n$. ∎

(Compare Proposition 3.2.4 and Corollary 3.2.6.)

It follows, just as in Theorem 3.2.7, that the set of equivalence classes $[\alpha]$ of edge-loops α in K, based at a vertex a^0, forms a group $\pi(K, a^0)$: the multiplication is defined by $[\alpha][\beta] = [\alpha . \beta]$, the identity element is $[a^0]$, and the inverse of $[\alpha]$ is $[\alpha^{-1}]$.

Definition 3.3.8 $\pi(K, a^0)$ is called the *edge-group* of K, based at a^0.

The resemblance between the definitions of $\pi(K, a^0)$ and $\pi_1(|K|, a^0)$ is of course no coincidence.

Theorem 3.3.9 $\pi(K, a^0) \cong \pi_1(|K|, a^0)$.

Proof. The theorem is proved by constructing a homomorphism $\theta\colon \pi(K, a^0) \to \pi_1(|K|, a^0)$, and then showing that θ is onto and (1-1).

Let $\alpha = a^0 a^1 \cdots a^n a^0$ be an edge-loop based at a^0. For each pair of vertices a^i, a^j that span a simplex of K, let $u_{ij}\colon |L| \to |K|$ be the simplicial map that sends 0 to a^i and 1 to a^j ($L = K(0, 1)$). Then u_{ij} is a path in $|K|$ from a^i to a^j, and we may define

$$\theta[\alpha] = [u_{01} . u_{12} \dots u_{n0}] \in \pi_1(|K|, a^0).$$

It is first necessary to check that θ is well-defined, that is, that $\theta[\alpha] = \theta[\beta]$ if $\alpha \sim \beta$. We need only consider the case where β is obtained from α by a single operation of type (a) or (b) in Definition 3.3.6, and by Corollary 3.2.6(b) operations of type (a) give no trouble, since u_{rr} is the 'constant path' at a^r. As for operations of type (b), we merely remark that if a^{r-1}, a^r, a^{r+1} span a simplex of K, then $u_{r-1,r}.u_{r,r+1} \simeq u_{r-1,r+1}$ rel 0, 1 by an obvious homotopy.

It is easy to see that θ is a homomorphism. For if $\beta = a^0a^{n+1}\ldots a^{n+m}a^0$ is another edge-loop based at a^0, we have

$$\begin{aligned}\theta[\alpha]\theta[\beta] &= [u_{01}\ldots u_{n0}][u_{0,n+1}\ldots u_{n+m,0}]\\ &= [u_{01}\ldots u_{n0}.u_{0,n+1}\ldots u_{n+m,0}]\\ &= \theta[\alpha.\beta].\end{aligned}$$

Next, θ is onto, since if $[u] \in \pi_1(|K|, a^0)$, we may assume by the Simplicial Approximation Theorem that $u\colon |L^{(r)}| \to |K|$ is a simplicial map for some $r \geqslant 0$. If the vertices of $|L^{(r)}|$ are $0 = b^0 < b^1 < \cdots < b^n = 1$, define $\alpha = u(b^0).u(b^1)\ldots u(b^n)$. Then $\theta[\alpha] = [u]$, so that θ is onto.

Lastly, θ is (1-1). For if $\alpha = a^0a^1\ldots a^na^0$ is any edge-loop, $\theta[\alpha]$ is represented by a simplicial map $u\colon |M| \to |K|$, where M is a triangulation of I with vertices $0 = c^0 < c^1 < \cdots < c^{n+1} = 1$, and $u(c^r) = a^r$ $(0 \leqslant r \leqslant n)$, $u(c^{n+1}) = a^0$. So if $\theta[\alpha] = 1$ in $\pi_1(|K|, a^0)$, there is a homotopy $F\colon I \times I \to |K|$, such that

$$F(t, 0) = u(t),$$

and

$$F(t, 1) = F(0, t) = F(1, t) = a^0 \qquad (t \in I).$$

Now $I \times I$ can be triangulated as shown in Fig. 3.5 by a complex N, the four sides of the square forming a subcomplex P.

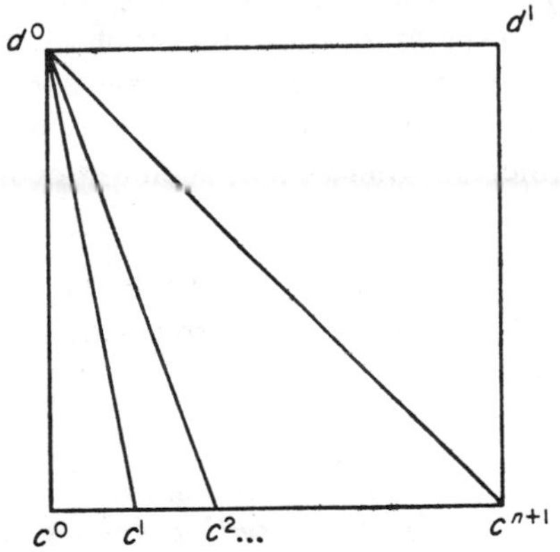

Fig. 3.5

Since $F \mid |P|$ is simplicial, by the Simplicial Approximation Theorem we may assume that F is a simplicial map $|(N, P)^{(r)}| \to |K|$ for some $r \geqslant 0$. We also know, by Corollary 3.3.4 and an argument similar to that used in Example 3.3.2, that $(N, P)^{(r)} \searrow c^0$. It follows that in N the edge-loop $\beta = c^0c^1\ldots c^{n+1}d^1d^0c^0$ is equivalent to the edge-loop c^0. For if (d^i, d^j, d^k) is a 2-simplex with free face (d^i, d^j), the edge-paths $\ldots d^id^j\ldots$ and $\ldots d^id^kd^j\ldots$ are equivalent, and if (d^i, d^j) is a 1-simplex with free vertex d^i, the edge-paths $\ldots d^jd^id^j\ldots$ and $\ldots d^j\ldots$ are equivalent: hence the sequence of elementary collapses of $(N, P)^{(r)}$ defines a sequence of equivalent edge-loops starting with β and ending with c^0. Finally we have

$$\begin{aligned}\alpha &= a^0a^1\ldots a^na^0\\ &\sim a^0a^1\ldots a^na^0a^0a^0a^0\\ &= F(c^0)F(c^1)\ldots F(c^{n+1})F(d^1)F(d^0)F(c^0)\\ &\sim F(c^0), \quad \text{since } F \text{ is simplicial,}\\ &= a^0,\end{aligned}$$

so that $[\alpha] = 1$, and so θ is (1-1). ∎

An obvious corollary of Theorem 3.3.9 is

Corollary 3.3.10 *$\pi_1(|K|, a^0)$ depends only on $|K^2|$.* ∎

In particular, S^n is simply-connected for $n > 1$. For by Example 2.3.13, S^n may be triangulated as $\dot{\sigma}$, where σ is an $(n + 1)$-simplex; and if $n > 1$ this has the same 2-skeleton as $K(\sigma)$, whose polyhedron is contractible.

On the face of it Theorem 3.3.9 does not tell us very much more about $\pi_1(|K|, a^0)$ than the original definition. Surprisingly enough, however, it is quite easy to give a finite set of generators and relations for $\pi(K, a^0)$, the trick being to ignore those parts of each edge-loop that are contained in a subcomplex whose polyhedron is contractible. If $|K|$ is path-connected, there exists such a subcomplex that contains all the vertices of K.

Definition 3.3.11 A 1-dimensional subcomplex L of K is called a *tree* if $|L|$ is contractible. Clearly trees are partially ordered by inclusion; a tree is *maximal* if it is not contained in a strictly larger tree. (Since K has only a finite number of simplexes, maximal trees certainly exist.)

Proposition 3.3.12 *If $|K|$ is path-connected, and L is a maximal tree, then L contains all the vertices of K.*

Proof. Suppose, if possible, that a is a vertex of K that is not in L. Since $|K|$ is path-connected, there is a path in $|K|$ from a vertex b, say, of L to a; hence, as in the proof of Theorem 3.3.9, there is an edge-path $ba^0 \dots a^n a$ in K. If a^r is the last vertex of this edge-path that is in L, (a^r, a^{r+1}) is a 1-simplex not in L (we may assume that $a^r \neq a^{r+1}$). Thus $\bar{L} = L \cup (a^r, a^{r+1}) \cup (a^{r+1})$ is a subcomplex strictly larger than L; moreover $|\bar{L}| \simeq |L|$, since the simplex (a^r, a^{r+1}) can be contracted to a^r without disturbing $|L|$. Hence $|\bar{L}|$ is contractible, and L is not a maximal tree, contrary to hypothesis. ∎

If $|K|$ is path-connected, and $|L|$ is a contractible subpolyhedron that contains all the vertices of K, we can now construct a group G with a finite number of generators and relations, which is isomorphic to $\pi(K, a^0)$, and hence to $\pi_1(|K|, a^0)$. Totally order the vertices of K, in the form $a^0 < a^1 < \cdots < a^m$; thus each simplex of K can be written in the form $(a^{i_0}, a^{i_1}, \dots, a^{i_n})$, where $i_0 < i_1 < \cdots < i_n$: a simplex written in this way is called an *ordered* simplex. Let G be the group generated by the symbols g_{ij}, one for each ordered 1-simplex (a^i, a^j) of $K - L$, subject to the relations $g_{ij}g_{jk}g_{ik}^{-1}$, one for each ordered 2-simplex (a^i, a^j, a^k) of $K - L$ (if, say, a^i, a^j span a simplex of L, g_{ij} is to be interpreted as 1).

Theorem 3.3.13 $G \cong \pi(K, a^0)$.

Proof. This time the theorem is proved by constructing homomorphisms $\theta\colon G \to \pi(K, a^0)$ and $\phi\colon \pi(K, a^0) \to G$, such that the composites $\phi\theta$ and $\theta\phi$ are identity isomorphisms.

To define θ, choose an edge-path α_r in L from a^0 to a^r, for each vertex a^r (we take $\alpha_0 = a^0$), and set $\theta(g_{ij}) = [\alpha_i . a^i a^j . \alpha_j^{-1}]$. Then for each ordered 2-simplex (a^i, a^j, a^k) of $K - L$, we have

$$\begin{aligned}\theta(g_{ij})\theta(g_{jk})(\theta(g_{ik}))^{-1} &= [\alpha_i . a^i a^j . \alpha_j^{-1}][\alpha_j . a^j a^k . \alpha_k^{-1}][\alpha_k . a^k a^i . \alpha_i^{-1}] \\ &= [\alpha_i . a^i a^j a^j a^k a^k a^i . \alpha_i^{-1}] \\ &= [\alpha_i . a^i a^j a^k a^i . \alpha_i^{-1}] \\ &= [\alpha_i . a^i a^k a^i . \alpha_i^{-1}] \\ &= [\alpha_i . \alpha_i^{-1}] \\ &= 1\end{aligned}$$

(if, say, a^i, a^j span a simplex of L, we still have $\theta(g_{ij}) = [\alpha_i . a^i a^j . \alpha_j^{-1}]$, since by Theorem 3.3.9 all edge-loops in L based at a^0 are equivalent). Thus by Proposition 1.3.20 θ extends in a unique way to a homomorphism $\theta\colon G \to \pi(K, a^0)$.

The homomorphism $\phi\colon \pi(K, a^0) \to G$ is defined as follows. Given any pair of vertices a^i, a^j of K, that span a simplex , let

$$h_{ij} = \begin{cases} g_{ij}, & \text{if } (a^i, a^j) \text{ is an ordered 1-simplex of } K - L, \\ g_{ji}^{-1}, & \text{if } (a^j, a^i) \text{ is an ordered 1-simplex of } K - L, \\ 1, & \text{otherwise.} \end{cases}$$

Then if $\alpha = a^0a^ia^j \ldots a^ka^0$ is an edge-loop in K, define

$$\phi[\alpha] = h_{0i}h_{ij} \ldots h_{k0} \in G.$$

It is easy to see that ϕ is an unambiguously defined homomorphism.

Now $\phi\theta(g_{ij}) = \phi[\alpha_i . a^ia^j . \alpha_j^{-1}] = g_{ij}$, so that $\phi\theta$ is the identity isomorphism of G. Morever if $\alpha = a^0a^ia^j \ldots a^ka^0$ is an edge-loop in K,

$$\begin{aligned} \theta\phi[\alpha] &= \theta\phi([\alpha_0 . a^0a^i . \alpha_i^{-1}] \ldots [\alpha_k . a^ka^0 . \alpha_0^{-1}]) \\ &= \theta\phi[\alpha_0 . a^0a^i . \alpha_i^{-1}] \ldots \theta\phi[\alpha_k . a^ka^0 . \alpha_0^{-1}]. \end{aligned}$$

But $[\alpha_r . a^ra^s . \alpha_s^{-1}] = 1$ unless a^r, a^s span a 1-simplex of $K - L$, and in any case $\theta\phi[\alpha_r . a^ra^s . \alpha_s^{-1}] = [\alpha_r . a^ra^s . \alpha_s^{-1}]$. Hence $\theta\phi[\alpha] = [\alpha]$, so that $\theta\phi$ is the identity isomorphism of $\pi(K, a^0)$. Thus θ and ϕ are themselves isomorphisms. ∎

Examples 3.3.14

(a) $\pi_1(S^1) \cong Z$, the additive group of integers. To see this, triangulate S^1 as the boundary $\dot{\sigma}$ of a 2-simplex $\sigma = (a^0, a^1, a^2)$, and take for L the subcomplex $\dot{\sigma} - (a^0, a^2)$. Certainly $|L|$ is contractible and contains all the vertices, so that $\pi_1(S^1) \cong \text{Gp}\,\{g_{02}\} \cong Z$.

(b) By Proposition 1.4.40(b) the real projective plane RP^2 can be obtained from a square $ABCD$ by identifying the sides AB and CD,

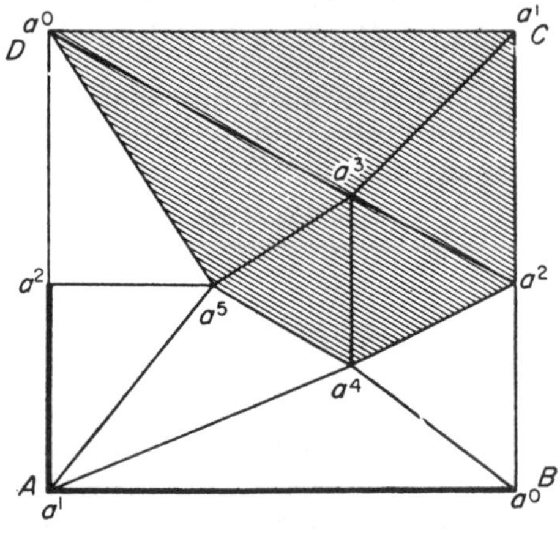

Fig. 3.6

and BC and DA (compare Chapter 2, Exercise 8). A triangulation of RP^2 is shown in Fig. 3.6, in which the shaded area represents a contractible subpolyhedron that contains all the vertices, and the vertices are totally ordered by their superfixes.

Thus $\pi_1(RP^2)$ is the group generated by g_{02}, g_{04}, g_{14}, g_{15} and g_{25}, subject to the relations $g_{01}g_{14}g_{04}^{-1}$, $g_{02}g_{24}g_{04}^{-1}$, $g_{02}g_{25}g_{05}^{-1}$, $g_{12}g_{25}g_{15}^{-1}$ and $g_{14}g_{45}g_{15}^{-1}$. Thus in $\pi_1(RP^2)$, $g_{14} = g_{04}$, $g_{02} = g_{04}$, $g_{02}g_{25} = 1$, $g_{25} = g_{15}$ and $g_{14} = g_{15}$. These relations imply that all five generators are equal, and $(g_{02})^2 = 1$, that is, $\pi_1(RP^2)$ is isomorphic to Z_2, the group of integers mod 2. ▌

The reader will see from this last example that, although Theorem 3.3.13 guarantees a finite set of generators and relations for $\pi_1(|K|)$, it may well produce far more generators and relations than are necessary. Indeed, for more complicated spaces than spheres or the real projective plane Theorem 3.3.13 may give such an unwieldy description of the fundamental group that it is useless for practical calculations. The trouble is that even comparatively simple spaces may need a large number of simplexes to triangulate them: for example, the torus (see Chapter 2, Exercise 8) cannot be triangulated with less than 7 0-simplexes, 21 1-simplexes and 14 2-simplexes.

We therefore seek a method of improving Theorem 3.3.13 so as to produce as few generators and relations as possible. The first step in this direction is to establish a theorem that expresses the fundamental group of the union of two polyhedra in terms of the fundamental groups of the two polyhedra and of their intersection. However, the result of this theorem is stated in terms of the *free product* of two groups, and so we must first define this.

Definition 3.3.15 Given two groups G and H, the *free product* $G * H$ is the group generated by all the elements of G and all the elements of H, subject to the relations $g_1g_2g_3^{-1}$, for all $g_1, g_2, g_3 \in G$ such that $g_1g_2 = g_3$, and $h_1h_2h_3^{-1}$, for all $h_1, h_2, h_3 \in H$ such that $h_1h_2 = h_3$.

Example 3.3.16 If G and H are each free groups generated by single elements a, b respectively, then $G * H = \text{Gp}\{a, b\}$. ▌

In fact the set of generators and relations for $G * H$ given in Definition 3.3.15 is in general unnecessarily large. $G * H$ can be described in terms of any finite sets of generators and relations for G and H, as follows.

Proposition 3.3.17 *If* $G = \text{Gp}\{a_1, \ldots, a_m; \alpha_1, \ldots, \alpha_n\}$ *and* $H = \text{Gp}\{b_1, \ldots, b_p; \beta_1, \ldots, \beta_q\}$, *then*

$$G * H \cong \text{Gp}\,\{a_1, \ldots, a_m, b_1, \ldots, b_p; \alpha_1, \ldots, \alpha_n, \beta_1, \ldots, \beta_q\}.$$

Proof. Let $G{:}H$ be the group

$$\text{Gp}\,\{a_1, \ldots, a_m, b_1, \ldots, b_p; \alpha_1, \ldots, \alpha_n, \beta_1, \ldots, \beta_q\},$$

and let $\theta\colon G{:}H \to G * H$ be the obvious homomorphism, that sends a word in a's and b's to itself; θ is unambiguously defined since each α or β is sent to 1. Similarly, let $\phi\colon G * H \to G{:}H$ be the obvious homomorphism: again ϕ is unambiguous, because each relation of the form $g_1g_2g_3^{-1}$, for example, must be a word in conjugates of α's. Moreover the composites $\theta\phi$ and $\phi\theta$ are both identity isomorphisms, so that θ and ϕ are isomorphisms. ∎

We can now state the theorem on the union of two polyhedra. For this, let L and M be subcomplexes of a simplicial complex K, such that $L \cup M = K$, and let $N = L \cap M$. Write λ, μ for the inclusion maps $|N| \subset |L|$, $|N| \subset |M|$ respectively.

Theorem 3.3.18 *If $|L|$, $|M|$ and $|N|$ are path-connected, and a^0 is a vertex of N, then $\pi_1(|K|, a^0)$ is the group obtained from $\pi_1(|L|, a^0) * \pi_1(|M|, a^0)$ by adding extra relations $(\lambda_*c)(\mu_*c)^{-1}$, one for each element c of $\pi_1(|N|, a^0)$.* (*As in Proposition* 3.3.17, *it suffices to add the relations $(\lambda_*c)(\mu_*c)^{-1}$, one for each element c in a finite set of generators for $\pi_1(|N|, a^0)$.*)

Proof. Let T_N be a maximal tree in N. As in Proposition 3.3.12, T_N can be extended to trees T_L in L, containing all the vertices of L, and T_M in M, containing all the vertices of M, in such a way that $T_L \cap N = T_N = T_M \cap N$ and $T_K = T_L \cup T_M$ is a tree containing all the vertices of K.

Now order the vertices of K: in doing so, the vertices of L, M and N are also ordered in an obvious way. By Theorem 3.3.13, $\pi_1(|K|, a^0)$ is generated by the symbols g_{ij}, one for each ordered 1-simplex of $K - T_K$, subject to the relations $g_{ij}g_{jk}g_{ik}^{-1}$, one for each ordered 2-simplex of $K - T_K$. This is clearly the same as the group generated by the symbols g_{ij}, h_{ij}, one for each ordered 1-simplex of $L - T_L$, $M - T_M$ respectively, with relations of the form $g_{ij}g_{jk}g_{ik}^{-1}$, $h_{ij}h_{jk}h_{ik}^{-1}$, together with $g_{ij}h_{ij}^{-1}$, whenever $g_{ij} = h_{ij}$ in K. But this is exactly $\pi_1(|L|, a^0) * \pi_1(|M|, a^0)$, with extra relations $(\lambda_*g_{ij})(\mu_*g_{ij})^{-1}$, one for each generator g_{ij} of $\pi_1(|N|, a^0)$. ∎

There are two important special cases of Theorem 3.3.18. First, if $|N|$ is simply-connected (in particular if $|N|$ is contractible or just a point), then $\pi_1(|K|, a^0) = \pi_1(|L|, a^0) * \pi_1(|M|, a^0)$. A more important corollary, however, refers to the following situation. Let $|K|$ be a path-connected polyhedron, and let $\alpha = a^0a^1 \ldots a^na^0$ ($n \geqslant 2$) be an edge-loop in K, in which no two consecutive vertices are the same. Let $|L|$ be a regular polygon of $(n + 1)$ sides in R^2, triangulated as shown in Fig. 3.7 (b is the centre of $|L|$).

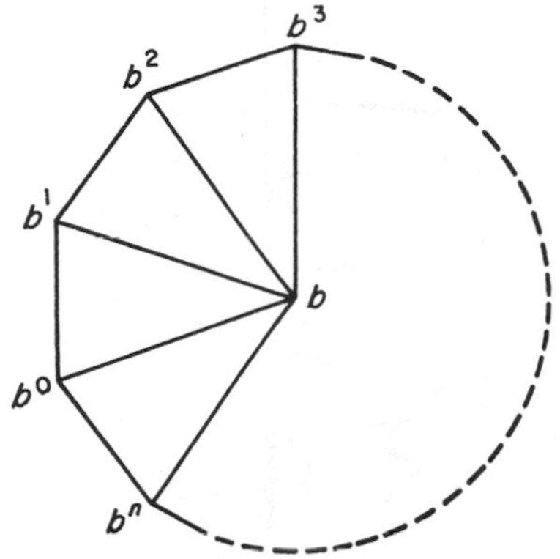

Fig. 3.7

Now (L, M) is a triangulation of (E^2, S^1), where M is the 'boundary' of L. Moreover α determines a (simplicial) map $f: S^1 \to |K|$ by the rule $f(b^r) = a^r$ ($0 \leqslant r \leqslant n$): let X be the adjunction space $|K| \cup_f E^2$.

Theorem 3.3.19 *$\pi_1(X, a^0)$ is obtained from $\pi_1(|K|, a^0)$ by adding the relation $\theta[\alpha]$, where $\theta: \pi(K, a^0) \to \pi_1(|K|, a^0)$ is the isomorphism of Theorem 3.3.9.*

Proof. In order to apply Theorem 3.3.18, it is first necessary to triangulate X. To do so, let $\mathcal{N}$ be the abstract simplicial complex formed from the abstractions of K and $(L, M)'$ by identifying the vertices b^r and a^r, for $0 \leqslant r \leqslant n$. If we identify b^r with a^r, we automatically also identify the abstract simplex (b^r, b^{r+1}) with (a^r, a^{r+1}) for $0 \leqslant r < n$, and (b^n, b^0) with (a^n, a^0), but no further identification of simplexes takes place. This is because

(a) since consecutive vertices of α are distinct, and each simplex of $(L, M)'$ meets M in a face, no simplex is reduced in dimension by the identification;

(b) two distinct 1-simplexes (b^{i_1}, c^1), (b^{i_2}, c^2) of $(L, M)' - M$ cannot be identified unless $c^1 = c^2$; but then b^{i_1} and b^{i_2} must be consecutive vertices of M;

(c) given two distinct 2-simplexes of $(L, M)'$, there must be a vertex, not in M, that is in one simplex but not the other.

(The reader may find Fig. 3.8 helpful in following the above argument.)

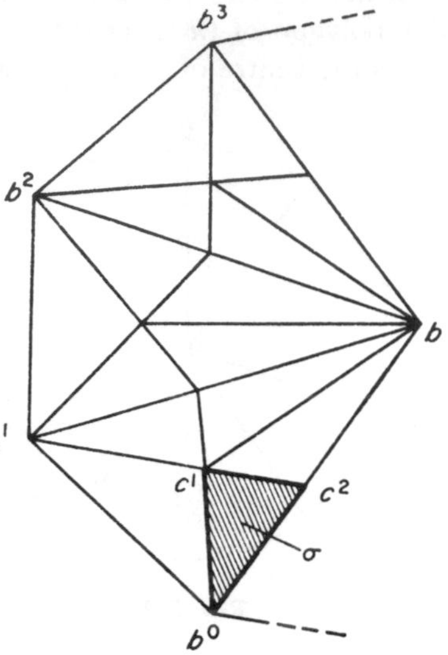

Fig. 3.8

It follows that if N is a geometric realization of $\mathscr{N}$, then $|N|$ is homeomorphic to X (it has the correct topology by Corollary 2.3.9).

Now choose a 2-simplex $\sigma = (b^0, c^1, c^2)$ in $(L, M)'$ that has b^0 as a vertex (see Fig. 3.8), and write $Y = |N - \sigma|$. By Theorem 3.3.18, $\pi_1(X, a^0)$ is obtained from $\pi_1(Y, a^0) * \pi_1(\sigma, a^0)$ by adding relations $(\lambda_* d)(\mu_* d)^{-1}$ for each generator d of $\pi_1(|\dot{\sigma}|, a^0)$, where λ and μ are the inclusion maps. But $\pi_1(\sigma, a^0) = 0$, since σ is contractible, and by Example 3.3.14(a) $\pi_1(|\dot{\sigma}|, a^0)$ is a free group generated by the single element $\theta[\beta]$, where $\beta = b^0c^1c^2b^0$. It follows that $\pi_1(X, a^0)$ is obtained from $\pi_1(Y, a^0)$ by adding the relation $\theta[\beta]$.

To complete the proof, note that, since $|L|$ is convex, radial projection from the barycentre of σ is a (strong) deformation retraction of $|(L, M)' - \sigma|$ onto $|M|$, and so can be extended to a deformation retraction $\rho\colon Y \to |K|$. So $\rho_*\colon \pi_1(Y, a^0) \to \pi_1(|K|, a^0)$ is an isomorphism, and clearly $\rho_*\theta[\beta] = \theta[\alpha]$. ∎

A somewhat surprising corollary of Theorem 3.3.19 is that any group with a finite set of generators and relations can be realized as the fundamental group of some polyhedron.

Theorem 3.3.20 *Let $G = \mathrm{Gp}\{b_1, \ldots, b_m; \beta_1, \ldots, \beta_n\}$. There exists a polyhedron $|K|$ and a vertex a^0 of K such that $\pi_1(|K|, a^0) \cong G$.*

Proof. We first construct K, and then show that it has the right properties.

Let Y be a 'wedge' of m circles $S_1^1, S_2^1, \ldots, S_m^1$. More precisely, take m copies $S_1^1, S_2^1, \ldots, S_m^1$ of S^1, where a point of S_r^1 is denoted by $(x_1, x_2)_r$, and let Y be the space obtained from the disjoint union by identifying together all the points $(1, 0)_1, (1, 0)_2, \ldots, (1, 0)_m$: call this point a^0.

X is now formed from Y by attaching 2-cells by maps obtained from the relations $\beta_1, \ldots, \beta_n$. Now each such β_r is a word in b's: say $\beta_r = b_{i_1}^{\epsilon_1} \cdots b_{i_p}^{\epsilon_p}$, where each ϵ is ± 1. Corresponding to β_r, define a map $f_r: S^1 \to Y$ by the rule

$$f_r(\cos\theta, \sin\theta) = \begin{cases} (\cos(p\theta - 2(q-1)\pi), \sin(p\theta - 2(q-1)\pi))_{i_q}, & \text{if } \epsilon_q = 1, \\ (\cos(2q\pi - p\theta), \sin(2q\pi - p\theta))_{i_q}, & \text{if } \epsilon_q = -1, \end{cases}$$

for $2(q-1)\pi/p \leqslant \theta \leqslant 2q\pi/p$, $1 \leqslant q \leqslant p$. In other words, S^1 is divided into p equal parts, and the qth segment is wrapped round $S_{i_q}^1$, forwards or backwards according as ϵ_q is 1 or -1. Now attach a 2-cell E_r^2 to Y by each of the maps f_r $(1 \leqslant r \leqslant n)$, and call the resulting identification space X.

The fact that X has the required properties is now an easy corollary of Theorem 3.3.19. By radial projection from the origin, each S_r^1 in Y may be triangulated as the boundary of an equilateral triangle (a_r^0, a_r^1, a_r^2), where $a_r^0 = (1, 0)_r$, $a_r^1 = (-\frac{1}{2}, \sqrt{3/2})_r$ and $a_r^2 = (-\frac{1}{2}, -\sqrt{3/2})_r$; a triangulation of Y results if we identify $a_1^0, a_2^0, \ldots, a_m^0$ to a single point a^0. Similarly, each E_r^2 may be triangulated as a regular polygon of $3p$ sides, where p is the number of segments into which the boundary S^1 of E_r^2 is divided in the definition of f_r: see Fig. 3.9.

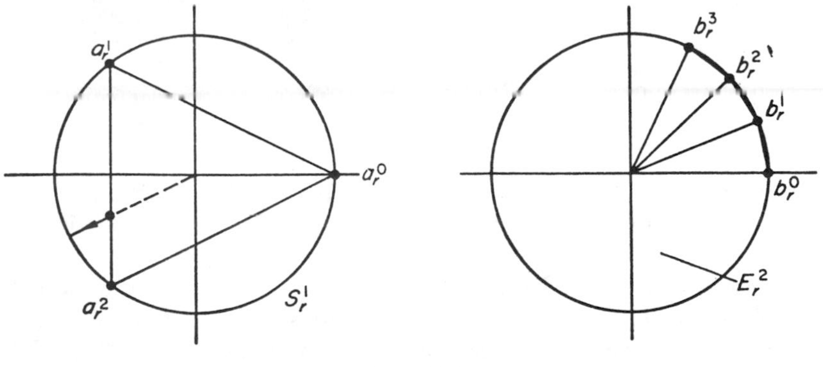

Fig. 3.9

The point of doing this is that now each 2-cell E_r^2 is attached to Y as in Theorem 3.3.19 by means of the edge-loop

$$\beta_r = (a^0_{i_1} a^1_{i_1} a^2_{i_1} a^0_{i_1})^{\epsilon_1} \cdots (a^0_{i_p} a^1_{i_p} a^2_{i_p} a^0_{i_p})^{\epsilon_p}.$$

Thus X is triangulable, and we might as well assume that X is a polyhedron $|K|$. Since $\pi_1(Y, a^0)$ is the free group generated by $\bar{b}_1, \ldots, \bar{b}_m$, where $\bar{b}_r = [a^0_r a^1_r a^2_r a^0_r]$, it follows at once that $\pi_1(X, a^0) \cong G$. ∎

This theorem is not only of interest in itself, but, used 'in reverse', it provides a very practical method of calculating the fundamental groups of certain spaces. By Corollary 3.3.10, if the 2-skeleton of a simplicial complex K is a triangulation of a space X constructed as in Theorem 3.3.20, then we can immediately write down a set of generators and relations for $\pi_1(|K|)$; and this method will usually yield a much smaller set of generators and relations than would be obtained by using Theorem 3.3.13 directly.

Examples 3.3.21

(a) Consider the real projective plane RP^2 again. By Proposition 1.4.40(b) this is the space $S^1 \cup_f E^2$, where $f: S^1 \to S^1$ is defined by $f(\cos\theta, \sin\theta) = (\cos 2\theta, \sin 2\theta)$: see Fig. 3.10, where α represents the generator of $\pi_1(S^1, a^0)$.

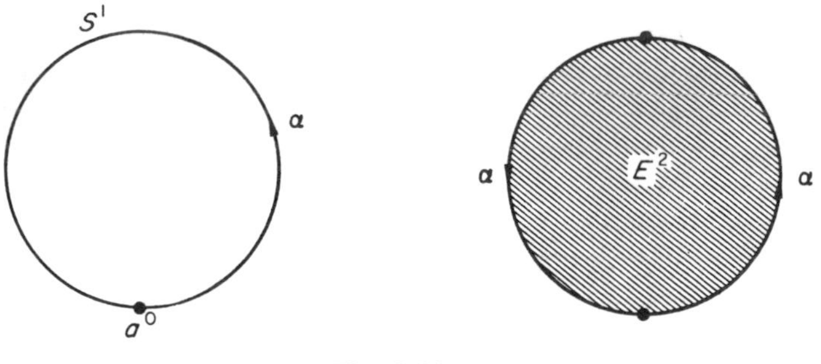

Fig. 3.10

It follows immediately that $\pi_1(RP^2, a^0) = \text{Gp}\,\{\alpha;\, \alpha^2\} \cong Z_2$. The reader should compare the ease of this proof with the messy calculations of Example 3.3.14(b).

(b) As in Chapter 2, Exercise 8, the torus T is the space obtained from a square $ABCD$ by identifying the sides AD and BC, and then AB and DC. By making the identifications on the boundary of $ABCD$ first, we see that this is the same as starting with the wedge Y of two circles S_1^1 and S_2^1, so that $\pi_1(Y)$ is the free group generated by b_1, b_2, say, and then attaching a 2-cell as in Theorem 3.3.20 by a map $f: S^1 \to Y$ corresponding to the word $b_1 b_2 b_1^{-1} b_2^{-1}$: see Fig. 3.11.

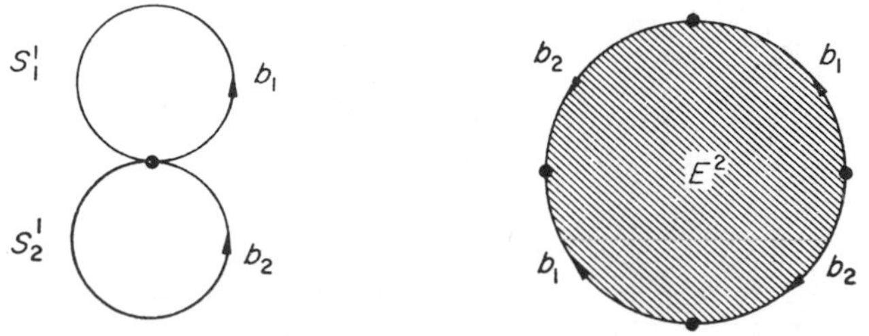

Fig. 3.11

Hence the fundamental group of the torus is Gp $\{b_1, b_2; b_1 b_2 b_1^{-1} b_2^{-1}\}$; in other words $\pi_1(T)$ is a free abelian group with two generators. ▌

We end this section with an example of the calculation of the fundamental group of a more complicated space. This example will also be needed in Chapters 5 and 8.

Example 3.3.22 Let X be the space obtained from (the surface of) a dodecahedron by identifying opposite faces after a twist through an angle $\pi/5$. By stereographic projection from the mid-point of one face, the dodecahedron can be drawn as in Fig. 3.12, in which the vertices and faces are labelled according to the identifications.

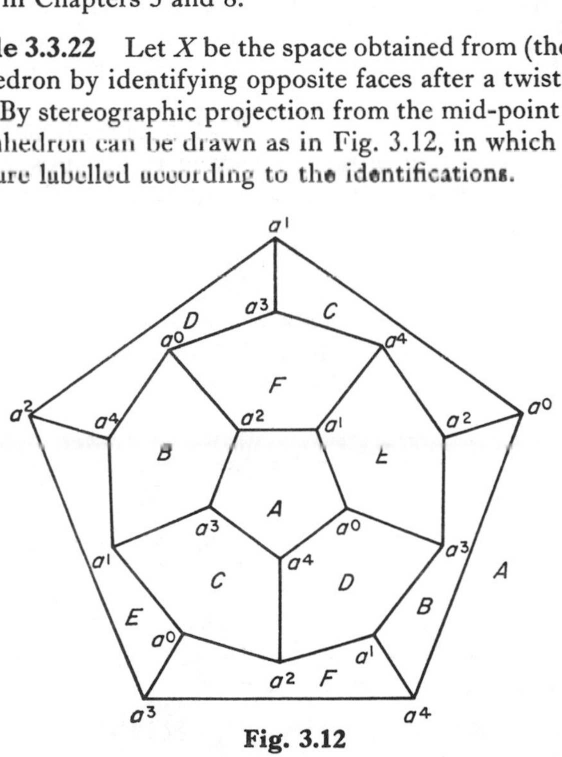

Fig. 3.12

It will be seen that the vertices and edges of the dodecahedron become, after identification, the space Y obtained from the five points a^0, a^1, a^2, a^3, a^4 by joining each pair of points by a line: see Fig. 3.13.

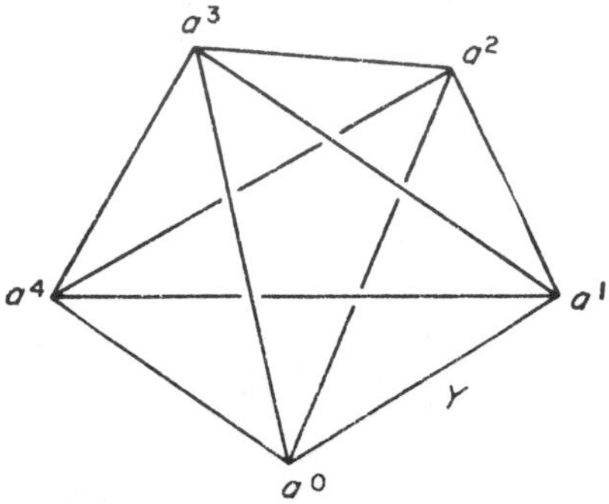

Fig. 3.13

Now Y is plainly triangulable, and X is the space obtained from Y by attaching six 2-cells A, B, C, D, E and F by the edge-loops $a^0a^1a^2a^3a^4a^0$, $a^0a^4a^1a^3a^2a^0$, $a^0a^2a^4a^3a^1a^0$, $a^0a^4a^2a^1a^3a^0$, $a^0a^3a^2a^4a^1a^0$ and $a^0a^2a^1a^4a^3a^0$ respectively. Thus $\pi_1(X, a^0)$ can be calculated by Theorem 3.3.19.

The first task is to calculate $\pi_1(Y, a^0)$. This is easily done by using Theorem 3.3.13: a maximal tree consists of the 1-simplexes (a^0, a^1), (a^0, a^2), (a^0, a^3) and (a^0, a^4), so that $\pi_1(Y, a^0)$ is the free group generated by

$$\begin{aligned} \alpha &= [a^0a^1a^2a^0], \\ \beta &= [a^0a^1a^3a^0], \\ \gamma &= [a^0a^1a^4a^0], \\ \delta &= [a^0a^2a^3a^0], \\ \epsilon &= [a^0a^2a^4a^0] \end{aligned}$$

and $$\zeta = [a^0a^3a^4a^0].$$

So by Theorem 3.3.19, $\pi_1(X, a^0)$ has these generators, subject to the following six relations, given by the faces A, B, C, D, E and F:

$$\alpha\delta\zeta, \quad \gamma^{-1}\beta\delta^{-1}, \quad \epsilon\zeta^{-1}\beta^{-1}, \quad \epsilon^{-1}\alpha^{-1}\beta, \quad \delta^{-1}\epsilon\gamma^{-1}, \quad \alpha^{-1}\gamma\zeta^{-1}$$

(we write $[a^0a^1a^2a^3a^4a^0]$, for instance, in the equivalent form $[a^0a^1a^2a^0][a^0a^2a^3a^0][a^0a^3a^4a^0] = \alpha\delta\zeta$). The first, third and fifth of these relations give $\alpha = \zeta^{-1}\delta^{-1}$, $\beta = \epsilon\zeta^{-1}$, $\gamma = \delta^{-1}\epsilon$; and the remaining three relations then become

$$\epsilon^{-1}\delta\epsilon\zeta^{-1}\delta^{-1}, \quad \epsilon^{-1}\delta\zeta\epsilon\zeta^{-1}, \quad \delta\zeta\delta^{-1}\epsilon\zeta^{-1}.$$

From the first of these, $\zeta = \delta^{-1}\epsilon^{-1}\delta\epsilon$, so that $\pi_1(X, a^0)$ now has two generators δ and ϵ, and two relations

$$\epsilon^{-2}\delta\epsilon\delta^{-1}\epsilon\delta, \qquad \epsilon^{-1}\delta\epsilon\delta^{-2}\epsilon\delta.$$

The second of these can be replaced by the product of itself and the inverse of the first (or rather by the conjugate of this element by ϵ^{-2}):

$$\epsilon^2(\epsilon^{-1}\delta\epsilon\delta^{-2}\epsilon\delta)\delta^{-1}\epsilon^{-1}\delta\epsilon^{-1}\delta^{-1} = \epsilon\delta\epsilon\delta^{-1}\epsilon^{-1}\delta^{-1};$$

and the first relation can then be replaced by this new relation multiplied by its inverse (with conjugation by $\epsilon\delta$ before and after the multiplication):

$$\delta^{-1}\epsilon^{-1}(\epsilon\delta\epsilon\delta^{-1}\epsilon^{-1}\delta^{-1})\delta\epsilon^{-1}\delta^{-1}\epsilon^2 = \epsilon\delta^{-1}\epsilon^{-2}\delta^{-1}\epsilon^2.$$

Now write $\delta = \eta\epsilon^{-1}$, so that the generators are now ϵ and η, and the relations are

$$\epsilon\eta\epsilon^{-1}\epsilon\epsilon\eta^{-1}\epsilon^{-1}\epsilon\eta^{-1} = \epsilon\eta\epsilon\eta^{-2}$$

and

$$\epsilon\epsilon\eta^{-1}\epsilon^{-2}\epsilon\eta^{-1}\epsilon^2 = \epsilon^2\eta^{-1}\epsilon^{-1}\eta^{-1}\epsilon^2.$$

These in turn are equivalent to

$$\eta^2(\epsilon\eta\epsilon)^{-1}, \qquad \epsilon^4(\eta\epsilon\eta)^{-1},$$

or to

$$\eta^3(\eta\epsilon)^{-2}, \qquad \epsilon^5(\eta\epsilon)^{-2},$$

and this gives a concise expression for $\pi_1(X, a^0)$ in terms of generators and relations. It is not immediately obvious that the group is non-trivial, but in the group of permutations of 1, 2, 3, 4, 5 the permutations

$$x: (1, 2, 3, 4, 5) \to (4, 2, 1, 3, 5)$$

and

$$y: (1, 2, 3, 4, 5) \to (2, 3, 4, 5, 1)$$

can easily be seen to satisfy $x^3 = (xy)^2 = y^5 = 1$, and so generate a group isomorphic to a quotient group of $\pi_1(X, a^0)$, which is therefore non-trivial. ∎

3.4 Classification of triangulable 2-manifolds

As an example of the application of the fundamental group to geometric problems, we shall show in this section that the theorems of Section 3.3 allow easy calculation of the fundamental groups of certain polyhedra known as *2-manifolds* or *surfaces*. By using geometric arguments as well, this leads to a complete classification of these 2-manifolds, up to homeomorphism. This is a good illustration of a typical procedure of algebraic topology: one first uses a geometric argument to show that every 2-manifold is homeomorphic to one of a standard set, and then shows that the 'standard' manifolds are all topologically distinct, by showing that their fundamental groups are all distinct.

For completeness, we shall first define manifolds in general, and then specialize to 2-manifolds. Roughly speaking, an ***n*-manifold** is a topological space that is locally 'like' Euclidean space R^n.

Definition 3.4.1 A Hausdorff space M is called an ***n*-manifold** if each point of M has a neighbourhood homeomorphic to an open set in R^n.

Notice that any space homeomorphic to an n-manifold is itself an n-manifold, as also is any open subset of an n-manifold.

Example 3.4.2 R^n itself is clearly an n-manifold, as also is S^n. To prove this, let e^n be the open unit disc in R^n, of points x such that $\|x\| < 1$, and note that the standard map $\theta: E^n \to S^n$ restricts to a homeomorphism $\theta: e^n \to S^n - (-1, 0, \ldots, 0)$. Thus every point of S^n other than $(-1, 0, \ldots, 0)$ is certainly contained in an open set homeomorphic to e^n, and we can deal with the exceptional point by constructing a similar homeomorphism from e^n to the complement of $(1, 0, \ldots, 0)$ in S^n.

Other examples of manifolds are the torus and the real projective plane: both of these are 2-manifolds, as can readily be proved from the definition.

Lastly, consider the space X obtained from two copies S^1_1, S^1_2 of the circle S^1 by identifying each point $(x_1, x_2)_1$ with the corresponding point $(x_1, x_2)_2$, except for the points $(1, 0)_1$ and $(1, 0)_2$, which remain distinct: see Fig. 3.14.

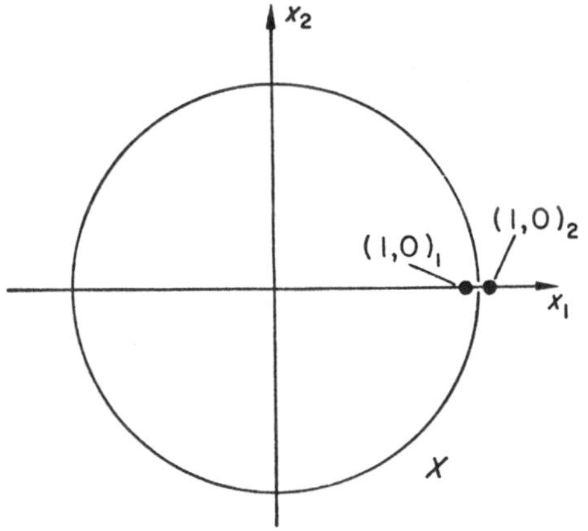

Fig. 3.14

Now $(1, 0)_1$ has an open neighbourhood consisting of $(1, 0)_1 \cup C$, where C is the complement in S^1 of $(1, 0)$ and $(-1, 0)$; and this neighbourhood is homeomorphic to an open interval in R^1. Similarly $(1, 0)_2$ has a neighbourhood homeomorphic to an open interval, and this property is clearly true for all other points of X. However, X fails to be a manifold, since open neighbourhoods of $(1, 0)_1$ and $(1, 0)_2$ always intersect, so that X is not Hausdorff. ∎

The last example shows the reason for insisting on the Hausdorff condition in Definition 3.4.1: we wish to exclude such freak spaces.

In order to apply the theorems of Section 3.3, we shall consider only *triangulable* n-manifolds in this chapter. Since we are particularly interested in 2-manifolds, this is only a mild restriction, for it can be shown that every compact 2-manifold is triangulable, although the proof of this is beyond the scope of this book. However, before attempting to prove the classification theorem for triangulable 2-manifolds, we need a few results about simplicial complexes whose polyhedra are manifolds: these are based on Theorem 2.4.5.

Proposition 3.4.3 *Let K be a simplicial complex whose polyhedron is an n-manifold. Then for each $x \in |K|$, $|\text{Lk}(x)| \simeq S^{n-1}$.*

Proof. By definition, there exists an open set U in R^n, and a homeomorphism h of U onto a subset $h(U)$ of $|K|$ that contains x; let $y = h^{-1}(x)$. Since U is open, there exists ϵ such that the set B of points z such that $d(y, z) \leqslant \epsilon$ is contained in U: then h is a homeomorphism of B onto a subset of $|K|$, and x is in an open set contained in $h(B)$.

But B can be triangulated as in Example 2.3.13, with y as the vertex a_0. Hence by Theorem 2.4.5 $|\text{Lk}(x)| \simeq |\text{Lk}(y)|$, which is homeomorphic to S^{n-1}. ∎

Corollary 3.4.4 *If $|K|$ is a 2-manifold, then*

(a) *$\dim K = 2$;*
(b) *each 1-simplex of K is a face of just two 2-simplexes.*

Proof. K cannot have a simplex σ of dimension $n > 2$, for if x were in the interior of such a simplex, then by Example 2.3.13 $|\text{Lk}(x)|$ would be homeomorphic to S^{n-1}. But $\pi_1(S^{n-1}) = 0$ if $n > 2$, whereas $\pi_1(S^1) \cong Z$, so that S^{n-1} is not homotopy-equivalent to S^1, and hence σ cannot exist.

Now let x be a point in the interior of a 1-simplex τ, and suppose that τ is a face of r 2-simplexes. Then $\text{Lk}(x)$ is the subcomplex shown in Fig. 3.15, with r 'strings' joining a^0 and a^1 (we must have $r > 0$, since otherwise $|\text{Lk}(x)| = a^0 \cup a^1$, which is not homotopy-equivalent to S^1).

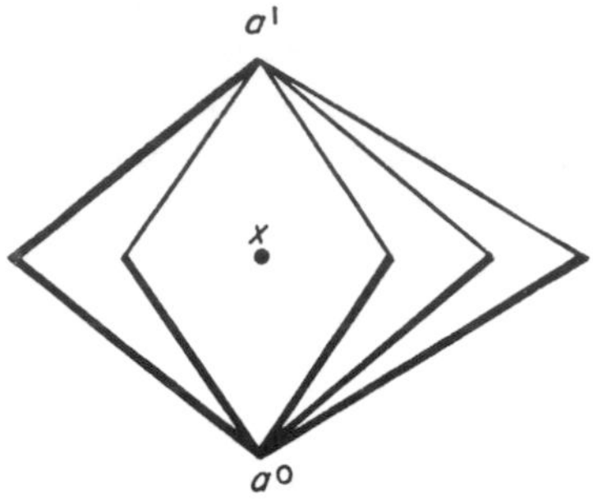

Fig. 3.15

A maximal tree of Lk (x) is shown in thick lines in Fig. 3.15. Thus by Theorem 3.3.13 $\pi_1(|\text{Lk}(x)|, a^0)$ is a free group generated by $(r - 1)$ elements. This contradicts Proposition 3.4.3 unless $r = 2$. ∎

We now start work on the classification theorem for triangulable 2-manifolds. As a first step, we prove the following result on path-connected polyhedra (there is no loss of generality in supposing that the polyhedra are path-connected, for otherwise, by Exercises 2 and 10, we merely consider the path-components separately).

Theorem 3.4.5 *Let K be a simplicial complex whose polyhedron is a path-connected 2-manifold. Then $|K|$ is homeomorphic to the space obtained from a regular polygon of 2n sides in R^2 by identifying the edges in pairs.*

Proof. We can construct a space homeomorphic to $|K|$ as follows. Choose any 2-simplex σ_1 of K: this is (linearly) homeomorphic to an equilateral triangle in R^2. Now choose any 1-face τ of σ_1; by Corollary 3.4.4(b) τ is a face of just one other 2-simplex, σ_2, say. The subspace $\sigma_1 \cup \sigma_2$ of $|K|$ is (simplicially) homeomorphic to the equilateral triangle with another triangle attached along one edge, and this in turn is simplicially homeomorphic to a square in R^2: see Fig. 3.16.

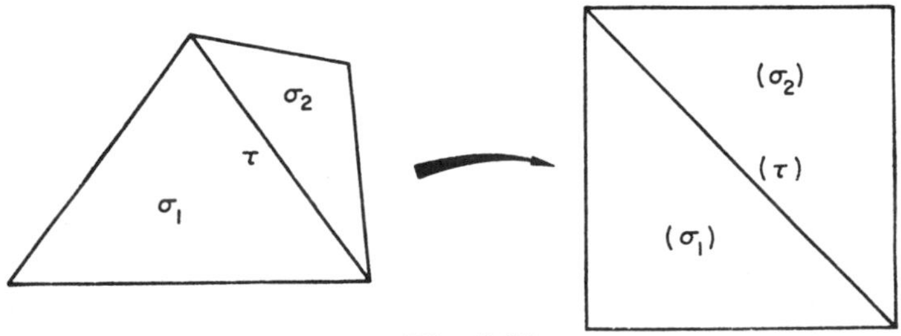

Fig. 3.16

This process can be continued: at the general stage we have $\sigma_1 \cup \cdots \cup \sigma_m$, simplicially homeomorphic to a regular $(m + 2)$-sided polygon in R^2, although some pairs of edges in this polygon may have to be identified: each edge corresponds to a 1-simplex of K that faces two 2-simplexes, and if both these simplexes are already represented in the polygon, then the edge must be identified with another edge of the polygon (it must be another *edge*, and not an 'interior' 1-simplex, since otherwise there would be a 1-simplex of K facing more than two 2-simplexes). If on the other hand only one of the two 2-simplexes is already represented in the polygon, call the other one σ_{m+1}, and construct a regular $(m + 3)$-sided polygon by attaching a triangle corresponding to σ_{m+1} along the appropriate edge, and taking a simplicial homeomorphism of the resulting space onto a regular polygon in R^2.

It is clear that we can continue attaching triangles and deforming into regular polygons, until we finally reach a regular polygon P of $2n$ sides in which each edge is identified with one other edge (this is why P must have an even number of edges). This is the result we want, provided every 2-simplex of K is now represented in P.

To prove that no 2-simplex has been left out, suppose on the contrary that P, with its appropriate identifications, is homeomorphic to $|L|$, where L is a subcomplex of K. Choose a vertex of L and a vertex of $K - L$, and join them by an edge-path (since $|K|$ is path-connected); let a be the last vertex in L and b be the next vertex, so that (a, b) is a 1-simplex of $K - L$. We can obtain a contradiction by showing that $|\mathrm{Lk}_K(a)|$ is not path-connected, and so certainly not homotopy-equivalent to S^1. For suppose, if possible, that some vertex

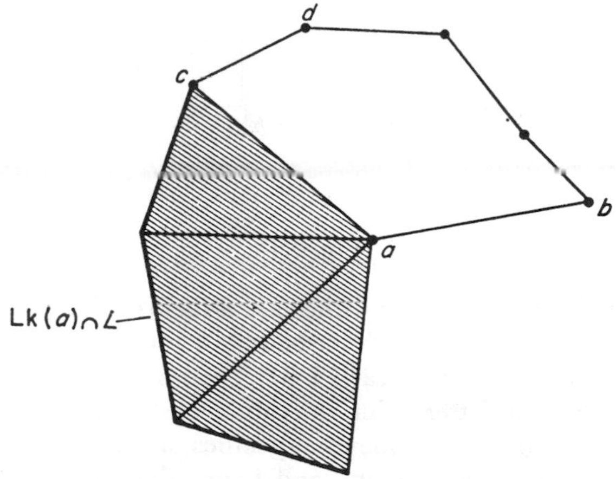

Fig. 3.17

in Lk$(a) \cap L$ can be joined to b by an edge-path in Lk(a). Once again, let c be the last vertex in Lk$(a) \cap L$, and d be the next vertex: see Fig. 3.17.

Now $(c, d) \in$ Lk(a), so that (a, c, d) is a 2-simplex of $K - L$. But it is clear from the construction of L that each 1-simplex of L, in particular (a, c), faces two 2-simplexes of L. Hence (a, c) faces at least three 2-simplexes of K, which contradicts Corollary 3.4.4(b). Thus $|$Lk$(a)|$ is not path-connected, which is again a contradiction, so that L must be the whole of K. ∎

Let the vertices of a regular $2n$-sided polygon be $b^0, b^1, \ldots, b^{2n-1}, b^0$, in order as we go round the boundary. Now if the edges of P are identified in pairs, an edge (c, d) is identified with one other edge (c', d'), say, where c is identified with c' and d with d'. For each such pair of edges, denote both by a symbol such as x, and denote the 'reversed' edges (d, c) and (d', c') by x^{-1}; of course, different symbols are to be used for different pairs of edges. In this way P can be specified, with its identifications, by the sequence of symbols such as x or x^{-1} corresponding to the sequence of edges $(b^0, b^1), (b^1, b^2), \ldots, (b^{2n-1}, b^0)$. For example, the torus can be specified in this way by the sequence $xyx^{-1}y^{-1}$, and the real projective plane by $xy^{-1}xy^{-1}$: see Fig. 3.18.

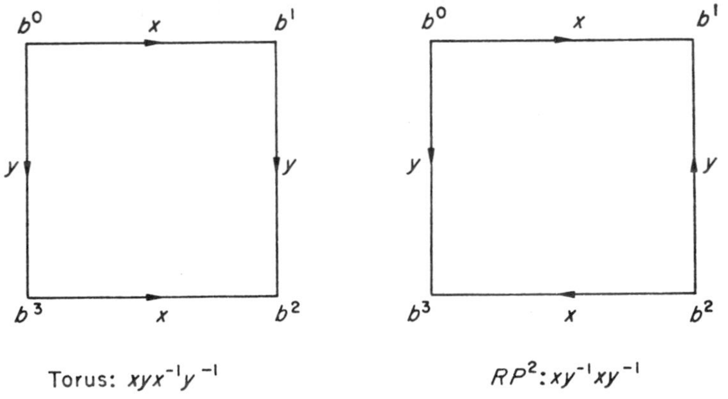

Fig. 3.18

Theorem 3.4.5 shows, then, that a polyhedron $|K|$ that is a path-connected 2-manifold can be described by a finite sequence of symbols such as x or x^{-1}, in which each letter occurs twice and there are at least two different letters: let us call such a sequence *admissible*. Clearly any admissible sequence specifies a regular polygon with identifications of edges in pairs, and hence determines a topological space. Unfortunately this is not yet a classification theorem, since it is quite possible for two different admissible sequences to specify homeomorphic spaces. The next step is to resolve this difficulty, by giving three rules for changing an admissible sequence, while altering the corresponding space only by a homeomorphism. To state these rules, denote (possibly empty) sequences of symbols by capital letters, and if say $A = \alpha_1\alpha_2\cdots\alpha_n$, where each α_r is of the form x or x^{-1}, write A^{-1} for the sequence $\alpha_n^{-1}\alpha_{n-1}^{-1}\cdots\alpha_1^{-1}$ (by convention $(x^{-1})^{-1} = x$).

Rule 1. Replace $ABxCDxE$ by $AyDB^{-1}yC^{-1}E$, where y is a new symbol.

Rule 2. Replace $ABxCDx^{-1}E$ by $AyDCy^{-1}BE$.

Rule 3. Replace $Axx^{-1}B$ or $Ax^{-1}xB$ by AB, provided AB contains at least two letters (each occurring twice, of course).

To justify these changes, we prove

Theorem 3.4.6 *The application of Rules 1–3 to an admissible sequence gives a new admissible sequence whose corresponding space is homeomorphic to the space corresponding to the original sequence.*

Proof. It is clearly sufficient to prove this for a single application of Rule 1, 2 or 3.

Rule 1. In the regular polygon corresponding to $ABxCDxE$, join the 'end-point' of A to the end-point of C by a straight line, denoted by y. Cut the polygon in two along y, and join the two pieces together again by identifying the edges corresponding to x: see Fig. 3.19.

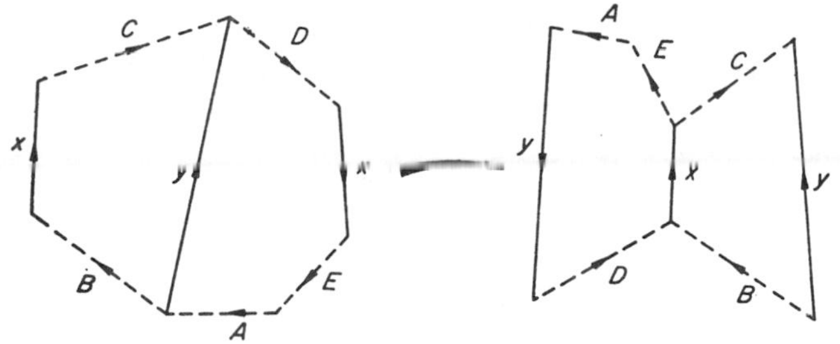

Fig. 3.19

The new space can be deformed into a regular polyhedron again, since it can be triangulated by joining the boundary edges to the mid-point of x. When corresponding edges of the new polygon are identified, we obtain a space homeomorphic to the original one; and the new sequence of symbols is $AyDB^{-1}yC^{-1}E$.

The reader should notice, however, that this proof is valid only if BC and ADE are non-empty sequences, since otherwise the polygon is not cut into two pieces. However, if BC is empty there is nothing to prove, whereas if ADE is empty, the replacement of $BxCx$ by $yB^{-1}yC^{-1}$ corresponds merely to going round the boundary of the polygon in the opposite direction (and putting $y = x^{-1}$). And BC and ADE cannot both be empty, since an admissible sequence contains at least two letters.

Rule 2. This admits a similar proof.

Rule 3. Consider the regular polygon corresponding to $Axx^{-1}B$. By starting from a different vertex in the boundary, if necessary, we may assume that each of A and B represents at least two edges. Make a cut y from the end-point of B to the end-point of x, and deform each of the two pieces into regular polygons in which the two edges x and y are made into just one edge; finally join the two polygons together by identifying the edges corresponding to xy^{-1}, and deform the result into a regular polygon again: see Fig. 3.20.

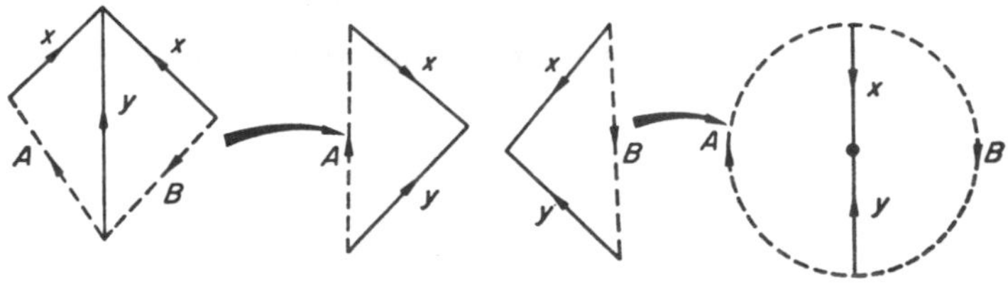

Fig. 3.20

As in Rule 1, when corresponding edges of this polygon are identified, we obtain a space homeomorphic to the original one, and the new sequence of symbols is AB. ∎

Rules 1–3 allow the reduction of admissible sequences to certain standard forms. Now each letter x in an admissible sequence occurs twice; call these two occurrences a *similar pair* if the sequence is of the form $\cdots x \cdots x \cdots$ or $\cdots x^{-1} \cdots x^{-1} \cdots$, and call them a *reversed pair* if the sequence is of the form $\cdots x \cdots x^{-1} \cdots$ or $\cdots x^{-1} \cdots x \cdots$. The following four steps can be applied to an admissible sequence, where each step is a combination of applications of Rules 1–3.

Step 1. Replace the sequence by AB, where A is of the form $x_1x_1x_2x_2 \cdots x_rx_r$ and B contains only reversed pairs (of course, A or B

may be empty). This is justified by the following sequence of operations, using only Rule 1, where C is supposed to be already of the form $x_1x_1x_2x_2\cdots$

$$CDxExF \to CyD^{-1}yE^{-1}F$$
$$\to CzzDE^{-1}F.$$

(Each similar pair may be assumed to be of the form $\cdots x\cdots x\cdots$, by replacing x^{-1} by a new symbol y if necessary.)

Step 2. Now replace AB by ACD, where C is of the form $y_1z_1y_1^{-1}z_1^{-1}\cdots y_sz_sy_s^{-1}z_s^{-1}$, and D contains only non-interlocking reversed pairs (two reversed pairs are said to *interlock* if they occur in the form $\cdots y\cdots z\cdots y^{-1}\cdots z^{-1}\cdots$). This is justified by Rule 2, where E is assumed to be already of the required form.

$$EFaGbHa^{-1}Ib^{-1}J \to EcGbHc^{-1}FIb^{-1}J \quad \text{(here } a \text{ is the `}x\text{' of Rule 2)}$$
$$\to EcGdFIHc^{-1}d^{-1}J \quad \text{(with } b \text{ as `}x\text{')}$$
$$\to EeFIHGde^{-1}d^{-1}J \quad \text{(with } c \text{ as `}x\text{')}$$
$$\to Eefe^{-1}f^{-1}FIHGJ \quad \text{(with } d \text{ as `}x\text{').}$$

Step 3. If A is non-empty, replace ACD by ED, where E is of the form $x_1x_1x_2x_2\cdots$, that is, convert all interlocking reversed pairs to similar pairs. This uses Rule 1, but in reverse:

$$Fxxaba^{-1}b^{-1}G \leftarrow Fyb^{-1}a^{-1}ya^{-1}b^{-1}G$$
$$\leftarrow Fyay^{-1}accG$$
$$\leftarrow FyyddccG.$$

Step 4. Finally, consider D, which consists only of non-interlocking reversed pairs. Let the closest pair in D be $\cdots x\cdots x^{-1}\cdots$; then there can be no symbols between x and x^{-1}, since if both members of a pair lie between x and x^{-1} they form a closer pair, whereas if just one member of a pair is between x and x^{-1} there is an interlocking pair in D. Thus we can 'cancel' xx^{-1} by Rule 3, and similarly cancel the rest of D, provided what remains always contains at least two letters.

The final result of Steps 1–4 is that the admissible sequence now has one of the forms

$$x_1y_1x_1^{-1}y_1^{-1}\cdots x_gy_gx_g^{-1}y_g^{-1} \qquad (g \geqslant 1)$$

or

$$x_1x_1x_2x_2\cdots x_hx_h \qquad (h \geqslant 2),$$

with three remaining special cases $xxyy^{-1}$, $xx^{-1}yy^{-1}$ and $xyy^{-1}x^{-1}$: these cannot be further reduced by Rule 3, although it will be noticed that $xyy^{-1}x^{-1}$ represents the same space as $x^{-1}xyy^{-1}$, and hence as $xx^{-1}yy^{-1}$, since a cyclic permutation of symbols merely corresponds to taking a different starting point for the boundary of the corresponding square.

Let M_g ($g \geqslant 1$) be the space obtained from a regular $4g$-sided polygon by identifying the edges according to the sequence $x_1y_1x_1^{-1}y_1^{-1}\cdots x_gy_gx_g^{-1}y_g^{-1}$, and let N_h ($h \geqslant 2$) be defined similarly using $x_1x_1\cdots x_hx_h$; also let N_1 and M_0 be the special cases defined by $xxyy^{-1}$ and $xx^{-1}yy^{-1}$ respectively. We have so far proved

Theorem 3.4.7 *A path-connected triangulable 2-manifold is homeomorphic to one of the spaces M_g ($g \geqslant 0$) or N_h ($h \geqslant 1$).* ▌

Examples 3.4.8

(a) M_0 is homeomorphic to S^2. For S^2 can be triangulated as the boundary of a 3-simplex (A, B, C, D), and the process described in the proof of Theorem 3.4.5 yields the square shown in Fig. 3.21, with corresponding sequence $xx^{-1}yy^{-1}$.

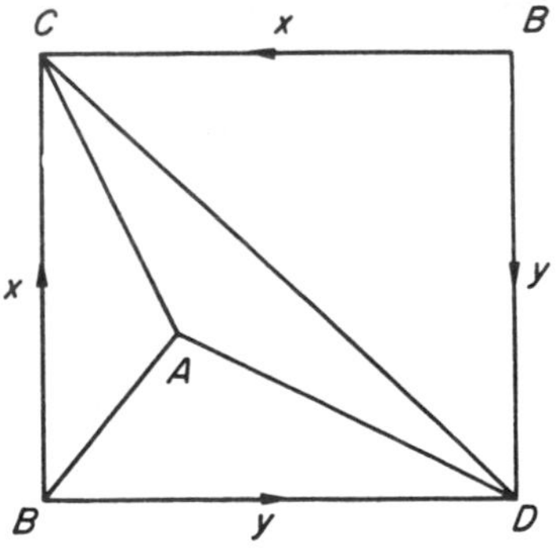

Fig. 3.21

(b) N_1 is the real projective plane RP^2. For Fig. 3.18 shows that RP^2 is the space defined by the sequence $xy^{-1}xy^{-1}$, and a single application of Rule 1 reduces this to $zzyy^{-1}$. ▌

In fact Theorem 3.4.7 is exactly the classification theorem for triangulable 2-manifolds, although it still remains to prove that each of M_g and N_h are topologically distinct, and that each of these spaces is a triangulable 2-manifold. The first point is established by calculating the fundamental groups of M_g and N_h, by using Theorem 3.3.20 (compare Examples 3.3.21).

Theorem 3.4.9

(a) $\pi_1(M_g) \cong \text{Gp}\,\{x_1, y_1, \ldots, x_g, y_g;\, x_1y_1x_1^{-1}y_1^{-1}\cdots x_gy_gx_g^{-1}y_g^{-1}\}$ (*this is to be interpreted as* 0 *if* $g = 0$).

(b) $\pi_1(N_h) \cong \text{Gp}\,\{x_1, \ldots, x_h;\, x_1^2 \ldots x_h^2\}$.

Proof.

(a) For $g \geqslant 1$, M_g is obtained by identifying edges in a regular $4g$-sided polygon P. Now all $4g$ vertices of P are identified together in M_g, since

$$\begin{aligned}\text{initial point of } x_1 &= \text{end point of } y_1\\ &= \text{end point of } x_1\\ &= \text{initial point of } y_1\\ &= \text{initial point of } x_2,\end{aligned}$$

and so on. Thus the boundary of P becomes, after identification, a 'wedge' of $2g$ circles, one for each letter x_r or y_r, and M_g is the space obtained by attaching a single 2-cell according to the word $x_1y_1x_1^{-1}y_1^{-1}\cdots x_gy_gx_g^{-1}y_g^{-1}$. Theorem 3.3.20 immediately yields (a), at least if $g \geqslant 1$. However, if $g = 0$, $M_0 = S^2$, and $\pi_1(S^2) = 0$ by Corollary 3.3.10.

(b) Again, for $h \geqslant 2$, N_h is obtained by identifying edges in a regular $2h$-sided polygon P. As in the proof of (a), all $2h$ vertices of P are identified together in N_h, and the boundary of P becomes a wedge of h circles, one for each letter x_r. So N_h is the space obtained by attaching a 2-cell according to the word $x_1^2\cdots x_h^2$, which proves (b) if $h \geqslant 2$. But for $h = 1$ $N_h = RP^2$, and $\pi_1(RP^2) = \text{Gp}\,\{x;\, x^2\}$ by Example 3.3.21(a). ∎

Corollary 3.4.10 *The spaces M_g and N_h are all topologically distinct.*

Proof. It is sufficient to show that their fundamental groups are not isomorphic. Now in general the problem of deciding whether two groups given by generators and relations are isomorphic is difficult, and may even be insoluble. However, it is sufficient here to remark that, by Proposition 1.3.24, if two groups G and H are isomorphic, then so are their 'abelianizations' $G/[G, G]$ and $H/[H, H]$.

Now by Proposition 1.3.28, $\pi_1(M_g)/[\pi_1(M_g), \pi_1(M_g)]$ is

$$\text{Ab}\,\{x_1, y_1, \ldots, x_g, y_g\},$$

and $\pi_1(N_h)/[\pi_1(N_h), \pi_1(N_h)]$ is

$$\text{Ab}\,\{x_1, \ldots, x_h; 2(x_1 + \cdots + x_h)\}.$$

By setting $y = x_1 + \cdots + x_h$, the latter is the same as

$$\text{Ab}\,\{x_1, \ldots, x_{h-1}, y; 2y\},$$

which is the direct sum of a free abelian group with $(h - 1)$ generators and a group isomorphic to Z_2. So by Theorem 1.3.30 the groups $\pi_1(M_g)$ and $\pi_1(N_h)$ are all distinct, so that no two of M_g or N_h can be homeomorphic. ∎

It follows also, of course, that no two of M_g or N_h are homotopy-equivalent, so that for triangulable 2-manifolds the classification up to homeomorphism is the same as the classification up to homotopy equivalence. In particular, any manifold homotopy-equivalent to S^2 must actually be homeomorphic to S^2; this result is of especial interest, since it remains an unsolved problem whether or not the corresponding result for 3-manifolds and S^3 is true: this is the famous 'Poincaré conjecture'.

To complete the classification of triangulable 2-manifolds, it now remains only to prove

Theorem 3.4.11 *Each of M_g and N_h is a triangulable 2-manifold.*

Proof. As in the proof of Theorem 3.4.9, Theorem 3.3.20 shows that each of M_g and N_h is triangulable. Hence it is sufficient to prove that they are 2-manifolds.

Consider the $4g$-sided polygon P corresponding to M_g, for $g \geqslant 1$. It is clear that a point of P not on the boundary has a neighbourhood homeomorphic to an open set in R^2. Also a point A on the boundary of P, other than a vertex, occurs in just two edges, say the edges corresponding to the symbol x_1: see Fig. 3.22.

Choose ϵ so that the two 'ϵ-neighbourhoods' of A (the shaded areas in Fig. 3.22) intersect the boundary of P only in the edges x_1. After identification, these ϵ-neighbourhoods fit together to make a neighbourhood of A that is clearly homeomorphic to an open disc in R^2.

Lastly, consider the point B of M_g corresponding to the $4g$ vertices of P. This too has a neighbourhood homeomorphic to an open set in R^2, obtained by piecing together ϵ-neighbourhoods, although this time there are $4g$ pieces instead of only 2. In detail, choose ϵ less than

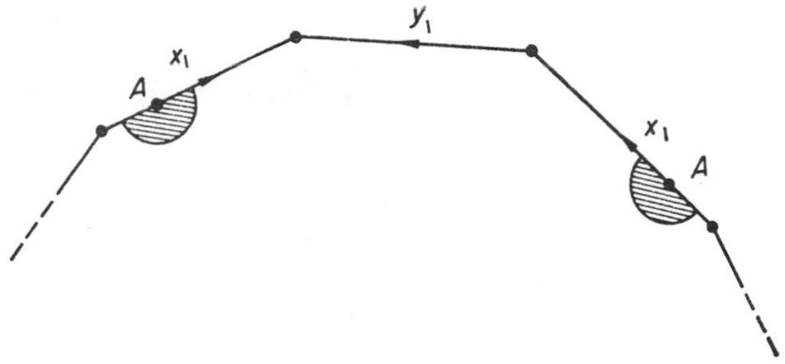

Fig. 3.22

half the length of an edge of P, so that the ϵ-neighbourhoods of the vertices are disjoint segments of an open disc: see Fig. 3.23.

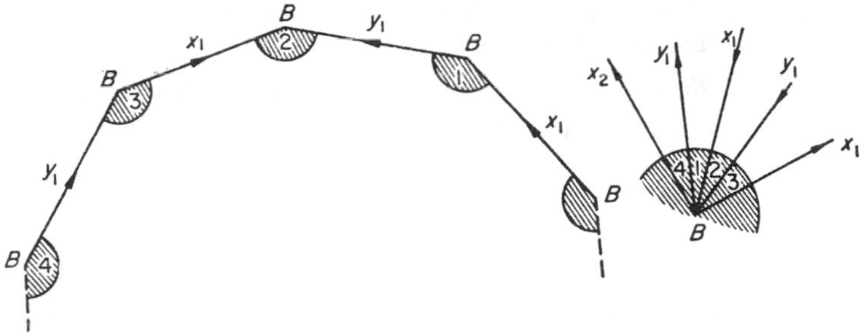

Fig. 3.23

After identification of edges, the numbered ϵ-neighbourhoods fit together as shown in Fig. 3.23 to make a segment of an open disc bounded by the beginning of edge x_1 and the beginning of edge x_2; this fits onto the segment between x_2 and x_3, and so on. Thus the $4g$ segments in P fit together to make a neighbourhood of B that is homeomorphic to an open disc in R^2. Hence M_g is a 2-manifold.

The reader should have no difficulty in adapting this proof to deal with N_h, for $h \geqslant 2$, and the special case $N_1 = RP^2$. And of course $M_0 = S^2$, which we have already seen in Example 3.4.2 is a 2-manifold. ▮

EXERCISES

1. Give an example of a space X, with two base points x_0 and x_1 such that $\pi_1(X, x_0)$ and $\pi_1(X, x_1)$ are not isomorphic.
2. Show that a connected open set in R^n is path-connected, and that a connected polyhedron is path-connected. (*Hint:* show that each path

component is a subpolyhedron.) Show also that a contractible space is path-connected.

3. Show by the following example that two spaces may be homotopy-equivalent without being of the same based homotopy type. Let X be the set of all points in R^2 on straight-line segments joining (0, 1) to $(x_1, 0)$, where x_1 runs through all points $1/n$, for each positive integer n, together with 0. Then X is contractible, but if $x_0 = (0, 0)$ is the base point, X and x_0 are not of the same based homotopy type. (Suppose that $F: X \times I \to X$ is a homotopy starting with the identity map, such that $F(x, 1) = F(x_0, t) = x_0$ for all $x \in X$, $t \in I$; obtain a contradiction to the continuity of F.)

4. Given spaces X and Y, with base points x_0 and y_0 respectively, show that $\pi_1(X \times Y, (x_0, y_0))$ is isomorphic to the direct sum of $\pi_1(X, x_0)$ and $\pi_1(Y, y_0)$. (This provides another proof that the fundamental group of a torus is a free abelian group with two generators, since the torus is homeomorphic to $S^1 \times S^1$.)

5. Prove the following generalization of Proposition 3.3.3. Let K be a simplicial complex that collapses onto a subcomplex L; then for any subcomplex M, $(K, M)' \searrow L'$. Prove also that $|L|$ is a strong deformation retract of $|K|$.

6. Show that a 1-dimensional complex whose polyhedron is simply-connected is collapsible onto a vertex. Show also that a 2-dimensional simplicial complex K in R^2 is collapsible onto a 1-dimensional subcomplex, and hence that K is collapsible to a vertex if $|K|$ is simply-connected. (However, not all contractible 2-dimensional simplicial complexes are collapsible to vertices: see for example Chapter 8, Exercise 5.)

7. Show that real projective n-space RP^n can be triangulated by identifying antipodal points in L', where L is the triangulation of S^n in Example 2.3.13; more precisely, by forming a geometric realization of the abstract complex formed from the abstraction of L' by identifying each vertex $(x_1, \ldots, x_{n+1})$ with $(-x_1, \ldots, -x_{n+1})$. Let σ be an n-simplex of the resulting simplicial complex K that has $(0, \ldots, 0, 1) = (0, \ldots, 0, -1)$ as a vertex; prove that RP^{n-1} is homeomorphic to a deformation retract of $|K - \sigma|$, and deduce that $\pi_1(RP^n) \cong \pi_1(RP^2) \cong Z_2$, for all $n \geqslant 2$.

8. Let $|K|$ be a path-connected polyhedron. Show that $\pi_1(|K| * S^0) = 0$.

9. A *topological group* G is a group that is also a topological space, such that the functions $m: G \times G \to G$ and $u: G \to G$ are continuous, where m is the multiplication and $u(g) = g^{-1}$ for all $g \in G$. Given loops v, w in G, based at the identity element e, define $v * w$ by $(v * w)(t) = m(v(t), w(t))$ $(t \in I)$. Prove that $v.w \simeq v * w \simeq w.v$ rel 0, 1, and deduce that $\pi_1(G, e)$ is abelian.

10. Show that the path components of an n-manifold are themselves n-manifolds, and that a connected n-manifold is path-connected.

11. Let X be the space obtained from an equilateral triangle by identifying edges as shown in Fig. 3.24.

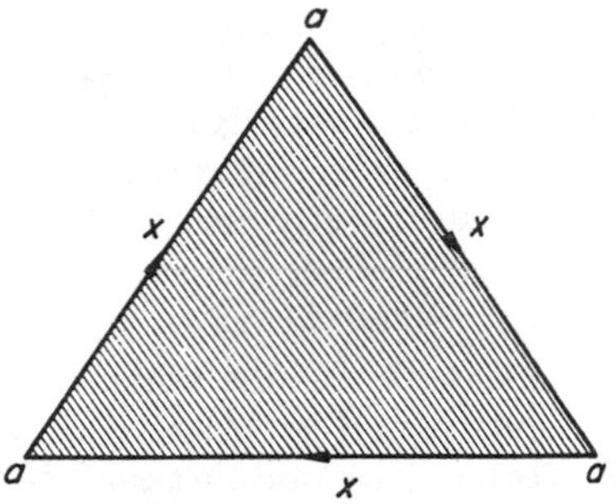

Fig. 3.24

Show that X is not a 2-manifold.

12. Show that if K is a triangulation of a connected 2-manifold, it cannot have a subcomplex (other than itself) whose polyhedron is also a 2-manifold.

13. If K is a simplicial complex such that $|\text{Lk}\,(a)|$ is connected for each vertex a, and each 1-simplex of K is a face of just two 2-simplexes, show that $|K|$ is a 2-manifold.

14. Let X and Y be triangulable 2-manifolds, and let $f: E^2 \to X$, $g: E^2 \to Y$ be embeddings, that is, homeomorphisms onto subspaces. Let e^2 be the subspace of E^2 of points x such that $\|x\| < 1$, and define the *connected sum* of X and Y, $X \# Y$, to be the space obtained from $X - f(e^2)$ and $Y - g(e^2)$ by identifying $f(s)$ with $g(s)$ for each point s of S^1 (with a little more care, this definition can be made independent of the particular embeddings f and g). Prove that

(a) $M_g \# M_1$ is homeomorphic to M_{g+1};
(b) $N_h \# M_1$ and $N_h \# N_2$ are both homeomorphic to N_{h+2};
(c) $N_h \# N_1$ is homeomorphic to N_{h+1}.

This shows, for example, that M_g can be thought of as the space obtained by 'sticking g toruses together', as in Fig. 3.25.

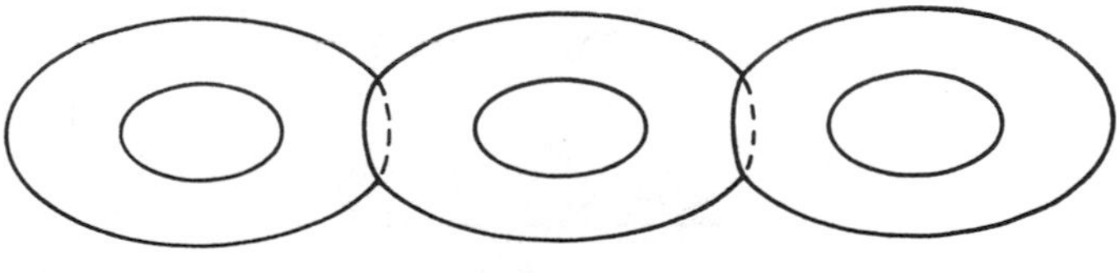

Fig. 3.25

15. A *2-manifold with boundary* is a Hausdorff space M in which each point has a neighbourhood homeomorphic to an open set in the half-plane $x_1 \geqslant 0$ in R^2, and the *boundary* of M, ∂M, is the subspace of M of those points that have neighbourhoods homeomorphic to open sets that meet the line $x_1 = 0$. If K is a simplicial complex whose polyhedron is a 2-manifold with boundary, show that for each point $x \in |K|$, $|\text{Lk}(x)|$ is homotopy-equivalent either to S^1 or to a point, and deduce that $\dim K = 2$. Show also that each 1-simplex of K faces either one or two 2-simplexes, and that if L is the subcomplex of K of those 1-simplexes that face exactly one 2-simplex, together with their vertices, then $|L| = \partial|K|$. (*Hint:* show that $\partial|K|$ is closed in $|K|$.) Prove also that $|L|$ is a 1-manifold.

16. Let K be a simplicial complex whose polyhedron is a path-connected 2-manifold with boundary, and let $|L|$ be a path component of $\partial|K|$; by subdividing, if necessary, assume that each 2-simplex of K meets L, if at all, in a face. Show that the subpolyhedron of $|K|$ consisting of those 2-simplexes that meet $|L|$ is homeomorphic to the space obtained from a regular polygon in R^2 by identifying edges according to a sequence of symbols of the form $aBa^{-1}C$, where B and C consist of single letters (and C may be empty). By using the polygons corresponding to the path components of $\partial|K|$, together with the remaining 2-simplexes of K, in the way that the 2-simplexes were used in the proof of Theorem 3.4.5, and then applying Rules 1–3, deduce that, if $\partial|K| \neq \varnothing$, $|K|$ is homeomorphic to the space corresponding to a sequence of symbols of the form

$$a_1B_1a_1^{-1}\cdots a_rB_ra_r^{-1}.x_1y_1x_1^{-1}y_1^{-1}\cdots x_gy_gx_g^{-1}y_g^{-1} \quad (g \geqslant 0, r \geqslant 1)$$

or

$$a_1B_1a_1^{-1}\cdots a_rB_ra_r^{-1}.x_1x_1\cdots x_hx_h \quad (h \geqslant 1, r \geqslant 1),$$

where the B's are sequences of single letters. Denote these spaces by M_g^r, N_h^r respectively, so that M_g^r, for example, is M_g with r discs removed: see Fig. 3.26 in the case of M_1^1 (torus with one hole).

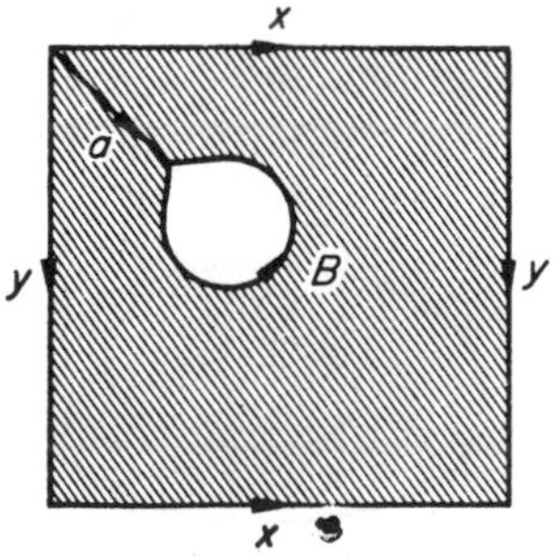

Fig. 3.26

Classify the triangulable 2-manifolds with boundary, up to homeomorphism, by establishing the following four propositions.

(a) Each of M_g^r, N_h^r is a triangulable 2-manifold with boundary.

(b) Any two of M_g^r, N_h^r, M_g or N_h that are homeomorphic must both have empty boundary, or both have the same r.

(c) By considering abelianizations of fundamental groups, the spaces M_g^r are all topologically distinct, as also are the spaces N_h^r.

(d) If M_g^r and N_{2g}^r were homeomorphic, so also would be M_g and N_{2g}.

NOTES ON CHAPTER 3

The fundamental group. The definition of the fundamental group is due to Poincaré [116], who also gave many examples of its calculation and applications, and introduced the term 'simply-connected'. The notation $\pi_1(X, x_0)$ may seem unnecessarily complicated, but is intended to emphasize that the fundamental group is just one example of the more general *homotopy groups*, $\pi_n(X, x_0)$, which will be studied in Chapters 6 and 7.

Collapsing. This idea is due to J. H. C. Whitehead [156], though see also Newman [109]. Whitehead's paper contains many more examples and applications. Collapsing plays an important role in combinatorial topology: an excellent survey will be found in Zeeman [167].

Calculation theorems. Theorem 3.3.9 was first proved by Tietze [144]. Theorem 3.3.18, usually known as van Kampen's Theorem, was in fact originally proved by Seifert, and only later (independently) by van Kampen [83], whose paper, however, also contains a proof of Theorem 3.3.19. That van Kampen's Theorem is not true for arbitrary topological spaces is shown by an example due to Griffiths [61]; but there are nevertheless generalizations, due to Olum [112] and R. Brown [35]. Example 3.3.22 is due to Poincaré [118].

Triangulated 2-manifolds. For the proof that compact 2-manifolds are triangulable, see Radó [120] or Ahlfors and Sario [6], Chapter 1. The original proof of the classification theorem is that of Dehn and Heegaard [13], but we follow more closely the proof of Brahana [24].

The Poincaré conjecture. Although this is an unsolved problem for 3- (and 4-)manifolds, curiously enough the corresponding result in dimensions greater than 4 is known to be true: see Smale [127], Stallings [132] and Zeeman [165, 166].

CHAPTER 4

HOMOLOGY THEORY

4.1 Introduction

In the previous chapter we defined and investigated the fundamental group, and saw that it was quite a powerful topological invariant; for example, it was strong enough to prove the classification theorem for triangulable 2-manifolds. However, the fundamental group yields no information at all in a large class of obvious problems: this is hardly surprising when we recall that the fundamental group of a polyhedron depends only on the 2-skeleton, and even fails to distinguish between S^2 and S^3. This chapter is concerned with setting up more algebraic invariants for a space X, called the (singular) ***homology groups*** $H_n(X)$. Like the fundamental group, these are homotopy-type invariants of X; and if X is triangulable the Simplicial Approximation Theorem yields effective calculation theorems: we shall see that the homology groups of a polyhedron $|K|$ can be calculated directly from the simplicial structure of K.

The homology groups (and the closely related cohomology groups) are useful in a large number of topological problems, and are in practice the standard tools of algebraic topology. In this chapter and the next we shall give several examples of their use, in particular the 'fixed-point' theorem of Lefschetz and the Alexander–Poincaré duality theorem for triangulable manifolds.

The plan of this chapter is as follows. Section 4.2 contains the definition and elementary properties of the homology groups, including the proof that they are homotopy-type invariants, and in Section 4.3 we shall see how to calculate the homology groups of a polyhedron; some applications to the topology of Euclidean spaces and spheres are given. In Section 4.4 we prove some more calculation theorems, and finally homology groups with arbitrary coefficients are defined in Section 4.5: this leads to a proof of the Lefschetz Fixed-Point Theorem.

4.2 Homology groups

Like the fundamental group, the homology groups of a space X are based on the set of maps of certain fixed spaces into X. This time the fixed spaces are the *standard n-simplexes* Δ_n, and we start by giving their definition.

Write a^n for the point $(0, \ldots, 0, 1)$ in R^n $(n \geqslant 1)$: by means of the standard identification of R^n as the subspace $R^n \times 0$ of $R^n \times R^m = R^{n+m}$, a^n may be regarded as a point of R^{n+m} for any $m \geqslant 0$. Write also a^0 for the point $(0, \ldots, 0)$ in any R^n. It is obvious that the points $a^0, \ldots, a^n$ are independent, and so may be taken to be the vertices of an n-simplex.

Definition 4.2.1 For $n \geqslant 0$, the *standard n-simplex* Δ_n is the simplex $(a^0, \ldots, a^n)$ in R^n (or in R^{n+m}, $m \geqslant 0$). When necessary, Δ_n is regarded as the polyhedron of $K(\Delta_n)$.

Definition 4.2.2 Given a space X, a *singular n-simplex* λ in X is a map $\lambda: \Delta_n \to X$.

Thus a singular 1-simplex in X is just a path in X, in the sense of Definition 3.2.1, so that it would appear that we could generalize the fundamental group by taking homotopy classes of singular n-simplexes in X, and making an appropriate definition of the 'product' of two singular simplexes. This can indeed be done, but the resulting groups are the *homotopy groups* $\pi_n(X)$ (compare Propositions 7.2.1 and 7.2.2). To define the *homology groups*, on the other hand, we construct groups from the sets of singular simplexes in a rather different, and more algebraic, fashion.

Definition 4.2.3 Given a space X, and an integer n, the nth *singular chain group* of X, $S_n(X)$, is defined to be the free abelian group with the singular n-simplexes in X as generators (we take $S_n(X) = 0$ if $n < 0$).

The groups $S_n(X)$ are of course not independent, since for example the restriction of a singular n-simplex $\lambda: \Delta_n \to X$ to Δ_{n-1} is a singular $(n-1)$-simplex. The relationships that arise by restricting singular simplexes in this way to faces of Δ_n can be formally described by the *boundary homomorphisms* $\partial: S_n(X) \to S_{n-1}(X)$, whose definition depends in turn on the *face maps* $F^r: \Delta_{n-1} \to \Delta_n$.

Now if K is a simplicial complex, a singular n-simplex $\lambda: \Delta_n \to |K|$ that happens to be a *simplicial* map is completely determined by the set of vertices $(\lambda a^0, \ldots, \lambda a^n)$, which span a (geometric) simplex of K (possibly with repeats). In this situation the singular simplex will often be denoted by $(\lambda a^0, \ldots, \lambda a^n)$; we hope that the context will always make clear whether the notation refers to the singular simplex λ in $|K|$ or to the geometric simplex $\lambda(\Delta_n)$ of K.

This notation allows us to specify certain elements of $S_{n-1}(\Delta_n)$, by taking the various $(n-1)$-dimensional faces of Δ_n.

Definition 4.2.4 The rth *face map* $F^r : \Delta_{n-1} \to \Delta_n$ is the element $(a^0, \ldots, \hat{a}^r, \ldots, a^n)$ of $S_{n-1}(\Delta_n)$, where the notation $\hat{a}^r$ means that the vertex a^r has been omitted.

The boundary homomorphism $\partial : S_n(X) \to S_{n-1}(X)$ is defined by associating with each singular n-simplex λ the set of composites $\lambda F^r : \Delta_{n-1} \to X$. Now ∂ must of course be defined to be the zero homomorphism if $n \leqslant 0$, but otherwise, by Proposition 1.3.20, a unique homomorphism ∂ can be defined by specifying its value on each generator of $S_n(X)$, that is, on each singular n-simplex.

Definition 4.2.5 Let λ be a singular n-simplex in X $(n \geqslant 1)$. Define

$$\partial(\lambda) = \sum_{r=0}^{n} (-1)^r \lambda F^r.$$

Observe that if X is a polyhedron $|K|$, and $\lambda : \Delta_n \to |K|$ is a simplicial map, then $\partial(\lambda) = \sum (-1)^r (\lambda a^0, \ldots, \widehat{\lambda a^r}, \ldots, \lambda a^n)$.

Example 4.2.6 If $\Delta_2 = (a^0, a^1, a^2)$ is regarded as an element of $S_2(\Delta_2)$, we have

$$\partial(a^0, a^1, a^2) = (a^1, a^2) - (a^0, a^2) + (a^0, a^1),$$

so that $\partial(a^0, a^1, a^2)$ is associated with the sum of the 1-simplexes in the boundary of (a^0, a^1, a^2), at least if these simplexes are given appropriate signs. Observe also that

$$\begin{aligned}
\partial^2(a^0, a^1, a^2) &= \partial(a^1, a^2) - \partial(a^0, a^2) + \partial(a^0, a^1) \\
&= (a^2) - (a^1) - (a^2) + (a^0) + (a^1) - (a^0) \\
&= 0,
\end{aligned}$$

as one might expect, since $\partial(a^0, a^1, a^2)$ represents a closed loop, which ought to have no 'boundary'. ∎

The property that $\partial^2 = 0$ holds quite generally for the singular chain groups and boundary homomorphisms of a space X.

Proposition 4.2.7 *Given a space X,*

$$\partial^2 = 0 : S_n(X) \to S_{n-2}(X),$$

for all integers n.

Proof. It is clearly sufficient to prove this for $n \geqslant 2$, and even then we need check it only for one typical singular n-simplex λ. Now

$$\begin{aligned}\partial^2(\lambda) &= \partial\left[\sum_r (-1)^r \lambda F^r\right] \\ &= \sum_{r,s} (-1)^{r+s} \lambda F^r F^s.\end{aligned}$$

But it is easy to see that $F^r F^s = F^s F^{r-1}$ if $s < r$, so that

$$\begin{aligned}\partial^2(\lambda) &= \sum_{s<r} (-1)^{r+s} \lambda F^s F^{r-1} + \sum_{s \geqslant r} (-1)^{r+s} \lambda F^r F^s \\ &= 0,\end{aligned}$$

since each expression $\lambda F^i F^j$ occurs twice, once with sign $(-1)^{i+j+1}$ in $\sum_{s<r}$, and once with sign $(-1)^{i+j}$ in $\sum_{s \geqslant r}$. ∎

Thus a space X gives rise to a set of abelian groups $S_n(X)$, one for each n, and homomorphisms $\partial\colon S_n(X) \to S_{n-1}(X)$ such that $\partial^2 = 0$. It is often convenient to consider this algebraic situation in the abstract.

Definition 4.2.8 A *chain complex* C is a direct sum $\bigoplus_n C_n$ of abelian groups C_n, one for each integer n, together with a homomorphism $\partial\colon C \to C$ such that $\partial^2 = 0$ and $\partial(C_n) \subset C_{n-1}$ for each n (each C_n is regarded as a subgroup of C).

In particular, the singular chain groups and boundary homomorphisms of a space X give rise to the *singular chain complex* of X, $S(X) = \bigoplus S_n(X)$. Sometimes, however, it is convenient as a technical device to introduce a fictitious 'singular (-1)-simplex', and hence define the 'reduced' singular chain complex of X, $\tilde{S}(X)$.

Definition 4.2.9 The *reduced singular chain complex* of X, $\tilde{S}(X)$, is defined by setting $\tilde{S}_n(X) = S_n(X)$ for $n \neq -1$, but by taking $\tilde{S}_{-1}(X)$ to be the free abelian group with a single generator $*$. The boundary homomorphism ∂ is the same as that of $S(X)$, except that $\partial(*) = 0$ and $\partial(\lambda) = *$ for each singular 0-simplex λ; clearly $\partial^2 = 0$, since if λ is a singular 1-simplex,

$$\partial^2(\lambda) = \partial(\lambda F^0 - \lambda F^1) = * - * = 0.$$

Yet a third chain complex arises if we consider a pair of spaces (X, Y).

Definition 4.2.10 If (X, Y) is a pair of spaces, the *relative singular chain complex* $S(X, Y) = \bigoplus S_n(X, Y)$ is defined by setting

$S_n(X, Y) = S_n(X)/S_n(Y)$, with the obvious identification of $S_n(Y)$ as a subgroup of $S_n(X)$. (Thus $S_n(X, Y)$ may be thought of as the free abelian group with generators those singular n-simplexes $\lambda: \Delta_n \to X$ whose image is not completely contained in Y.) The boundary homomorphism $\partial: S_n(X, Y) \to S_{n-1}(X, Y)$ is the homomorphism induced by $\partial: S_n(X) \to S_{n-1}(X)$, in the sense of Theorem 1.3.11: obviously $\partial^2 = 0$ once again.

There is no question of defining a 'reduced' relative singular chain complex, since the singular (-1)-simplex $*$ is supposed to be common to all spaces, and hence $S_n(X)/S_n(Y) = \tilde{S}_n(X)/\tilde{S}_n(Y)$.

The reader will notice that $S(X, Y) = S(X)$ if Y happens to be empty. Thus any theorem about relative chain complexes immediately specializes to 'non-relative' chain complexes on putting $Y = \varnothing$.

Although they are in fact topological invariants, the various chain complexes that we have constructed are too unwieldy for practical purposes. However, the fact that $\partial^2 = 0$ in a chain complex allows us to construct other groups, namely the homology groups, that turn out to be much easier to handle, and are actually homotopy-type invariants. As before, it is convenient first to consider the abstract algebraic situation.

Definition 4.2.11 Given a chain complex $C = \bigoplus C_n$, the group of *n-boundaries* $B_n(C)$ is defined to be the image of $\partial: C_{n+1} \to C_n$, and the group of *n-cycles* $Z_n(C)$ is the kernel of $\partial: C_n \to C_{n-1}$. Clearly $B_n(C) \subset Z_n(C)$, since $\partial^2 = 0$; the nth *homology group* $H_n(C)$ is defined to be the quotient group $Z_n(C)/B_n(C)$. We write $B(C)$ for $\bigoplus B_n(C) = \operatorname{Im} \partial$, $Z(C)$ for $\bigoplus Z_n(C) = \operatorname{Ker} \partial$, and $H(C)$ for $\bigoplus H_n(C)$; by Proposition 1.3.33, $H(C) \cong Z(C)/B(C)$.

In particular, we write $B_n(X)$, $Z_n(X)$, $H_n(X)$ and $H_*(X)$ for $B_n(C)$, $Z_n(C)$, $H_n(C)$ and $H(C)$ respectively, if $C = S(X)$: $H_n(X)$ is the nth (singular) *homology group* of X. Similarly if $C = \tilde{S}(X)$ we write $\tilde{B}_n(X)$, $\tilde{Z}_n(X)$, $\tilde{H}_n(X)$ (the nth *reduced* homology group of X) and $\tilde{H}_*(X)$, and if $C = S(X, Y)$ we write $B_n(X, Y)$, $Z_n(X, Y)$, $H_n(X, Y)$ and $H_*(X, Y)$; $H_n(X, Y)$ is the nth *relative* homology group of (X, Y). Notice that all these homology groups are trivial if $n < 0$, except that $\tilde{H}_{-1}(\varnothing) \cong Z$.

Example 4.2.12 Let P be a single point. Clearly for each $n \geqslant 0$ we have $S_n(P) \cong Z$, generated by λ_n, the only possible map from Δ_n to P. Moreover $\partial: S_n(P) \to S_{n-1}(P)$ is an isomorphism if n is even and is zero if n is odd. Thus for n even, $n \geqslant 2$, we have $Z_n(P) = 0$ so that $H_n(P) = 0$; for n odd, $n \geqslant 1$, we have $B_n(P) = Z_n(P) = S_n(P)$, so that $H_n(P) = 0$. On the other hand $Z_0(P) \cong Z$ and $B_0(P) = 0$, so that $H_0(P) \cong Z$. To sum up,

$$H_n(P) \cong \begin{cases} 0, & n \neq 0 \\ Z, & n = 0. \end{cases}$$

A similar calculation shows that $\tilde{H}_n(P) = 0$ for all n. ▌

Example 4.2.13 If X is any path-connected space, $H_0(X) \cong Z$. For the generators of $S_0(X) = Z_0(X)$ may be taken to be the points of X, and since a singular 1-simplex is just a path, $B_0(X)$ is the free abelian group generated by all $x - y$, where x, y are points of X. Thus $H_0(X) \cong Z$, and a generator is the coset $[x]$, for any point $x \in X$.

A similar argument shows that, if X is not path-connected, $H_0(X)$ is a free abelian group with one generator for each path component. Also $\tilde{H}_0(X)$ is a free abelian group with one fewer generators than there are path components: that is, $H_0(X) \cong \tilde{H}_0(X) \oplus Z$. ▌

Having defined the homology groups, the next step, as in Chapter 3, is to show that a continuous map $f\colon X \to Y$ gives rise to homomorphisms $f_*\colon H_n(X) \to H_n(Y)$. As usual, we do this in two stages, first considering the algebraic situation.

Definition 4.2.14 Given chain complexes $C = \bigoplus C_n$ and $D = \bigoplus D_n$, a *chain map* $\theta\colon C \to D$ is a homomorphism from C to D such that $\theta\partial = \partial\theta$ and $\theta(C_n) \subset D_n$ (we write ∂ indiscriminately for the boundary homomorphism in either C or D). If θ is also an isomorphism (of groups), θ is called a *chain isomorphism*.

Notice that if $\phi\colon D \to E$ is also a chain map, then so is the composite $\phi\theta\colon C \to E$.

Proposition 4.2.15 *A chain map $\theta\colon C \to D$ gives rise to homomorphisms $\theta_*\colon H_n(C) \to H_n(D)$, one for each n; with the following properties:*

(a) *If $1\colon C \to C$ is the identity chain isomorphism, 1_* is the identity isomorphism for each n.*

(b) *If θ is a chain isomorphism, each θ_* is an isomorphism.*

(c) *If $\phi\colon D \to E$ is another chain map, then $(\phi\theta)_* = \phi_*\theta_*$.*

Proof. Since $\theta\partial = \partial\theta$, it is clear that $\theta(B_n(C)) \subset B_n(D)$ and $\theta(Z_n(C)) \subset Z_n(D)$. Thus by Theorem 1.3.11 θ induces homomorphisms $\theta_*\colon H_n(C) \to H_n(D)$. The proofs of properties (a)–(c) are very easy, and are left to the reader. ▌

Thus, given a continuous map of pairs $f: (X, Y) \to (A, B)$, in order to make f induce homology homomorphisms we must construct from f a chain map $f_\cdot: S(X, Y) \to S(A, B)$.

Proposition 4.2.16 *Let $f: (X, Y) \to (A, B)$ be a map of pairs. Then f induces a chain map $f_\cdot: S(X, Y) \to S(A, B)$, with the properties:*

(a) *if f is the identity map $1: (X, Y) \to (X, Y)$, then $f_\cdot$ is the identity chain isomorphism;*
(b) *if $g: (A, B) \to (C, D)$ is another map of pairs, then $(gf)_\cdot = g_\cdot f_\cdot$.*
Also a map $f: X \to A$ induces a chain map $f_\cdot: \tilde{S}(X) \to \tilde{S}(A)$, with similar properties.

Proof. First define $f_\cdot: S(X) \to S(A)$, by sending the singular n-simplex $\lambda: \Delta_n \to X$ to the composite $f\lambda: \Delta_n \to A$. This clearly defines a chain map, and $f_\cdot S(Y) \subset S(B)$, so that $f_\cdot$ induces a chain map $f_\cdot: S(X, Y) \to S(A, B)$ as well. Properties (a) and (b) are trivial, and the modification to reduced chain complexes is made by setting $f_\cdot(*) = *$. ∎

Corollary 4.2.17 *f induces homomorphisms $f_*: H_n(X, Y) \to H_n(A, B)$, $f_*: \tilde{H}_n(X) \to \tilde{H}_n(A)$, all n, with the properties:*

(a) *if f is the identity map, f_* is the identity isomorphism;*
(b) $(gf)_* = g_* f_*$. ∎

Thus homeomorphic spaces have isomorphic homology groups. Indeed, more than this is true, since the homology groups are homotopy-type invariants. As in the case of the fundamental group, the proof consists in showing that homotopic maps induce the same homology homomorphisms, though in the spirit of the present chapter we first consider the analogous algebraic situation.

Definition 4.2.18 Given chain complexes C and D, and chain maps $\theta, \phi: C \to D$, a *chain homotopy* h between θ and ϕ is a homomorphism $h: C \to D$ such that $\phi(c) - \theta(c) = \partial h(c) + h\partial(c)$ for each $c \in C$. In this case, the chain maps θ and ϕ are said to be *chain-homotopic*.

Proposition 4.2.19 *If $\theta, \phi: C \to D$ are chain-homotopic chain maps, then $\theta_* = \phi_*: H(C) \to H(D)$.*

Proof. A typical element of $H(C)$ is a coset $[z]$, where $z \in Z(C)$, and $\theta_*[z] = [\theta(z)]$. But if h is the chain homotopy between θ and ϕ, we have

$$\begin{aligned}\phi(z) - \theta(z) &= \partial h(z) + h\partial(z) \\ &= \partial h(z),\end{aligned}$$

since $\partial(z) = 0$. Hence $[\phi(z)] - [\theta(z)] = [\partial h(z)] = 0$. ∎

Thus it only remains to show that homotopic maps induce chain-homotopic chain maps. Now a homotopy G between two maps $f, g: X \to Y$ induces a chain map $G_.: S(X \times I) \to S(Y)$, and so to construct a chain homotopy between $f_.$ and $g_.$ it is sufficient to consider $(i_0)_.$ and $(i_1)_.$, where $i_0, i_1: X \to X \times I$ are the inclusions as $X \times 0$, $X \times 1$ respectively. The chain homotopy here is defined by sending $\lambda: \Delta_n \to X$ to $\lambda \times 1: \Delta_n \times I \to X \times I$, composed with a certain element of $S_{n+1}(\Delta_n \times I)$, which in turn is obtained from a triangulation of $\Delta_n \times I$.

Now if K is any simplicial complex in R^p, $|K| \times I \subset R^p \times R^1 = R^{p+1}$ can be regarded as a polyhedron by the following method, which is similar to the definition of the derived complex in Definition 2.5.7. Suppose, as an inductive hypothesis, that for each $r < n$, $|K^r| \times I$ is the polyhedron of a simplicial complex $K^r \times I$, such that

(a) $i_0, i_1: |K^r| \to |K^r \times I|$ are simplicial maps;

(b) if L is a subcomplex of K^r, there is a subcomplex $L \times I$ of $K^r \times I$ such that $|L \times I| = |L| \times I$.

(If $n = 0$ the hypothesis is vacuous.) Now if σ is an n-simplex of K, the 'boundary' $\sigma \times 0 \cup |\dot\sigma| \times I \cup \sigma \times 1$ is already the polyhedron of $K(\sigma) \times 0 \cup \dot\sigma \times I \cup K(\sigma) \times 1$. Thus we may define

$$K^n \times I = K^{n-1} \times I \cup \{\bar\sigma\tau\} \cup \{(\bar\sigma)\},$$

where σ runs through all n-simplexes of K, τ through all simplexes in the 'boundary' of each $\sigma \times I$, and $\bar\sigma$ denotes the point $(\hat\sigma, \frac{1}{2})$. This definition is justified in exactly the same way as the definition of the derived complex; we omit the details. Finally, $K \times I$ is defined to be $K^m \times I$, where $m = \dim K$.

Example 4.2.20 If $K = K(\Delta_1)$, $K \times I$ is the simplicial complex in Fig. 4.1 overleaf. ∎

In particular, $\Delta_n \times I$ is the polyhedron of $K(\Delta_n) \times I$. By using its simplicial structure, we can pick an element of $S_{n+1}(\Delta_n \times I)$ and hence construct the required chain homotopy between the chain maps induced by $i_0, i_1: X \to X \times I$.

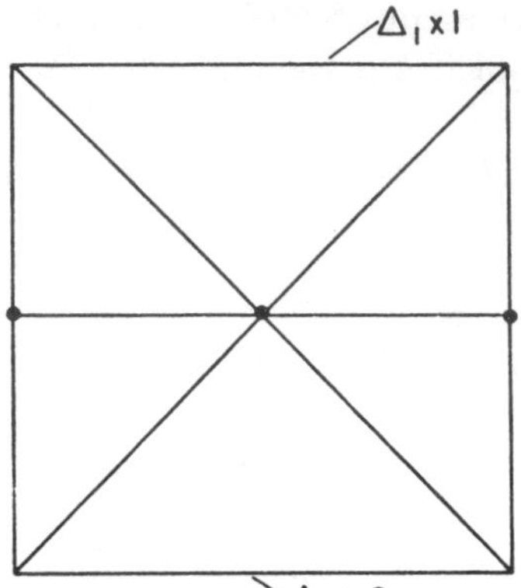

Fig. 4.1

Proposition 4.2.21 *Given any space X, there exists a chain homotopy $h\colon S(X) \to S(X \times I)$ between $(i_0)_\cdot$ and $(i_1)_\cdot$.*

Proof. Suppose that we have already defined $h\colon S_r(X) \to S_{r+1}(X \times I)$, for all spaces X, and for all $r < n$ (if $r < 0$, take $h = 0$). Let λ be a singular n-simplex in X, and define

$$h(\lambda) = (\lambda \times 1)_\cdot(a[(c^0, \ldots, c^n) - (b^0, \ldots, b^n) - h\partial(a^0, \ldots, a^n)]),$$

where $a = (\hat{\Delta}_n, \frac{1}{2})$, $c^r = (a^r, 1)$, $b^r = (a^r, 0)$, and $a[\ \ ]$ is defined by the rule $a(d^0, \ldots, d^r) = (a, d^0, \ldots, d^r)$, extended linearly (by induction, $h\partial(a^0, \ldots, a^n)$ is a linear combination of *simplicial* maps, since $F^r \times 1\colon \Delta_{n-1} \times I \to \Delta_n \times I$ is simplicial for each r). Then

$$\begin{aligned}\partial h(\lambda) = (\lambda \times 1)_\cdot((c^0, \ldots, c^n) - (b^0, \ldots, b^n) - h\partial(a^0, \ldots, a^n) \\ - a\partial[(c^0, \ldots, c^n) - (b^0, \ldots, b^n) - h\partial(a^0, \ldots, a^n)]).\end{aligned}$$

But by the inductive hypothesis we have

$$\begin{aligned}\partial h\partial(a^0, \ldots, a^n) &= \partial(c^0, \ldots, c^n) - \partial(b^0, \ldots, b^n) - h\partial^2(a^0, \ldots, a^n) \\ &= \partial(c^0, \ldots, c^n) - \partial(b^0, \ldots, b^n),\end{aligned}$$

so that

$$\begin{aligned}\partial h(\lambda) &= (\lambda \times 1)_\cdot((c^0, \ldots, c^n) - (b^0, \ldots, b^n) - h\partial(a^0, \ldots, a^n)) \\ &= (i_1)_\cdot(\lambda) - (i_0)_\cdot(\lambda) - h\partial(\lambda),\end{aligned}$$

since obviously $(\lambda \times 1)_\cdot h(\mu) = h\lambda_\cdot(\mu)$.

This completes the inductive step and hence the definition of h. ∎

Notice that h can be extended to a chain homotopy from $\tilde{S}(X)$ to $\tilde{S}(X \times I)$ by putting $h(*) = 0$.

Corollary 4.2.22 *If* $f \simeq g: (X, Y) \to (A, B)$, *then*

$$f_* = g_*: H_*(X, Y) \to H_*(A, B).$$

Similarly, if $f \simeq g: X \to A$, *then*

$$f_* = g_*: \tilde{H}_*(X) \to \tilde{H}_*(A).$$

Proof. The chain homotopy h clearly induces a chain homotopy $h: S(X, Y) \to S(X \times I, Y \times I)$. Thus if $G: (X \times I, Y \times I) \to (A, B)$ is the homotopy between f and g, then for each $x \in S(X, Y)$ we have

$$\begin{aligned} \partial G_. h(x) + G_. h\partial(x) &= G_.(\partial h(x) + h\partial(x)) \\ &= G_.((i_1)_. x - (i_0)_. x) \\ &= g_.(x) - f_.(x). \end{aligned}$$

Hence $f_* = g_*$, by Proposition 4.2.19. The proof for reduced homology is similar. ∎

Corollary 4.2.23 *If* $(X, Y) \simeq (A, B)$, *then* $H_n(X, Y) \cong H_n(A, B)$ *for each* n. *Similarly* $\tilde{H}_n(X) \cong \tilde{H}_n(A)$ *if* $X \simeq A$.

Proof. Let $f: (X, Y) \to (A, B)$ be a homotopy equivalence, and $g: (A, B) \to (X, Y)$ be a homotopy inverse to f. Then

$$g_* f_* = (gf)_* = 1,$$

the identity isomorphism of $H_n(X, Y)$. Similarly $f_* g_*$ s the identity isomorphism of $H_n(A, B)$, so that f_* and g_* are isomorphisms. ∎

4.3 Methods of calculation: simplicial homology

As in Chapter 3, having defined the homology groups and proved that they are homotopy-type invariants, we now face the problem of calculation. Once again the Simplicial Approximation Theorem can be used to reduce the problem considerably in the case of polyhedra (or indeed spaces homotopy-equivalent to polyhedra), and we shall see that it is sufficient to consider those singular simplexes that are actually *simplicial* maps from the standard simplexes: this is the analogue for homology of Theorem 3.3.9.

Suppose then that (K, L) is a simplicial pair. Write $\Delta_n(K)$ for the subgroup of $S_n(|K|)$ generated by the simplicial maps $\lambda: \Delta_n \to |K|$, and let $\Delta_n(K, L) = \Delta_n(K)/\Delta_n(L)$ (which may be regarded as a subgroup of $S_n(|K|, |L|)$). It is clear that $\partial(\Delta_n(K)) \subset \Delta_{n-1}(K)$, so that $\Delta(K) = \bigoplus \Delta_n(K)$ and $\Delta(K, L) = \bigoplus \Delta_n(K, L)$ are sub-chain complexes of $S(|K|)$, $S(|K|, |L|)$ respectively. Similarly, $\tilde{\Delta}(K) \subset \tilde{S}(K)$ is defined by setting $\tilde{\Delta}_{-1}(K) = \tilde{S}_{-1}(|K|)$. Finally, write $H_n(K)$, $H_n(K, L)$ and $\tilde{H}_n(K)$ for $H_n(\Delta(K))$, $H_n(\Delta(K, L))$ and $H_n(\tilde{\Delta}(K))$ respectively. (Compare Definition 3.3.8.)

In fact $H_n(K) \cong H_n(|K|)$, $H_n(K, L) \cong H_n(|K|, |L|)$ and $\tilde{H}_n(K) \cong \tilde{H}_n(|K|)$. Now $\Delta(K) \subset S(|K|)$, and the method of proof is to construct a chain map that is an inverse, to within chain homotopy, of the inclusion chain map. The idea here is a very simple one: the Simplicial Approximation Theorem is used to replace each singular n-simplex in $|K|$ by a simplicial map of some triangulation of Δ_n, in a coherent way.

Proposition 4.3.1 *For each n, and for each singular n-simplex λ in $|K|$, there exists a simplicial complex M_λ such that $|M_\lambda| = \Delta_n$, a simplicial map $g_\lambda: |M_\lambda| \to |K|$, and a homotopy G_λ between λ and g_λ. Moreover*

(a) *if λ is already simplicial, then $M_\lambda = K(\Delta_n)$, $g_\lambda = \lambda$, and G_λ is the constant homotopy;*

(b) *for each face map $F^r: \Delta_{n-1} \to \Delta_n$, $F^r: |M_{\lambda F^r}| \to |M_\lambda|$ is simplicial, $g_\lambda F^r = g_{\lambda F^r}$, and $G_\lambda(F^r \times 1) = G_{\lambda F^r}$.*

Proof. Suppose, as an inductive hypothesis, that we have already constructed M_μ, g_μ and G_μ for all singular m-simplexes μ in $|K|$, for $m < n$; suppose also that each G_μ is a semi-constant homotopy in the sense of Definition 2.5.21. The induction starts, since the hypothesis is vacuous for $n = 0$.

Consider a singular n-simplex λ. If λ is simplicial, then so is each λF^r, and we may take $M_\lambda = K(\Delta_n)$, $g_\lambda = \lambda$, and G_λ to be the constant homotopy. If λ is not simplicial, on the other hand, the inductive hypothesis ensures that we already have the required maps and homotopies on each face of Δ_n. Moreover these fit together where the faces overlap, so that we have a simplicial complex N such that $|N| = |\dot{\Delta}_n|$, together with a simplicial map $h: |N| \to |K|$ and a semi-constant homotopy H between $\lambda \mid |N|$ and h. Now define

$$M = N \cup \{a\sigma\} \cup (a),$$

where $a = \hat{\Delta}_n$ and σ runs through all simplexes of N; the usual argument shows that $|M| = \Delta_n$. Corollary 2.5.22 now yields a simplicial map $g_\lambda: |(M, N)^{(s)}| \to |K|$, for some s, such that $g_\lambda = h$ on $|N|$, and $\lambda \simeq g_\lambda$ by a homotopy G_λ that extends H (and, as remarked after Corollary 2.5.22, G_λ may be taken to be semi-constant). Thus if we set $M_\lambda = (M, N)^{(s)}$, the inductive step is complete. ▌

Observe that if (K, L) is a simplicial pair, and $\lambda(\Delta_n) \subset |L|$, then g_λ and G_λ may be taken to be maps into $|L|$ as well.

The required chain map from $S(|K|)$ to $\Delta(K)$ is constructed by sending the singular n-simplex λ to $(g_\lambda)_. x_\lambda$, where x_λ is a suitable element of $\Delta_n(M_\lambda)$. If $M_\lambda = K(\Delta_n)$ we take x_λ to be the identity map of Δ_n; otherwise suppose that we already have $x_{\lambda F^r}$ for each r, and hence $\sum(-1)^r(F^r)_. x_{\lambda F^r} \in \Delta_{n-1}(N)$. We cannot quite take x_λ to be $a\sum(-1)^r(F^r)_. x_{\lambda F^r}$, since M_λ is $(M, N)^{(s)}$, not M; instead, we take its image under a standard chain map $\phi: \Delta(M) \to \Delta((M, N)^{(s)})$.

Definition 4.3.2 Given a simplicial pair (M, N), the *subdivision chain map* $\phi: \Delta(M) \to \Delta((M, N)')$ is defined inductively as follows. Suppose that $\phi: \Delta_m(M) \to \Delta_m((M, N)')$ has been defined for all $m < n$, such that ϕ is the identity for $m = 0$ and on all singular simplexes in $|N|$. Take a simplicial map $\lambda: \Delta_n \to |M|$, whose image is not contained in $|N|$, and define $\phi(\lambda) = a\phi\partial(\lambda)$, where a is the barycentre of $\lambda(\Delta_n)$. Certainly $\partial\phi(\lambda) = \phi\partial(\lambda)$ if $\lambda \in \Delta(N)$; otherwise,

$$\begin{aligned} \partial\phi(\lambda) &= \partial[a\phi\partial(\lambda)] \\ &= \phi\partial(\lambda) - a[\partial\phi\partial(\lambda)] \\ &= \phi\partial(\lambda), \end{aligned}$$

since $\partial\phi\partial(\lambda) = \phi\partial^2(\lambda) = 0$. Thus ϕ is indeed a chain map.

If $\lambda \in S_n(|K|)$ is not simplicial, define

$$x_\lambda = \phi^s\left[a\sum(-1)^r(F^r)_. x_{\lambda F^r}\right] \in \Delta_n(M_\lambda),$$

and hence define $\alpha: S(|K|) \to \Delta(K)$ by setting

$$\alpha(\lambda) = (g_\lambda)_. x_\lambda,$$

for each singular simplex λ in $|K|$.

Proposition 4.3.3 *α is a chain map.*

Proof. We first show that $\partial x_\lambda = \sum(-1)^r(F^r)_. x_{\lambda F^r}$, by induction. For

$$\partial x_\lambda = \phi^s\left[\sum(-1)^r(F^r)_. x_{\lambda F^r} - a\partial\sum(-1)^r(F^r)_. x_{\lambda F^r}\right].$$

But $\partial \sum (-1)^r (F^r)_. x_{\lambda F^r} = \sum (-1)^r (F^r)_. \partial (x_{\lambda F^r}) = 0$ as in Proposition 4.2.7, so that

$$\begin{aligned} \partial x_\lambda &= \phi^s [\sum (-1)^r (F^r)_. x_{\lambda F^r}] \\ &= \sum (-1)^r (F^r)_. x_{\lambda F^r}, \end{aligned}$$

since ϕ is the identity on N.

It follows that

$$\begin{aligned} \partial \alpha(\lambda) &= (g_\lambda)_. \partial x_\lambda \\ &= (g_\lambda)_. \sum (-1)^r (F^r)_. x_{\lambda F^r} \\ &= \sum (-1)^r (g_{\lambda F^r})_. x_{\lambda F^r} \\ &= \alpha \partial (\lambda). \quad \blacksquare \end{aligned}$$

It is clear that $\alpha\beta = 1$, where $\beta: \Delta(K) \to S(|K|)$ is the inclusion chain map, and (from the remark after Proposition 4.3.1) that α induces a chain map $\alpha: \Delta(K, L) \to S(|K|, |L|)$, with a similar property. To complete the proof that $H_*(K) \cong H_*(|K|)$, we construct a chain homotopy between $\beta\alpha$ and the identity chain isomorphism of $S(|K|)$. This is very similar to Corollary 4.2.22, and the chain homotopy is defined by sending the singular n-simplex λ to $(G_\lambda)_. y_\lambda$, where y_λ is a suitable element of $S_{n+1}(\Delta_n \times I)$ and G_λ is the homotopy obtained in Proposition 4.3.1.

To define y_λ, we need a triangulation N_λ of $\Delta_n \times I$ that has $K(\Delta_n)$ at the '0 end' and M_λ at the '1 end', and such that $F^r \times 1: |N_{\lambda F^r}| \to |N_\lambda|$ is simplicial for each face map $F^r: \Delta_{n-1} \to \Delta_n$. If we suppose that this has already been done for all singular m-simplexes with $m < n$, then we have a suitable triangulation P of $\Delta_n \times 0 \cup |\dot{\Delta}_n \times I| \cup \Delta_n \times 1$, so that we may take

$$N_\lambda = P \cup \{a\sigma\} \cup (a),$$

where $a = (\hat{\Delta}_n, \frac{1}{2})$ and σ runs through all simplexes of P.

Now suppose that we have constructed $y_\mu \in \Delta_{m+1}(N_\mu)$ (regarded as a subgroup of $S_{m+1}(\Delta_m \times I)$) for all singular m-simplexes μ with $m < n$, satisfying

$$\partial y_\mu = (i_1)_. x_\mu - i_0 - \sum (-1)^r (F^r \times 1)_. y_{\mu F^r},$$

where $i_0, i_1: \Delta_m \to \Delta_m \times I$ are the inclusions as $\Delta_m \times 0$ and $\Delta_m \times 1$, and $y_{\mu F^r}$ is taken to be zero if $m = 0$ (compare the proof of Proposition 4.2.21). Take a singular n-simplex λ, and define

$$y_\lambda = a[(i_1)_. x_\lambda - i_0 - \sum (-1)^r (F^r \times 1)_. y_{\lambda F^r}].$$

Then

$$\partial y_\lambda = (i_1)_. x_\lambda - i_0 - \sum (-1)^r (F^r \times 1)_. y_{\lambda F^r}$$

as required, since

$$\begin{aligned}\partial[(i_1)_. x_\lambda - i_0 - \textstyle\sum (-1)^r (F^r \times 1)_. y_{\lambda F^r}] &= \textstyle\sum (-1)^r (F^r \times 1)_. [(i_1)_. x_{\lambda F^r} - i_0 - \partial y_{\lambda F^r}] \\ &= \textstyle\sum (-1)^{r+s} (F^r \times 1)_. (F^s \times 1)_. y_{\lambda F^r F^s} \\ &= 0 \quad \text{as in Proposition 4.2.7.}\end{aligned}$$

Finally, define $h\colon S(|K|) \to S(|K|)$ by $h(\lambda) = (G_\lambda)_. y_\lambda$.

Proposition 4.3.4 *h is a chain homotopy between* $1\cdot$ *and* $\beta\alpha$.

Proof.

$$\begin{aligned}\partial h(\lambda) &= (G_\lambda)_. \partial y_\lambda \\ &= (G_\lambda)_. [(i_1)_. x_\lambda - i_0 - \textstyle\sum (-1)^r (F^r \times 1)_. y_{\lambda F^r}] \\ &= (g_\lambda)_. x_\lambda - \lambda - \textstyle\sum (-1)^r (G_{\lambda F^r})_. y_{\lambda F^r} \\ &= \beta\alpha(\lambda) - \lambda - h\partial(\lambda). \quad \blacksquare\end{aligned}$$

Corollary 4.3.5 *$\beta_*\colon H_*(K, L) \to H_*(|K|, |L|)$ is an isomorphism, as also is $\beta_*\colon \tilde{H}_*(K) \to \tilde{H}_*(|K|)$.*

Proof. The chain homotopy h induces a similar chain homotopy

$$h\colon S(|K|, |L|) \to S(|K|, |L|),$$

and we can extend h to $\tilde{S}(|K|)$ by setting $h(*) = 0$. ∎

Observe that if $f\colon (|K|, |L|) \to (|M|, |N|)$ is a simplicial map of pairs, $f_*\colon H_*(|K|, |L|) \to H_*(|M|, |N|)$ may be taken to be the homomorphism from $H_*(K, L)$ to $H_*(M, N)$ induced by the restriction of $f_.$ to a chain map from $\Delta(K, L)$ to $\Delta(M, N)$. Indeed, even if f is not simplicial, f_* may be identified with the homomorphism $f_*\colon H_*(K, L) \to H_*(M, N)$ induced by the composite chain map

$$\Delta(K, L) \xrightarrow{\beta} S(|K|, |L|) \xrightarrow{f_.} S(|M|, |N|) \xrightarrow{\alpha} \Delta(M, N);$$

for certainly the diagram

$$\begin{array}{ccc} H_*(K, L) & \xrightarrow{(\alpha f_. \beta)_*} & H_*(M, N) \\ \beta_* \downarrow & & \downarrow \beta_* \\ H_*(|K|, |L|) & \xrightarrow[f_*]{} & H_*(|M|, |N|) \end{array}$$

is commutative, because $f_.\beta$ and $\beta\alpha f_.\beta$ are chain-homotopic.

Example 4.3.6 Let L and M be subcomplexes of a simplicial complex K, where $K = L \cup M$, and let $i: (|M|, |L \cap M|) \to (|K|, |L|)$ be the inclusion map. Then $i_*: H_n(|M|, |L \cap M|) \to H_n(|K|, |L|)$ is an isomorphism. For, since i is simplicial, it is sufficient to consider

$$i_{\cdot}: \Delta(M, L \cap M) \to \Delta(K, L);$$

but this is an isomorphism since a simplex of K is in $M - (L \cap M)$ if and only if it is in $K - L$. ▌

This result is known as the Excision Theorem, since it expresses the fact that the 'excision' of the simplexes in $K - M$ from both K and L does not affect $H_*(|K|, |L|)$. Indeed, an analogous result holds for arbitrary topological spaces: see Theorem 8.2.1.

Corollary 4.3.5 is the analogue for homology of Theorem 3.3.9 for the fundamental group, and like that theorem does not by itself provide a practical method of calculation. Even though each $\Delta_n(K, L)$ is a finitely generated group, there are many more generators than necessary: for example, if P is a single point then $\Delta_n(P) \cong Z$ for each $n \geqslant 0$. What we should like to do now is to reduce the chain complex still further until there is just one generator for each (geometric) simplex of $K - L$ (compare Theorem 3.3.13).

This is achieved by taking the quotient of $\Delta(K)$, for example, by the sub-chain complex $\Delta^0(K)$ generated by all expressions of the form

$$(b^0, \ldots, b^n) - \epsilon_\rho(b^{\rho 0}, \ldots, b^{\rho n}),$$

(together with all $(b^0, \ldots, b^n)$ containing a repeated vertex), where $(b^0, \ldots, b^n)$ is a singular n-simplex of $\Delta(K)$, ρ is a permutation of $0, 1, \ldots, n$, and ϵ_ρ is $+1$ or -1 according as ρ is even or odd. It is not quite obvious that $\partial\Delta^0(K) \subset \Delta^0(K)$, but this is easy to prove.

Proposition 4.3.7 $\partial\Delta^0(K) \subset \Delta^0(K)$.

Proof. It is sufficient to consider the action of ∂ on $(b^0, \ldots, b^n) + (b^0, \ldots, b^{r+1}, b^r, \ldots, b^n)$, since every permutation is a composite of transpositions of this form. But

$$\partial[(b^0, \ldots, b^n) + (b^0, \ldots, b^{r+1}, b^r, \ldots, b^n)]$$
$$= \sum_{s \neq r, r+1} (-1)^s[(b^0, \ldots, \hat{b}^s, \ldots, b^n) + (b^0, \ldots, \hat{b}^s, \ldots, b^{r+1}, b^r, \ldots, b^n)],$$

since the terms involving $\hat{b}^r$ and $\hat{b}^{r+1}$ occur with opposite sign, and so cancel. ▌

Definition 4.3.8 If K is a simplicial complex, the *simplicial chain complex* of K, $C(K)$, is defined by setting $C(K) = \Delta(K)/\Delta^0(K)$. There are similar definitions of $C(K, L)$ and $\tilde{C}(K)$ (where $\tilde{C}_{-1}(K) = \tilde{\Delta}_{-1}(K)$).

We shall write $[b^0, \ldots, b^n]$ for the coset of $(b^0, \ldots, b^n)$: observe that this coset is zero if the set of vertices $b^0, \ldots, b^n$ contains a repeat, so that $C(K)$, for example, has one generator for each geometric simplex of K.

Let $\alpha: \Delta(K, L) \to C(K, L)$ ($\alpha: \tilde{\Delta}(K) \to \tilde{C}(K)$) be the quotient chain map. We prove that α induces isomorphisms between $H_n(K, L)$ and the *simplicial homology groups* $H_n(C(K, L))$.

Theorem 4.3.9 *$\alpha_*: H_n(K, L) \to H_n(C(K, L))$ and $\alpha_*: \tilde{H}_n(K) \to H_n(\tilde{C}(K))$ are isomorphisms.*

Proof. As in Corollary 4.3.5, we construct a chain map $\beta: C(K) \to \Delta(K)$ which is an inverse to α, to within chain homotopy. To define β, totally order the vertices of K, in the form $b^0 < b^1 < \cdots < b^m$, write each generator of $C(K)$ in the form $[b^{i_0}, \ldots, b^{i_n}]$, where $i_0 < \cdots < i_n$, and define

$$\beta[b^{i_0}, \ldots, b^{i_n}] = (b^{i_0}, \ldots, b^{i_n}).$$

Clearly $\alpha\beta$ is the identity chain isomorphism of $C(K)$, so that it remains to produce a chain homotopy $h: \Delta(K) \to \Delta(K)$ between $\beta\alpha$ and the identity.

Suppose as an inductive hypothesis that we have constructed $h: \Delta_m(K) \to \Delta_{m+1}(K)$ for all $m < n$, such that

$$\beta\alpha(\mu) - \mu = \partial h(\mu) + h\partial(\mu)$$

for all generators μ of $\Delta_m(K)$; suppose further that $h(\mu)$ actually lies in $\Delta_{m+1}(K(\mu(\Delta_m)))$ (if $m \leqslant 0$ we may take $h = 0$). If now λ is a generator of $\Delta_n(K)$, $n \geqslant 1$, then h has already been defined on $\partial(\lambda)$, and

$$\begin{aligned}\partial[\beta\alpha(\lambda) - \lambda - h\partial(\lambda)] &= \beta\alpha\partial(\lambda) - \partial(\lambda) - \partial h\partial(\lambda) \\ &= 0,\end{aligned}$$

since $h\partial^2(\lambda) = 0$. Moreover all elements involved are in $\Delta(K(\lambda(\Delta_n)))$; but $\lambda(\Delta_n)$ is a simplex, and so is contractible; so $H_n(\lambda(\Delta_n)) = 0$. Thus there must exist an element $h(\lambda) \in \Delta_{n+1}(K(\lambda(\Delta_n))) \subset \Delta_{n+1}(K)$, such that $\partial h(\lambda) = \beta\alpha(\lambda) - \lambda - h\partial(\lambda)$, as required.

It follows that $\alpha_*: H_n(K) \to H_n(C(K))$ is an isomorphism; for

(K, L) we observe that β and h induce corresponding homomorphisms for (K, L), and for reduced homology we set $\beta(*) = *$, $h(*) = 0$. ∎

Corollary 4.3.10 *$H_n(K, L)$ depends only on the $(n+1)$-skeleton, and is zero if $n >$ dim K.* ∎

Observe that if $f\colon (|K|, |L|) \to (|M|, |N|)$ is a simplicial map, then $f_{.}\Delta^0(K, L) \subset \Delta^0(M, N)$, so that $f_{.}$ induces a chain map $f_{.}\colon C(K, L) \to C(M, N)$. The corresponding homology homomorphism may be identified with f_*, since the diagram

$$\begin{array}{ccc} H_*(K, L) & \xrightarrow{f_*} & H_*(M, N) \\ \downarrow{\scriptstyle \alpha_*} & & \downarrow{\scriptstyle \alpha_*} \\ H(C(K, L)) & \xrightarrow[f_*]{} & H(C(M, N)) \end{array}$$

is obviously commutative. Similarly $\phi\Delta^0(K) \subset \Delta^0(K')$, where ϕ is the subdivision chain map of Definition 4.3.2, so that ϕ induces $\phi\colon C(K) \to C(K')$, whose induced homology homomorphism is the same as that induced by the original ϕ. Moreover an obvious induction argument shows that ϕ sends each generator of $C(K)$, considered as an n-simplex σ of K, to the sum of the n-simplexes in $(K(\sigma))$ (with appropriate signs). Thus if $h\colon |K'| \to |K|$ is a simplicial approximation to the identity map, the remark after Proposition 2.5.13 shows that $h_{.}\phi\colon C(K) \to C(K)$ sends each generator to plus or minus itself. Hence, by an obvious adaptation of the argument in Theorem 4.3.9, $h_{.}\phi$ is chain-homotopic to the identity, and so $h_*\phi_*$ is the identity isomorphism. But h_* is an isomorphism because h is a homotopy equivalence; thus ϕ_* is the inverse isomorphism to h_*. Similar remarks apply to relative homology.

We end this section with some calculations and examples.

Example 4.3.11 S^1 may be triangulated as the boundary $\dot\sigma$ of a 2-simplex $\sigma = (b^0, b^1, b^2)$. It follows immediately that $H_n(S^1) = 0$ if $n > 1$ (and also if $n < 0$); moreover $H_0(S^1) \cong Z$ by Example 4.2.13. Thus it remains only to calculate $H_1(S^1) \cong H_1(C(\dot\sigma))$.

Now $C_1(\dot\sigma)$ has three generators, $[b^0, b^1]$, $[b^0, b^2]$ and $[b^1, b^2]$, where $\partial[b^0, b^1] = [b^1] - [b^0]$, $\partial[b^0, b^2] = [b^2] - [b^0]$ and $\partial[b^1, b^2] = [b^2] - [b^1]$. It is easy to see that $Z_1(C(\dot\sigma))$ is isomorphic to Z, with generator $[b^0, b^1] + [b^1, b^2] - [b^0, b^2]$, and since clearly $B_1(C(\dot\sigma)) = 0$ we have $H_1(S^1) \cong H_1(C(\dot\sigma)) \cong Z$. ∎

Example 4.3.12 Let us now calculate $H_*(S^n)$ for all $n \geqslant 0$, by triangulating S^n as $\dot\sigma$, where σ is the $(n+1)$-simplex $(b^0, \ldots, b^n)$.

Now it is possible, of course, to calculate $H(C(\dot{\sigma}))$ directly as in Example 4.3.11, but this method is tedious and complicated. Instead, we use a trick, based on the fact that $\dot{\sigma}$ and $K(\sigma)$ differ by only one simplex, and σ is contractible. It is convenient also to do the work in terms of reduced homology.

Consider $\tilde{C}(\dot{\sigma})$ and $\tilde{C}(K(\sigma))$. Obviously $\tilde{C}_m(\dot{\sigma}) = 0$ if $m > n$ or $m < -1$, and the inclusion (simplicial) map $i\colon |\dot{\sigma}| \to |K(\sigma)|$ induces isomorphisms $i_{.}\colon \tilde{C}_m(\dot{\sigma}) \to \tilde{C}_m(K(\sigma))$ for all $m \leqslant n$. Moreover the following diagram is commutative, since $i_{.}$ is a chain map.

$$\begin{array}{ccccccccccc} & & 0 & \longrightarrow & \tilde{C}_n(\dot{\sigma}) & \xrightarrow{\partial} & \tilde{C}_{n-1}(\dot{\sigma}) & \longrightarrow \cdots \xrightarrow{\partial} & \tilde{C}_{-1}(\dot{\sigma}) & \longrightarrow & 0 \\ & & & & \downarrow i_{.} & & \downarrow i_{.} & & \downarrow i_{.} & & \\ 0 \to & \tilde{C}_{n+1}(K(\sigma)) & \xrightarrow[\partial]{} & & \tilde{C}_n(K(\sigma)) & \xrightarrow[\partial]{} & \tilde{C}_{n-1}(K(\sigma)) & \to \cdots \xrightarrow[\partial]{} & \tilde{C}_{-1}(K(\sigma)) & \to & 0. \end{array}$$

Now $|K(\sigma)|$, being homeomorphic to E^{n+1}, is contractible, and hence the reduced homology groups of $K(\sigma)$ are the same as those of a point, which are all zero. It follows at once that $H_m(\tilde{C}(\dot{\sigma})) = 0$ for $m < n$. Also $H_n(\tilde{C}(\dot{\sigma})) \cong Z_n(\tilde{C}(\dot{\sigma})) \cong Z_n(\tilde{C}(K(\sigma))) \cong B_n(\tilde{C}(K(\sigma))) \cong \tilde{C}_{n+1}(K(\sigma))$, so that $H_n(\tilde{C}(\sigma)) \cong Z$, and a generator is the coset $[\partial[b^0, \ldots, b^n]]$. To sum up,

$$\tilde{H}_m(S^n) \cong \begin{cases} Z, & \text{if } m = n \\ 0, & \text{otherwise.} \end{cases}$$

(Notice that this result is true even if $n = -1$, if we interpret S^{-1} as the empty set!)

Since $K(\sigma) - \dot{\sigma}$ has only the one simplex σ, $C(K(\sigma), \dot{\sigma})$ has just one generator $[b^0, \ldots, b^n]$, and so

$$H_m(E^{n+1}, S^n) \cong \begin{cases} Z, & m = n + 1 \\ 0, & \text{otherwise.} \end{cases} \quad \blacksquare$$

An immediate consequence is that S^n and S^m are not homotopy-equivalent if $m \neq n$, and so are certainly not homeomorphic. This result is particularly useful since it can be used with Theorem 2.4.5 to prove some important theorems on homeomorphisms and possible triangulations of certain spaces.

Theorem 4.3.13 *If $|K|$ and $|L|$ are homeomorphic polyhedra, then* $\dim K = \dim L$.

Proof. Let $\dim K = m$, $\dim L = n$, and suppose if possible that $n > m$. Take a point x in the interior of an n-simplex σ of L, so that $\text{Lk}\,(x) = \dot{\sigma}$, and $\tilde{H}_{n-1}(\text{Lk}\,(x)) \cong Z$. But if y is any point of $|K|$, $\text{Lk}\,(y)$

is a subcomplex of dimension at most $(m - 1)$, so that $\tilde{H}_{n-1}(\text{Lk}(y)) = 0$. Hence $|\text{Lk}(x)| \not\simeq |\text{Lk}(y)|$, which contradicts Theorem 2.4.5. ∎

Corollary 4.3.14 *If K is a triangulation of S^n or E^n, then* $\dim K = n$.

Proof. By Example 2.3.13, both S^n and E^n have triangulations by n-dimensional complexes. ∎

We can actually say rather more about possible triangulations of E^n.

Proposition 4.3.15 *If K is a triangulation of E^n, with homeomorphism $h\colon |K| \to E^n$, there exists an $(n - 1)$-dimensional subcomplex L of K such that $h|L| = S^{n-1}$.*

Proof. In any case E^n can be triangulated as $K(\sigma)$, where σ is an n-simplex, and then $\dot{\sigma}$ triangulates S^{n-1}. Moreover, by rotating E^n about the origin if necessary, we can arrange that any given point x of S^{n-1} is in the interior of an $(n - 1)$-simplex τ of $\dot{\sigma}$. It follows that $\text{Lk}(x) = \dot{\sigma} - \tau$, and it is easy to see that $|\dot{\sigma} - \tau|$ is contractible to the vertex of σ that is not in τ (the homotopy involved is linear). On the other hand if $x \in E^n - S^{n-1}$, then x is in the interior of σ and so $\text{Lk}(x) = \dot{\sigma}$. Hence S^{n-1} is the set of points of E^n such that $|\text{Lk}(x)|$ is contractible, and by Theorem 2.4.5 this is true however E^n is triangulated.

Thus if $h(x) \in S^{n-1}$ and x is in the interior of a simplex τ, then every point in the interior of τ, having the same link as x, will be mapped into S^{n-1}. Hence $h(\tau) \subset S^{n-1}$, since S^{n-1} is closed. That is to say, $S^{n-1} = h|L|$, where L is the subcomplex of K of those simplexes τ such that $h(\tau) \subset S^{n-1}$. And $\dim L = n - 1$, since L is a triangulation of S^{n-1}. ∎

Another important consequence of Example 4.3.12 is

Theorem 4.3.16 *Let U, V be open sets in R^m, R^n respectively. If U and V are homeomorphic, then $m = n$.*

Proof. Let $h\colon U \to V$ be the homeomorphism, and let x be any point of U. Choose ϵ so that B, the ϵ-neighbourhood of $h(x)$, is contained in V, and then choose η so that B', the η-neighbourhood of x, is contained in $h^{-1}(B)$. Thus h is a homeomorphism of B' onto a subset of B, and $h(x)$ is contained in an open set in $h(B')$. But B and B' are homeomorphic to cells, and so are triangulable, so that by Theorem 2.4.5 we must have $|\text{Lk}(x)| \simeq |\text{Lk}(h(x))|$. But $|\text{Lk}(x)|$ is homeomorphic to S^{m-1} and $|\text{Lk}(h(x))|$ to S^{n-1}, so that $m = n$. ∎

In particular, R^m and R^n cannot be homeomorphic unless $m = n$.

Finally, let us calculate the homology groups of the real projective plane RP^2.

Example 4.3.17 Consider the triangulation of RP^2 by the simplicial complex K shown in Fig. 4.2.

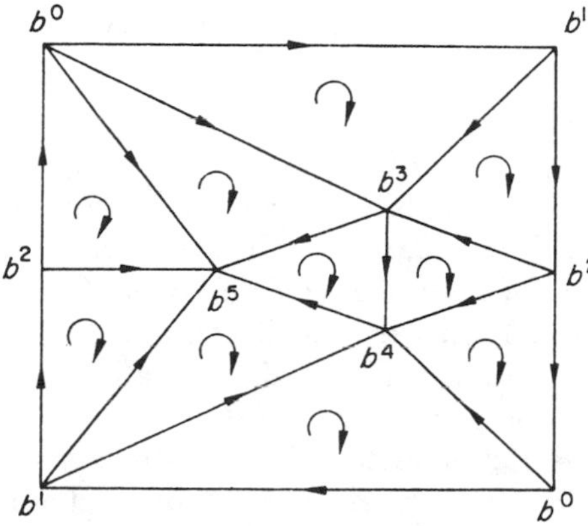

Fig. 4.2

As in the case of S^n, it would be possible, though very laborious, to calculate $H(C(K))$ directly from the definition. We prefer instead to compare K with two subcomplexes: L, consisting of K without the 2-simplex (b^3, b^4, b^5), and M, consisting only of (b^0, b^1), (b^1, b^2), (b^0, b^2) and their vertices.

Write $i: |M| \to |L|$, $j: |L| \to |K|$ for the inclusion (simplicial) maps. Now $|M|$ is a strong deformation retract of $|L|$, so that i is a homotopy equivalence and $i_*: H_n(C(M)) \to H_n(C(L))$ is an isomorphism for each n. Also M is a triangulation of S^1, so that $H_n(C(M))$ is Z for $n = 0$, 1 and zero otherwise; hence the same is true for L. Moreover, a generator of $H_1(C(M)) \cong H_1(C(L))$ is $[z]$, where $z = [b^0, b^1] + [b^1, b^2] - [b^0, b^2]$.

Now consider the commutative diagram

$$\begin{array}{ccccccccc} 0 & \longrightarrow & C_2(L) & \xrightarrow{\partial} & C_1(L) & \xrightarrow{\partial} & C_0(L) & \longrightarrow & 0 \\ & & \downarrow j_. & & \downarrow j_. & & \downarrow j_. & & \\ 0 & \longrightarrow & C_2(K) & \xrightarrow[\partial]{} & C_1(K) & \xrightarrow[\partial]{} & C_0(K) & \longrightarrow & 0, \end{array}$$

in which $j_.$ is the identity on $C_1(L)$ and $C_0(L)$. Let c be the element of $C_2(L)$ (or $C_2(K)$) defined by

$$c = [b^0, b^1, b^3] + [b^1, b^2, b^3] + [b^2, b^4, b^3] + [b^0, b^4, b^2] + [b^0, b^1, b^4] + [b^1, b^5, b^4] + [b^1, b^2, b^5] + [b^0, b^5, b^2] + [b^0, b^3, b^5];$$

direct calculation shows that $\partial[b^3, b^4, b^5] = 2z - \partial(c)$ in $C(K)$ (each 1-simplex in $\partial(c + [b^3, b^4, b^5])$ occurs twice, with opposite signs if it does not occur in z). So if $r[b^3, b^4, b^5] + d \in Z_2(C(K))$, with $d \in C_2(L)$, we have

$$\begin{aligned} 0 &= \partial(r[b^3, b^4, b^5] + d) \\ &= 2rz + \partial(d - c) \end{aligned}$$

in $C_1(L)$. That is, $[2rz] = 0$ in $H_1(C(L)) \cong Z$, so that $r = 0$. But $H_2(C(L)) = 0$, so that $Z_2(C(L)) = 0$ and hence $d = 0$ as well. It follows that $Z_2(C(K)) = 0$, or $H_2(C(K)) = 0$.

Now $C_1(K) = C_1(L)$ and $Z_1(C(K)) = Z_1(C(L))$; also $B_1(C(K))$ differs from $B_1(C(L))$ only in that it contains extra elements $2nz$ for each integer n. Since $H_1(C(L)) \cong Z$, generated by $[z]$, it follows that $H_1(C(K)) \cong Z_2$, also generated by $[z]$.

Finally, $H_0(RP^2) \cong Z$ since RP^2 is path-connected. To sum up, $H_1(RP^2) \cong Z_2$, $H_0(RP^2) \cong Z$, and all other homology groups are zero. ∎

4.4 Methods of calculation: exact sequences

The examples at the end of Section 4.3 will no doubt have convinced the reader that calculation directly from Theorem 4.3.9 would be extremely laborious for general polyhedra. In the case of S^n and RP^2 we were able to perform the calculation by various tricks, but these had to be invented separately for each space, and gave no insight into any sort of general procedure. One object of this section is to prove a theorem enabling the homology groups of polyhedra to be calculated from still further simplified chain complexes, in which the generators, instead of corresponding to single simplexes, correspond to certain subcomplexes called 'blocks'. The situation in homology is thus once again similar to Section 3.3: there, Theorem 3.3.13 theoretically gave a method of calculation of the fundamental group of a polyhedron, but it was much quicker in practice to use the result of Theorem 3.3.20.

In order to prove the calculation theorem, and also for its independent interest, we shall first show that, if (X, Y) is any pair of spaces, the homology groups of Y, X and (X, Y) can be fitted into an exact sequence, called the *exact homology sequence* of the pair (X, Y). In establishing this exact sequence, it is convenient, in the spirit of

Section 4.2, first to consider the abstract algebraic situation. This approach has the incidental advantage of yielding some other useful exact sequences. In order to state the fundamental theorem, one definition is necessary.

Definition 4.4.1 A sequence of chain complexes and chain maps $0 \to C \to D \to E \to 0$ is an *exact sequence of chain complexes* if it is exact considered as a sequence of abelian groups and homomorphisms (we write 0 for the chain complex C in which each $C_n = 0$).

Theorem 4.4.2 *Given an exact sequence of chain complexes*

$$0 \longrightarrow C \xrightarrow{f} D \xrightarrow{g} E \longrightarrow 0$$

there exists a homomorphism $\partial_*: H_n(E) \to H_{n-1}(C)$ for each n, such that the sequence

$$\cdots \longrightarrow H_n(C) \xrightarrow{f_*} H_n(D) \xrightarrow{g_*} H_n(E) \xrightarrow{\partial_*} H_{n-1}(C) \longrightarrow \cdots$$

is an exact sequence of abelian groups and homomorphisms. Moreover, given a commutative diagram of chain complexes and chain maps

$$\begin{array}{ccccccccc} 0 & \longrightarrow & C & \xrightarrow{f} & D & \xrightarrow{g} & E & \longrightarrow & 0 \\ & & \downarrow{\scriptstyle\alpha} & & \downarrow{\scriptstyle\beta} & & \downarrow{\scriptstyle\gamma} & & \\ 0 & \longrightarrow & C' & \xrightarrow[f']{} & D' & \xrightarrow[g']{} & E' & \longrightarrow & 0 \end{array}$$

in which the rows are exact sequences, the corresponding diagram

$$\begin{array}{ccccccccc} \cdots \longrightarrow & H_n(C) & \xrightarrow{f_*} & H_n(D) & \xrightarrow{g_*} & H_n(E) & \xrightarrow{\partial_*} & H_{n-1}(C) & \longrightarrow \cdots \\ & \downarrow{\scriptstyle\alpha_*} & & \downarrow{\scriptstyle\beta_*} & & \downarrow{\scriptstyle\gamma_*} & & \downarrow{\scriptstyle\alpha_*} & \\ \cdots \longrightarrow & H_n(C') & \xrightarrow[f'_*]{} & H_n(D') & \xrightarrow[g'_*]{} & H_n(E') & \xrightarrow[\partial'_*]{} & H_{n-1}(C') & \longrightarrow \cdots \end{array}$$

is a commutative diagram of abelian groups and homomorphisms.

Proof. We must first define ∂_*. Now a typical element of $H_n(E)$ is a coset $[z]$, where $z \in Z_n(E)$, and since g is onto, $z = g(d)$ for some $d \in D_n$. Thus $g\partial(d) = \partial g(d) = \partial(z) = 0$, so that by exactness $\partial(d) = f(c)$ for a unique element $c \in C_{n-1}$. Moreover $f\partial(c) = \partial f(c) = \partial^2(d) = 0$, so that $\partial(c) = 0$ (since f is (1-1)) and $c \in Z_{n-1}(C)$. Define $\partial_*[z] = [c] \in H_{n-1}(C)$; this appears to depend on the choice of z and d, but if z', d', c' is another choice of z, d, and c, then $g(d - d') \in B_n(E)$, so that $g(d - d') = \partial g(d'') = g\partial(d'')$ for some $d'' \in D$, since g is onto. Hence $g(d - d' - \partial(d'')) = 0$, so that $d - d' - \partial(d'') =$

$f(c'')$, for some $c'' \in C$, and so $\partial(d) - \partial(d') = \partial f(c'') = f\partial(c'')$, and $c - c' = \partial(c'')$, so that $[c] = [c'] \in H_{n-1}(C)$.

It is easy to see that ∂_* is a homomorphism. For if $[g(d_1)] = [z_1]$ and $[g(d_2)] = [z_2]$ in $H_n(E)$, then $[g(d_1 + d_2)] = [z_1 + z_2]$, so that $\partial_*([z_1] + [z_2])$ is given by $\partial(d_1 + d_2) = \partial(d_1) + \partial(d_2)$, and hence

$$\partial_*([z_1] + [z_2]) = \partial_*[z_1] + \partial_*[z_2].$$

The proof that the sequence of homology groups is exact proceeds in three stages.

(a) $\operatorname{Im} f_* = \operatorname{Ker} g_*$. Certainly $\operatorname{Im} f_* \subset \operatorname{Ker} g_*$, since $gf = 0$ implies $g_*f_* = 0$. Conversely if $[z] \in \operatorname{Ker} g_*$ then $g(z) = \partial(e)$ for some $e \in E$; but $e = g(d)$ for some $d \in D$, so that $g(z) = \partial g(d) = g\partial(d)$, and $g(z - \partial(d)) = 0$ so that $z - \partial(d) \in \operatorname{Im} f$. Hence $[z] = [z - \partial(d)] \in \operatorname{Im} f_*$, and $\operatorname{Ker} g_* \subset \operatorname{Im} f_*$.

(b) $\operatorname{Im} g_* = \operatorname{Ker} \partial_*$. It is clear from the definition that $\partial_* g_* = 0$, for an element of $\operatorname{Im} g_*$ is $[g(d)]$, where $\partial(d) = 0$. On the other hand if $[z] \in \operatorname{Ker} \partial_*$, then $z = g(d)$, where $\partial(d) = f\partial(c)$ for some $c \in C$, so that $z = g(d - f(c))$, where $\partial(d - f(c)) = 0$. Hence $[z] \in \operatorname{Im} g_*$.

(c) $\operatorname{Im} \partial_* = \operatorname{Ker} f_*$. Again, it is clear that $f_*\partial_* = 0$, so that $\operatorname{Im} \partial_* \subset \operatorname{Ker} f_*$. But if $[z] \in \operatorname{Ker} f_*$, then $f(z) = \partial(d)$ for some $d \in D$, so that $\partial g(d) = gf(z) = 0$, and $[g(d)] \in H(E)$ satisfies $\partial_*[g(d)] = [z]$. Hence $\operatorname{Ker} f_* \subset \operatorname{Im} \partial_*$.

Finally, Proposition 4.2.15(c) shows that $\beta_* f_* = f'_* \alpha_*$ and $\gamma_* g_* = g'_* \beta_*$. And if $[g(d)] \in H(E)$, then $\partial_*[g(d)] = [c]$, where $\partial(d) = f(c)$: thus $\alpha_* \partial_*[g(d)] = [\alpha(c)]$. But $f'\alpha(c) = \beta f(c) = \beta\partial(d) = \partial\beta(d)$, so that $[\alpha(c)] = \partial_*[g'\beta(d)] = \partial'_*[\gamma g(d)] = \partial'_* \gamma_*[g(d)]$. ∎

The exact homology sequence of a pair (X, Y) follows immediately.

Theorem 4.4.3 *Given a pair of spaces (X, Y), let $i: Y \to X$ be the inclusion map, and let $j: S(X) \to S(X, Y)$ be the quotient chain map. There is an exact sequence*

$$\cdots \longrightarrow H_n(Y) \xrightarrow{i_*} H_n(X) \xrightarrow{j_*} H_n(X, Y) \xrightarrow{\partial_*} H_{n-1}(Y) \longrightarrow \cdots,$$

such that if $f: (X, Y) \to (A, B)$ is a map of pairs, the diagram

$$\begin{array}{ccccccccc}
\cdots \longrightarrow & H_n(Y) & \xrightarrow{i_*} & H_n(X) & \xrightarrow{j_*} & H_n(X, Y) & \xrightarrow{\partial_*} & H_{n-1}(Y) & \longrightarrow \cdots \\
& \downarrow{\scriptstyle f_*} & & \downarrow{\scriptstyle f_*} & & \downarrow{\scriptstyle f_*} & & \downarrow{\scriptstyle f_*} & \\
\cdots \longrightarrow & H_n(B) & \xrightarrow[i'_*]{} & H_n(A) & \xrightarrow[j'_*]{} & H_n(A, B) & \xrightarrow[\partial'_*]{} & H_{n-1}(B) & \longrightarrow \cdots
\end{array}$$

is commutative. ∎

The groups $H_n(X)$ and $H_n(Y)$ in Theorem 4.4.3 can of course be replaced by the corresponding reduced groups $\tilde{H}_n(X)$ and $\tilde{H}_n(Y)$: the resulting exact sequence is called the *reduced* homology sequence of the pair (X, Y).

Example 4.4.4 Let x be any point of the space X. In the reduced homology sequence of the pair (X, x), the groups $\tilde{H}_n(x)$ are all zero; hence $\tilde{H}_n(X) \cong H_n(X, x)$ for all n. ▮

It is sometimes useful to have a form of Theorem 4.4.3 that involves only relative homology groups.

Theorem 4.4.5 *Given a triple* (X, Y, Z), *let* $i: (Y, Z) \to (X, Z)$ *and* $j: (X, Z) \to (X, Y)$ *be the inclusion maps. There is an exact sequence*

$$\cdots \to H_n(Y, Z) \xrightarrow{i_*} H_n(X, Z) \xrightarrow{j_*} H_n(X, Y) \xrightarrow{\partial_*} H_{n-1}(Y, Z) \to \cdots,$$

called the exact homology sequence of the triple (X, Y, Z).

Proof. $0 \to S(Y, Z) \xrightarrow{i} S(X, Z) \xrightarrow{j} S(X, Y) \to 0$ is clearly an exact sequence of chain complexes. ▮

Of course, a continuous map of triples gives rise to a commutative diagram involving two exact sequences, just as in Theorem 4.4.3. It should also be noticed that in fact Theorem 4.4.3 is just the special case of Theorem 4.4.5 obtained by putting $Z = \varnothing$.

Consider now a simplicial pair (K, L). By applying Theorem 4.4.2 to the exact sequence of chain complexes

$$0 \longrightarrow \Delta(L) \xrightarrow{i} \Delta(K) \xrightarrow{j} \Delta(K, L) \longrightarrow 0,$$

we obtain an exact sequence

$$\cdots \longrightarrow H_n(L) \xrightarrow{i_*} H_n(K) \xrightarrow{j_*} H_n(K, L) \xrightarrow{\partial_*} H_{n-1}(L) \longrightarrow \cdots,$$

which might at first sight be different from the exact sequence obtained from Theorem 4.4.3. However, there is a commutative diagram

$$\begin{array}{ccccccccc}
0 & \longrightarrow & \Delta(L) & \xrightarrow{i} & \Delta(K) & \xrightarrow{j} & \Delta(K, L) & \longrightarrow & 0 \\
 & & \downarrow{\scriptstyle\beta} & & \downarrow{\scriptstyle\beta} & & \downarrow{\scriptstyle\beta} & & \\
0 & \longrightarrow & S(|L|) & \xrightarrow[i]{} & S(|K|) & \xrightarrow[j]{} & S(|K|, |L|) & \longrightarrow & 0,
\end{array}$$

which by Theorem 4.4.2 gives rise to a commutative diagram involving the two exact sequences; since each chain map β induces isomorphisms in homology, the two exact sequences may therefore be identified. Similar remarks apply to the reduced homology sequence of (K, L) and to the exact sequence of a simplicial triple (K, L, M). Moreover a similar proof shows that the exact sequence of (K, L), for example, coincides also with that obtained by Theorem 4.4.2 from the exact sequence of chain complexes

$$0 \longrightarrow C(L) \xrightarrow{i_\cdot} C(K) \xrightarrow{j} C(K, L) \longrightarrow 0.$$

Another useful exact sequence arises if we have a simplicial complex K with two subcomplexes L and M such that $K = L \cup M$. Write $i_1\colon |L \cap M| \to |L|$, $i_2\colon |L \cap M| \to |M|$, $i_3\colon |L| \to |K|$ and $i_4\colon |M| \to |K|$ for the various inclusion maps.

Theorem 4.4.6 *There is an exact sequence, called the Mayer–Vietoris sequence of the triad $(K; L, M)$:*

$$\cdots \longrightarrow H_n(L \cap M) \xrightarrow{\eta_*} H_n(L) \oplus H_n(M) \xrightarrow{\xi_*} H_n(K) \xrightarrow{\partial_*} H_{n-1}(L \cap M) \longrightarrow \cdots,$$

where $\eta_*(x) = (i_1)_*x \oplus (-(i_2)_*x)$ and $\xi_*(x \oplus y) = (i_3)_*x + (i_4)_*y$. Moreover if $(P; Q, R)$ is another triad, a continuous map $f\colon |K| \to |P|$ such that $f|L| \subset |Q|$ and $f|M| \subset |R|$ gives rise to a commutative diagram involving the two Mayer–Vietoris sequences.

Proof. $\Delta(L) \oplus \Delta(M) = \oplus(\Delta_n(L) \oplus \Delta_n(M))$ can be made into a chain complex by taking as its boundary homomorphism $\partial_L \oplus \partial_M$, where ∂_L and ∂_M are the boundary homomorphisms in $\Delta(L)$ and $\Delta(M)$ respectively. Moreover, by Proposition 1.3.33,

$$H(\Delta(L) \oplus \Delta(M)) \cong H_*(L) \oplus H_*(M).$$

Now consider the sequence

$$0 \longrightarrow \Delta(L \cap M) \xrightarrow{\eta} \Delta(L) \oplus \Delta(M) \xrightarrow{\xi} \Delta(K) \longrightarrow 0,$$

where $\eta(x) = (i_1)_.x \oplus (-(i_2)_.x)$ and $\xi(x \oplus y) = (i_3)_.x + (i_4)_.y$. It is easy to see that η and ξ are chain maps, and that the sequence is exact. Hence the Mayer–Vietoris sequence is exact, by Theorem 4.4.2.

To prove the last part, recall that $f_*\colon H_*(K) \to H_*(P)$, for example, is the homomorphism induced by the composite

$$\Delta(K) \xrightarrow{\beta} S(|K|) \xrightarrow{f_\cdot} S(|P|) \xrightarrow{\alpha} \Delta(P).$$

Moreover, by the remark after Proposition 4.3.1 the chain map α may be chosen so as to restrict correctly to the corresponding chain maps α for Q, R and $Q \cap R$. It follows that the diagram

$$\begin{array}{ccccccccc} 0 & \longrightarrow & \Delta(L \cap M) & \xrightarrow{\eta} & \Delta(L) \oplus \Delta(M) & \xrightarrow{\xi} & \Delta(K) & \longrightarrow & 0 \\ & & \downarrow{\scriptstyle \alpha f_{\cdot}\beta} & & \downarrow{\scriptstyle \alpha f_{\cdot}\beta \oplus \alpha f_{\cdot}\beta} & & \downarrow{\scriptstyle \alpha f_{\cdot}\beta} & & \\ 0 & \longrightarrow & \Delta(Q \cap R) & \xrightarrow[\eta]{} & \Delta(Q) \oplus \Delta(R) & \xrightarrow[\xi]{} & \Delta(P) & \longrightarrow & 0 \end{array}$$

is commutative, which by Theorem 4.4.2 completes the proof. ∎

By an argument similar to that used for the exact sequence of a pair, the chain complexes $C(K)$, etc., may be used instead of $\Delta(K)$, etc., in setting up the Mayer–Vietoris sequence. It should also be noted that there is a corresponding theorem for arbitrary topological spaces, whose proof, however, is more complicated than that of Theorem 4.4.6 (see the notes at the end of the chapter).

An obvious modification of the proof of Theorem 4.4.6 shows that the homology groups could all be replaced by the corresponding reduced homology groups, or alternatively by relative homology groups: given a triad $(K; L, M)$ and a subcomplex N of K, there is an exact sequence

$$\cdots \to H_n(L \cap M, L \cap M \cap N) \to H_n(L, L \cap N) \oplus H_n(M, M \cap N) \\ \to H_n(K, N) \to H_{n-1}(L \cap M, L \cap M \cap N) \to \cdots,$$

called the *relative Mayer–Vietoris sequence*. Once again, a continuous map gives rise to a commutative diagram of reduced or relative Mayer–Vietoris sequences.

Example 4.4.7 Let L be the triangulation of S^n in Example 2.3.13 ($n \geqslant 0$), with vertices $a_1, a_1', \ldots, a_{n+1}, a_{n+1}'$. Let M be the subcomplex obtained by omitting a_{n+1}' and N the subcomplex obtained by omitting a_{n+1}, so that $M \cap N$ is a triangulation of S^{n-1}: see Fig. 4.3 overleaf. By Theorem 4.4.6, we have an exact sequence

$$\cdots \longrightarrow \tilde{H}_m(S^{n-1}) \xrightarrow{\eta_*} \tilde{H}_m(M) \oplus \tilde{H}_m(N) \xrightarrow{\xi_*} \tilde{H}(S^n) \xrightarrow{\partial_*} \\ \tilde{H}_{m-1}(S^{n-1}) \longrightarrow \cdots.$$

Since $|M|$ and $|N|$ are obviously contractible, ∂_* is an isomorphism, and so we recover the results of Example 4.3.12 by induction, starting with the trivial observation that

$$\tilde{H}_m(S^0) \cong \begin{cases} Z, & m = 0, \\ 0, & \text{otherwise.} \end{cases} \quad ∎$$

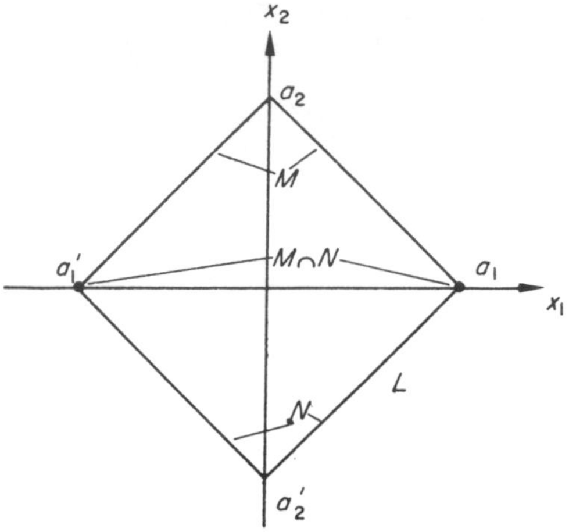

Fig. 4.3

The isomorphism ∂_* in Example 4.4.7 is a special case of a more general isomorphism, between the homology groups of a polyhedron and its 'suspension'.

Definition 4.4.8 Given a simplicial complex K, the *suspension* SK is $K * L$, where L consists only of two vertices a and b. By the remarks at the end of Section 2.3, homeomorphic polyhedra have homeomorphic suspensions, so that we can unambiguously write $S|K|$ for $|K| * S^0$, and even SX for a triangulated space X. Moreover a continuous map $f\colon |K| \to |M|$ gives rise to a map $Sf\colon S|K| \to S|M|$, defined by $Sf = f * 1$, where 1 is the identity map of S^0.

Example 4.4.9 If $\bar{S}^{n-1}$ is as in Example 2.3.13, $S(\bar{S}^{n-1})$ may be identified with $\bar{S}^n$, by taking a and b in Definition 4.4.8 to be a_{n+1} and a'_{n+1} respectively. (This is often rather loosely expressed by saying that $S(S^{n-1})$ 'is' S^n.) ▮

Theorem 4.4.10 *For each n, there is an isomorphism*

$$s_*\colon \tilde{H}_n(K) \to \tilde{H}_{n+1}(SK),$$

called the suspension isomorphism. Moreover if $f\colon |K| \to |M|$ is any continuous map, then $s_* f_* = (Sf)_* s_*$.

Proof. Define $s\colon \tilde{\Delta}(K) \to \tilde{\Delta}(SK)$ by $s(\lambda) = \lambda a - \lambda b$, for each generator λ of $\Delta(K)$ (this is interpreted as $a - b$ if $\lambda = *$). Of course,

s is not really a chain map, since $s(\tilde{\Delta}_n(K)) \subset \tilde{\Delta}_{n+1}(K)$, but it is nevertheless true that $\partial s = s\partial$, so that s induces homomorphisms $s_*: \tilde{H}_n(K) \to \tilde{H}_{n+1}(SK)$.

To complete the proof, we compare s_* with the homomorphism ∂_* in the reduced Mayer–Vietoris sequence of the triad $(SK; K * a, K * b)$. Now given $[z] \in \tilde{H}_n(K)$, where $z \in Z_n(\tilde{\Delta}(K))$, we have $s(z) = za - zb = \xi(za \oplus -zb)$, where

$$(\partial \oplus \partial)(za \oplus -zb) = (-1)^{n+1}(z \oplus (-z)) = (-1)^{n+1}\eta(z).$$

Hence $\partial_* s_*[z] = (-1)^{n+1}[z]$. But since $|K * a|$ and $|K * b|$ are contractible, they have zero reduced homology, so that ∂_*, and hence s_*, is an isomorphism.

That $s_* f_* = (Sf)_* s_*$ follows now from Theorem 4.4.6. ∎

Observe that $s\Delta^0(K) \subset \Delta^0(SK)$, so that s induces a chain map $s: \tilde{C}(K) \to \tilde{C}(SK)$, whose induced homology homomorphism may be identified with s_*.

Example 4.4.11 It is useful to define 'standard generators' σ_n of $H_n(\bar{S}^n) \cong Z$ $(n \geqslant 0)$, by setting $\sigma_0 = [(a_1) - (a_1')]$ and $\sigma_n = s_*\sigma_{n-1}$ $(n \geqslant 1)$, using the identification of $S(\bar{S}^{n-1})$ with $\bar{S}^n$ in Example 4.4.9. Thus, for example, $\sigma_1 = [(a_1, a_2) - (a_1, a_2') - (a_1', a_2) + (a_1', a_2')]$, and in general σ_n has a representative cycle z_n, that contains $(a_1, \ldots, a_{n+1})$. A corresponding generator of $H_n(\bar{E}^n, \bar{S}^{n-1})$ is $\bar{\sigma}_n$, where $\sigma_{n-1} = \partial_*\bar{\sigma}_n$, and ∂_* is the homomorphism in the exact homology sequence of $(\bar{E}^n, \bar{S}^{n-1})$: thus a representative cycle for $\bar{\sigma}_n$ is $a_0 z_{n-1}$, which contains $(a_0, a_1, \ldots, a_{n+1})$. ∎

Example 4.4.11 has interesting consequences concerned with fixed points of maps of S^n to itself (compare Theorem 2.5.23).

Proposition 4.4.12 *Let $f: S^n \to S^n$ be a continuous map without fixed points. Then $f_*(\sigma_n) = (-1)^{n+1}\sigma_n$.*

Proof. By Corollary 2.2.4 $f \simeq g$, where $g(x) = -x$ for all $x \in S^n$. But (regarding g as a map of $\bar{S}^n$) g is the simplicial map that exchanges a_i and a_i' $(1 \leqslant i \leqslant n)$, and it is easy to see that therefore $g_*(\sigma_n) = (-1)^{n+1}\sigma_n$. ∎

Corollary 4.4.13 *If $f: S^n \to S^n$ is a map homotopic to the identity, and n is even, then f has a fixed point.* ∎

This result, in the special case $n = 2$, is popularly known as the 'Hairy Ball Theorem': if one imagines a hair growing out of each point of the surface of a ball, it is impossible to brush them flat without a parting, since otherwise there would be a map homotopic to the identity (along the hairs), without a fixed point.

We turn our attention now to the calculation theorem for the homology groups of polyhedra, mentioned at the beginning of this section. Since $H_*(K, \varnothing) = H_*(K)$ and $\tilde{H}_*(K) \cong H_*(K, a)$ for any vertex a, it will be sufficient to consider only the case of relative homology. Roughly speaking, the method is to construct a sequence of subcomplexes of K that in some sense generalize the skeletons of K. Now if K^n denotes the n-skeleton, for any subcomplex L the set $K^n \cup L - K^{n-1} \cup L$ consists of the n-simplexes in $K - L$, so that $H_r(K^n \cup L, K^{n-1} \cup L) = 0$ unless $r = n$, and

$$H_n(K^n \cup L, K^{n-1} \cup L) \cong C_n(K, L).$$

This property of the skeletons is generalized by forming a sequence of subcomplexes $K = M^p \supset M^{p-1} \supset \cdots \supset M^0 \supset L$ with the property that $H_r(M^n, M^{n-1}) = 0$ unless $r = n$: we then define a new chain complex C by setting $C_n \cong H_n(M^n, M^{n-1})$, and it turns out that $H(C) \cong H_*(K, L)$ for any such sequence of subcomplexes.

It is particularly convenient to work with a sequence of subcomplexes constructed by dividing K into certain generalizations of simplexes called 'blocks'. However, we must first prove the general result on the homology of chain complexes constructed as above.

Suppose, then, that $K = M^p \supset M^{p-1} \supset \cdots \supset M^0 \supset L$ is a sequence of subcomplexes such that $H_r(M^n, M^{n-1}) = 0$ unless $r = n$, for all integers r and n. Let C be the chain complex $\bigoplus C_n$, where $C_n = H_n(M^n, M^{n-1})$ (M^n is to be interpreted as K if $n > p$ and L if $n < 0$), and the boundary homomorphism $d: C_n \to C_{n-1}$ is defined to be the composite

$$H_n(M^n, M^{n-1}) \xrightarrow{\partial_*} H_{n-1}(M^{n-1}, L) \xrightarrow{j_*} H_{n-1}(M^{n-1}, M^{n-2}),$$

where ∂_* and j_* are homomorphisms in the exact sequences of the triples (M^n, M^{n-1}, L) and (M^{n-1}, M^{n-2}, L) respectively (in fact $d = \partial_*$, the homomorphism in the exact sequence of the triple (M^n, M^{n-1}, M^{n-2}): see Exercise 6). Certainly C is a chain complex, since d^2 involves a composite of two successive homomorphisms (j_* and ∂_*) in the exact sequence of the same triple, so that $d^2 = 0$.

Theorem 4.4.14 *For each n, $H_n(C) \cong H_n(K, L)$.*

Proof. This is really just an exercise in handling exact sequences. First consider the exact sequence of the triple (M^n, M^{n-1}, L):

$$\cdots \longrightarrow H_{r+1}(M^n, M^{n-1}) \xrightarrow{\partial_*} H_r(M^{n-1}, L) \xrightarrow{i_*} H_r(M^n, L) \xrightarrow{j_*} H_r(M^n, M^{n-1}) \longrightarrow \cdots.$$

Since $H_r(M^n, M^{n-1}) = 0$ unless $r = n$, $i_*: H_r(M^{n-1}, L) \to H_r(M^n, L)$ is an isomorphism for $r \neq n - 1$, n. So for $r > n$, $H_r(M^n, L) \cong H_r(M^{n-1}, L) \cong \cdots \cong H_r(M^{-1}, L) = H_r(L, L) = 0$, and in particular it follows that $j_*: H_n(M^n, L) \to H_n(M^n, M^{n-1})$ is (1-1)

Now consider part of the chain complex C. By definition the following diagram is commutative (the superscripts to ∂_* and j_* are merely for identification purposes).

$$\begin{array}{ccc}
H_{n+1}(M^{n+1}, M^n) & \xrightarrow{=} & C_{n+1} \\
\downarrow{\scriptstyle \partial_*^1} & & \\
H_n(M^n, L) & & \downarrow{\scriptstyle d} \\
\downarrow{\scriptstyle j_*^1} & & \\
H_n(M^n, M^{n-1}) & \xrightarrow{=} & C_n \\
\downarrow{\scriptstyle \partial_*^2} & & \\
H_{n-1}(M^{n-1}, L) & & \downarrow{\scriptstyle d} \\
\downarrow{\scriptstyle j_*^2} & & \\
H_{n-1}(M^{n-1}, M^{n-2}) & \xrightarrow{=} & C_{n-1}
\end{array}$$

Now

$$\begin{aligned} Z_n(C) &= \operatorname{Ker} \partial_*^2, \quad \text{since } j_*^2 \text{ is (1-1)} \\ &= \operatorname{Im} j_*^1 \\ &\cong H_n(M^n, L), \quad \text{since } j_*^1 \text{ is (1-1).} \end{aligned}$$

It follows that

$$\begin{aligned} H_n(C) &= Z_n(C)/B_n(C) \\ &\cong H_n(M^n, L)/\operatorname{Im} \partial_*^1. \end{aligned}$$

But $\operatorname{Im} \partial_*^1 = \operatorname{Ker} [i_*: H_n(M^n, L) \to H_n(M^{n+1}, L)]$, so that

$$\begin{aligned} H_n(C) &\cong \operatorname{Im} [i_*: H_n(M^n, L) \to H_n(M^{n+1}, L)], \quad \text{by Proposition 1.3.12} \\ &= H_n(M^{n+1}, L), \quad \text{since } H_n(M^{n+1}, M^n) = 0. \end{aligned}$$

But $H_n(M^{n+1}, L) \cong H_n(M^{n+2}, L) \cong \cdots \cong H_n(M^p, L) = H_n(K, L)$, and hence $H_n(C) \cong H_n(K, L)$. ∎

Of course, if $M^n = K^n \cup L$, then $H_r(M^n, M^{n-1}) = 0$ unless $r = n$, and $H_n(M^n, M^{n-1}) \cong C_n(K, L)$. It is easy to see that the boundary in $C(K, L)$ defined in Theorem 4.4.14 is the same as the ordinary boundary homomorphism in this case, so that we recover the chain complex $C(K, L)$.

The next step is to define *blocks*: these are generalizations of simplexes, and the corresponding 'block skeletons' form a particularly convenient sequence of subcomplexes to which Theorem 4.4.14 can be applied.

Definition 4.4.15 An *n-block* in a simplicial complex K is a pair of subcomplexes $(e, \dot{e})$, such that $\dim e = n$ and

$$H_r(e, \dot{e}) \cong \begin{cases} Z, & r = n \\ 0, & r \neq n. \end{cases}$$

$\dot{e}$ is called the *boundary* of e, and the *interior* of e is the set of simplexes in $e - \dot{e}$.

For example, if σ is an n-simplex of K, then $(K(\sigma), \dot{\sigma})$ is an n-block, by Example 4.3.12. The interior of this block consists of the single simplex σ.

We next wish to divide K into blocks, in such a way as to generalize the structure of K as a simplicial complex.

Definition 4.4.16 A *block dissection* of K is a set of blocks such that

(a) each simplex is in the interior of just one block;

(b) the boundary of each n-block is in the union of the m-blocks, for $m < n$.

A subcomplex L of K is a *block subcomplex* if it is a union of blocks. In particular, the *block n-skeleton* of K, K_n, is the union of all the m-blocks for $m \leqslant n$.

Example 4.4.17 Given any simplicial complex K, the set of pairs $(K(\sigma), \dot{\sigma})$, for all simplexes of K, forms a block dissection of K. For this block dissection, every subcomplex is a block subcomplex. Similarly, the set of all pairs $(K(\sigma)', (\dot{\sigma})')$ forms a block dissection of K'. ∎

Thus a block dissection of K is indeed a generalization of the dissection of K as a simplicial complex.

Example 4.4.18 The torus T can be triangulated as shown in Fig. 4.4.

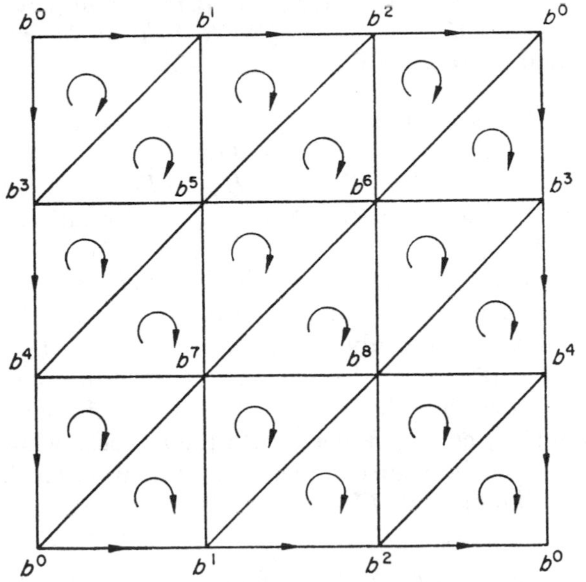

Fig. 4.4

A block dissection consists of the blocks

e^2, the whole simplicial complex,

$e_1^1 = (b^0, b^1), (b^1, b^2), (b^2, b^0), (b^0), (b^1), (b^2),$

$e_2^1 = (b^0, b^3), (b^3, b^4), (b^4, b^0), (b^0), (b^3), (b^4),$

$e^0 = (b^0),$

where $\dot{e}^2 = e_1^1 \cup e_2^1$, $\dot{e}_1^1 = \dot{e}_2^1 = e^0$, and $\dot{e}^0 = \varnothing$. This certainly satisfies (a) and (b) of Definition 4.4.16, though it remains to prove that these really are blocks. We postpone the proof until after Proposition 4.4.19. ∎

Example 4.4.18 illustrates the practical difficulty that usually arises in constructing a block dissection: it is easy enough to find subcomplexes that fit together in the right way, but less easy to show that they are genuine blocks. However, the following proposition will often resolve this difficulty.

Proposition 4.4.19 *If (M, N) is a triangulation of (E^n, S^{n-1}), and $f\colon |M| \to |K|$ is a simplicial map that is* (1-1) *on $|M| - |N|$, then*

$(f(M), f(M))$ is an n-block, where $f(M)$ is the subcomplex of K of all simplexes $f(\sigma)$, $\sigma \in M$. (If $n = 0$, (E^0, S^{-1}) is to be interpreted as $(0, \varnothing)$, where 0 is the origin.)

Proof. Since f is (1-1) on $|M| - |N|$, it is (1-1) on each simplex of $M - N$, and hence $f_{.}: \Delta(M, N) \to \Delta(f(M), f(N))$ is a chain isomorphism: thus $f_*: H_*(E^n, S^{n-1}) \to H_*(f(M), f(N))$ is an isomorphism. Since, by Corollary 4.3.14, $\dim f(M)$ is clearly n, the proof is complete. ∎

Thus, for example, e^2, e_1^1, e_2^1 and e^0 in Example 4.4.18 are blocks.

The usefulness of a block dissection lies in the fact that if L is a block subcomplex, the subcomplexes $M^n = K_n \cup L$ satisfy the hypotheses of Theorem 4.4.14. This is the next theorem, in which we shall also prove that $H_n(M^n, M^{n-1})$ is a free abelian group with generators in (1-1)-correspondence with the n-blocks whose interiors are in $K - L$; thus the chain complex formed from the block dissection is a generalization of $C(K, L)$.

Theorem 4.4.20 *Let K be a simplicial complex with a block dissection, and let L be a block subcomplex. If $M^n = K_n \cup L$ for $n \geqslant 0$ (and $M^n = L$ for $n < 0$), then $H_r(M^n, M^{n-1}) = 0$ unless $r = n$, and (if $n \geqslant 0$) $H_n(M^n, M^{n-1})$ is a free abelian group with generators in (1-1)-correspondence with the n-blocks whose interiors are in $K - L$.*

Proof. $M^n - M^{n-1}$ is the union of the interiors of those n-blocks e_i whose interiors are contained in $K - L$. Since each simplex is in the interior of a unique block, it follows that $\Delta_r(M^n, M^{n-1}) = \bigoplus \Delta_r(e_i, \dot{e}_i)$, for those e_i whose interiors are in $K - L$. But $\partial(\lambda)$ has the same value whether λ is regarded as a generator of $\Delta_r(M^n, M^{n-1})$ or of some $\Delta_r(e_i, \dot{e}_i)$; hence $H_r(M^n, M^{n-1}) \cong \bigoplus H_r(e_i, \dot{e}_i)$, which is zero unless $r = n$, and if $r = n$ is a free abelian group with generators in (1-1)-correspondence with those e_i whose interiors are in $K - L$. ∎

It follows from Theorem 4.4.14 that $H_n(K, L) \cong H_n(C)$ for each n, where $C_n = H_n(M^n, M^{n-1})$. The only practical problem that remains before we can use C to compute $H_*(K, L)$ is the determination of the boundary homomorphisms d, and these can be described as follows.

Since $M^n - M^{n-1}$ has no simplex of dimension greater than n, $B_n(C(M^n, M^{n-1})) = 0$, and we may identify $H_n(M^n, M^{n-1})$ with $Z_n(C(M^n, M^{n-1}))$, which is a subgroup of $C_n(M^n, M^{n-1}) = C_n(K, L)$. Let θ: $C_n \to C_n(K, L)$ be the inclusion homomorphism so defined; the boundary homomorphism d is thus completely determined by

Proposition 4.4.21 *θ is a chain map.*

Proof. By definition, d is the composite

$$H_n(M^n, M^{n-1}) \xrightarrow{\partial_*} H_{n-1}(M^{n-1}, L) \xrightarrow{j_*} H_{n-1}(M^{n-1}, M^{n-2}),$$

where, by the remarks after Theorem 4.4.5, we may use simplicial chain complexes throughout. Now if

$$x \in H_n(M^n, M^{n-1}) = Z_n(C(M^n, M^{n-1})),$$

$j_*\partial_*(x)$ is given by regarding x as an element of $C(M^n, L)$, taking its boundary $\partial(x)$ in $C(M^n, L)$, and noting that in fact $\partial(x) \in Z(C(M^n, L))$: then

$$j_*\partial_*(x) = j_{.}\partial(x) \in Z_{n-1}(C(M^{n-1}, M^{n-2})) = H_{n-1}(M^{n-1}, M^{n-2}).$$

However, $\partial(x)$ has the same value whether x is regarded as an element of $C(M^n, L)$ or of $C(K, L)$, and $j_{.}$ has no effect on $\partial(x)$, since it is a linear combination of $(n - 1)$-simplexes that are not in L, and so not in M^{n-2}. Hence $\theta d(x) = \partial\theta(x)$. ▌

In particular θ induces a homomorphism $\theta_*\colon H(C) \to H_*(K, L)$.

Proposition 4.4.22 *θ_* is the isomorphism of Theorem* 4.4.14.

Proof. The isomorphism of Theorem 4.4.14 is given by choosing a representative cycle x in $C_n = H_n(M^n, M^{n-1})$, 'lifting' to $H_n(M^n, L)$, and mapping to $H_n(M^{n+1}, L) \cong H_n(K, L)$ by i_*. But if $j_*\partial_*(x) = 0$, then by Proposition 4.4.21 we have $\partial\theta(x) = 0$ in $C(K, L)$, so that $\theta(x) \in Z(C(M^n, L))$. That is, x becomes the coset $[\theta(x)] = \theta_*(x)$ in $H_n(M^{n+1}, L) \cong H_n(K, L)$. ▌

Example 4.4.23 Consider the triangulation and block dissection of the torus T obtained in Example 4.4.18. If we identify the resulting chain complex C with a sub-chain complex of the simplicial chain complex, using the chain map θ, we may take the groups C_n to be the free abelian groups generated by the following elements.

$$C_2\colon \quad z^2 = [b^0, b^1, b^3] + [b^1, b^5, b^3] + \cdots$$

(that is, z^2 is the sum of all the 2-simplexes of T, where each 2-simplex is identified with a generator of the simplicial chain complex according to the arrows in Fig. 4.4: the arrow on (b^0, b^1, b^3), for example, indicates that we are to take $[b^0, b^1, b^3]$ rather than, say, $[b^0, b^3, b^1]$).

$$C_1: \quad z_1^1 = [b^0, b^1] + [b^1, b^2] + [b^2, b^0]$$
$$z_2^1 = [b^0, b^3] + [b^3, b^4] + [b^4, b^0]$$

(these are also indicated on Fig. 4.4 by arrows on the corresponding simplexes).

$$C_0: \quad z^0 = [b^0].$$

It is easy to see that these will do as generators: for example, $\partial(z^2) = 0$, since $\partial(z^2)$ contains every 1-simplex twice, with opposite signs, and hence $z^2 \in Z_2(C(e^2, e_1^1 \cup e_2^1))$; on the other hand, z^2 cannot be a multiple of any other cycle. Also $\partial(z_1^1) = \partial(z_2^1) = 0$ and $\partial(z^0) = 0$, so that the homology groups of T are

$$H_0(T) \cong Z, \quad H_1(T) \cong Z \oplus Z, \quad H_2(T) \cong Z, \quad H_n(T) = 0 \text{ otherwise.}$$

As an example of the calculation of relative homology groups, consider the (block) subcomplex e_1^1 of T. To calculate $H_*(T, e_1^1)$, we use the blocks e^2 and e_2^1: again $\partial(z^2) = \partial(z_2^1) = 0$, so that $H_1(T, e_1^1) \cong Z$, $H_2(T, e_1^1) \cong Z$ and $H_n(T, e_1^1) = 0$ otherwise. ∎

The same method can be used to calculate the homology groups of the triangulable 2-manifolds of Chapter 3.

Theorem 4.4.24 *The homology groups of M_g, N_h $(g \geqslant 0, h \geqslant 1)$ are given by:*

(a) $H_0(M_g) \cong H_2(M_g) \cong Z$, $H_1(M_g) \cong 2gZ$, $H_n(M_g) = 0$ *otherwise;*
(b) $H_0(N_h) \cong Z$, $H_1(N_h) \cong (h - 1)Z \oplus Z_2$, $H_n(N_h) = 0$ *otherwise,*
where rZ denotes the direct sum of r copies of Z.

Proof. As usual, we first dispose of the special cases M_0 and N_1. Now $M_0 = S^2$ and $N_1 = RP^2$, and the homology groups of these have been calculated in Examples 4.3.12 and 4.3.17. Also the case of M_1 (the torus) has just been done in Example 4.4.23; the general case is done in a similar way, with a block dissection of M_g, for example, having one 2-block, $2g$ 1-blocks and one 0-block.

Consider the triangulation of M_g $(g \geqslant 1)$ obtained as in Theorem 3.4.11: take the $4g$-sided polygon P corresponding to M_g, divide each side into three equal parts, join the resulting vertices to the centre, and then subdivide the resulting triangulation of P relative to the boundary. A block dissection of M_g is then obtained as follows, where the boundary edges of P correspond to the sequence of symbols $x_1 y_1 x_1^{-1} y_1^{-1} \cdots x_g y_g x_g^{-1} y_g^{-1}$.

Let e^2 = all simplexes in M_g,

e_r^1 = all simplexes in the edge x_r,

$\bar{e}_r^1$ = all simplexes in the edge y_r,

e^0 = a, the point to which all the original vertices of P are identified.

If we also take $\dot{e}^2 = \bigcup_r (e_r^1 \cup \bar{e}_r^1)$, $\dot{e}_r^1 = \dot{\bar{e}}_r^1 = a$, Proposition 4.4.19 ensures that these are all blocks (see Fig. 4.5).

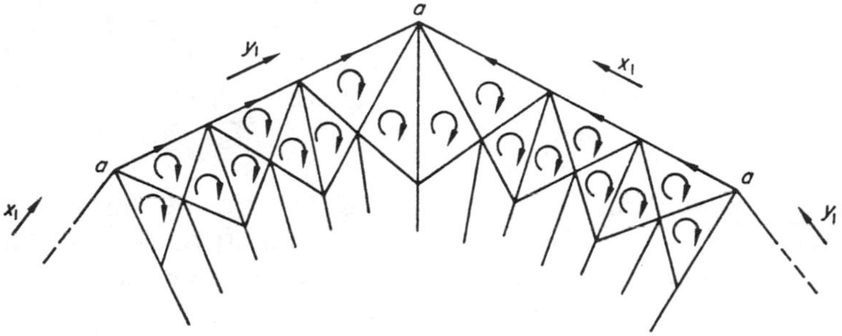

Fig. 4.5

The resulting chain complex C may be taken to be that generated by

z^2 = the sum of all the 2-simplexes,

z_r^1 = the sum of the three 1-simplexes in e_r^1,

$\bar{z}_r^1$ = the sum of the three 1-simplexes in $\bar{e}_r^1$,

and $z^0 = [a]$,

where these simplexes are identified with generators of the simplicial chain complex according to the arrows in Fig. 4.5. These will certainly do as generators: for example $\partial(z^2) = 0$ in $C(e^2, \dot{e}^2)$ because $\partial(z^2)$ contains each 'interior' 1-simplex twice, with opposite signs; and z^2 cannot be a multiple of another cycle.

Finally,

$$\partial(z_r^1) = \partial(\bar{z}_r^1) = 0;$$

and

$$\begin{aligned}\partial(z^2) &= (z_1^1 + \bar{z}_1^1 - z_1^1 - \bar{z}_1^1) + \cdots + (z_g^1 + \bar{z}_g^1 - z_g^1 - \bar{z}_g^1) \\ &= 0.\end{aligned}$$

This proves (a), and the reader should have no difficulty in adapting the proof to deal with (b). ∎

As a final example, let us calculate the homology groups of RP^n, for each n.

Example 4.4.25 We may as well regard RP^n as the space obtained from $\bar{S}^n$ by identifying antipodal points, where as in Example 2.3.13 $\bar{S}^n$ is the polyhedron of a simplicial complex L_n in R^{n+1}. A triangulation M_n of RP^n is then obtained by identifying antipodal points of L'_n.

By Proposition 1.4.40, M_n is also the simplicial complex obtained by identifying antipodal points of L'_{n-1} in $(L_n^+)'$, where L_n^+ is the subcomplex of L_n of those simplexes lying in the region $x_{n+1} \geqslant 0$. Since L_n^+ is a triangulation of E^n, by Proposition 4.4.19 there is a block dissection of M_n with just one n-block $e^n = M_n$, the boundary $\dot{e}^n$ being M_{n-1}, which in turn is a single $(n-1)$-block e^{n-1} whose boundary is M_{n-2}, and so on, until $e^0 = M_0$, the single point $a_1 = a'_1$.

The corresponding chain complex C has $C_r = 0$ for $r > n$ or $r < 0$, and otherwise C_r has just one generator. It remains to choose these generators and calculate the boundary homomorphisms. Write z_n for the 'standard' generating cycle of $Z_n(L_n)$, as in Example 4.4.11; then $z_{n-1}a_{n+1}$ is a generating cycle for $Z_n(L_n^+, L_{n-1})$. Write also z_n, $z_{n-1}a_{n+1}$ for the images of these elements in the corresponding simplicial chain complexes (they are also generating cycles). By the remarks after Corollary 4.3.10, $\phi(z_{n-1}a_{n+1})$ is a generating cycle for $Z_n(C((L_n^+)', L'_{n-1}))$, and hence if $p\colon |L_n^+| \to |M_n|$ is the identification map, the proof of Proposition 4.4.19 shows that $p_.\phi(z_{n-1}a_{n+1})$ is a generating cycle for $Z_n(C(M_n, M_{n-1})) = C_n$. Moreover in $C(M_n)$

$$\begin{aligned} \partial p_.\phi(z_{n-1}a_{n+1}) &= p_.\phi\partial(z_{n-1}a_{n+1}) \\ &= (-1)^n p_.\phi(z_{n-1}) \\ &= (-1)^n p_.\phi(z_{n-2}a_n - z_{n-2}a'_2). \end{aligned}$$

But if z'_{n-2} denotes z_{n-2} with a_r and a'_r interchanged for each r, we have $z_{n-2} = (-1)^{n-1}z'_{n-2}$, and $p_.\phi(z'_{n-2}a'_n) = p_.\phi(z_{n-2}a_n)$. Hence

$$\begin{aligned} \partial p_.\phi(z_{n-1}a_{n+1}) &= (-1)^n p_.\phi(z_{n-2}a_n) + p_.\phi(z_{n-2}a_n) \\ &= (1 + (-1)^n)p_.\phi(z_{n-2}a_n). \end{aligned}$$

It follows that the homology groups of RP^n are

$$\begin{aligned} H_0(RP^n) &\cong Z, \\ H_r(RP^n) &= 0 \quad \text{if } r < 0,\ r > n, \text{ or } r \text{ is even}, \\ H_r(RP^n) &\cong Z_2 \quad \text{if } r \text{ is odd and } 0 < r < n, \\ H_n(RP^n) &\cong Z \quad \text{if } n \text{ is odd.} \end{aligned}$$

∎

4.5 Homology groups with arbitrary coefficients, and the Lefschetz Fixed-Point Theorem

So far in this chapter we have been working with homology groups of chain complexes in which the chain groups are free abelian groups. Thus for example, if X is a topological space, an element of $S(X)$ is a formal linear combination $\sum_i r_i\lambda_i$, where the λ_i are singular simplexes and the r_i are integers. However, it is often useful to consider a generalization in which the r_i, instead of being integers, are elements of an arbitrary abelian group G. The new chain complex that results is written $S(X; G)$, and the corresponding homology groups are $H_n(X; G)$, the homology groups of X *with coefficients in* G; thus $H_n(X)$ appears as the special case $H_n(X; Z)$. In fact the groups $H_n(X; G)$ are completely determined by the groups $H_n(X)$, so that this generalization cannot be expected to yield new information about X; the idea is rather that $H_*(X; G)$ may be simpler and easier to handle than $H_*(X)$. For example, if K is a simplicial complex and G is a field, we shall see that each $H_n(K; G)$ is a finite-dimensional vector space over G, and so is determined up to isomorphism by its dimension. Moreover the homology homomorphisms induced by continuous maps are linear maps of vector spaces, and this fact can be used to give algebraic conditions for a continuous map $f: |K| \to |K|$ to have a fixed point.

Since the definition of $H_*(X; G)$ is purely algebraic, we shall follow our usual procedure and consider first the abstract situation. The idea of 'taking coefficients in G instead of in Z' is formalized by the notion of the *tensor product* of two abelian groups.

Definition 4.5.1 Given abelian groups A and B, the *tensor product* $A \otimes B$ is the abelian group generated by all symbols of the form $a \otimes b$, for each element $a \in A$ and each $b \in B$, subject to the relations $a_1 \otimes (b_1 + b_2) - a_1 \otimes b_1 - a_1 \otimes b_2$ and $(a_1 + a_2) \otimes b_1 - a_1 \otimes b_1 - a_2 \otimes b_1$, for each $a_1, a_2 \in A$ and $b_1, b_2 \in B$.

Example 4.5.2 If R is any ring, then by Proposition 1.3.20 the multiplication $m: R \times R \to R$ may be regarded as a homomorphism $m: R \otimes R \to R$. ∎

Example 4.5.3 For any abelian group G, $G \otimes Z \cong G$. To prove this, define a homomorphism $\theta: G \otimes Z \to G$ (using Proposition 1.3.20) by the rule $\theta(g \otimes n) = ng$ ($n \in Z, g \in G$), where ng means

$g + g + \cdots + g$ (n times) if $n > 0$ and $-(-ng)$ if $n < 0$ (and of course $0g = 0$). Clearly θ is onto, and it is also (1-1), since

$$\begin{aligned}\theta\left[\sum \lambda_i(g_i \otimes n_i)\right] = 0 &\Rightarrow \sum_i \lambda_i n_i g_i = 0 \\ &\Rightarrow \left(\sum \lambda_i n_i g_i\right) \otimes 1 = 0 \\ &\Rightarrow \sum (\lambda_i n_i g_i \otimes 1) = 0 \\ &\Rightarrow \sum \lambda_i (g_i \otimes n_i) = 0.\end{aligned}$$

Hence θ is an isomorphism (indeed, we shall often identify the groups $G \otimes Z$ and G, using the isomorphism θ). ∎

Example 4.5.4 If p and q are positive integers, $Z_p \otimes Z_q \cong Z_{(p,q)}$, where (p, q) is the highest common factor of p and q. For we can define a homomorphism $\theta\colon Z_p \otimes Z_q \to Z_{(p,q)}$ by $\theta(r \otimes s) = rs \pmod{(p, q)}$ ($r \in Z_p$, $s \in Z_q$). Again θ is clearly onto, and to prove that it is also (1-1), note that $\sum \lambda_i(r_i \otimes s_i) = \sum \lambda_i r_i s_i(1 \otimes 1)$. Thus

$$\begin{aligned}\theta\left[\sum \lambda_i(r_i \otimes s_i)\right] = 0 &\Rightarrow \sum \lambda_i r_i s_i = 0 \quad \text{mod } (p, q) \\ &\Rightarrow \sum \lambda_i r_i s_i = ap + bq,\end{aligned}$$

for some integers a and b, so that

$$\begin{aligned}\sum \lambda_i(r_i \otimes s_i) &= ap(1 \otimes 1) + bq(1 \otimes 1) \\ &= a(p \otimes 1) + b(1 \otimes q) \\ &= 0. \ ∎\end{aligned}$$

Example 4.5.5 If p is any positive integer, then $Z_p \otimes Q = 0$, where Q is the additive group of rationals. For if $r \in Z_p$ and $q \in Q$, then $r \otimes q = (pr) \otimes (q/p) = 0$. ∎

The process of forming the tensor product can be applied to homomorphisms as well as to groups.

Proposition 4.5.6 *Homomorphisms $f\colon A \to A'$, $g\colon B \to B'$ give rise to a homomorphism $f \otimes g\colon A \otimes B \to A' \otimes B'$, such that given further homomorphisms $f'\colon A' \to A''$, $g'\colon B' \to B''$, we have $(f' \otimes g')(f \otimes g) = (f'f) \otimes (g'g)$.*

Proof. Define $f \otimes g\colon A \otimes B \to A' \otimes B'$ by the rule

$$(f \otimes g)(a \otimes b) = f(a) \otimes g(b):$$

this defines a homomorphism by Proposition 1.3.20. The property $(f' \otimes g')(f \otimes g) = (f'f) \otimes (g'g)$ is immediate from the definition. ▮

Finally, before applying tensor products to chain complexes, it is useful to have two general rules for manipulating tensor products of abelian groups.

Proposition 4.5.7

(a) $A \otimes B \cong B \otimes A$.

(b) *If* $A = \bigoplus_i A_i$ *and* $B = \bigoplus_j B_j$, *then* $A \otimes B \cong \bigoplus_{i,j} (A_i \otimes B_j)$.

Proof. (a) is trivial. As for (b), let $p_i, q_j: A_i, B_j \to \oplus A_i, \oplus B_j$ be the obvious inclusion homomorphisms, and define homomorphisms

$$\theta: A \otimes B \to \oplus(A_i \otimes B_j), \qquad \phi: \oplus (A_i \otimes B_j) \to A \otimes B$$

by $\theta[(a_i) \otimes (b_j)] = (a_i \otimes b_j)$, $\phi((x_{ij})) = \sum (p_i \otimes q_j)x_{ij}$ $(x_{ij} \in A_i \otimes B_j)$. (Note that the latter sum is only a finite sum, since by definition of the direct sum, all but a finite number of the x_{ij} are zero.) Now

$$\begin{aligned} \phi\theta[(a_i) \otimes (b_j)] &= \phi((a_i \otimes b_j)) \\ &= \sum (p_i \otimes q_j)(a_i \otimes b_j) \\ &= \sum p_i a_i \otimes q_j b_j \\ &= (a_i) \otimes (b_j), \end{aligned}$$

so that $\phi\theta$ is the identity, and

$$\begin{aligned} \theta\phi((a_i \otimes b_j)) &= \theta\left[\sum p_i a_i \otimes q_j b_j\right] \\ &= \theta[(a_i) \otimes (b_j)] \\ &= (a_i \otimes b_j), \end{aligned}$$

so that $\theta\phi$ is also the identity (it is clearly sufficient to check $\theta\phi$ on an element of the form $(a_i \otimes b_j)$, since every element of $\oplus(A_i \otimes B_j)$ is a finite sum of such elements). Thus θ and ϕ are inverse isomorphisms. ▮

We are particularly interested, of course, in applying the tensor product construction to chain complexes. Suppose that C is a chain complex with boundary homomorphisms $\partial: C_n \to C_{n-1}$, and that G is any abelian group.

Proposition 4.5.8 *$C \otimes G$ is a chain complex. Moreover if $f: C \to D$ is a chain map, then so is $f \otimes 1: C \otimes G \to D \otimes G$.*

Proof. By Proposition 4.5.7(b), $C \otimes G \cong \oplus (C_n \otimes G)$, so we may set $(C \otimes G)_n = C_n \otimes G$. Since $(\partial \otimes 1)(\partial \otimes 1) = (\partial^2 \otimes 1) = 0$,

$C \otimes G$, with the boundary homomorphisms $\partial \otimes 1$, is certainly a chain complex. Moreover if f is a chain map, then so is $f \otimes 1$, since

$$\begin{aligned}(f \otimes 1)(\partial \otimes 1) &= (f\partial \otimes 1)\\ &= (\partial f \otimes 1)\\ &= (\partial \otimes 1)(f \otimes 1),\end{aligned}$$

and $(f \otimes 1)(C_n \otimes G) \subset D_n \otimes G$. ∎

Definition 4.5.9 Let (X, Y) be a pair of spaces. The homology groups of (X, Y) *with coefficients in* G are defined by

$$H_n(X, Y; G) = H_n(S(X, Y) \otimes G).$$

We write $H_*(X, Y; G)$ for $\bigoplus H_n(X, Y; G)$, and $H_n(X; G)$ if Y is empty. Similarly the reduced homology groups of X *with coefficients in* G are defined by

$$\tilde{H}_n(X; G) = H_n(\tilde{S}(X) \otimes G).$$

Given a continuous map $f: (X, Y) \to (A, B)$, we obtain a chain map $f_. \otimes 1: S(X, Y) \otimes G \to S(A, B) \otimes G$, and the induced homology homomorphisms $f_*: H_n(X, Y; G) \to H_n(A, B; G)$ are defined by $f_* = (f_. \otimes 1)_*$. Just as in Section 4.2, we can prove that $1_* = 1$, that $(gf)_* = g_* f_*$, and that $f_* = g_*$ if $f \simeq g$. Indeed, these results follow immediately from Proposition 4.2.16 and 4.2.21, in virtue of Proposition 4.5.6. Similarly, we can deduce from the results of Section 4.3 that, if (K, L) is a simplicial pair, then $H_*(|K|, |L|; G)$ is also the homology of the chain complexes $\Delta(K, L) \otimes G$ and $C(K, L) \otimes G$.

By Example 4.5.3, $S(X, Y) \otimes Z$ may be identified with $S(X, Y)$, so that $H_*(X, Y; Z)$ is what we have previously called $H_*(X, Y)$: we shall continue to omit the coefficient group if it happens to be Z. Moreover, since $S(X, Y)$ is a direct sum of copies of Z, one for each singular simplex in X whose image is not contained in Y, it follows using Proposition 4.5.7 as well that, for any abelian group G, $S(X, Y) \otimes G$ is the corresponding direct sum of copies of G; hence in particular the sequence of chain complexes

$$0 \longrightarrow S(Y) \otimes G \xrightarrow{i_. \otimes 1} S(X) \otimes G \xrightarrow{j \otimes 1} S(X, Y) \otimes G \longrightarrow 0$$

is exact. Thus the results of Section 4.4 all hold in the corresponding versions for homology with coefficients in G (though see Exercise 13 in the case of the 'block dissection' calculation theorem).

The reader is warned, however, that although $H_n(X, Y; G)$ is defined to be $H_n(S(X, Y) \otimes G)$, it does not follow that $H_n(X, Y; G) = H_n(X, Y) \otimes G$. For example, it was proved in Example 4.3.17 that $H_0(RP^2) \cong Z$, $H_1(RP^2) \cong Z_2$, and $H_2(RP^2) = 0$; but the same method (or that of Theorem 4.4.24 or Example 4.4.25) will show that $H_0(RP^2; Z_2) \cong H_1(RP^2; Z_2) \cong H_2(RP^2; Z_2) \cong Z_2$. Nevertheless it is possible to calculate $H_*(X, Y; G)$ by a purely algebraic process from $H_*(X, Y)$. Since it is rather complicated, we shall not prove the general theorem here (though see Exercise 16), but will confine our attention to the most useful special cases $G = Q$ or Z_p. In this context, it is worth noting that there are particular advantages in taking coefficients in a field.

Proposition 4.5.10 *Let C be a chain complex and let F be a field. Then $H(C \otimes F)$ is a vector space over F. Moreover if $g: C \to D$ is a chain map, the induced homology homomorphism $g_*: H(C \otimes F) \to H(D \otimes F)$ is a linear map of vector spaces.*

Proof. We show first that $C \otimes F$ is a vector space over F. It is necessary only to define an action of F on $C \otimes F$, and this can be done by setting $f(c \otimes f') = c \otimes (ff')$. This obviously makes $C \otimes F$ into a vector space, and also

$$\begin{aligned} f[(\partial \otimes 1)(c \otimes f')] &= f(\partial c \otimes f') \\ &= \partial c \otimes ff' \\ &= (\partial \otimes 1)[f(c \otimes f')], \end{aligned}$$

so that $\partial \otimes 1$ is a linear map, and hence $H(C \otimes F)$, being a quotient of a subspace of $C \otimes F$, is also a vector space over F. A similar argument shows that $g \otimes 1$ is a linear map, so that the same is true of g_*. ∎

Corollary 4.4.11 *Given a pair (X, Y) and a field F, then each $H_n(X, Y; F)$ is a vector space over F. Moreover if $g: (X, Y) \to (A, B)$ is a continuous map, then $g_*: H_*(X, Y; F) \to H_*(A, B; F)$ is a linear map.* ∎

We next establish the results on the relation between homology with coefficients in Q or Z_p and ordinary homology, that is, homology with coefficients in Z. As usual, the abstract situation is considered first.

In order to deal with coefficients in Q, an algebraic lemma is necessary.

Lemma 4.5.12 *Let g be an element of an abelian group G. If $g \otimes 1 = 0$ in $G \otimes Q$, there exists an integer $n \neq 0$ such that $ng = 0$ in G.*

Proof. In the *free* abelian group generated by the symbols $g \otimes q$ ($g \in G$, $q \in Q$), $g \otimes 1$ is a finite sum of elements of the form

$$g_1 \otimes (q_1 + q_2) - g_1 \otimes q_1 - g_1 \otimes q_2$$

or

$$(g_1 + g_2) \otimes q_1 - g_1 \otimes q_1 \otimes g_2 \otimes q_1.$$

Thus if G_0 is the subgroup of G generated by the elements g_1 and g_2 that occur in this finite sum, G_0 is a finitely generated abelian group that contains g, and $g \otimes 1 = 0$ in $G_0 \otimes Q$. But now Theorem 1.3.30 can be applied to G_0, and then Examples 4.5.3 and 4.5.5 show that there exists n such that $ng = 0$ in G_0, and hence in G. ∎

Theorem 4.5.13 *Let C be any chain complex. Then for each n, $H_n(C \otimes Q) \cong H_n(C) \otimes Q$.*

Proof. We first show that if $F \xrightarrow{\alpha} G \xrightarrow{\beta} H$ is an exact sequence of abelian groups, then so is

$$F \otimes Q \xrightarrow{\alpha \otimes 1} G \otimes Q \xrightarrow{\beta \otimes 1} H \otimes Q.$$

To prove this, note first that $(\beta \otimes 1)(\alpha \otimes 1) = 0$, so that $\operatorname{Im}(\alpha \otimes 1) \subset \operatorname{Ker}(\beta \otimes 1)$. Conversely, let $\sum (g_i \otimes q_i) \in \operatorname{Ker}(\beta \otimes 1)$, so that $\sum \beta g_i \otimes q_i = 0$. Since this is a finite sum, there exists an integer $m \neq 0$ such that each mq_i is an integer, so that $\sum \beta(mq_i g_i) \otimes 1/m = 0$, or $\sum \beta(mq_i g_i) \otimes 1 = 0$. Hence by Lemma 4.5.12 there exists an integer n such that $\beta(\sum nmq_i g_i) = 0$ in H, so that $\sum nmq_i g_i = \alpha(f)$ for some $f \in F$. But then

$$\begin{aligned} \sum g_i \otimes q_i &= \sum nmq_i g_i \otimes 1/nm \\ &= \alpha(f) \otimes 1/nm \\ &\in \operatorname{Im}(\alpha \otimes 1). \end{aligned}$$

Thus $\operatorname{Ker}(\beta \otimes 1) \subset \operatorname{Im}(\alpha \otimes 1)$, and the sequence is exact.

To return to the chain complex C, the exact sequence

$$0 \longrightarrow Z_n \longrightarrow C_n \xrightarrow{\partial} C_{n-1}$$

yields the exact sequence

$$0 \longrightarrow Z_n \otimes Q \longrightarrow C_n \otimes Q \xrightarrow{\partial \otimes 1} C_{n-1} \otimes Q,$$

so that $Z_n(C \otimes Q) = \text{Ker}\,(\partial \otimes 1) = Z_n \otimes Q$. Hence the exact sequence

$$0 \longrightarrow Z_n \longrightarrow C_n \xrightarrow{\partial} B_{n-1} \longrightarrow 0$$

yields

$$0 \longrightarrow Z_n(C \otimes Q) \longrightarrow C_n \otimes Q \xrightarrow{\partial \otimes 1} B_{n-1} \otimes Q \longrightarrow 0,$$

so that $B_{n-1}(C \otimes Q) \cong B_{n-1} \otimes Q$. Finally, a similar argument applied to the exact sequence

$$0 \to B_n \to Z_n \to H_n(C) \to 0$$

shows that $H_n(C \otimes Q) \cong H_n(C) \otimes Q$. ∎

In particular $H_n(X, Y; Q) \cong H_n(X, Y) \otimes Q$ for a pair (X, Y). Thus if (K, L) is a simplicial pair, and $H_n(K, L)$ is a direct sum of m copies of Z and a finite group, $H_n(K, L; Q)$ is a direct sum of m copies of Q. That is, $H_n(K, L; Q)$ measures the 'free part' of $H_n(K, L)$.

The case of coefficients Z_p is dealt with by a rather different method. Once again, an algebraic lemma is necessary first.

Lemma 4.5.14 *Let G be an abelian group, and let $\alpha: G \to G$ be the homomorphism defined by $\alpha(g) = pg$. There is an exact sequence*

$$0 \longrightarrow \text{Ker}\,\alpha \longrightarrow G \xrightarrow{\alpha} G \xrightarrow{\beta} G \otimes Z_p \longrightarrow 0,$$

where $\beta(g) = g \otimes 1$.

Proof. Clearly β is onto, so it is necessary only to show that $\text{Im}\,\alpha = \text{Ker}\,\beta$. Now $\beta\alpha(g) = pg \otimes 1 = g \otimes p = 0$, so that $\text{Im}\,\alpha \subset \text{Ker}\,\beta$. On the other hand we can define a homomorphism $\gamma: G \otimes Z_p \to G/\text{Im}\,\alpha$ by $\gamma(g \otimes n) =$ the coset $[ng]$, where $n \in Z_p$; this is unambiguous, since $[pg] = 0$. Moreover $\gamma\beta$ is just the 'quotient homomorphism' $G \to G/\text{Im}\,\alpha$, so that if $g \in \text{Ker}\,\beta$, $\gamma\beta(g) = 0$ in $G/\text{Im}\,\alpha$, and hence $g \in \text{Im}\,\alpha$. Thus $\text{Ker}\,\beta \subset \text{Im}\,\alpha$, and the sequence is exact. ∎

It is usual to write $\text{Tor}\,(G, Z_p)$ for $\text{Ker}\,[\alpha: G \to G]$. The reader should have no difficulty in showing that $\text{Tor}\,(Z, Z_p) = 0$, that $\text{Tor}\,(Z_p, Z_q) \cong Z_{(p,q)}$, and that $\text{Tor}\,(\bigoplus_i G_i, Z_p) \cong \bigoplus_i \text{Tor}\,(G_i, Z_p)$; this suffices to calculate $\text{Tor}\,(G, Z_p)$ for any finitely generated abelian group G.

If C is a chain complex in which each C_n is a free abelian group (as it is, of course, if $C = S(X, Y)$), there is an exact sequence of chain complexes

$$0 \longrightarrow C \xrightarrow{\alpha} C \xrightarrow{\beta} C \otimes Z_p \longrightarrow 0.$$

By Theorem 4.4.2 this leads to an exact sequence

$$\cdots \longrightarrow H_n(C) \xrightarrow{\alpha_*} H_n(C) \xrightarrow{\beta_*} H_n(C \otimes Z_p) \xrightarrow{\partial_*} H_{n-1}(C) \longrightarrow \cdots,$$

where again $\alpha_*(x) = px$ for each $x \in H_n(C)$. This sequence is called the *exact coefficient sequence* associated with Z_p, and ∂_* is known as a *Bockstein boundary homomorphism.* The exact coefficient sequence can be broken up into short exact sequences, one for each n:

$$0 \to H_n(C)/\text{Ker}\,\beta_* \to H_n(C \otimes Z_p) \to \text{Im}\,\partial_* \to 0,$$

where the homomorphisms are induced by β_* and ∂_*. Now $\text{Ker}\,\beta_* = \text{Im}\,\alpha_*$, so that $H_n(C)/\text{Ker}\,\beta_* \cong H_n(C) \otimes Z_p$, by Lemma 4.5.14; also $\text{Im}\,\partial_* = \text{Ker}\,\alpha_* = \text{Tor}\,(H_{n-1}(C), Z_p)$. So we have (almost) proved

Theorem 4.5.15 *Let C be a chain complex in which each C_n is a finitely-generated free abelian group, and let p be any positive integer. Then*

$$H_n(C \otimes Z_p) \cong H_n(C) \otimes Z_p \oplus \text{Tor}\,(H_{n-1}(C), Z_p).$$

Proof. Let r be any divisor of p, so that $p = rs$, say. There is a commutative diagram

$$\begin{array}{ccccccccc}
0 & \longrightarrow & C & \xrightarrow{\alpha} & C & \xrightarrow{\beta} & C \otimes Z_p & \longrightarrow & 0 \\
 & & \downarrow{\scriptstyle \lambda} & & \downarrow{\scriptstyle 1} & & \downarrow{\scriptstyle \mu} & & \\
0 & \longrightarrow & C & \xrightarrow[\bar{\alpha}]{} & C & \xrightarrow[\bar{\beta}]{} & C \otimes Z_r & \longrightarrow & 0
\end{array}$$

in which $\lambda(c) = sc$, $\bar{\alpha}(c) = rc$, and μ may be regarded as the 'quotient homomorphism' $C/\text{Im}\,\alpha \to C/\text{Im}\,\bar{\alpha}$. This gives rise to a commutative diagram

$$\begin{array}{ccccccccc}
\cdots \longrightarrow & H_n(C) & \xrightarrow{\alpha_*} & H_n(C) & \xrightarrow{\beta_*} & H_n(C \otimes Z_p) & \xrightarrow{\partial_*} & H_{n-1}(C) & \longrightarrow \cdots \\
 & \downarrow{\scriptstyle \lambda_*} & & \downarrow{\scriptstyle 1} & & \downarrow{\scriptstyle \mu_*} & & \downarrow{\scriptstyle \lambda_*} & \\
\cdots \longrightarrow & H_n(C) & \xrightarrow[\bar{\alpha}_*]{} & H_n(C) & \xrightarrow[\bar{\beta}_*]{} & H_n(C \otimes Z_r) & \xrightarrow[\bar{\partial}_*]{} & H_{n-1}(C) & \longrightarrow \cdots,
\end{array}$$

and hence, by the above discussion, to a commutative diagram (of exact sequences)

$$\begin{array}{ccccccccc}
0 & \longrightarrow & H_n(C) \otimes Z_p & \xrightarrow{\beta_*} & H_n(C \otimes Z_p) & \xrightarrow{\partial_*} & \text{Tor}\,(H_{n-1}(C), Z_p) & \longrightarrow & 0 \\
 & & \downarrow{\scriptstyle \nu} & & \downarrow{\scriptstyle \mu_*} & & \downarrow & & \\
0 & \longrightarrow & H_n(C) \otimes Z_r & \xrightarrow[\bar{\beta}_*]{} & H_n(C \otimes Z_r) & \xrightarrow[\bar{\partial}_*]{} & \text{Tor}\,(H_{n-1}(C), Z_r) & \longrightarrow & 0,
\end{array}$$

where again ν may be regarded as the quotient homomorphism $H_n(C)/\text{Im}\, \alpha_* \to H_n(C)/\text{Im}\, \bar{\alpha}_*$. Now it is clear from Lemma 4.5.14 that the order of every element of Tor $(H_{n-1}(C), Z_p)$ and $C_n \otimes Z_p$ (and hence $H_n(C \otimes Z_p)$) is a divisor of p. In particular, by Theorem 1.3.30, Tor $(H_{n-1}(C), Z_p)$ is a direct sum of groups Z_r, for various divisors r of p; and if x is a generator of one of these, say Z_r, we can write $x = \partial_*(y)$ for some $y \in H_n(C \otimes Z_p)$. Then $\partial_*(ry) = rx = 0$, so that $ry = \beta_*(z)$ for some $z \in H_n(C) \otimes Z_p$. So $\beta_*\nu(z) = \mu_*\beta_*(z) = r\mu_*(y) = 0$, and hence $\nu(z) = 0$ and $z = rt$ for some $t \in H_n(C) \otimes Z_p$. To sum up, we have $x = \partial_*(y - \beta_*(t))$, where $r(y - \beta_*(t)) = 0$. Hence by defining $\gamma(x) = y - \beta_*(t)$, and making this construction for each generator of Tor $(H_{n-1}(C), Z_p)$, there results a homomorphism γ: Tor $(H_{n-1}(C), Z_p) \to H_n(C \otimes Z_p)$ such that $\partial_*\gamma = 1$. Consequently, by Proposition 1.3.36,

$$H_n(C \otimes Z_p) \cong H_n(C) \otimes Z_p \oplus \text{Tor}\,(H_{n-1}(C), Z_p). \quad \blacksquare$$

Note. The restriction in Theorem 4.5.15, that each C_n should be finitely generated, is not really necessary, but is included in order to simplify the proof. In fact the result of Theorem 4.5.15 is true in much greater generality: see Exercise 16.

It follows from Theorem 4.5.15 that, for any simplicial pair (K, L) and any integer $p \geqslant 2$, we have

$$H_n(K, L; Z_p) \cong H_n(K, L) \otimes Z_p \oplus \text{Tor}\,(H_{n-1}(K, L), Z_p).$$

As an example of the use of homology with coefficients other than the integers, we end this section with a proof of the Lefschetz Fixed-Point Theorem. Suppose given a simplicial complex K, and a continuous map $f: |K| \to |K|$; then by Corollary 4.5.11 each

$$f_*^n: H_n(K; Q) \to H_n(K; Q)$$

is a linear map of finite-dimensional vector spaces (the superfix n on f_* indicates the n-dimensional component).

Definition 4.5.16 The *Lefschetz number* $L(f)$ of the map $f: |K| \to |K|$ is defined by $L(f) = \sum_{n=0}^{\infty} (-1)^n \,\text{tr}\,(f_*^n)$. (Recall Proposition 1.3.59. This is only a finite sum, since $f_*^n = 0$ for $n > \dim K$.)

The main result about the Lefschetz number is that, if $L(f) \neq 0$, then f has a fixed point. A lemma is necessary here.

Lemma 4.5.17 *If $f_n: C_n(K) \otimes Q \to C_n(K) \otimes Q$ are the components of a chain map that induces f_*, then $L(f) = \sum (-1)^n \,\text{tr}\,(f_n)$.*

Proof. Write C_n for $C_n(K) \otimes Q$, and Z_n, B_n for the corresponding groups of cycles and boundaries. Now $f_n: C_n \to C_n$ restricts to $f'_n: Z_n \to Z_n$, $f''_n: B_n \to B_n$, and induces $\bar{f}_n: C_n/Z_n \to C_n/Z_n$, $f^n_*: Z_n/B_n \to Z_n/B_n$. By Proposition 1.3.60, we have

$$\begin{aligned} \operatorname{tr}(f_n) &= \operatorname{tr}(f'_n) + \operatorname{tr}(\bar{f}_n) \\ &= \operatorname{tr}(f''_n) + \operatorname{tr}(f^n_*) + \operatorname{tr}(\bar{f}_n). \end{aligned}$$

But $\partial \otimes 1$ induces an isomorphism $\bar{\partial}: C_n/Z_n \to B_{n-1}$, and $f''_{n-1}\partial = \bar{\partial}f_n$. Hence $\operatorname{tr}(\bar{f}_n) = \operatorname{tr}(f''_{n-1})$, so that

$$L(f) = \sum(-1)^n \operatorname{tr}(f^n_*) = \sum(-1)^n \operatorname{tr}(f_n). \blacksquare$$

In particular, Lemma 4.5.17 can be applied to the identity map $1: |K| \to |K|$, to show that $L(1) = \sum(-1)^n \alpha_n$, where α_n is the number of n-simplexes of K. Thus $\sum(-1)^n\alpha_n$ depends only on the homotopy type of $|K|$; it is usually called the *Euler–Poincaré characteristic* of K, and written $\chi(|K|)$.

Theorem 4.5.18 *Given a map $f: |K| \to |K|$ without fixed points, then $L(f) = 0$.*

Proof. Suppose that K is in some Euclidean space R^m, and let d be the metric in R^m. Since $|K|$ is compact and f has no fixed point, $d(x, f(x))$ attains a greatest lower bound $\delta > 0$, say, as x runs over all points of $|K|$. Take an integer n such that mesh $K^{(n)} < \frac{1}{2}\delta$, and let $g: |K^{(n+t)}| \to |K^{(n)}|$ be a simplicial approximation to $f: |K^{(n)}| \to |K^{(n)}|$. Now if $h: |K^{(n+t)}| \to |K^{(n)}|$ is a simplicial approximation to the identity, $g \simeq f \simeq fh$, so that $g_* = f_* h_*$. But it was remarked after Corollary 4.3.10 that h_* is the inverse isomorphism to ϕ^t_*, where ϕ is the subdivision chain map; hence $f_* = g_*\phi^t_*$, so that $g_{.}\phi^t: C(K^{(n)}) \to C(K^{(n)})$ is a chain map that induces f_*.

By Lemma 4.5.17, it is sufficient now to prove that, for each simplex σ of $K^{(n)}$, $g_{.}\phi^t(\sigma)$ is a linear combination of simplexes other than σ, for then each $\operatorname{tr}(f_r)$ is zero. Suppose, if possible, that σ is a simplex such that $g_{.}\phi^t(\sigma)$ does contain σ. Then since $\phi^t(\sigma)$ is a linear combination of simplexes that are all contained in σ, it follows that at least one of these must be mapped by g back to σ, and so there is a point $x \in \sigma$ such that $g(x) \in \sigma$ also, that is, $d(x, g(x)) \leqslant$ mesh $K^{(n)} < \frac{1}{2}\delta$. But the proof of Theorem 2.5.3 shows that $f(x)$ and $g(x)$ are both in some simplex of $K^{(n)}$, so that $d(f(x), g(x)) < \frac{1}{2}\delta$. Hence $d(x, f(x)) \leqslant d(x, g(x)) + d(f(x), g(x)) < \delta$, which contradicts the definition of δ. $\blacksquare$

Corollary 4.5.19 *If $L(f) \neq 0$, f has a fixed point.* ∎

We conclude with some examples of the use of Theorem 4.5.18.

Proposition 4.5.20 *Let K be a simplicial complex such that $|K|$ is path-connected, and $H_n(K)$ is a finite group for each $n > 0$, and let $f\colon |K| \to |K|$ be a continuous map. Then f has a fixed point.*

Proof. By Theorem 4.5.13 and Example 4.5.5,

$$H_n(K; Q) \cong H_n(K) \otimes Q \cong \begin{cases} Q, & n = 0 \\ 0, & \text{otherwise.} \end{cases}$$

And since any vertex of K will do as a representative for a generator of $H_0(K)$, $f_*\colon H_0(K; Q) \to H_0(K; Q)$ is the identity isomorphism. Hence $L(f) = 1$. ∎

Proposition 4.5.20 provides another proof of Theorem 2.5.23; but a rather more interesting example is that any map $f\colon RP^n \to RP^n$ has a fixed point if n is even: for by Example 4.4.25 $H_r(RP^n)$ is either 0 or Z_2 $(r > 0)$ if n is even. That this result need not be true if n is odd is shown, for example, by identifying RP^1 with S^1, and taking the map of S^1 that just rotates it through an angle π.

Finally, Theorem 4.5.18 provides an alternative proof of Proposition 4.4.12: if $f\colon S^n \to S^n$ $(n > 0)$ satisfies $f_*(\sigma_n) \to d\sigma_n$, then

$$f_*\colon H_n(S^n; Q) \to H_n(S^n; Q)$$

is multiplication by d, whereas $f_*\colon H_0(S^n; Q) \to H_0(S^n; Q)$ is the identity isomorphism. Hence $L(f) = 1 + (-1)^n d$, which is zero if and only if $d = (-1)^{n+1}$.

EXERCISES

1. Show that the relation of being chain-homotopic is an equivalence relation on the set of chain maps from a chain complex C to a chain complex D. C and D are said to be *chain-equivalent*, and $f\colon C \to D$ is a *chain equivalence*, if there exists a chain map $g\colon D \to C$ such that gf and fg are chain-homotopic to the respective identity chain isomorphisms; prove that this sets up an equivalence relation on any set of chain complexes.
2. Given chain complexes C and D, write $[C, D]$ for the set of chain homotopy classes of chain maps from C to D. Show that $[C, D]$ can be made into an abelian group by defining $(f + g)(c) = f(c) + g(c)$ for chain maps f and g, and extending this definition to equivalence classes. Prove also

(a) $[C, D]$ depends, up to isomorphism, only on the chain equivalence classes of C and D;
(b) if X and Y are spaces then $[S(X), S(Y)]$ depends only on the homotopy types of X and Y;
(c) if C is a chain complex in which C_n is a free abelian group with one generator, and $C_r = 0$ if $r \neq n$, then $[C, D] \cong H_n(D)$.

3. Let X be a path-connected space, with base point x_0. Show that a homomorphism $h: \pi_1(X, x_0) \to H_1(X)$ can be defined by sending a loop u to the corresponding singular 1-simplex u. Prove that h is onto, and that its kernel is the commutator subgroup $[\pi, \pi]$ of $\pi_1(X, x_0)$: thus $H_1(X)$ is isomorphic to $\pi_1(X, x_0)$ 'made abelian'. (*Hint:* first prove that h is onto, and $[\pi, \pi] \subset \operatorname{Ker} h$, so that h induces a homomorphism $\bar{h}: \pi_1(X, x_0)/[\pi, \pi] \to H_1(X)$; now show that $\bar{h}$ is (1-1), by showing that, if u is a loop corresponding to $\partial(c) \in B_1(X)$, the singular 2-simplexes in c can be used to construct a loop v such that $[v] = 1$ in $\pi_1(X, x_0)$ and $[u] = [v]$ in $\pi_1(X, x_0)/[\pi, \pi]$.)
4. Show that an m-manifold cannot be homeomorphic to an n-manifold unless $m = n$.
5. Given a simplicial pair (K, L), let $K \cup CL$ be the simplicial complex obtained by 'adding a cone' to L, that is, by forming the union of K and $CL = L * a$, where a is a single vertex. Show that there is an isomorphism $\alpha: H_*(K, L) \to \tilde{H}^*(K \cup CL)$, such that, if $f: (|K|, |L|) \to (|M|, |N|)$ is a map of pairs and $\bar{f}: |K \cup CL| \to |M \cup CN|$ is the obvious map formed from f and $f * 1$, then the diagram

$$\begin{array}{ccc} H_*(K, L) & \xrightarrow{f_*} & H_*(M, N) \\ {\scriptstyle\alpha}\downarrow & & \downarrow{\scriptstyle\alpha} \\ \tilde{H}_*(K \cup CL) & \xrightarrow[\bar{f}_*]{} & \tilde{H}_*(M \cup CN) \end{array}$$

is commutative.
6. Given a 'quadruple' of spaces (X, Y, Z, W), show that the following diagram is commutative, where the homomorphisms are those in the exact homology sequences of (X, Y, Z), (X, Y, W) and (Y, Z, W).

$$\begin{array}{ccccccccc} & & & & H_n(X, W) & & & & \\ & & & {\scriptstyle j_*}\swarrow & & \searrow{\scriptstyle i_*} & & & \\ H_n(Y, Z) & \xrightarrow{i_*} & H_n(X, Z) & & \xrightarrow{j_*} & & H_n(X, Y) & \xrightarrow{\partial_*} & H_{n-1}(Y, Z) \\ & {\scriptstyle\partial_*}\searrow & & \swarrow{\scriptstyle\partial_*} & & & & {\scriptstyle\partial_*}\searrow \quad \nearrow{\scriptstyle j_*} & \\ & & H_{n-1}(Z, W) & & & & & H_{n-1}(Y, W) & \end{array}$$

7. Let (K, L) be a simplicial pair. Show that there is a suspension isomorphism $s_*: H_n(K, L) \to H_{n+1}(SK, SL)$, such that if $f: (|K|, |L|) \to$

$(|M|, |N|)$ is a map of pairs, then $s_* f_* = (Sf)_* s_*$. Prove also that the diagram

$$\begin{array}{ccccccccc}
\cdots \longrightarrow & \tilde{H}_n(L) & \xrightarrow{i_*} & \tilde{H}_n(K) & \xrightarrow{j_*} & H_n(K, L) & \xrightarrow{\partial_*} & \tilde{H}_{n-1}(L) & \rightarrow \cdots \\
& \downarrow s_* & & \downarrow s_* & & \downarrow s_* & & \downarrow s_* & \\
\cdots \rightarrow & \tilde{H}_{n+1}(SL) & \xrightarrow[i_*]{} & \tilde{H}_{n+1}(SK) & \xrightarrow[j_*]{} & H_{n+1}(SK, SL) & \xrightarrow[\partial_*]{} & \tilde{H}_n(SL) & \rightarrow \cdots,
\end{array}$$

where the rows are the exact homology sequences of the pairs (K, L) and (SK, SL), is commutative.

8. Let (K, L, M) be a simplicial triple, in which K has a block dissection and L and M are block subcomplexes. Show that in the exact homology sequence of the triple

$$\cdots \rightarrow H_n(L, M) \xrightarrow{i_*} H_n(K, M) \xrightarrow{j_*} H_n(K, L) \xrightarrow{\partial_*} H_{n-1}(L, M) \rightarrow \cdots,$$

the homomorphisms i_*, j_* and ∂_* may be calculated by using the blocks and the 'block chain complexes', in exactly the same way that these homomorphisms are defined using the simplicial chain complexes.

9. Let p and q be coprime integers, with $p \geqslant 2$. The *Lens space* $L(p, q)$ is the space obtained from E^3 by making identifications on the boundary S^2, as follows. Divide the equator S^1 into p equal parts by vertices $a^0, a^1, \ldots, a^{p-1}$, and by joining to the 'poles' $a = (0, 0, 1)$, $b = (0, 0, -1)$, divide S^2 into $2p$ 'triangles': see Fig. 4.6.

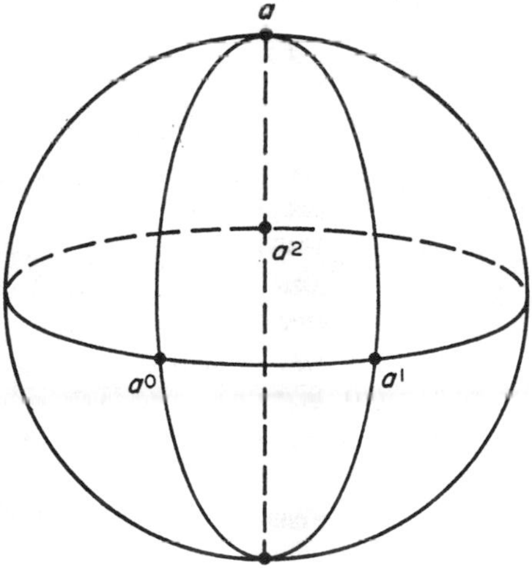

Fig. 4.6

$L(p, q)$ is the space obtained by identifying each triangle $aa^r a^{r+1}$ with $ba^{r+q}a^{r+q+1}$, so that a and b are identified, as also are a^r and a^{r+q},

a^{r+1} and a^{r+q+1} (the superfixes r, $r + q$, etc., are interpreted as elements of Z_p).

Show that $L(p, q)$ is triangulable, and that $L(p, q)$ is homeomorphic to $L(p, q')$ if $q = -q'$ or if $q - q'$ is divisible by p; also that $L(2, 1)$ is homeomorphic to RP^3. Show also that $L(p, q)$ is homeomorphic to $L(p, q')$ if $qq' = 1 \pmod p$. (*Hint:* cut E^3 into 'tetrahedra' $aa^r a^{r+1} b$, and reassemble by identifying the triangles on S^2 as above: this produces E^3 again, but with the line formed by identifying together all edges $a^r a^{r+1}$ taking the place of ab. $L(p, q)$ is still the space obtained by making certain identifications on the boundary of the new E^3.)

By using a suitable block dissection, show that

$$H_1(L(p, q)) \cong Z_p, \quad H_2(L(p, q)) = 0 \quad \text{and} \quad H_3(L(p, q)) \cong Z.$$

10. Show that $(A \otimes B) \otimes C \cong A \otimes (B \otimes C)$ for any three abelian groups A, B and C.
11. Show that a homomorphism $\theta: G \to H$ between abelian groups gives rise to homomorphisms $\theta_*: H_n(X, Y; G) \to H_n(X, Y; H)$, for any pair (X, Y), such that if $f: (X, Y) \to (A, B)$ is a map of pairs, $\theta_* f_* = f_* \theta_*$.
12. Given an exact sequence of abelian groups

$$0 \to F \xrightarrow{\alpha} G \xrightarrow{\beta} H \to 0,$$

and a pair (X, Y), show that there is an exact sequence

$$\cdots \to H_n(X, Y; F) \xrightarrow{\alpha_*} H_n(X, Y; G) \xrightarrow{\beta_*} H_n(X, Y; H) \xrightarrow{\partial_*} H_{n-1}(X, Y; F) \to \cdots$$

(this is the *exact coefficient sequence*, and ∂_* is the *Bockstein boundary homomorphism*, associated with the exact sequence $0 \to F \xrightarrow{\alpha} G \xrightarrow{\beta} H \to 0$).

13. Let K be a simplicial complex with a block dissection, and let L be a block subcomplex. Let C be the chain complex $\bigoplus C_n$, where $C_n = H_n(M^n, M^{n-1})$, and $M^n = K_n \cup L$; and for any abelian group G, let $C(G) = \bigoplus C_n(G)$ be the chain complex similarly defined by $C_n(G) = H_n(M^n, M^{n-1}; G)$. Show that $C(G)$ and $C \otimes G$ are chain-isomorphic, so that although there appear to be two generalizations to arbitrary coefficients of the method of calculating homology from block dissections; in fact these generalizations coincide. (*Hint:* define a homomorphism $\alpha: H_n(X, Y) \otimes G \to H_n(X, Y; G)$, which is an isomorphism if $(X, Y) = (E^n, S^{n-1})$.)
14. Given an exact sequence $0 \to A \xrightarrow{\alpha} B \xrightarrow{\beta} C \to 0$ of abelian groups, and another abelian group G, show that the sequence

$$A \otimes G \xrightarrow{\alpha \otimes 1} B \otimes G \xrightarrow{\beta \otimes 1} C \otimes G \longrightarrow 0$$

is exact, but (for example, by taking $A = B = Z$, $C = G = Z_2$) the homomorphism $\alpha \otimes 1$ need not be (1-1). Show, however, that if the sequence $0 \to A \xrightarrow{\alpha} B \xrightarrow{\beta} C \to 0$ is a split exact sequence, in the sense of Proposition 1.3.36, then

$$0 \longrightarrow A \otimes G \xrightarrow{\alpha \otimes 1} B \otimes G \xrightarrow{\beta \otimes 1} C \otimes G \longrightarrow 0$$

is also a split exact sequence. (*Hint:* to show that $\text{Ker}(\beta \otimes 1) \subset \text{Im}(\alpha \otimes 1)$, construct a homomorphism $\gamma: C \otimes G \to B \otimes G/\text{Im}(\alpha \otimes 1)$, such that $\gamma\bar{\beta} = 1$, where $\bar{\beta}: B \otimes G/\text{Im}(\alpha \otimes 1) \to C \otimes G$ is the homomorphism induced by $\beta \otimes 1$; hence $\bar{\beta}$ is (1-1) and hence an isomorphism.)

15. Given abelian groups A and B, write A in the form F/R, where F is a free abelian group, so that there is an exact sequence

$$0 \to R \xrightarrow{\alpha} F \xrightarrow{\beta} A \to 0.$$

Define $\text{Tor}(A, B) = \text{Ker}(\alpha \otimes 1)$, so that by Exercise 14 there is an exact sequence

$$0 \longrightarrow \text{Tor}(A, B) \longrightarrow R \otimes B \xrightarrow{\alpha \otimes 1} F \otimes B \xrightarrow{\beta \otimes 1} A \otimes B \longrightarrow 0.$$

Establish the following properties of $\text{Tor}(A, B)$.

(a) Given another abelian group $A' = F'/R'$, and a homomorphism $f: A \to A'$, there exists a unique homomorphism $\bar{f}: \text{Tor}(A, B) \to \text{Tor}(A', B)$ such that the diagram

$$\begin{array}{ccccccccc}
0 \longrightarrow & \text{Tor}(A, B) & \longrightarrow & R \otimes B & \xrightarrow{\alpha \otimes 1} & F \otimes B & \xrightarrow{\beta \otimes 1} & A \otimes B & \longrightarrow 0 \\
& \downarrow{\scriptstyle \bar{f}} & & \downarrow{\scriptstyle f_2 \otimes 1} & & \downarrow{\scriptstyle f_1 \otimes 1} & & \downarrow{\scriptstyle f \otimes 1} & \\
0 \longrightarrow & \text{Tor}(A', B) & \longrightarrow & R' \otimes B & \xrightarrow[\alpha' \otimes 1]{} & F' \otimes B & \xrightarrow[\beta' \otimes 1]{} & A' \otimes B & \longrightarrow 0
\end{array}$$

is commutative, where f_1 and f_2 are any homomorphisms that make the diagram

$$\begin{array}{ccccccc}
0 \longrightarrow & R & \xrightarrow{\alpha} & F & \xrightarrow{\beta} & A & \longrightarrow 0 \\
& \downarrow{\scriptstyle f_2} & & \downarrow{\scriptstyle f_1} & & \downarrow{\scriptstyle f} & \\
0 \longrightarrow & R' & \xrightarrow[\alpha']{} & F' & \xrightarrow[\beta']{} & A' & \longrightarrow 0
\end{array}$$

commutative (such homomorphisms exist by Proposition 1.3.37).

(b) $\text{Tor}(A, B)$ depends only, up to isomorphism, on the groups A and B, and not on the particular representation $A = F/R$.

(c) $\text{Tor}(\bigoplus_i A_i, \bigoplus_j B_j) \cong \bigoplus_{i,j} \text{Tor}(A_i, B_j)$.

(For the proof that this definition of $\text{Tor}(A, B)$ coincides with that of Section 4.5 if $B = Z_p$, see Exercise 17.)

16. Let C be a chain complex in which each C_n is a free abelian group, and let G be any abelian group. Prove that

$$H_n(C \otimes G) \cong H_n(C) \otimes G \oplus \text{Tor}\,(H_{n-1}(C), G).$$

(*Hint:* use the theorem that any subgroup of a free abelian group is a free abelian group to deduce from Exercise 14 that

$$0 \longrightarrow Z_n(C) \otimes G \xrightarrow{\alpha_n \otimes 1} C_n \otimes G \xrightarrow{\partial \otimes 1} B_{n-1}(C) \otimes G \longrightarrow 0$$

is a split exact sequence, and deduce that $H_n(C \otimes G) \cong Z_n(C) \otimes G/\text{Im}\,(\beta_n \otimes 1) \oplus \text{Ker}\,(\beta_{n-1} \otimes 1)$, where $\beta_n: B_n(C) \to Z_n(C)$ is the inclusion homomorphism. Then use the exact sequence

$$0 \longrightarrow B_{n-1}(C) \xrightarrow{\beta_{n-1}} Z_{n-1}(C) \longrightarrow H_{n-1}(C) \longrightarrow 0$$

to define $\text{Tor}\,(H_{n-1}(C), G)$.)

17. Deduce from Exercise 16 and the exact sequence just before Theorem 4.5.15 that $\text{Tor}\,(Z_p, H_n(C)) \cong \text{Tor}\,(H_n(C), Z_p)$, for any p, and any chain complex C in which each C_n is a free abelian group. By constructing a suitable chain complex, hence prove that $\text{Tor}\,(A, Z_p) \cong \text{Tor}\,(Z_p, A)$ for any abelian group A. ($\text{Tor}\,(Z_p, A)$, in the sense of Exercise 15, is clearly what was called $\text{Tor}\,(A, Z_p)$ in Section 4.5.)

18. Given a simplicial complex K and a field F, let h_n be the dimension of $H_n(K; F)$, as a vector space over F. The *Euler–Poincaré characteristic* of K, with coefficients in F, $\chi(|K|; F)$, is defined to be $\sum_{n=0}^{\infty} (-1)^n h_n$; show that $\chi(|K|; F) = \chi(|K|)$.

NOTES ON CHAPTER 4

Homology groups. The homology groups of a polyhedron were first introduced by Poincaré [116], and the generalization to the singular homology groups of an arbitrary topological space was made by Lefschetz [90] and Eilenberg [50] (although the basic idea is contained in Veblen [147]). Relative homology groups were introduced by Lefschetz [88]. The proof of the homotopy-type invariance of homology was first given by Alexander [7, 9] and Veblen [147], although of course their work was done in terms of the simplicial homology groups of a polyhedron.

There is another way of defining the homology groups of a space, which in general yields different groups from the singular homology groups (although the two theories coincide on polyhedra): these groups are the *Čech homology groups* of Čech [37] (following ideas of Alexandroff [13]). Apparently different definitions were given by Vietoris [148] and Alexander [11], but Dowker [47] proved that these definitions are equivalent to Čech's. A good exposition of Čech homology theory will be found in Eilenberg and Steenrod [56], Chapter 9.

A different approach to homology theory, by means of an axiomatic definition, has much to recommend it: see Eilenberg and Steenrod [55], or [56], Chapter 1.

Exact sequences. The exact homology sequence of a pair was formalized by Eilenberg and Steenrod [55], although the idea seems to be due to Hurewicz: see [75]. The Mayer–Vietoris sequence has a rather longer history: a formula for the homology groups of the union of two polyhedra was given by Mayer [103] and Vietoris [149], but the form of the result given in Theorem 4.4.6 is due to Eilenberg and Steenrod [56], Chapter 1. In fact Eilenberg and Steenrod prove the more general version involving arbitrary topological spaces.

Fixed points of maps of S^n. Proposition 4.4.12 was first proved by Brouwer [25].

Homology with arbitrary coefficients. Homology with coefficients Z_2 was first used by Tietze [144] and Alexander and Veblen [12], the generalization to coefficients Z_p, for various integers p, being made by Alexander [9]. Čech [39] defined homology with coefficients in an arbitrary abelian group, and established the result of Exercise 16 (although our formulation is closer to that of Eilenberg and MacLane [51]).

The Lefschetz Fixed-Point Theorem and Euler–Poincaré characteristic. Lefschetz's original proof of Theorem 4.5.18 can be found in [86, 87], though see also Hopf [67, 68]. These papers, and [89], Chapter 6, also contain a generalization in the form of an equality between $L(f)$ and the sum of the 'indices' of the fixed points of f: this is the Lefschetz Fixed-Point Formula.

In essence the Euler–Poincaré characteristic is due to Euler, whose definition was extended by Cauchy [36], and then by Poincaré [116].

Lens spaces. Tietze [144] first defined Lens spaces, and established many of their properties, including the fact that they are 3-manifolds, Reidemeister [122] proved that $L(p, q)$ and $L(p, q')$ are homeomorphic if and only if $q' = \pm q^{\pm 1} \pmod p$, and J. H. C. Whitehead [158] showed that a necessary and sufficient condition for $L(p, q)$ and $L(p, q')$ to be homotopy-equivalent is that qq' or $-qq'$ should be a quadratic residue mod p; thus $L(7, 1)$ and $L(7, 2)$ are homotopy-equivalent 3-manifolds that are not homeomorphic.

CHAPTER 5

COHOMOLOGY AND DUALITY THEOREMS

5.1 Introduction

We have seen in Section 4.5 how the idea of the homology groups of a pair can be generalized by taking coefficients in an arbitrary abelian group G. This process of generalization was purely algebraic, and bore no relation to the topology: a chain complex C gave rise to a new chain complex $C \otimes G$, whose homology, in the case where $C = S(X, Y)$, was defined to be $H_*(X, Y; G)$.

There is, however, another way of using a chain complex C and an abelian group G to yield a new chain complex. This process is in a sense dual to that of passing from C to $C \otimes G$, and will be the concern of this chapter. The idea is that, given abelian groups A and B, the set of homomorphisms from A to B can be given the structure of an abelian group, for which the notation $A \pitchfork B$ is used (the reader should notice the resemblance to the notion of a dual space in vector space theory). Just as in the case of the tensor product, this construction can be applied to the chain complex C to yield another chain complex $C \pitchfork G$; and if in particular $C = S(X, Y)$, the homology groups of $S(X, Y) \pitchfork G$ are called the *cohomology groups* of (X, Y), with coefficients in G: these are usually written $H^n(X, Y; G)$. The behaviour of cohomology groups resembles that of homology groups, but with one important difference: this time, given a continuous map $f\colon (X, Y) \to (A, B)$, we obtain corresponding cohomology homomorphisms $f^*\colon H^n(A, B; G) \to H^n(X, Y; G)$, that is, cohomology 'reverses the direction of maps'.

At first sight this definition seems rather pointless, particularly since the groups $H^n(X, Y; G)$, like $H_n(X, Y; G)$, are completely determined by the groups $H_n(X, Y)$, and indeed if (K, L) is a simplicial pair and F is a field, $H_n(K, L; F)$ and $H^n(K, L; F)$ are dual vector spaces over F. However, the language of cohomology allows a neat statement of the duality theorems of Sections 5.3 and 5.4, and in any case cohomology has a great advantage over homology, in that it is possible to define a product between elements of $H^*(X, Y)$, and thus make $H^*(X, Y)$ into a ring. This increases the power of cohomology: it may be that two spaces X and Y have isomorphic homology groups (and hence isomorphic cohomology groups), although $H^*(X)$ and $H^*(Y)$ are not isomorphic as rings, and so X and Y are not of the same homotopy type. However, the cohomology product will not be defined in this chapter, since it is easier to set up in the context of Chapter 8; but the reader should be aware of its existence, as a powerful reason for the study of cohomology.

The definitions will be given in Section 5.2, which also contains some calculation theorems for cohomology groups. The rest of this chapter is concerned with some duality theorems for triangulable manifolds: in Section 5.3 we shall prove the duality theorems of Poincaré and Alexander, which relate the homology and cohomology groups of triangulable manifolds, and in Section 5.4 we shall define manifolds with boundary, and prove the corresponding duality theorem, due to Lefschetz.

5.2 Definitions and calculation theorems

We start by considering the algebraic situation, which should be compared with Section 4.5. Given abelian groups A and B, write $A \pitchfork B$ for the set of homomorphisms from A to B (many authors use the notation Hom (A, B) instead of $A \pitchfork B$).

Proposition 5.2.1 *$A \pitchfork B$ can be given the structure of an abelian group.*

Proof. Given homomorphisms $f, g\colon A \to B$, define $f + g$ by the rule

$$(f + g)(a) = f(a) + g(a) \qquad (a \in A).$$

It is a trivial exercise to prove that $f + g$ is another homomorphism, and that $f + g = g + f$. Moreover $(f + g) + h = f + (g + h)$ if $h\colon A \to B$ is another homomorphism. Finally, define $0\colon A \to B$ by $0(a) = 0$ for all a, and $-f\colon A \to B$ by $(-f)(a) = -f(a)$; clearly

$$0 + f = f = f + 0$$

and

$$f + (-f) = (-f) + f = 0$$

for all $f \in A \pitchfork B$, so that $A \pitchfork B$ is an abelian group. ∎

Examples 5.2.2

(a) For any abelian group G, $Z \pitchfork G \cong G$. For we can define a homomorphism $\theta\colon Z \pitchfork G \to G$ by $\theta(f) = f(1)$ for all $f\colon Z \to G$, which is (1-1) since $f(1) = 0$ implies $f(n) = 0$ for all $n \in Z$, and onto, since given any $g \in G$ we can define $f\colon Z \to G$ by $f(n) = ng$, so that $\theta(f) = g$.

(b) If p and q are positive integers, $Z_p \pitchfork Z_q \cong Z_{(p,q)}$. For just as in (a), there is an isomorphism between $Z_p \pitchfork Z_q$ and the subgroup of Z_q consisting of possible values r of $f(1)$ for the homomorphisms f from Z_p to Z_q. But such r's are characterized by the property $pr = 0 \pmod q$ or $pr = qs$ for some s. If $p = a(p, q)$ and $q = b(p, q)$, then $ar = bs$ and $(a, b) = 1$, so that the possible values of r are just the (p, q) multiples of b, and these form a subgroup of Z_q isomorphic to $Z_{(p,q)}$.

(c) If p is a positive integer, then $Z_p \pitchfork Z = Z_p \pitchfork Q = 0$. For in either case, given a homomorphism f, let $f(1) = r$. Then

$$0 = pf(1) = pr,$$

which is impossible unless $r = 0$. ∎

As in the case of the tensor product, the process of forming $A \pitchfork B$ can be applied to homomorphisms as well as to groups (compare also the idea of a dual linear map between dual vector spaces).

Proposition 5.2.3 *Homomorphisms $f\colon A' \to A$, $g\colon B \to B'$ give rise to a homomorphism $f \pitchfork g\colon A \pitchfork B \to A' \pitchfork B'$, such that if $f'\colon A'' \to A'$ and $g'\colon B' \to B''$ are further homomorphisms, $(f' \pitchfork g')(f \pitchfork g) = (ff') \pitchfork (g'g)$.* (The reader should take careful note of the behaviour of composites here.)

Proof. If α is an element of $A \pitchfork B$, define $(f \pitchfork g)(\alpha) = g\alpha f$; this is certainly a homomorphism from A' to B', and clearly

$$(f \pitchfork g)(\alpha + \beta) = (f \pitchfork g)(\alpha) + (f \pitchfork g)(\beta).$$

Moreover

$$\begin{aligned}(f' \pitchfork g')(f \pitchfork g)(\alpha) &= (f' \pitchfork g')(g\alpha f)\\ &= g'g\alpha ff'\\ &= (ff' \pitchfork g'g)(\alpha).\end{aligned}$$ ∎

Finally, before applying this construction to chain complexes, we need a result analogous to Proposition 4.5.7(b) (there is, of course, no analogue of Proposition 4.5.7(a), since for example $Z \pitchfork Z_2 \cong Z_2$, but $Z_2 \pitchfork Z = 0$). Unfortunately it is not true that $\pitchfork$ is distributive for arbitrary direct sums: for example, let $A_i = Z$ for each integer i, and let $A = \bigoplus_i A_i$; then the homomorphism that sends 1 in each A_i to 1 in Z is a perfectly good element of $A \pitchfork Z$, but is not an element of $\bigoplus_i (A_i \pitchfork Z)$, since if it were it would have to be zero for all but a finite set of values of i. However, as long as we stick to finite direct sums, there is no difficulty.

Proposition 5.2.4 *If* $A = \bigoplus_{i=1}^{m} A_i$ *and* $B = \bigoplus_{j=1}^{n} B_j$, *then* $A \pitchfork B \cong \bigoplus_{i,j} (A_i \pitchfork B_j)$.

Proof. It is clearly sufficient to prove the two propositions

(a) $(A \oplus B) \pitchfork C \cong (A \pitchfork C) \oplus (B \pitchfork C)$;
(b) $A \pitchfork (B \oplus C) \cong (A \pitchfork B) \oplus (A \pitchfork C)$.

To prove (a), define $\theta: (A \oplus B) \pitchfork C \to (A \pitchfork C) \oplus (B \pitchfork C)$ by

$$\theta(f) = (fi_A) \oplus (fi_B),$$

where $f \in (A \oplus B) \pitchfork C$ and $i_A, i_B: A, B \to A \oplus B$ are the obvious inclusion homomorphisms. Now θ is a homomorphism, since

$$\begin{aligned}\theta(f + g) &= (f + g)i_A \oplus (f + g)i_B \\ &= (fi_A + gi_A) \oplus (fi_B + gi_B) \\ &= (fi_A \oplus fi_B) + (gi_A \oplus gi_B) \\ &= \theta(f) + \theta(g).\end{aligned}$$

Also θ is (1-1), since $f = 0$ if $fi_A = fi_B = 0$, and onto, since given $f: A \to C$ and $g: B \to C$ we can define $h: A \oplus B \to C$ by

$$h(a \oplus b) = f(a) + g(b);$$

then $\theta(h) = f \oplus g$. ∎

In extending the construction '$\pitchfork G$' to chain complexes, some care is necessary in view of the behaviour of direct sums: the group $C \pitchfork G$ need not be isomorphic to the group $\bigoplus (C_n \pitchfork G)$.

Definition 5.2.5 Given a chain complex C, with boundary homomorphism $\partial: C_n \to C_{n-1}$, and an abelian group G, the chain complex $C \pitchfork G$ is defined to be $\bigoplus_n (C \pitchfork G)_n$, where $(C \pitchfork G)_n = C_{-n} \pitchfork G$. The boundary homomorphism in $C \pitchfork G$ is $\delta = \partial \pitchfork 1$; clearly $\delta^2 = (\partial \pitchfork 1)(\partial \pitchfork 1) = \partial^2 \pitchfork 1 = 0$, by Proposition 5.2.3. ($(C \pitchfork G)_n$ is defined to be $C_{-n} \pitchfork G$, and not $C_n \pitchfork G$, so that δ sends $(C \pitchfork G)_n$ to $(C \pitchfork G)_{n-1}$.)

Notice that $C \pitchfork G = \bigoplus_n (C \pitchfork G)_n$ is not the same as the *group*

$C \pitchfork G$, unless only a finite number of the C_n are non-zero, as is the case, for example, if $C = C(K, L)$ for a simplicial pair (K, L). However, if C is a chain complex we shall always take $C \pitchfork G$ to mean $\bigoplus (C \pitchfork G)_n$.

Proposition 5.2.6 *A chain map $f\colon C \to D$ gives rise to a chain map $f \pitchfork 1\colon D \pitchfork G \to C \pitchfork G$, such that*

(a) *if $1\colon C \to C$ is the identity chain isomorphism, then $1 \pitchfork 1$ is also the identity chain isomorphism;*

(b) *if $g\colon D \to E$ is another chain map, then $(f \pitchfork 1)(g \pitchfork 1) = (gf) \pitchfork 1$.*

Proof. Let $f \pitchfork 1\colon D \pitchfork G \to C \pitchfork G$ be what was previously called $f \pitchfork 1$, on each $(D \pitchfork G)_n$: this certainly defines a homomorphism from $D \pitchfork G$ to $C \pitchfork G$. Properties (a) and (b) follow immediately from Proposition 5.2.3, and finally $f \pitchfork 1$ is a chain map, since

$$\begin{aligned} \delta(f \pitchfork 1) &= (\partial \pitchfork 1)(f \pitchfork 1) \\ &= (\partial f) \pitchfork 1 \\ &= (f\partial) \pitchfork 1 \\ &= (f \pitchfork 1)(\partial \pitchfork 1) \\ &= (f \pitchfork 1)\delta. \quad \blacksquare \end{aligned}$$

If C is the singular chain complex of a pair (X, Y), the homology groups of $C \pitchfork G$ are called the *cohomology groups* of (X, Y).

Definition 5.2.7 Given a pair (X, Y), the *nth cohomology group of* (X, Y), with coefficients in the abelian group G, is defined by $H^n(X, Y; G) = H_{-n}(S(X, Y) \pitchfork G)$. Of course, if $Y = \varnothing$ we write $H^n(X; G)$; and also $H^*(X, Y; G)$, $H^*(X; G)$ for $\bigoplus H^n(X, Y; G)$, $\bigoplus H^n(X; G)$ respectively.

Similarly, the *reduced* cohomology groups of X are defined by

$$\tilde{H}^n(X; G) = H_{-n}(\tilde{S}(X) \pitchfork G), \qquad \tilde{H}^*(X; G) = \bigoplus \tilde{H}^n(X; G).$$

The word 'cohomology' is used, of course, to prevent confusion with the homology groups. We have defined $H^n(X, Y; G)$ to be $H_{-n}(S(X, Y) \pitchfork G)$, rather than $H_n(S(X, Y) \pitchfork G)$, so that $H^n(X, Y; G)$ is zero for $n < 0$ (and also, as we shall see, so that $H^n(K, L; F)$ and $H_n(K, L; F)$ are dual spaces over F, if (K, L) is a simplicial pair and F is a field). By analogy with homology, we shall usually write $H^n(X, Y)$ instead of $H^n(X, Y; Z)$.

Just as in Section 4.5, a continuous map $f: (X, Y) \to (A, B)$ gives rise to a chain map $f_{.} \pitchfork 1: S(A, B) \pitchfork G \to S(X, Y) \pitchfork G$, whose induced homology homomorphisms are called the *cohomology homomorphisms* $f^*: H^n(A, B; G) \to H^n(X, Y; G)$. Once again, Propositions 4.2.16 and 4.2.21, together with Proposition 5.2.3, show that $1^* = 1$, $(gf)^* = f^*g^*$, and $f^* = g^*$ if $f \simeq g$. Also, if (K, L) is a simplicial pair, then $H^*(|K|, |L|; G)$ may be identified with the homology of the chain complexes $\Delta(K, L) \pitchfork G$ and $C(K, L) \pitchfork G$. In particular $\phi^*: H^*(K') \to H^*(K)$ and $h^*: H^*(K) \to H^*(K')$ are inverse isomorphisms, where h is a simplicial approximation to the identity map.

We can also take over the exact sequence theorems of Section 4.4, in virtue of the following Proposition.

Proposition 5.2.8 *Given an exact sequence of abelian groups*

$$0 \longrightarrow A \xrightarrow{\alpha} B \xrightarrow{\beta} C \longrightarrow 0,$$

where C is a free abelian group, and another abelian group G, then

$$0 \longrightarrow C \pitchfork G \xrightarrow{\beta \pitchfork 1} B \pitchfork G \xrightarrow{\alpha \pitchfork 1} A \pitchfork G \longrightarrow 0$$

is an exact sequence.

Proof. By Corollary 1.3.38, the exact sequence $0 \to A \to B \to C \to 0$ is a split exact sequence, and so $B \cong A \oplus C$. Hence by Proposition 5.2.4 $B \pitchfork G \cong (C \pitchfork G) \oplus (A \pitchfork G)$, and it is easy to conclude that the sequence

$$0 \longrightarrow C \pitchfork G \xrightarrow{\beta \pitchfork 1} B \pitchfork G \xrightarrow{\alpha \pitchfork 1} A \pitchfork G \longrightarrow 0$$

is therefore also a split exact sequence. ∎

In particular, if $0 \to C \xrightarrow{f} D \xrightarrow{g} E \to 0$ is an exact sequence of chain complexes, and each E_n is a free abelian group, then

$$0 \longrightarrow E \pitchfork G \xrightarrow{g \pitchfork 1} D \pitchfork G \xrightarrow{f \pitchfork 1} C \pitchfork G \longrightarrow 0$$

is an exact sequence of chain complexes. We immediately obtain, for example, the *exact cohomology sequence of the triple* (X, Y, Z):

$$\cdots \longrightarrow H^n(X, Y; G) \xrightarrow{j^*} H^n(X, Z; G) \xrightarrow{i^*} H^n(Y, Z; G) \xrightarrow{\delta^*} H^{n+1}(X, Y; G) \longrightarrow \cdots.$$

Similarly, there is a cohomology Mayer–Vietoris sequence of a simplicial triad $(K; L, M)$, and a relative Mayer–Vietoris sequence given another subcomplex N. As in homology, continuous maps give rise to commutative diagrams involving these exact sequences. In particular the homomorphism $s\colon \tilde{\Delta}(K) \to \tilde{\Delta}(SK)$ induces the *cohomology suspension isomorphism* $s^*\colon \tilde{H}^{n+1}(SK) \to \tilde{H}^n(K)$, where if $f\colon |K| \to |L|$ is a continuous map $f^*s^* = s^*(Sf)^*$. And as a consequence of the exact sequence theorems, cohomology can be calculated directly from a block dissection (though as in Section 4.5 there is an apparent ambiguity about how to do this: see Exercise 1 which resolves this ambiguity).

As before, the next step is to establish results that connect cohomology groups with various coefficients, and also cohomology with homology. The latter is done by constructing a pairing between elements of $H(C \pitchfork G)$ and $H(C \otimes G)$, given a chain complex C and a *ring* (not merely an abelian group) G.

Proposition 5.2.9 *Given a chain complex C and a ring G, there is a homomorphism*

$$H_{-n}(C \pitchfork G) \otimes H_n(C \otimes G) \to G,$$

called the Kronecker product, *where the image of $x \otimes y$ is written $\langle x, y\rangle$. Moreover if $f\colon C \to D$ is a chain map, and $x \in H_{-n}(D \pitchfork G)$, $y \in H_n(C \otimes G)$, then $\langle (f \pitchfork 1)_*(x), y\rangle = \langle x, (f \otimes 1)_*(y)\rangle$.*

Proof. Given $\alpha \in (C \pitchfork G)_{-n}$ and $\sum c_i \otimes g_i \in C_n \otimes G$, define

$$\left\langle \alpha, \sum c_i \otimes g_i \right\rangle = \sum \alpha(c_i) g_i$$

(using the multiplication in G). It is easy to see that this defines a homomorphism $(C \pitchfork G)_{-n} \otimes (C \otimes G)_n \to G$. Now given $x \in H_{-n}(C \pitchfork G)$ and $y \in H_n(C \otimes G)$, take representative cycles α, c for x, y respectively, and define $\langle x, y\rangle = \langle \alpha, c\rangle$. This does not depend on the choice of α and c, since, for example, if $\beta \in (C \pitchfork G)_{-n+1}$, then

$$\langle \delta(\beta), c\rangle = \langle \beta, (\partial \pitchfork 1)(c)\rangle = 0.$$

Finally, represent $x \in H_{-n}(D \pitchfork G)$, $y \in H_n(C \otimes G)$ by cycles α, c respectively. Then

$$\begin{aligned} \langle (f \pitchfork 1)_*(x), y\rangle &= \langle (f \pitchfork 1)(\alpha), c\rangle \\ &= \langle \alpha, (f \otimes 1)(c)\rangle \\ &= \langle x, (f \otimes 1)_*(y)\rangle. \quad \blacksquare \end{aligned}$$

It follows that if $f\colon (X, Y) \to (A, B)$ is a continuous map, and $x \in H^n(A, B; G)$, $y \in H_n(X, Y; G)$, then

$$\langle f^*(x), y\rangle = \langle x, f_*(y)\rangle.$$

If X is path-connected, $G \simeq H_0(X; G)$, so that the Kronecker product $H^n(X; G) \otimes H_n(X; G) \to G$ may be regarded as a homomorphism into $H_0(X; G)$. It is therefore possible to make the following slight generalization, at least for a polyhedron $|K|$.

Proposition 5.2.10 *If G is a ring, there exists a homomorphism*

$$H^r(K; G) \otimes H_n(K; G) \to H_{n-r}(K; G),$$

called the cap product, *where the image of $x \otimes y$ is written $x \cap y$.*

Proof. Totally order the vertices of K. Given a generator $\sigma = [b^0, \ldots, b^n]$ of $C(K)$, with its vertices in the correct order, let $\sigma' = [b^0, \ldots, b^{n-r}]$ and $\sigma'' = [b^{n-r}, \ldots, b^n]$. Given also $\alpha \in C_r(K) \pitchfork G$, define

$$\alpha \cap \left(\sum \sigma_i \otimes g_i\right) = \sum \sigma_i' \otimes (\alpha(\sigma_i'')g_i).$$

It is easy to see that

$$(\partial \otimes 1)(\alpha \cap c) = \alpha \cap (\partial \otimes 1)(c) + (-1)^{n-r}\delta(\alpha) \cap c,$$

so that if $x \in H^r(K; G)$, $y \in H_n(K; G)$ are represented by cycles α, c respectively, we can unambiguously define $x \cap y = [\alpha \cap c]$. And this certainly defines a homomorphism from $H^r(K; G) \otimes H_n(K; G)$ to $H_{n-r}(K; G)$. ∎

In fact the cap product can be defined for arbitrary spaces: see Exercise 2.

If in particular $G = F$, a field, Proposition 5.2.9 leads to the following analogue of Proposition 4.5.10.

Proposition 5.2.11 *Let C be a chain complex in which each C_n is a finitely-generated abelian group, and let F be a field. Then the Kronecker product makes $H_{-n}(C \pitchfork F)$ and $H_n(C \otimes F)$ into dual vector spaces over F. Moreover if $g: C \to D$ is a chain map, $(g \pitchfork 1)_*$ and $(g \otimes 1)_*$ are dual linear maps.*

Proof. If $\alpha \in (C \pitchfork F)_{-n}$ and $f \in F$, define $f\alpha$ by

$$(f\alpha)(c) = f\alpha(c) \qquad (c \in C_n).$$

This makes $(C \pitchfork F)_{-n}$ into a vector space over F. If the Kronecker product is regarded as a function

$$(C \pitchfork F)_{-n} \times (C \otimes F)_n \to F,$$

it is certainly linear in each variable. Moreover if $\langle \alpha, \sum c_i \otimes f_i \rangle = 0$ for all $c_i \in C_n$, $f_i \in F$, then in particular $\alpha(c) = 0$ for all $c \in C_n$, so that

$\alpha = 0$. On the other hand, if $\langle \alpha, \sum c_i \otimes f_i \rangle = 0$ for all $\alpha \in (C \pitchfork F)_{-n}$, but $\sum c_i \otimes f_i \neq 0$, there exists a linear map $\beta: C_n \otimes F \to F$ such that $\beta(\sum c_i \otimes f_i) \neq 0$. Thus if α is defined by $\alpha(c) = \beta(c \otimes 1)$ $(c \in C_n)$, we have

$$\begin{aligned}\langle \alpha, \sum c_i \otimes f_i \rangle &= \sum \alpha(c_i) f_i \\ &= \sum \beta(c_i \otimes 1) f_i \\ &= \beta(\sum c_i \otimes f_i) \\ &\neq 0,\end{aligned}$$

contrary to hypothesis: thus $\sum c_i \otimes f_i = 0$ and so $(C \pitchfork F)_{-n}$, $(C \otimes F)_n$ are dual vector spaces under the Kronecker product.

Now

$$\begin{aligned}\delta(\alpha) = 0 &\Leftrightarrow \langle \delta(\alpha), c \rangle = 0 \quad \text{for all } c \in C_{n+1} \otimes F \\ &\Leftrightarrow \langle \alpha, (\partial \otimes 1)(c) \rangle = 0;\end{aligned}$$

hence $Z_{-n}(C \pitchfork F) = \mathscr{A}(B_n(C \otimes F))$. Similarly $Z_n(C \otimes F) = \mathscr{A}(B_{-n}(C \pitchfork F))$, so that, by Proposition 1.3.55, $B_{-n}(C \pitchfork F) = \mathscr{A}(Z_n(C \otimes F))$, and $H_{-n}(C \pitchfork F)$ and $H_n(C \otimes F)$ are dual vector spaces (under the Kronecker product, of course).

That $(g \pitchfork 1)_*$ and $(g \otimes 1)_*$ are dual linear maps is immediate from Proposition 5.2.9. ∎

Corollary 5.2.12 *Given a simplicial pair (K, L) and a field F, $H^n(K, L; F)$ and $H_n(K, L; F)$ are dual vector spaces over F. Moreover if $g: (|K|, |L|) \to (|M|, |N|)$ is a continuous map, g^* and g_* are dual linear maps.* ∎

There is a similar result to that of Proposition 5.2.11, relating $H_{-n}(C \pitchfork Z_p)$ and $H_n(C \otimes Z_p)$, even though p may not be a prime and so Z_p may not be a field.

Proposition 5.2.13 *Let C be a chain complex in which each C_n is a finitely-generated abelian group, and let p be any positive integer. Then $H_{-n}(C \pitchfork Z_p) \cong H_n(C \otimes Z_p) \pitchfork Z_p$.*

Proof. Each $(C \otimes Z_p)_n$ is a finitely-generated abelian group, in which each element has finite order dividing p; let us call such a group a *p-group*. Now given a p-group G, a subgroup H, and a homomorphism $\alpha: H \to Z_p$, there exists a homomorphism $\beta: G \to Z_p$ such that $\beta | H = \alpha$. For suppose g is in G but not H; then the set $S = \{s \in Z_p \mid sg \in H\}$ is clearly a (cyclic) subgroup of Z_p, and $f: S \to Z_p$, defined by $f(s) = \alpha(sg)$, is a homomorphism. So f must be multiplication by some $t \in Z_p$, that is, $\alpha(sg) = ts$ for all $s \in S$. It follows that we may extend α to the multiples of g by setting $\alpha(g) = t$; and by continuing in this way, after a finite number of steps we shall have extended α to the whole of G.

Now consider an exact sequence of p-groups $A \xrightarrow{f} B \xrightarrow{g} C$: we show that the sequence

$$C \pitchfork Z_p \xrightarrow{g \pitchfork 1} B \pitchfork Z_p \xrightarrow{f \pitchfork 1} A \pitchfork Z_p$$

is also exact. For certainly $(f \pitchfork 1)(g \pitchfork 1) = gf \pitchfork 1 = 0$, so that $\text{Im}\,(g \pitchfork 1) \subset \text{Ker}\,(f \pitchfork 1)$. But if $\alpha: B \to Z_p$ is in $\text{Ker}\,(f \pitchfork 1)$, then $\alpha f = 0$, so that $\alpha(\text{Ker}\, g) = \alpha(\text{Im}\, f) = 0$. Thus α induces $\bar{\alpha}: \text{Im}\, g \to Z_p$, such that $\bar{\alpha} g = \alpha$; but $\bar{\alpha}$ can be extended as above to the whole of C, so that $\alpha \in \text{Im}\,(g \pitchfork 1)$ and hence $\text{Im}\,(g \pitchfork 1) = \text{Ker}\,(f \pitchfork 1)$.

Now define $\theta: C_n \pitchfork Z_p \to (C_n \otimes Z_p) \pitchfork Z_p$ by $\theta(\alpha)(c) = \langle \alpha, c \rangle$, $\alpha \in C_n \pitchfork Z_p$, $c \in C_n \otimes Z_p$. As in the proof of Proposition 5.2.11, θ is (1-1) and onto, and so is a chain isomorphism; our result now follows by a proof similar to that of Theorem 4.5.13. ∎

Corollary 5.2.14 *Given a simplicial pair* (K, L), $H^n(K, L; Z_p) \cong H_n(K, L; Z_p) \pitchfork Z_p$. ∎

Theorems relating cohomology groups with various coefficients can be established by the methods of Section 4.5. Indeed, they can actually be deduced from the theorems of Section 4.5, in virtue of the following proposition.

Proposition 5.2.15 *Let C be a chain complex in which each C_n is a finitely-generated abelian group, and let G be any abelian group. Then there is a chain isomorphism*

$$\theta: (C \pitchfork Z) \otimes G \to C \pitchfork G.$$

Proof. Define $\theta: (C_n \pitchfork Z) \otimes G \to C_n \pitchfork G$ by the rule

$$[\theta(\alpha \otimes g)](c) = \alpha(c)g,$$

for $\alpha \in C_n \pitchfork Z, g \in G$, and $c \in C_n$. This certainly defines a homomorphism θ, and in fact θ is an isomorphism. For if the generators of C_n are $\sigma_1, \ldots, \sigma_r$, $C_n \pitchfork Z$ is a free abelian group with generators $s_1, \ldots, s_r$, where $s_i(\sigma_j) = 1$ if $i = j$ and 0 otherwise. Thus a homomorphism $\phi: C_n \pitchfork G \to (C_n \pitchfork Z) \otimes G$ can be defined by the rule

$$\phi(\beta) = \sum_{i=1}^{r} s_i \otimes \beta(\sigma_i) \qquad (\beta \in C_n \pitchfork G).$$

And then

$$[\theta\phi(\beta)](\sigma_j) = \sum_i s_i(\sigma_j).\beta(\sigma_i)$$
$$= \beta(\sigma_j),$$

and

$$\phi\theta(s_j \otimes g) = \sum_i s_i \otimes [\theta(s_j \otimes g)](\sigma_i)$$
$$= \sum_i s_i \otimes s_j(\sigma_i)g$$
$$= s_j \otimes g,$$

so that $\theta\phi = 1$, $\phi\theta = 1$, and θ and ϕ are inverse isomorphisms.

Lastly, to show that θ is a chain isomorphism, observe that for $c \in C_{n+1}$, $\alpha \in C_n \pitchfork Z$, and $g \in G$, we have

$$[\delta\theta(\alpha \otimes g)](c) = [\theta(\alpha \otimes g)](\partial c)$$
$$= \alpha(\partial c).g$$
$$= (\delta\alpha)(c).g$$
$$= [\theta(\delta\alpha \otimes g)](c)$$
$$= [\theta(\delta \otimes 1)(\alpha \otimes g)](c),$$

so that $\delta\theta = \theta(\delta \otimes 1)$. ∎

Corollary 5.2.16 *Let (K, L) be a simplicial pair. Then*

(a) $H^n(K, L; Q) \cong H^n(K, L) \otimes Q$;

(b) $H^n(K, L; Z_p) \cong H^n(K, L) \otimes Z_p \oplus \text{Tor}\,(H^{n+1}(K, L), Z_p)$, *for any positive integer p.* ∎

Rather surprisingly, Corollary 5.2.16 leads to a formula expressing the cohomology groups $H^n(K, L)$ in terms of the homology groups $H_n(K, L)$. To state this formula, given a finitely generated abelian group A, use Theorem 1.3.30 to write A in the form $FA \oplus TA$, where FA is a free abelian group and TA is a finite group.

Proposition 5.2.17 $H^n(K, L) \cong FH_n(K, L) \oplus TH_{n-1}(K, L)$.

Proof. By Theorem 4.5.13, Corollary 5.2.12 and Corollary 5.2.16(a),

$$H^n(K, L) \otimes Q \cong H_n(K, L) \otimes Q,$$

being dual (finite-dimensional) vector spaces over Q. Thus

$$FH^n(K, L) \cong FH_n(K, L).$$

But we also know, by Theorem 4.5.15, Corollary 5.2.14 and Corollary 5.2.16(b), that for any positive integer p,

$$H^n(K, L) \otimes Z_p \oplus \text{Tor}\,(H^{n+1}(K, L), Z_p)$$
$$\cong [H_n(K, L) \otimes Z_p \oplus \text{Tor}\,(H_{n-1}(K, L), Z_p)] \pitchfork Z_p.$$

By taking p to be the l.c.m. of the orders of the elements of $TH^n(K, L)$, $TH^{n+1}(K, L)$, $TH_n(K, L)$ and $TH_{n-1}(K, L)$, and using the fact that $FH^n(K, L) \cong FH_n(K, L)$, it follows that

$$TH^n(K, L) \oplus TH^{n+1}(K, L) \cong TH_n(K, L) \oplus TH_{n-1}(K, L).$$

But $H^0(K, L) = Z_0(\Delta(K, L) \pitchfork Z)$, since $(\Delta(K, L) \pitchfork Z)_1 = 0$, and so $TH^0(K, L) = 0$ and $TH^1(K, L) \cong TH_0(K, L)$. Proceeding by induction on n, we obtain $TH^n(K, L) \cong TH_{n-1}(K, L)$, so that

$$H^n(K, L) \cong FH^n(K, L) \oplus TH^n(K, L)$$
$$\cong FH_n(K, L) \oplus TH_{n-1}(K, L). \blacksquare$$

Example 5.2.18 We have already calculated $H_*(RP^n)$, in Example 4.4.25:

$H_0(RP^n) \cong Z,$

$H_r(RP^n) = 0$ if $r < 0$, $r > n$ or r is even,

$H_r(RP^n) \cong Z_2$ if r is odd and $0 < r < n$,

$H_n(RP^n) \cong Z$ if n is odd.

It follows that $H^*(RP^n)$ is given by:

$H^0(RP^n) \cong Z,$

$H^r(RP^n) = 0$ if $r < 0$, $r > n$ or r is odd (unless $r = n$),

$H^r(RP^n) \cong Z_2$ if r is even and $0 < r \leqslant n$,

$H^n(RP^n) \cong Z$ if n is odd.

Similarly, the homology and cohomology of RP^n, with Z_2 coefficients, are given by:

$H_r(RP^n; Z_2) = H^r(RP^n; Z_2) = 0$ if $r < 0$ or $r > n$,

$H_r(RP^n; Z_2) \cong H^r(RP^n; Z_2) \cong Z_2$ if $0 \leqslant r \leqslant n$. $\blacksquare$

Example 5.2.19 Consider the triangulable 2-manifolds M_g and N_h. By Theorem 4.4.24, we have

$H_0(M_g) \cong H_2(M_g) \cong Z,$ $H_1(M_g) \cong 2gZ,$ $H_r(M_g) = 0$ otherwise,

$H_0(N_h) \cong Z,$ $H_1(N_h) \cong (h - 1)Z \oplus Z_2,$ $H_r(N_h) = 0$ otherwise.

Hence

$H^0(M_g) \cong H^2(M_g) \cong Z$, $H^1(M_g) \cong 2gZ$, $H^r(M_g) = 0$ otherwise.

$H^0(N_h) \cong Z$, $H^1(N_h) \cong (h - 1)Z$, $H^2(N_h) \cong Z_2$,

$H^r(N_h) = 0$ otherwise.

Also

$$H_0(M_g; Z_2) \cong H_2(M_g; Z_2) \cong Z_2, \quad H_1(M_g; Z_2) \cong 2gZ_2,$$

$$H_0(N_h; Z_2) \cong H_2(N_h; Z_2) \cong Z_2, \quad H_1(N_h; Z_2) \cong hZ_2,$$

cohomology with coefficients in Z_2 being the same groups. ∎

The reader should notice the following facts about Examples 5.2.18 and 5.2.19.

(a) If $H_n(RP^n) \cong Z$, then $H_r(RP^n) \cong H^{n-r}(RP^n)$, for all r.
(b) In any case, $H_r(RP^n; Z_2) \cong H^{n-r}(RP^n; Z_2)$ for all r.
(c) Similarly, $H_2(M_g) \cong Z$ and $H_r(M_g) \cong H^{n-r}(M_g)$.
(d) $H_r(N_h; Z_2) \cong H^{n-r}(N_h; Z_2)$.

These are all special cases of the Poincaré Duality Theorem, which we shall prove in Section 5.3. Certain triangulable spaces X (generalizations of n-manifolds) will be called *homology n-manifolds*; the statement of Poincaré Duality is that, for such an X, $H_r(X) \cong H^{n-r}(X)$ if $H_n(X) \cong Z$, and in any case $H_r(X; Z_2) \cong H^{n-r}(X; Z_2)$.

5.3 The Alexander–Poincaré Duality Theorem

In this section we shall prove a rather more general theorem than the Poincaré Duality Theorem that has just been outlined, to the effect that, if K is a triangulation of a homology n-manifold and (L, M) is a pair of subcomplexes of K, then

$$H_r(L, M) \cong H^{n-r}(|K| - |M|, |K| - |L|).$$

If we put $M = \varnothing$ and $L = K$, we recover the original Poincaré Duality Theorem; on the other hand if K is a triangulation of S^n we obtain the Alexander Duality Theorem, which is a very useful one in dealing with subspaces of S^n: in particular it gives a proof of a generalization of the (piecewise-linear) Jordan Curve Theorem.

Recall from Proposition 3.4.3 that if K is a triangulation of an n-manifold, then for each $x \in |K|$, $|\mathrm{Lk}(x)| \simeq S^{n-1}$. Now we could prove the theorems of this section for triangulable manifolds only, but since the only property we shall use is that, for some triangulation K and each $x \in |K|$, $H_*(\mathrm{Lk}(x)) \cong H_*(S^{n-1})$, we may as well consider all spaces having this property: these are the *homology n-manifolds.*

Definition 5.3.1 A path-connected space X is a *homology n-manifold* if there exists a triangulation K of X, such that for each point $x \in |K|$, and for each r, $H_r(\mathrm{Lk}(x)) \cong H_r(S^{n-1})$.

In other words, for each x we have

$$\tilde{H}_r(\mathrm{Lk}(x)) \cong \begin{cases} Z, & r = n - 1 \\ 0, & \text{otherwise.} \end{cases}$$

By Theorem 2.4.5, if this property holds for one triangulation of X, then it holds for every triangulation. Also, as we have just remarked, any triangulable path-connected n-manifold is a homology n-manifold. See Exercise 9, however, for an example of a homology n-manifold that is *not* an n-manifold.

Examples 5.3.2 The triangulable 2-manifolds M_g and N_h are of course homology 2-manifolds. Also S^n and RP^n are homology n-manifolds: for by Example 3.4.2 S^n is an n-manifold; and as for RP^n, we know at least that it is triangulable, by Example 4.4.25. But since RP^n is formed from S^n by identifying antipodal points, it is very easy to see that RP^n is an n-manifold: given a pair x, x' of antipodal points of S^n, choose ϵ so that the ϵ-neighbourhoods of x and x' do not intersect; then after identification these ϵ-neighbourhoods become a single open set in RP^n, containing the point corresponding to x and x', and clearly homeomorphic to an open set in R^n. ∎

A triangulation of a homology n-manifold has several convenient properties, which can be obtained by using Theorem 2.4.5. The most important of these are collected together in the next theorem.

Theorem 5.3.3 *Let K be a triangulation of a homology n-manifold. Then K has the following properties.*

(a) $\dim K = n$.

(b) *Each point of $|K|$ is contained in at least one n-simplex.*

(c) *Each $(n - 1)$-simplex of K faces two n-simplexes.*

(d) *Given n-simplexes σ and τ in K, there exists a sequence of n-simplexes $\sigma = \sigma_1, \sigma_2, \ldots, \sigma_r = \tau$, such that each $\sigma_i \cap \sigma_{i+1}$ is an $(n - 1)$-simplex.*

Proof. We may assume that $n \geqslant 1$, since a homology 0-manifold is obviously just a point.

(a) Certainly $\dim K \geqslant n$, for otherwise $\dim \mathrm{Lk}(x)$ would be less than $(n - 1)$ for all $x \in |K|$, and so $H_{n-1}(\mathrm{Lk}(x))$ would be zero. On the other hand, if K had an m-simplex σ, for $m > n$, then for points x in the interior of σ, $|\text{Lk}(x)|$ would be homeomorphic to S^{m-1}, contradicting Definition 5.3.1.

(b) This is immediate: if x were in no n-simplex, dim Lk(x) would be less than $(n - 1)$.

(c) Let x be a point in the interior of an $(n - 1)$-simplex σ, and suppose that σ is a face of r n-simplexes ($r > 0$). Corresponding to each n-simplex τ that has σ as a face, there is a subcomplex $\dot{\tau} - \sigma$ of Lk(x); the union of these is Lk(x), and any two intersect in $\dot{\sigma}$: see Fig. 5.1.

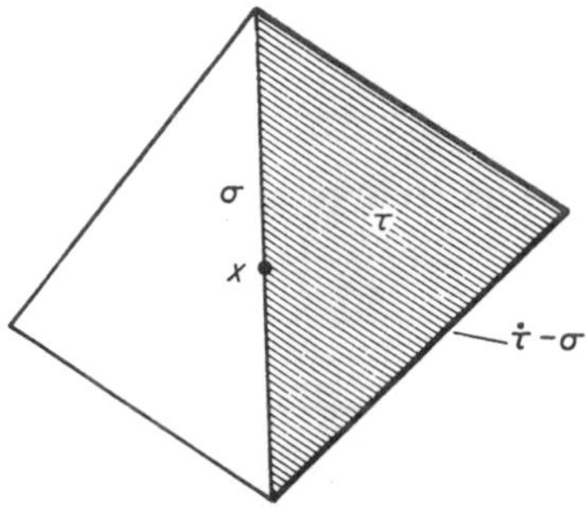

Fig. 5.1

An easy calculation by induction on r, using the reduced Mayer–Vietoris sequence, shows that $\tilde{H}_{n-1}(\text{Lk}(x))$ is a free abelian group with $(r - 1)$ generators, so that r must be 2.

(d) Choose a particular n-simplex σ, and let L be the set of n-simplexes of K that can be 'connected to σ' in this way (with their faces), and M be the set of n-simplexes that cannot be connected to σ (with their faces). Then L and M are subcomplexes, and $L \cup M = K$. Moreover, if we assume that M is non-empty, then $L \cap M$ is non-empty, since $|K|$ is path-connected, and $\dim (L \cap M) \leqslant n - 2$, since an $(n - 1)$-simplex of $L \cap M$ would have to be a face of an n-simplex of L and an n-simplex of M. This already contradicts the assumption that $M \neq \varnothing$ if $n = 1$, so we may assume from now on that $n \geqslant 2$.

Let a be a vertex of $L \cap M$, and consider Lk(a). a must be a vertex of an n-simplex of L and an n-simplex of M, so that both $\text{Lk}(a) \cap L$ and $\text{Lk}(a) \cap M$ contain $(n - 1)$-simplexes. Also $\dim(\text{Lk}(a) \cap L \cap M) \leqslant n - 3$, and by (c) every $(n - 2)$-simplex of Lk(a) is a face of two $(n - 1)$-simplexes; thus if $c_L = \sum \tau$, for all $(n - 1)$-simplexes τ of $\text{Lk}(a) \cap L$, and $c_M = \sum \tau$, for all $(n - 1)$-simplexes τ of $\text{Lk}(a) \cap M$, then c_L and c_M are linearly independent cycles of $C(\mathrm{Lk}(a)) \otimes Z_2$ (in $\partial(c_L)$, for example, each $(n - 2)$-simplex occurs twice). It follows that $H_{n-1}(\mathrm{Lk}(a); Z_2)$ has dimension at least 2, as a vector space over Z_2, so that by Theorem 4.5.15 $\mathrm{Lk}(a)$ cannot have the same homology as S^{n-1}. This contradiction shows that M must be empty, and so $L = K$. ∎

It follows from this theorem, and Chapter 3, Exercise 13, that every homology 2-manifold is actually a 2-manifold. A similar result holds for homology 3-manifolds, but not for manifolds of higher dimension: see Exercises 8 and 9.

It is clear from the remarks at the end of Section 5.2 that the Poincaré Duality Theorem is not true for all homology n-manifolds, unless coefficients Z_2 are used. Those homology manifolds for which the theorem is true for Z coefficients are exactly those that are *orientable*, in the sense of the next definition.

Definition 5.3.4 A homology n-manifold X is *orientable* if there exists a triangulation K of X, for which the n-simplexes can be identified with elements of $C_n(K)$ in such a way that, if σ is any $(n - 1)$-simplex, and τ_1, τ_2 are the two n-simplexes that have σ as a face, then σ occurs with opposite signs in $\partial(\tau_1)$ and $\partial(\tau_2)$.

Notice that a homology 0-manifold (a point) is certainly orientable.

Example 5.3.5 S^1 is orientable, since it can be triangulated as shown in Fig. 5.2, and the 1-simplexes identified with generators of the simplicial chain group according to the arrows in Fig. 5.2.

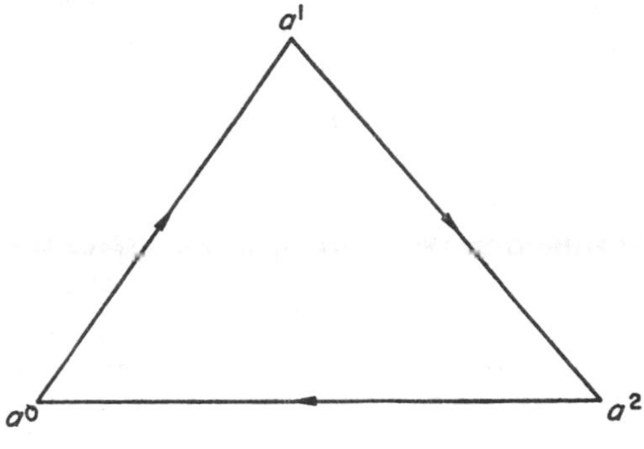

Fig. 5.2

As it stands Definition 5.3.4 is not much use, since it is not clear that the definition is independent of the particular triangulation. In order to clear up this point, and indeed to provide a practical test for orientability, we prove

Proposition 5.3.6 *Let X be a homology n-manifold. Then* $H_n(X) \cong Z$ *if X is orientable, and* $H_n(X) = 0$ *otherwise. In any case,* $H_n(X; Z_2) \cong Z_2$.

Proof. Let K be a triangulation of X, and suppose that the n-simplexes are identified with elements of $C_n(K)$ as in Definition 5.3.4. Then z, the sum of the n-simplexes of K, is an element of $Z_n(C(K))$, and hence so also is any integer multiple of z. On the other hand if z' is an element of $Z_n(C(K))$ that contains $r\sigma$ for some n-simplex σ, then z' must contain $r\tau$ for every n-simplex τ that meets σ in an $(n - 1)$-simplex. And so z' contains $r\tau$ for every n-simplex τ that can be connected to σ as in Theorem 5.3.3(d), that is, for every τ in K; hence $z' = rz$, and $H_n(K) \cong Z$.

Conversely, the same argument shows that, however the n-simplexes of K are identified with elements of $C_n(K)$, any element $z' \in Z_n(C(K))$ must be of the form rz, where $z = \sum \pm\sigma$, and σ runs through the n-simplexes of K. If $H_n(K) \cong Z$, we must have $\partial(z) = 0$, so that it is possible to change the identification of n-simplexes so as to satisfy Definition 5.3.4.

Finally, the argument used to prove that $H_n(K) \cong Z$ if X is orientable shows that in any case $H_n(K; Z_2) \cong Z_2$. But by Theorem 4.5.15, $H_n(K)$, being a finitely generated free abelian group, must be isomorphic either to Z or 0; hence $H_n(K) = 0$ if X is not orientable. ∎

Corollary 5.3.7 *Two homotopy-equivalent homology n-manifolds are either both orientable or both non-orientable.* ∎

Example 5.3.8 By Theorem 4.4.24, each M_g is orientable, and each N_h is non-orientable. S^n is orientable, and by Example 4.4.25 RP^n is orientable if and only if n is odd. ∎

We turn now to the duality theorem, which states that, if $|K|$ is an orientable homology n-manifold, and (L, M) is a pair of subcomplexes of K, then $H_r(L, M) \cong H^{n-r}(|K| - |M|, |K| - |L|)$ for each r. Our proof will be given in terms of simplicial homology and cohomology, and since $|K| - |L|$ and $|K| - |M|$ are not polyhedra the first step is to replace them by the supplements $\bar{L}$ and $\bar{M}$, in the sense of Definition 2.5.18.

Proposition 5.3.9 *Let K be a simplicial complex, and let (L, M) be a pair of subcomplexes. There is a commutative diagram*

$$\begin{array}{ccccccccc}
\cdots \longrightarrow & H_r(\bar{L}) & \xrightarrow{i_*} & H_r(\overline{M}) & \xrightarrow{j_*} & H_r(\overline{M},\bar{L}) & \xrightarrow{\partial_*} & H_{r-1}(\bar{L}) & \longrightarrow \cdots \\
& \downarrow f_* & & \downarrow f_* & & \downarrow f_* & & \downarrow f_* & \\
\cdots \longrightarrow & H_r(|K|-|L|) & \xrightarrow{i_*} & H_r(|K|-|M|) & \xrightarrow{j_*} & H_r(|K|-|M|,|K|-|L|) & \xrightarrow{\partial_*} & H_{r-1}(|K|-|L|) & \longrightarrow \cdots
\end{array}$$

in which each f_* is induced by an inclusion map, and is an isomorphism. A similar result holds for cohomology.

Proof. Certainly the diagram is commutative, by Theorem 4.4.3. The proof is completed by showing that $f\colon |\bar{L}| \to |K| - |L|$ and $f\colon |\overline{M}| \to |K| - |M|$ are homotopy equivalences: thus the induced homomorphisms f_* are isomorphisms, and then $f_*\colon H_*(\overline{M}, \bar{L}) \to H_*(|K| - |M|, |K| - |L|)$ is an isomorphism as well, by Proposition 1.3.35.

To show that $f\colon |\overline{M}| \to |K| - |M|$, for example, is a homotopy equivalence, we prove that $|\overline{M}|$ is a strong deformation retract of $|K| - |M|$. Now if σ is a simplex of $K' - (M' \cup \overline{M})$, then each vertex of σ is in either M' or $\overline{M}$; moreover σ has a vertex in M' because it is not in $\overline{M}$, and a vertex in $\overline{M}$ because if σ had all its vertices in M' it would be a simplex of M', by Corollary 2.5.11. It follows that σ is of the form $(a^0, \ldots, a^n)$, where $a^0, \ldots, a^r \in \overline{M}$, $a^{r+1}, \ldots, a^n \in M'$, and $0 \leqslant r < n$; thus the face $(a^0, \ldots, a^r)$ is in $\overline{M}$ since it cannot meet M', and $(a^{r+1}, \ldots, a^n)$ is in M' since all its vertices are in M': see Fig. 5.3, in which $n = 2$ and $r = 1$.

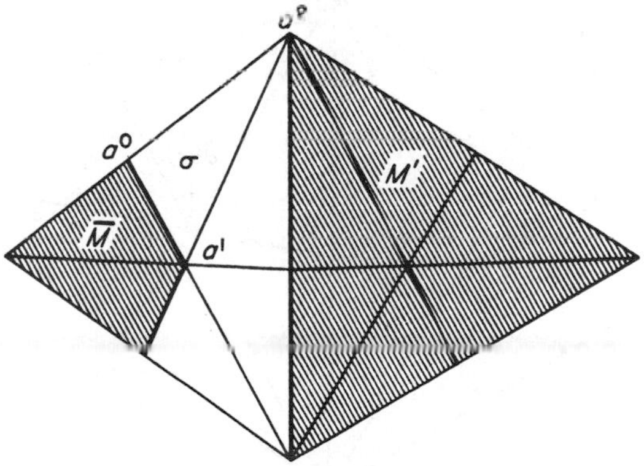

Fig. 5.3

It is now easy to define a (strong) deformation retraction $\rho\colon |K| - |M| \to |\overline{M}|$: if $x \in |\overline{M}|$, set $\rho(x) = x$, and if $x \in \sigma - (\sigma \cap |M|)$, then $x = \sum_{i=0}^{n} \lambda_i a^i$, where $\sum_{i=0}^{r} \lambda_i > 0$ and $\sum_{i=0}^{n} \lambda_i = 1$; put

$\rho(x) = (\sum_{i=0}^{r} \lambda_i a^i)/(\sum_{i=0}^{r} \lambda_i)$ (this represents radial projection from a^2 in Fig. 5.3). Then ρ is continuous on each simplex, and the definitions of ρ coincide on the intersection of two simplexes, so that ρ is continuous by Proposition 1.4.15. And ρ is a strong deformation retraction, since $\rho f = 1$ and $f\rho \simeq 1$ by a linear homotopy. ∎

The main tool in the proof of the duality theorem is the existence of a block dissection of a triangulation of a homology n-manifold, that is 'dual' to the ordinary simplicial dissection, in the sense that there is a (1-1)-correspondence between the r-blocks and the $(n - r)$-simplexes. We shall build up as much of this theory as possible for simplicial complexes in general, and specialize to homology manifolds only when necessary.

Suppose then that K is any simplicial complex. For each simplex σ of K, define subcomplexes

$$e(\sigma) = \text{all simplexes of } K' \text{ of form } (\hat{\sigma}_n, \ldots, \hat{\sigma}_0), \text{ where}$$

$$\sigma_n > \cdots > \sigma_0 > \sigma,$$

$$\dot{e}(\sigma) = \text{all simplexes of } e(\sigma) \text{ not having } \hat{\sigma} \text{ as a vertex.}$$

Clearly these *are* subcomplexes. As an example, see Fig. 5.4, in which σ is the simplex (a^1, a^2).

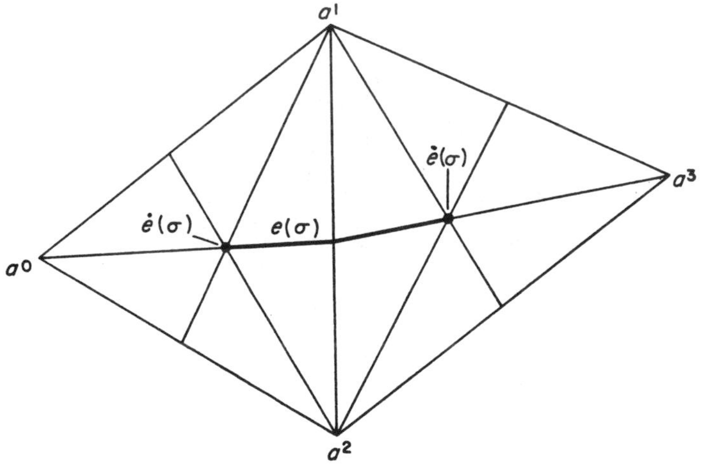

Fig. 5.4

In general the pair $(e(\sigma), \dot{e}(\sigma))$ need not be a block: see, for example, Fig. 5.5, in which $\sigma = (a^0)$, so that $H_1(e(\sigma), \dot{e}(\sigma)) \cong Z \oplus Z$.

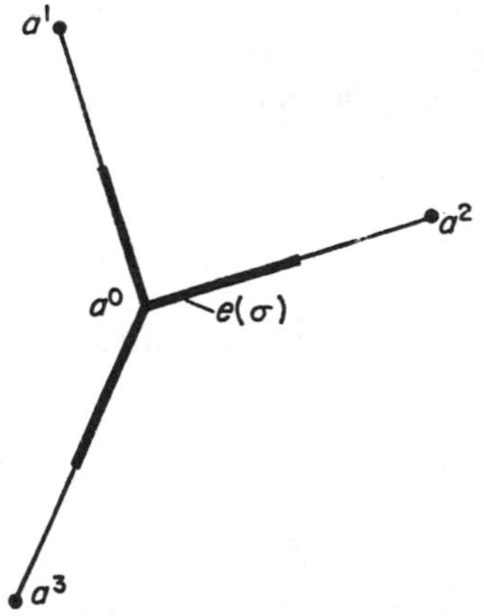

Fig. 5.5

However, these subcomplexes do have some convenient properties which we list in the next proposition (we shall refer to the set of simplexes in $e(\sigma) - \dot{e}(\sigma)$ as the *interior* of $e(\sigma)$, just as if $e(\sigma)$ were a block).

Proposition 5.3.10

(a) *Each simplex of K' is in the interior of just one $e(\sigma)$.*
(b) *$\dot{e}(\sigma)$ is the union of all the $e(\tau)$ for which σ is a (proper) face of τ.*
(c) *For each $\sigma \in K$, $\mathrm{Lk}_{K'}(\hat{\sigma}) = \dot{e}(\sigma) * (\dot{\sigma})'$.*

Proof. Consider the simplex $(\hat{\sigma}_n, \ldots, \hat{\sigma}_0)$ of K', where $\sigma_n > \cdots > \sigma_0$. This simplex is in the interior of $e(\sigma_0)$, and cannot be in the interior of any other $e(\sigma)$, which proves (a). As for (b), notice that

$$(\hat{\sigma}_n, \ldots, \hat{\sigma}_0) \in \dot{e}(\sigma) \Leftrightarrow \sigma_0 > \sigma,$$

but $\sigma_0 \neq \sigma$. Finally

$$\begin{aligned}(\hat{\sigma}_n, \ldots, \hat{\sigma}_0) \in \mathrm{Lk}(\hat{\sigma}) &\Leftrightarrow \hat{\sigma}_n, \ldots, \hat{\sigma}_0, \hat{\sigma} \text{ are the vertices of a simplex of } K' \\ &\Leftrightarrow \sigma_r > \sigma > \sigma_{r-1} \text{ for some } r \\ &\Leftrightarrow (\hat{\sigma}_n, \ldots, \hat{\sigma}_r) \in \dot{e}(\sigma) \quad \text{and} \quad (\hat{\sigma}_{r+1}, \ldots, \hat{\sigma}_0) \in (\dot{\sigma})',\end{aligned}$$

which proves (c). ∎

It follows from (a) and (b) that the set of all pairs $(e(\sigma), \dot{e}(\sigma))$ will form a block dissection of K', provided each is a block. It is at this point that we need to know that $|K|$ is a homology manifold.

Corollary 5.3.11 *Let K be a triangulation of a homology n-manifold. Then for each r-simplex σ of K, $(e(\sigma), \dot{e}(\sigma))$ is an $(n-r)$-block, and the set of all $(e(\sigma), \dot{e}(\sigma))$ forms a block dissection of K'.*

Proof. For each $\sigma \in K$,

$$H_s(\mathrm{Lk}_{K'}(\hat{\sigma})) \cong \begin{cases} Z, & s = n-1 \\ 0, & \text{otherwise.} \end{cases}$$

But by Example 2.3.13 $(\dot{\sigma})'$ is a triangulation of S^{r-1}, so that

$$\begin{aligned} \tilde{H}_s(\mathrm{Lk}(\hat{\sigma})) &\cong \tilde{H}_s(|\dot{e}(\sigma)| * S^{r-1}) \\ &\cong \tilde{H}_{s-r}(\dot{e}(\sigma)), \quad \text{by Example 2.3.18 and Theorem 4.4.10,} \end{aligned}$$

so that

$$\tilde{H}_s(\dot{e}(\sigma)) \cong \begin{cases} Z, & s = n-r-1 \\ 0, & \text{otherwise.} \end{cases}$$

On the other hand $e(\sigma) = \dot{e}(\sigma) * \hat{\sigma}$, and so is contractible. Hence $\tilde{H}_s(e(\sigma)) = 0$ for all s, and $H_s(e(\sigma), \dot{e}(\sigma)) \cong \tilde{H}_{s-1}(\dot{e}(\sigma))$ by the exact reduced homology sequence of the pair $(e(\sigma), \dot{e}(\sigma))$. It follows that

$$H_s(e(\sigma), \dot{e}(\sigma)) \cong \begin{cases} Z, & s = n-r \\ 0, & \text{otherwise,} \end{cases}$$

so that $(e(\sigma), \dot{e}(\sigma))$ is an $(n-r)$-block. Thus the set of all $(e(\sigma), \dot{e}(\sigma))$ forms a block dissection of K', by Proposition 5.3.10(a) and (b). ∎

In order to calculate homology from the blocks $e(\sigma)$, we must identify the corresponding 'block chain complex' with a sub-chain complex of $C(K)$, as in Proposition 4.4.21: this is done by choosing generators of each $Z_{n-r}(C(e(\sigma), \dot{e}(\sigma))) \subset C_{n-r}(K)$. Suppose now that $|K|$ is an orientable homology n-manifold, and that the n-simplexes of K are identified with elements of $C(K)$ as in Definition 5.3.4: thus $z \in C_n(K)$, the sum of the n-simplexes of K, is a representative cycle for a generator of $H_n(K)$. Totally order the vertices of K' so that $\hat{\sigma} < \hat{\tau}$ if $\dim \sigma > \dim \tau$; let $\phi\colon C(K) \to C(K')$ be the subdivision chain map, and let $h\colon |K'| \to |K|$ be a simplicial approximation to the identity. Finally, for each r-simplex σ of K (considered as an element of $C(K)$), let $s \in C_r(K) \pitchfork Z$ be the homomorphism that sends σ to 1 and all other r-simplexes to 0. Define $z(\sigma) = (h_{.} \pitchfork 1)(s) \cap \phi(z) \in C_{n-r}(K')$.

Proposition 5.3.12 *$z(\sigma)$ is a generator of $Z_{n-r}(C(e(\sigma), \dot{e}(\sigma))) \cong Z$.*

Proof. By the remarks after Corollary 4.3.10, each generator $[\hat{\sigma}_n, \ldots, \hat{\sigma}_0]$ of $C_n(K')$ occurs in $\phi(z)$, with coefficient ± 1. Now

$$(h_{.} \pitchfork 1)(s) \cap [\hat{\sigma}_n, \ldots, \hat{\sigma}_0] = [\hat{\sigma}_n, \ldots, \hat{\sigma}_r] s h_{.} [\hat{\sigma}_r, \ldots, \hat{\sigma}_0],$$

and this is zero except for just one simplex $(\hat{\sigma}_r, \ldots, \hat{\sigma}_0)$, contained in σ, when it is $\pm[\hat{\sigma}_n, \ldots, \hat{\sigma}_r]$. Hence $z(\sigma) \in C_{n-r}(e(\sigma))$.

Moreover, by the proof of Proposition 5.2.10

$$\begin{aligned} \partial z(\sigma) &= (-1)^{n-r}\delta(h_{.} \pitchfork 1)(s) \cap \phi(z) \quad (\text{since } \partial\phi(z) = \phi\partial(z) = 0) \\ &= (-1)^{n-r}(h_{.} \pitchfork 1)\delta(s) \cap \phi(z) \\ &\in C_{n-r-1}(\dot{e}(\sigma)), \text{ by Proposition 5.3.10(b).} \end{aligned}$$

Thus $z(\sigma) \in Z_{n-r}(C(e(\sigma), \dot{e}(\sigma)))$.

Lastly, $z(\sigma)$ is a generating cycle since, as we saw above, each simplex in $z(\sigma)$ has coefficient ± 1. ∎

We are now, at last, in a position to prove the Alexander–Poincaré Duality Theorem.

Theorem 5.3.13 *Let K be a triangulation of an orientable homology n-manifold, and let (L, M) be a pair of subcomplexes. Then for each r there exists an isomorphism $D\colon H^r(L, M) \to H_{n-r}(\bar{M}, \bar{L})$.*

Proof. Note first that $\bar{L}$ is the union of the blocks $e(\sigma)$, for all $\sigma \notin L$, so that $\bar{L}$ is a block subcomplex. For a simplex of $\bar{L}$ does not meet $|L|$, and so all its vertices are barycentres of simplexes not in L: so this simplex is in the interior of some $e(\sigma)$, where $\sigma \notin L$. Conversely, if $\sigma \notin L$, then no simplex having σ as a face can be in L, so no simplex in $e(\sigma)$ can have a vertex in $|L|$, and hence $e(\sigma) \subset \bar{L}$.

Similarly $\bar{M}$ is the union of the $e(\sigma)$, for all $\sigma \notin M$. Moreover, since for example $e(\sigma)$ is contained in $\bar{M}$ if and only if its interior is contained in $\bar{M}$, a simplex σ is in $L - M$ if and only if the interior of $e(\sigma)$ is in $\bar{M} - \bar{L}$.

Now consider the isomorphism $D\colon C_r(L, M) \pitchfork Z \to \bar{C}_{n-r}(\bar{M}, \bar{L})$, defined by $D(s) = z(\sigma)$, where $\bar{C}$ is the chain complex obtained from the block dissection into $e(\sigma)$'s, and s is the homomorphism that sends σ to 1 and all other r-simplexes to 0. If d is the boundary homomorphism in $\bar{C}$, $dD(s)$ can be calculated by Proposition 4.4.21:

$$\begin{aligned} \theta dD(s) &= \partial\theta D(s) \\ &= (-1)^{n-r}(h_{.} \pitchfork 1)\delta(s) \cap \phi(z) \end{aligned}$$

by the proof of Proposition 5.3.12, where δ is regarded as the boundary homomorphism in $C(L) \pitchfork Z$, and we omit simplexes in $\bar{L}$. Thus

$$dD(s) = \sum (-1)^{n-r}(\tau),$$

where the sum is taken over those $(r+1)$-simplexes τ such that $\partial(\tau) = \sigma + \cdots$ and the interior of $e(\tau)$ is not in $\bar{L}$. But as we saw above, the interior of $e(\tau)$ is in $\overline{M} - \overline{L}$ if and only if τ is in $L - M$, and so

$$dD(s) = (-1)^{n-r}D\delta(s),$$

where now δ is the boundary homomorphism in $C(L, M) \pitchfork Z$. Hence D induces an isomorphism

$$D: H^r(L, M) \to H_{n-r}(\overline{M}, \overline{L}). \blacksquare$$

Note that, by Proposition 4.4.22, D is the isomorphism induced by the homomorphism from $C(L, M) \pitchfork Z$ to $C(\overline{M}, \overline{L})$ given by sending s to $z(\sigma)$, for each σ in $L - M$.

Corollary 5.3.14 *With the notation of Proposition 5.2.17,*

$$FH^r(L, M) \cong FH_{n-r}(\overline{M}, \overline{L})$$

and

$$TH^r(L, M) \cong TH_{n-r-1}(\overline{M}, \overline{L}). \blacksquare$$

It follows, of course, that we may interchange homology and cohomology in Theorem 5.3.13: $H_r(L, M) \cong H^{n-r}(\overline{M}, \overline{L})$ for all r. Moreover, these isomorphisms remain valid if we replace integer coefficients by Q or Z_p, for any positive integer p.

We emphasize, however, that Theorem 5.3.13 has been proved only for *orientable* homology manifolds. Indeed, the theorem would be false if $|K|$ were non-orientable: for if $(L, M) = (K, \varnothing)$, then $(\overline{M}, \overline{L}) = (K', \varnothing)$, so that if Theorem 5.3.13 were true, we would have

$$H_n(K') \cong H^0(K) \cong Z,$$

since $|K|$ is path-connected. But this contradicts Proposition 5.3.6.

However, there is still a duality theorem for non-orientable homology manifolds, provided Z_2 coefficients are used throughout. On using $z \otimes 1 \in C_n(K) \otimes Z_2$ instead of z, the method of proof of Theorem 5.3.13 will prove

Theorem 5.3.15 *Let K be a triangulation of a homology n-manifold, not necessarily orientable, and let (L, M) be a pair of subcomplexes. Then for each r, the isomorphism $D: C(L, M) \pitchfork Z_2 \to C(\overline{M}, \overline{L}) \otimes Z_2$, given by $D(s) = (h_{\cdot} \pitchfork 1)(s) \cap (\phi \otimes 1)(z \otimes 1)$, induces an isomorphism $D: H^r(L, M; Z_2) \to H_{n-r}(\overline{M}, \overline{L}; Z_2)$.* Also

$$H_r(L, M; Z_2) \cong H^{n-r}(\overline{M}, \overline{L}; Z_2). \blacksquare$$

Before discussing corollaries and applications of the duality theorems, it is worth noting that Theorem 5.3.13 has the following converse.

Theorem 5.3.16 *Given a path-connected polyhedron $|K|$, and a positive integer n, such that for each pair of subcomplexes (L, M) of K and for each r, we have $H^r(L, M) \cong H_{n-r}(\bar{M}, \bar{L})$, then $|K|$ is an orientable homology n-manifold.*

Proof. Choose an r-simplex σ of K, and let $(L, M) = (K(\sigma), \dot{\sigma})$, so that (L, M) is a triangulation of (E^r, S^{r-1}), and by Proposition 5.2.17

$$H^s(L, M) \cong \begin{cases} Z, & s = r \\ 0, & \text{otherwise.} \end{cases}$$

On the other hand, as in the proof of Theorem 5.3.13, $\bar{L} = \bigcup e(\tau)$ for all $\tau \notin K(\sigma)$, and $\bar{M} = \bigcup e(\tau)$, for all $\tau \notin \dot{\sigma}$; hence $\bar{M} = \bar{L} \cup e(\sigma)$, and also $\bar{L} \cap e(\sigma) = \dot{e}(\sigma)$, by the definition of $\bar{L}$. It follows that

$$\begin{aligned} H_s(e(\sigma), \dot{e}(\sigma)) &\cong H_s(\bar{M}, L), \quad \text{by Example 4.3.6} \\ &\cong H^{n-s}(L, M) \\ &\cong \begin{cases} Z, & s = n - r \\ 0, & \text{otherwise,} \end{cases} \end{aligned}$$

so that $(e(\sigma), \dot{e}(\sigma))$ is an $(n - r)$-block.

To finish the proof, we just reverse the proof of Corollary 5.3.11:

$$\begin{aligned} \tilde{H}_s(\text{Lk}_{K'}(\hat{\sigma})) &\cong \tilde{H}_{s-r}(\dot{e}(\sigma)) \\ &\cong \tilde{H}_{s-r+1}(e(\sigma), \dot{e}(\sigma)) \\ &\cong \begin{cases} Z, & s = n - 1 \\ 0, & \text{otherwise.} \end{cases} \end{aligned}$$

But by Theorem 2.4.5 we can replace $\text{Lk}_{K'}$ by Lk_K; and clearly $\text{Lk}(\hat{\sigma}) = \text{Lk}(x)$ for all x in the interior of σ. That is, $\text{Lk}(x)$ has the correct reduced homology groups for each point $x \in |K|$, so that $|K|$ is a homology n-manifold. And it is orientable, since

$$H_n(K) \cong H^0(K) \cong Z. \quad ∎$$

The 'standard' Poincaré and Alexander duality theorems can easily be deduced from Theorems 5.3.13 and 5.3.15.

Theorem 5.3.17 (Poincaré duality.) *Let K be a triangulation of a homology n-manifold. If $|K|$ is orientable, there is an isomorphism $D: H^r(K) \to H_{n-r}(K')$ for all r; in any case there is an isomorphism $D: H^r(K; Z_2) \to H_{n-r}(K'; Z_2)$ for all r.*

Proof. Apply Theorems 5.3.13 and 5.3.15, with $(L, M) = (K, \varnothing)$. ∎

Notice that if $x \in H^r(K)$, then $D(x) = h^*(x) \cap \phi_*[z]$, where $h: |K'| \to |K|$ is a simplicial approximation to the identity map. Since h^* is an isomorphism, the Poincaré duality isomorphism may conveniently be regarded as the isomorphism $\bar{D}: H^r(K') \to H_{n-r}(K')$, given by $\bar{D}(x) = x \cap \phi_*[z]$.

Theorem 5.3.17 gives a useful sufficient condition for the orientability of $|K|$.

Corollary 5.3.18 *Let K be a triangulation of a homology n-manifold. If $H_1(K; Z_2) = 0$, then $|K|$ is orientable.*

Proof. We have $H_n(K; Z_2) \cong H^0(K; Z_2) \cong Z_2$ and $H_{n-1}(K; Z_2) \cong H^1(K; Z_2) \cong H_1(K; Z_2) \cong 0$. But by Theorem 4.5.15

$$H_n(K; Z_2) \cong H_n(K) \otimes Z_2 \oplus \operatorname{Tor}(H_{n-1}(K), Z_2)$$

and

$$H_{n-1}(K; Z_2) \cong H_{n-1}(K) \otimes Z_2 \oplus \operatorname{Tor}(H_{n-2}(K), Z_2).$$

Now $H_{n-1}(K)$ is a finitely generated abelian group, and so is a direct sum of groups isomorphic to Z or Z_p, for various integers p; but since $H_{n-1}(K) \otimes Z_2 = 0$ there are no Z's, and all p's are odd. It follows that $\operatorname{Tor}(H_{n-1}(K), Z_2) = 0$, and so

$$H_n(K) \otimes Z_2 \cong H_n(K; Z_2) \cong Z_2.$$

Hence $H_n(K) \cong Z$ by Theorem 5.3.6, and $|K|$ is orientable. ∎

Theorem 5.3.19 (Alexander duality.) *Let K be a triangulation of S^n, and let L be a subcomplex of K. Then $\tilde{H}^r(L) \cong \tilde{H}_{n-r-1}(S^n - |L|)$, for all r.*

Proof. Let a be a vertex of L. Since S^n is an orientable homology n-manifold, we have

$$\tilde{H}^r(L) \cong H^r(L, a) \cong H_{n-r}(\bar{a}, \bar{L}) \cong H_{n-r}(S^n - a, S^n - |L|),$$

using also Proposition 5.3.9. But it is easy to see that $S^n - a$ is homeomorphic to $E^n - S^{n-1}$ (use the standard map of Section 1.4), and so is contractible. Hence $\tilde{H}_*(S^n - a) = 0$, and the exact reduced homology sequence of $(S^n - a, S^n - |L|)$ shows that

$$H_{n-r}(S^n - a, S^n - |L|) \cong \tilde{H}_{n-r-1}(S^n - |L|). ∎$$

Naturally, Theorems 5.3.17 and 5.3.19 remain true if we interchange homology and cohomology.

The duality theorems have many interesting corollaries. Some of these depend on the ring structure of cohomology, and so will have to be postponed to Chapter 8, but we shall conclude this section with a few results on the Euler–Poincaré characteristic and inclusions of homology manifolds in each other.

Proposition 5.3.20 *Let X be a homology n-manifold, where n is odd. Then $\chi(X) = 0$.*

Proof. Whether or not X is orientable, we have

$$H_r(X; Z_2) \cong H^{n-r}(X; Z_2) \cong H_{n-r}(X; Z_2),$$

by Theorem 5.3.17 and Corollary 5.2.12. Thus if α_r is the dimension of $H_r(X; Z_2)$, as a vector space over Z_2,

$$\sum (-1)^r \alpha_r = 0, \quad \text{since } n \text{ is odd.}$$

On the other hand, Theorem 4.5.15 gives

$$\begin{aligned} H_r(X; Z_2) &\cong H_r(X) \otimes Z_2 \oplus \operatorname{Tor}(H_{r-1}(X), Z_2) \\ &\cong [FH_r(X) \oplus TH_r(X) \oplus TH_{r-1}(X)] \otimes Z_2. \end{aligned}$$

Thus $\sum (-1)^r \alpha_r = \sum (-1)^r \beta_r$, where β_r is the dimension of $FH_r(X) \otimes Z_2$, as a vector space over Z_2. But this is the same as the dimension of $FH_r(X) \otimes Q \cong H_r(X; Q)$ as a vector space over Q, so that

$$\sum (-1)^r \beta_r = \chi(X). \blacksquare$$

Proposition 5.3.21 *Let (K, L) be a simplicial pair, where both $|K|$ and $|L|$ are homology n-manifolds. Then $K = L$.*

Proof. By Theorem 5.3.15,

$$H_0(K', \bar{L}; Z_2) \cong H^n(L; Z_2) \cong H_0(L; Z_2) \cong Z_2.$$

Thus in the exact homology sequence of the pair (K, L):

$$\cdots \longrightarrow H_0(\bar{L}; Z_2) \xrightarrow{i_*} H_0(K'; Z_2) \xrightarrow{j_*} H_0(K', \bar{L}; Z_2) \longrightarrow 0,$$

since $H_0(K'; Z_2) \cong H_0(K', \bar{L}; Z_2) \cong Z_2$, i_* must be the zero homomorphism. But this is impossible unless $\bar{L} = \varnothing$, that is, $K = L$. $\blacksquare$

In other words, a homology n-manifold cannot be properly contained in another, as a subpolyhedron. Of course, it is essential for this result that the dimensions of the two homology manifolds should be the same; for we cannot have one contained in another of lower dimension, and on the other hand it is certainly possible to have one contained in another of higher dimension: for example, S^{n-1} in S^n.

The duality theorems allow us to say quite a lot about homology n-manifolds contained in S^{n+1}.

Proposition 5.3.22 *Let $|L|$ be a non-orientable homology n-manifold. Then L cannot be a subcomplex of a triangulation of S^{n+1}.*

Proof. Suppose, if possible, that K is a triangulation of S^{n+1}, having L as a subcomplex. Then $\tilde{H}_0(\bar{L}) \cong \tilde{H}^n(L) = 0$, by Theorem 5.3.19. But $\tilde{H}_0(\bar{L}; Z_2) \cong \tilde{H}^n(L; Z_2) \cong H^n(L; Z_2) \cong H_0(L; Z_2) \cong Z_2$ ($n > 0$, since otherwise L must be a point, and so orientable), which contradicts Theorem 4.5.15. ∎

That is, a homology n-manifold that is a subpolyhedron of S^{n+1} must be orientable. In particular, none of the 2-manifolds N_h can be a subpolyhedron of S^3.

Proposition 5.3.23 *Let $|L|$ be an orientable homology n-manifold ($n > 1$), and let L be a subcomplex of some triangulation of S^{n+1}. Then $S^{n+1} - |L|$ has two path components.*

Proof. By Theorem 5.3.19,

$$\begin{aligned}\tilde{H}_0(S^n - |L|) &\cong \tilde{H}^n(L)\\ &\cong H^n(L), \quad \text{since } n \geqslant 1\\ &\cong Z, \quad \text{since } |L| \text{ is orientable.}\end{aligned}$$

Thus $H_0(S^n - |L|) \cong Z \oplus Z$, so that by Example 4.2.13 $S^n - |L|$ has two path components. ∎

In particular the complement in S^{n+1} of any subpolyhedron homeomorphic to S^n must have two path components, and indeed, by Chapter 3, Exercise 2, these path components are connected sets. This result is a generalization of the Jordan Curve Theorem: the complement in S^2 of any subpolyhedron homeomorphic to S^1 has two connected components.

5.4 Manifolds with boundary and the Lefschetz Duality Theorem

In this section we shall generalize the duality theorems of Section 5.3 to manifolds 'with boundary' (compare Chapter 3, Exercises 15 and 16). These are spaces which are locally like either Euclidean space R^n or the half-plane $x_1 \geqslant 0$.

Definition 5.4.1 A Hausdorff space M is called an *n-manifold with boundary* ($n \geqslant 1$) if each point of M has a neighbourhood homeomorphic to an open set in the subspace $x_1 \geqslant 0$ of R^n.

Examples 5.4.2 E^n is an n-manifold with boundary. For each point of $E^n - S^{n-1}$ has a neighbourhood that is already an open set in R^n; moreover (E^n, S^{n-1}) is homeomorphic to $(\sigma, |\dot{\sigma}|)$ for an n-simplex σ, so that a point x of S^{n-1} has a neighbourhood in E^n that is homeomorphic to the intersection of an open set in R^n with σ, and this (if small enough) is of the required form (we can ensure that the given point x is mapped under the homeomorphism to an interior point of an $(n - 1)$-face of σ). See Fig. 5.6.

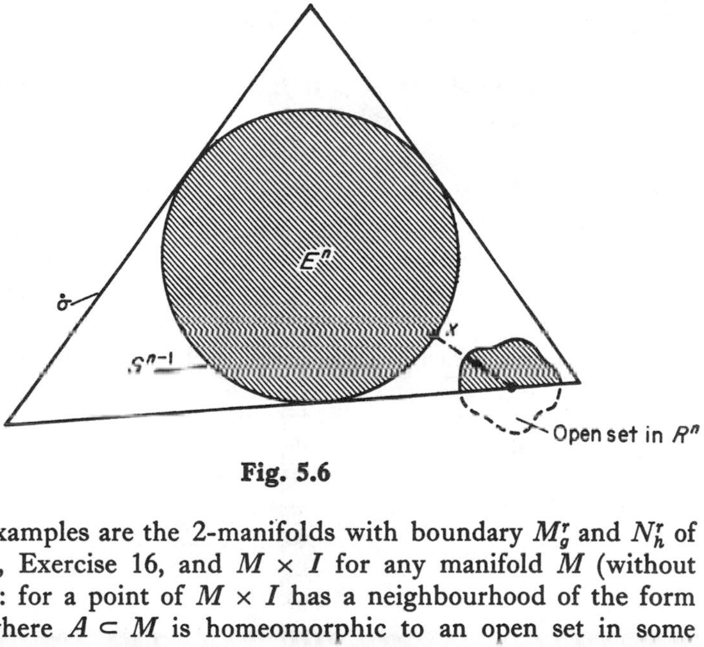

Fig. 5.6

Other examples are the 2-manifolds with boundary M_g^r and N_h^r of Chapter 3, Exercise 16, and $M \times I$ for any manifold M (without boundary): for a point of $M \times I$ has a neighbourhood of the form $A \times B$, where $A \subset M$ is homeomorphic to an open set in some Euclidean space, and B is an open set in I (and is not the whole of I). ∎

We are particularly interested, of course, in those manifolds with boundary that are triangulable. Information about possible triangulations can be obtained from the following proposition, which generalizes Proposition 3.4.33.

Proposition 5.4.3 *Let K be a triangulation of an n-manifold with boundary. Then for each $x \in |K|$, $|\mathrm{Lk}(x)|$ is homotopy-equivalent either to S^{n-1} or to a point.*

Proof. If x has a neighbourhood homeomorphic to an open set in $x_1 \geqslant 0$ that does not meet $x_1 = 0$, the argument of Proposition 3.4.3 applies, to show that $|\text{Lk}(x)| \simeq S^{n-1}$. Otherwise, there exists a point y in $x_1 = 0$, an $\epsilon > 0$, and a homeomorphism h of $B \cap (x_1 \geqslant 0)$ onto a subset of $|K|$ such that $h(y) = x$, where B is the ϵ-neighbourhood of y in R^n. But $B \cap (x_1 \geqslant 0)$ can be triangulated as $K(\sigma)$, where σ is an n-simplex with y in the interior of an $(n-1)$-face. Hence, by Theorem 2.4.5 $|\text{Lk}(x)| \simeq |\text{Lk}(y)|$, which is clearly contractible. ∎

It follows that if M is a triangulable manifold with boundary, the set of points of M having all neighbourhoods homeomorphic to open sets that meet $x_1 = 0$ is exactly the set of points x such that $|\text{Lk}(x)|$ is contractible. This subset of M is called the *boundary* of M, ∂M; notice that any homeomorphism of M onto another manifold with boundary, N, must map ∂M onto ∂N.

Proposition 5.4.4 *If K is a triangulation of an n-manifold with boundary, there exists a subcomplex L of K such that $|L| = \partial|K|$. Moreover, $|L|$ is an $(n-1)$-manifold.*

Proof. We show first that $\partial|K|$ is a closed subspace of $|K|$. Now each point x of $|K| - \partial|K|$ has a neighbourhood homeomorphic to an open set in R^n; the same is therefore true for each point in this neighbourhood, and so $|K| - \partial|K|$ is open, and hence $\partial|K|$ is closed.

If σ is a simplex of K that meets $\partial|K|$ at a point x in its interior, then $|\text{Lk}(x)|$ is contractible. But $\text{Lk}(x) = \text{Lk}(y)$ for all points y in the

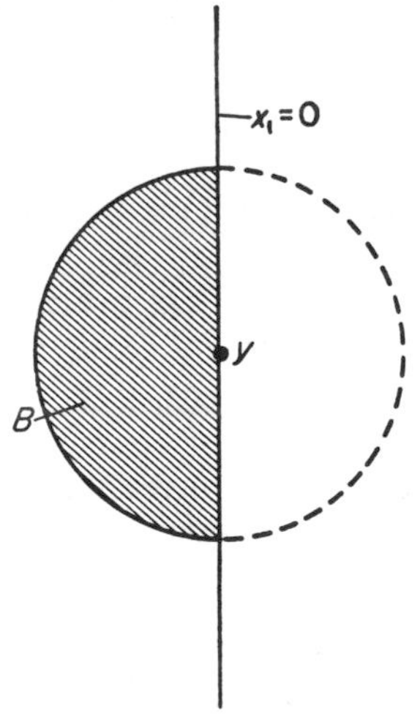

Fig. 5.7

interior of σ, so that the interior of σ is contained in $\partial|K|$. Hence $\sigma \subset \partial|K|$ since $\partial|K|$ is closed, and so if L is the subcomplex of K of those simplexes that are contained in $\partial|K|$, $|L| = \partial|K|$.

Lastly, consider a point $x \in |L|$. Then there exists a point y in $x_1 = 0$, an $\epsilon > 0$, and a homeomorphism h as in Proposition 5.4.3: see Fig. 5.7.

Now it is clear that points z in $x_1 \geqslant 0$ such that $d(y, z) < \epsilon$ are mapped by h to $|K| - |L|$ if they do not lie in $x_1 = 0$, and to $|L|$ otherwise (a point in $x_1 > 0$, for example, has a neighbourhood that is an open set in R^n, and is contained in $B \cap (x_1 \geqslant 0)$). Hence $x = h(y)$ has a neighbourhood in $|L|$ that is homeomorphic to the set of points z in $x_1 = 0$ such that $d(y, z) < \epsilon$, and this in turn is homeomorphic to an open set in R^{n-1}. ▮

So far in this section, we have considered triangulable manifolds with boundary, and it is perfectly possible to prove the Lefschetz Duality Theorem for these spaces only. However, in the spirit of Section 5.3, we prefer to work with rather more general spaces, the *homology* manifolds with boundary.

Definition 5.4.5 A path-connected space X is a *homology n-manifold with boundary* ($n \geqslant 1$) if there exists a triangulation K of X, such that for each point $x \in |K|$, $\tilde{H}_*(\mathrm{Lk}(x))$ is isomorphic either to $\tilde{H}_*(S^{n-1})$ or to 0. The *boundary* of X, ∂X, is the set of points x such that $\tilde{H}_*(\mathrm{Lk}(x)) = 0$; observe that $X - \partial X \neq \varnothing$, since a point in the interior of a simplex of maximum dimension cannot be in ∂X.

By Theorem 2.4.5, this property holds for all triangulations of X if it holds for one, and the definition of ∂X is independent of the particular triangulation; also, by Proposition 5.4.3, every path-connected triangulable n-manifold with boundary is a homology n-manifold with boundary. So, of course, is every homology n-manifold in the sense of Definition 5.3.1; we shall sometimes call such homology manifolds *closed*, if we wish to stress that their boundaries are empty.

Examples 5.4.6 E^n, M_g^r and N_h^r are all homology manifolds with boundary. ▮

We should like to be able to say also that $X \times I$ is a homology n-manifold with boundary if X is a (closed) homology $(n - 1)$-manifold. This is true, but is a little more difficult to prove than the corresponding result in Examples 5.4.2. The following lemma is necessary.

Lemma 5.4.7 *Let x be a vertex of a simplicial complex K, and let $L = \mathrm{Lk}_K(x)$. Let y be any point of $|L|$, and let z be the mid-point of xy* (see Fig. 5.8). *Then for each r, $\tilde{H}_r(\mathrm{Lk}_K(z)) \cong \tilde{H}_{r-1}(\mathrm{Lk}_L(y))$.*

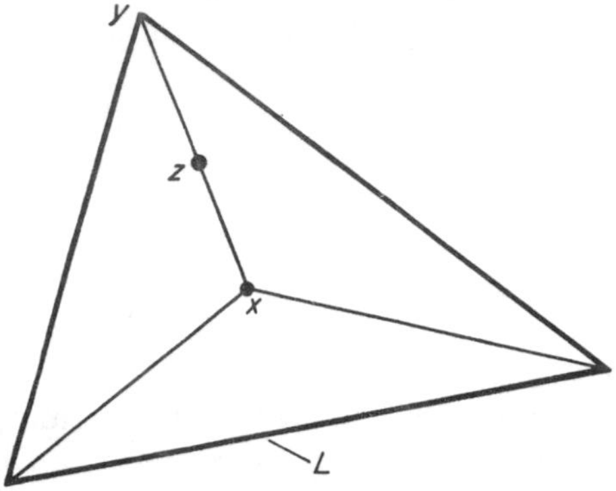

Fig. 5.8

Proof. Suppose first that y is a vertex of L, so that (x, y) is a 1-simplex. Then

$$\begin{aligned}\sigma \in \mathrm{Lk}_K(z) &\Leftrightarrow x, y \text{ and } \sigma \text{ are faces of a simplex of } K, \text{ but } \sigma \\ &\qquad \text{does not contain both } x \text{ and } y \\ &\Leftrightarrow \sigma \in \mathrm{Lk}_L(y) \cup \mathrm{Lk}_L(y) * x \cup \mathrm{Lk}_L(y) * y \\ &\qquad = \mathrm{Lk}_L(y) * (x \cup y)\end{aligned}$$

(if for example σ contains neither x nor y, then $\sigma \in L$ and $\sigma \in \mathrm{Lk}_K(y)$). Hence $\tilde{H}_r(\mathrm{Lk}_K(z)) \cong \tilde{H}_{r-1}(\mathrm{Lk}_L(y))$ by Theorem 4.4.10.

On the other hand if y is not a vertex of L, we may as well assume that y is a barycentre of some simplex of L. Thus y is a vertex of $(L') * x$, and by Theorem 2.4.5 this subdivision has not altered the homotopy types of either $|\mathrm{Lk}_K(z)|$ or $|\mathrm{Lk}_L(y)|$. ∎

Proposition 5.4.8 *If X is a closed homology $(n - 1)$-manifold, then $X \times I$ is a homology n-manifold with boundary, and $\partial X = X \times 0 \cup X \times 1$.*

Proof. Let K be a triangulation of X, and consider the 'cone' $CK = K * a$, where a is a single vertex. If $x \in |CK| - (|K| \cup a)$, x is an interior point of a straight-line segment ay, where y is a point of $|K|$: thus if z is the mid-point of ay, $\mathrm{Lk}(x) = \mathrm{Lk}(z)$, so that by Lemma 5.4.7

$$\tilde{H}_r(\mathrm{Lk}(x)) = \tilde{H}_r(\mathrm{Lk}(z)) \cong \tilde{H}_{r-1}(\mathrm{Lk}_K(y)) \cong \begin{cases} Z, & r = n - 1 \\ 0, & \text{otherwise.}\end{cases}$$

On the other hand, if $x \in |K|$, it is easy to see that $\mathrm{Lk}_{CK}(x) = \mathrm{Lk}_K(x) * a$, so that $\mathrm{Lk}_{CK}(x)$ is contractible, and $\tilde{H}_r(\mathrm{Lk}(x)) = 0$ for all r. That is to say, $|CK|$ is almost a homology n-manifold with boundary: each point except a satisfies the conditions of Definition 5.4.5 (a does not in general, since $\mathrm{Lk}(a) = K$).

Now $|K| \times I$ can be triangulated as $K \times I$, as in Section 4.2, and $K \times I$ has subcomplexes $K \times 0$ and $K \times 1$ that triangulate $|K| \times 0$ and $|K| \times 1$ respectively. Thus a simplicial complex M can be formed from the union of $K \times I$ and CK by identifying points of $K \times 1 = K$ with corresponding points of the subcomplex K of CK; and it is easy to see that M is another triangulation of $|CK|$ (see Fig. 5.9).

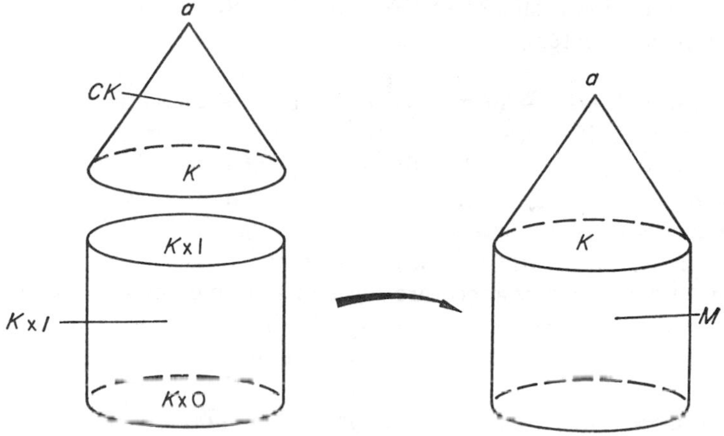

Fig. 5.9

It follows from Theorem 2.4.5 that, if $x \in |K| \times [0, 1)$, $|\mathrm{Lk}_{K \times I}(x)| = |\mathrm{Lk}_M(x)| \simeq |\mathrm{Lk}_{CK}(x)|$, so that

$$\tilde{H}_r(\mathrm{Lk}_{K \times I}(x)) \cong \begin{cases} Z, & \text{if } r = n - 1 \text{ and } x \in |K| \times (0, 1) \\ 0, & \text{otherwise.} \end{cases}$$

Similarly $\tilde{H}_r(\mathrm{Lk}_{K \times I}(x)) = 0$ for all r if $x \in |K| \times 1$, and certainly $|K| \times I$ is path-connected if $|K|$ is. Thus $|K| \times I$ is a homology n-manifold with boundary, and $\partial(|K| \times I) = |K| \times 0 \cup |K| \times 1$. ∎

The reader will have noticed that if X is a homology n-manifold with boundary, and X is a manifold with boundary or X is of the form $Y \times I$ for some (closed) homology $(n - 1)$-manifold, then ∂X is a subpolyhedron of X and each path component is a closed homology $(n - 1)$-manifold. This result is in fact true for all homology manifolds with boundary, though it is by no means an obvious consequence of Definition 5.4.5; indeed, to prove it we must first establish the

Lefschetz Duality Theorem. Let us call a homology n-manifold with boundary X *special* if, for each triangulation, ∂X is a subpolyhedron, and each path-component of ∂X is a closed homology $(n - 1)$-manifold. Our plan of action is first to prove the Lefschetz Duality Theorem for special homology manifolds with boundary, and then to deduce by induction on n that every homology n-manifold with boundary is special.

As in the case of the duality theorems of Section 5.3, the Lefschetz Duality Theorem takes two forms according as the manifold is orientable or not. Now the proof of Theorem 5.3.3 shows that if $|K|$ is a homology n-manifold with boundary, then $\dim K = n$ and every $(n - 1)$-simplex is a face of one or two n-simplexes. Definition 5.3.4 can therefore be extended as follows.

Definition 5.4.9 A homology n-manifold with boundary is said to be *orientable* if there exists a triangulation K, for which the n-simplexes can be identified with elements of $C_n(K)$ in such a way that, if σ is an $(n - 1)$-simplex that faces two n-simplexes τ_1 and τ_2, then σ occurs with opposite sign in $\partial(\tau_1)$ and $\partial(\tau_2)$.

Of course, this definition suffers from the same disadvantages as Definition 5.3.4. To resolve this difficulty (and indeed as the main tool in proving the Lefschetz Duality Theorem) we define the *double* of a special homology n-manifold with boundary.

Definition 5.4.10 Let X be a special homology n-manifold with (non-empty) boundary, and let X_0 and X_1 be two copies of X. The *double* of X, $2X$, is defined to be the space obtained from $X_0 \cup \partial X \times I \cup X_1$ by identifying points of ∂X_i with corresponding points of $\partial X \times i = \partial X$, for $i = 0, 1$.

Proposition 5.4.11 *$2X$ is a closed homology n-manifold, and is orientable if and only if X is.*

Proof. Let (K, L) be a triangulation of $(X, \partial X)$. $2X$ is certainly path-connected and triangulable, as $2K$, defined to be $K_0 \cup L \times I \cup K_1$ with appropriate identifications, where K_0 and K_1 are two copies of K. It is also clear from Proposition 5.4.8 that $\tilde{H}_r(\mathrm{Lk}(x)) \cong \tilde{H}_r(S^{n-1})$ for all points $x \in |2K|$ that do not lie in $|L \times 0|$ or $|L \times 1|$. On the other hand, if say $x \in |L \times 0|$, then $\mathrm{Lk}_{2K}(x) = \mathrm{Lk}_{L \times I}(x) \cup \mathrm{Lk}_K(x)$, and $\mathrm{Lk}_{L \times I}(x) \cap \mathrm{Lk}_K(x) = \mathrm{Lk}_L(x)$. But $\tilde{H}_*(\mathrm{Lk}_{L \times I}(x)) = \tilde{H}_*(\mathrm{Lk}_K(x)) = 0$, so that, by the reduced Mayer–Vietoris sequence,

$$\tilde{H}_r(\mathrm{Lk}_{2K}(x)) \cong \tilde{H}_{r-1}(\mathrm{Lk}_L(x)) \cong \tilde{H}_{r-1}(S^{n-2}) \cong \tilde{H}_r(S^{n-1}).$$

The same argument works for points of $|L \times 1|$, so that $|2K| = 2X$ is a closed homology n-manifold.

If $2X$ is orientable, identify the n-simplexes of $2K$ with elements of $C_n(2K)$, as in Definition 5.3.4. In particular this identifies the n-simplexes of K with elements of $C_n(K)$, and shows that X is orientable. To prove the converse, suppose that the n-simplexes of K are identified with elements of $C_n(K)$ as in Definition 5.4.9. Now an $(n - 1)$-simplex of K faces just one n-simplex of K if and only if it lies in L; so if z is the sum of the n-simplexes of K, $\partial^2(z) = 0$ and $\partial(z)$ is the sum of all the $(n - 1)$-simplexes of L, with appropriate signs. Since the path components of $|L|$ are obviously subpolyhedra and are homology $(n - 1)$-manifolds, this means that each path component of $|L|$ is orientable: they can be oriented by identifying each $(n - 1)$-simplex of L with the corresponding element in $\partial(z)$ (with its sign). To deduce that $|2K|$ is orientable, orient K_0 in the same way as K, K_1 in the opposite way (that is, σ in K_1 corresponds to $-\sigma$ in K), and $L \times I$ as follows. Let $h\colon S(|L|) \to S(|L| \times I)$ be the homomorphism of Proposition 4.2.21, which clearly restricts to $h\colon \Delta(L) \to \Delta(L \times I)$ and induces $h\colon C(L) \to C(L \times I)$. It is easy to see that each n-simplex τ of $L \times I$ occurs in just one expression $h(\sigma)$, where σ is an $(n - 1)$-simplex of L (already identified with an element of $C(L)$): identify τ with the corresponding element in $h(\sigma)$ (with its sign). Now Definition 5.3.4 is certainly satisfied for $(n - 1)$-simplexes of $K_0 - L$ or $K_1 - L$; on the other hand the formula

$$\partial h(\sigma) + h\partial(\sigma) = (i_1).\sigma - (i_0).\sigma$$

shows that Definition 5.3.4 is also satisfied for $(n - 1)$-faces of n-simplexes τ of $L \times I$: for if the face occurs in $\partial h(\sigma)$ the fact that L is oriented correctly will give us our result, and if the face is $(i_1).\sigma$ or $(i_0).\sigma$ the result follows because of the chosen orientation of K_0 and K_1 (other faces of τ must cancel in the expression for $\partial h(\sigma)$). Hence $2X$ is orientable. ▮

It follows that Definition 5.4.9 is independent of the triangulation of X, at least if X is special. It is also worth noting explicitly the following result, obtained in proving Proposition 5.4.11.

Corollary 5.4.12 *If X is orientable (and special), then each path component of ∂X is orientable.* ▮

The converse is not true: for example, each of the manifolds N_h^r is non-orientable (if N_h^r were orientable, it is easy to see that N_h would also be orientable); however ∂N_h^r is a disjoint union of S^1's, each of which is certainly orientable.

We are now in a position to prove the Lefschetz Duality Theorem.

Theorem 5.4.13 *Let X be a special homology n-manifold with boundary, and let (K, L) be a triangulation of $(X, \partial X)$. Then if X is orientable,*

$$H^r(K, L) \cong H_{n-r}(K) \qquad \text{(all } r\text{)},$$

and in any case

$$H^r(K, L; Z_2) \cong H_{n-r}(K; Z_2).$$

Proof. We may assume that L is not empty, for otherwise this is just Theorem 5.3.17. Suppose first that X is orientable, so that by Proposition 5.4.11 $2X$ is also orientable. Thus $2K$ is a triangulation of an orientable homology n-manifold, and Theorem 5.3.13 may be applied to the pair of subcomplexes $(2K, K_1)$, to obtain

$$H^r(2K, K_1) \cong H_{n-r}(\bar{K}_1).$$

But by Example 4.3.6 (applied to cohomology) we have

$$H^r(2K, K_1) \cong H^r(K_0 \cup L \times I, L \times 1).$$

However, in the exact cohomology sequence of the pair $(L \times I, L \times 1)$, the inclusion of $|L \times 1|$ in $|L \times I|$ is clearly a homotopy equivalence, so that $H_s(L \times I, L \times 1) = 0$ for all s. Thus in the exact cohomology sequence of the triple $(K_0 \cup L \times I, L \times I, L \times 1)$, we have

$$H^r(K_0 \cup L \times I, L \times 1) \cong H^r(K_0 \cup L \times I, L \times I).$$

Finally, using Example 4.3.6 again, we have

$$H^r(K_0 \cup L \times I, L \times I) \cong H^r(K_0, L \times 0) \cong H^r(K, L).$$

On the other hand,

$$\begin{aligned} |\bar{K}_1| &\simeq |2K| - |K_1|, \quad \text{by Proposition 5.3.9} \\ &= |K_0| \cup |L| \times [0, 1) \\ &\simeq |K_0|, \quad \text{by an obvious deformation retraction.} \end{aligned}$$

Hence

$$H_{n-r}(\bar{K}_1) \cong H_{n-r}(K_0) \cong H_{n-r}(K).$$

A similar proof works, using Z_2 coefficients, if X is not orientable. ∎

Of course, we can interchange homology and cohomology, and also use coefficients Q or Z_p, for any positive integer p, if X is orientable. Theorem 5.4.13 allows us now to justify our original definition of homology manifolds with boundary, by showing that they are all special.

Theorem 5.4.14 *Let X be a homology n-manifold with boundary. Then X is special.*

Proof. We shall prove this theorem by induction on n, supposing it to be true for all homology manifolds with boundary, of dimension $(n - 1)$. The induction starts, with $n = 1$, since in any triangulation of a homology 1-manifold X it is easy to see that ∂X is the set of vertices that are faces of just one 1-simplex.

Suppose then that X is a homology n-manifold with boundary $(n > 1)$, and that K is a triangulation of X. Let x be a vertex of K, and write $L = \text{Lk}(x)$. Now $\tilde{H}_r(L) \cong \tilde{H}_r(S^{n-1})$ or 0 according as x is in $X - \partial X$ or ∂X; hence $|L|$ is path-connected, and by Lemma 5.4.7 $|L|$ is a homology $(n - 1)$-manifold with boundary. Thus by the inductive hypothesis L contains a subcomplex M such that $|M| = \partial|L|$, and each path component of $|M|$ is a closed homology $(n - 2)$-manifold.

If $x \in X - \partial X$, Theorem 5.4.13 gives

$$H_0(L, M; Z_2) \cong H^{n-1}(L; Z_2) \cong Z_2.$$

If M were non-empty, the exact reduced homology sequence of the pair (L, M) shows this to be a contradiction, since $\tilde{H}_0(L) = 0$. Hence M is empty and so $|L|$ is a *closed* homology manifold. By Lemma 5.4.7 again, this means that for each point $y \in |L|$, the mid-point of the segment xy is in $X - \partial X$. But $\text{Lk}(z)$ is the same for all interior points z of xy, so that all points of these segments, except their end-points in $|L|$, lie in $X - \partial X$, and so x has a neighbourhood contained in $X - \partial X$. Now the same is true for any point x of $X - \partial X$, for we could take x to be a barycentre, and make it a vertex by subdividing. It follows that $X - \partial X$ is open; hence ∂X is closed; hence there is a subcomplex ∂K of K such that $|\partial K| = \partial|K|$, as in the proof of Proposition 5.4.4.

Now suppose that x is a vertex of ∂K. Again, Theorem 5.4.13 gives

$$H_r(L, M; Z_2) \cong H^{n-r-1}(L; Z_2) \cong \begin{cases} Z_2, & r = n - 1 \\ 0, & \text{otherwise.} \end{cases}$$

Thus the exact reduced homology sequence of the pair (L, M), and the fact that $\tilde{H}_r(L) = 0$, yields

$$\tilde{H}_r(M; Z_2) \cong \begin{cases} Z_2, & r = n - 2 \\ 0, & \text{otherwise.} \end{cases}$$

Since $\tilde{H}_{n-2}(M)$ must be a free abelian group, and $\tilde{H}_{n-3}(M; Z_2) = 0$, Theorem 4.5.15 shows that therefore $\tilde{H}_{n-2}(M) \cong Z$. Since $\tilde{H}_*(L) = 0$, it follows by the reduced Mayer–Vietoris sequence that $\tilde{H}_{n-1}(2L) \cong Z$; hence $2|L|$ is orientable, $|L|$ is orientable, and we may use Theorem 5.4.13 again to give $\tilde{H}_r(M) \cong \tilde{H}_r(S^{n-2})$, all r.

Now by Lemma 5.4.7, interior points z of segments xy ($y \in |L|$) lie in ∂X if and only if $y \in |M|$; indeed, since ∂X is closed, whole segments xy lie in ∂X if and only if $y \in |M|$. It follows that $\mathrm{Lk}_{\partial K}(x) = M$, so that $\tilde{H}_r(\mathrm{Lk}_{\partial K}(x)) \cong \tilde{H}_r(M) \cong \tilde{H}_r(S^{n-2})$. The same is true if x is a general point of ∂X by the usual trick: assume x is a barycentre and make it a vertex by subdividing. Thus the inductive step, and hence the proof of the theorem, is complete. ∎

We conclude this section with a short discussion of *cobordism*. As we have seen, the boundary of a homology n-manifold with boundary is a union of closed homology $(n-1)$-manifolds; cobordism is the study of the reverse problem: given a closed homology manifold (or a union of them), is it the boundary of a homology manifold with boundary? In particular, two closed $(n-1)$-manifolds X and Y are said to be *cobordant* if there exists an n-manifold Z such that ∂Z is the disjoint union of X and Y; and the problem can alternatively be stated: given X and Y, how do we tell whether or not they are cobordant? The problem remains unsolved in general for homology manifolds (though it has been done for *differentiable* manifolds: see the notes at the end of this chapter). However, it is sometimes possible to prove that X and Y are *not* cobordant by using the following result about the Euler–Poincaré characteristic.

Proposition 5.4.15 *Let X be a homology n-manifold with boundary. Then $\chi(\partial X)$ is even.*

Proof. Consider the Mayer–Vietoris sequence associated with the subpolyhedra $X_0 \cup (\partial X \times I)$ and $(\partial X \times I) \cup X_1$ of $2X$ (with coefficients Z_2):

$$\cdots \to H_r(\partial X \times I; Z_2) \to H_r(X_0 \cup \partial X \times I; Z_2) \oplus H_r(\partial X \times I \cup X_1; Z_2) \to H_r(2X; Z_2) \to H_{r-1}(\partial X \times I; Z_2) \to \cdots.$$

Now $\partial X \times I \simeq \partial X$, and $X_0 \cup \partial X \times I \simeq X \simeq \partial X \times I \cup X_1$, all by obvious deformation retractions. Thus the Mayer–Vietoris sequence can be amended so as to read

$$\cdots \to H_r(\partial X; Z_2) \to H_r(X; Z_2) \oplus H_r(X; Z_2) \to H_r(2X; Z_2) \to H_{r-1}(\partial X; Z_2) \to \cdots.$$

But an exact sequence of vector spaces over Z_2

$$0 \to V_1 \to V_2 \to \cdots \to V_m \to 0$$

may be regarded as a chain complex, so that, as in Lemma 4.5.17, we have $\sum(-1)^r \dim V_r = 0$. Applying this to the Mayer–Vietoris sequence, we obtain

$$\chi'(\partial X) - 2\chi'(X) + \chi'(2X) = 0,$$

where $\chi'(X) = \sum(-1)^r \dim H_r(X; Z_2)$. But it was proved in Proposition 5.3.20 that $\chi'(X) = \chi(X)$, so that

$$\chi(\partial X) - 2\chi(X) + \chi(2X) = 0.$$

If n is even, each path component of ∂X is an odd-dimensional closed homology manifold, so that by Proposition 5.3.20 $\chi(\partial X) = 0$. On the other hand if n is odd, then $\chi(2X) = 0$, so that $\chi(\partial X) = 2\chi(X)$, which is even. ∎

Corollary 5.4.16 *RP^n and S^n cannot be cobordant if n is even; neither can N_h and M_g be if h is odd (for any g), nor N_h and $N_{h'}$, if $h - h'$ is odd.*

Proof. If n is even, $\chi(RP^n) = 1$ and $\chi(S^n) = 2$. Hence

$$\chi(RP^n \cup S^n) = \chi(RP^n) + \chi(S^n) = 3,$$

and so $RP^n \cup S^n$ cannot be a boundary. The results about N_h and M_g follow similarly, since $\chi(M_g) = 2 - 2g$ and $\chi(N_h) = 2 - h$. ∎

In fact every M_g is a boundary, as also is N_h if h is even; on the other hand N_h is cobordant to N_1 if h is odd (see Exercise 17). Thus two 2-manifolds X and Y are cobordant (counting the empty set as a 2-manifold) if and only if $\chi(X) + \chi(Y)$ is even.

EXERCISES

1. Let K be a simplicial complex with a block dissection, and let L be a block subcomplex. Let C be the chain complex $\bigoplus C_n$, where $C_n = H_n(M^n, M^{n-1})$, and $M^n = K_n \cup L$; and for any abelian group G, let $C(G) = \bigoplus C_n(G)$ be the chain complex defined by $C_{-n}(G) = H^n(M^n, M^{n-1}; G)$. Show that $C \pitchfork G$ and $C(G)$ are chain-isomorphic, so that the two ways of generalizing to cohomology the method of calculating homology by a block dissection in fact coincide.
2. Show that a cap product $H^r(X; G) \otimes H_n(X; G) \to H_{n-r}(X; G)$ can be defined for an arbitrary space X, as follows. Given a generator λ of $S_n(X)$, let $\lambda' = \lambda F_1$ and $\lambda'' = \lambda F_2$, where $F_1: \Delta_{n-r} \to \Delta_n$ and $F_2: \Delta_r \to \Delta_n$ are defined by $F_1 = (a^0, \ldots, a^{n-r})$, $F_2 = (a^{n-r}, \ldots, a^n)$. Given also $\alpha \in S_r(X) \pitchfork G$, define $\alpha \cap (\sum \lambda_i \otimes g_i) = \sum \lambda_i' \otimes (\alpha(\lambda_i'')g_i)$;

show that this induces the required cap product between cohomology and homology. Prove also that this coincides with the cap product of Proposition 5.2.10 if X is a polyhedron, and that $f_*(f^*(x) \cap y) = x \cap f_*(y)$ if f is a continuous map.

3. Given an exact sequence $0 \to A \xrightarrow{\alpha} B \xrightarrow{\beta} C \to 0$ of abelian groups, and another abelian group G, show that the sequence

$$0 \longrightarrow C \pitchfork G \xrightarrow{\beta \pitchfork 1} B \pitchfork G \xrightarrow{\alpha \pitchfork 1} A \pitchfork G$$

is exact, but that $\alpha \pitchfork 1$ need not be onto.

4. Given abelian groups A and B, write A in the form F/R, where F is a free abelian group, so that there is an exact sequence

$$0 \to R \xrightarrow{\alpha} F \xrightarrow{\beta} A \to 0.$$

Define $\mathrm{Ext}\,(A, B) = R \pitchfork G/\mathrm{Im}\,(\alpha \pitchfork 1)$, so that by Exercise 2 there is an exact sequence

$$0 \longrightarrow A \pitchfork G \xrightarrow{\beta \pitchfork 1} F \pitchfork G \xrightarrow{\alpha \pitchfork 1} R \pitchfork G \longrightarrow \mathrm{Ext}\,(A, B) \longrightarrow 0.$$

Establish the following properties of $\mathrm{Ext}\,(A, B)$.

(a) $\mathrm{Ext}\,(A, B)$ depends only on A and B, and not on the particular representation $A = F/R$.

(b) $\mathrm{Ext}\,(\bigoplus_i A_i, \bigoplus_j B_j) \cong \bigoplus_{i,j} \mathrm{Ext}\,(A_i, B_j)$, provided both direct sums are finite.

(c) $\mathrm{Ext}\,(A, B) = 0$ if A is a free abelian group; $\mathrm{Ext}\,(Z_p, Z) \cong Z_p$, $\mathrm{Ext}\,(Z_p, Z_q) \cong Z_{(p,q)}$, $\mathrm{Ext}\,(Z_p, Q) = 0$, for positive integers p and q.

5. Let C be a chain complex in which each C_n is a free abelian group, and let G be any abelian group. Prove that

$$H_{-n}(C \pitchfork G) \cong H_n(C) \pitchfork G \oplus \mathrm{Ext}\,(H_{n-1}(C), G).$$

6. If in Exercise 5 each C_n is also finitely generated, show that

$$H_{-n}(C \pitchfork G) \cong H_{-n}(C \pitchfork Z) \otimes G \oplus \mathrm{Tor}\,(H_{-n-1}(C \pitchfork Z), G).$$

(*Hint:* use Proposition 5.2.15.)

7. Let M be a closed orientable homology n-manifold. Show that $H_{n-1}(M)$ is a free abelian group.

8. Show that a homology n-manifold (with or without boundary) is an n-manifold in the sense of Definitions 3.4.1 and 5.4.1, if $n \leqslant 3$. (*Hint:* prove this by induction on n, using Lemma 5.4.7 and the classification theorems for triangulable 2-manifolds.)

9. The result of Exercise 8 does not hold if $n \geqslant 4$. For example, let X be the space obtained from the (solid) dodecahedron by making identifications on the boundary, as in Example 3.3.22. Show that X is a triangulable 3-manifold, and that $\pi_1(X) \cong \mathrm{Gp}\,\{a, b;\ a^3(ab)^{-2}, b^5(ab)^{-2}\}$.

Show also that $H_3(X) \cong Z$, and deduce from Chapter 4, Exercise 3, and Theorem 5.3.17 that $H_1(X) = H_2(X) = 0$, so that X has the same homology groups as S^3.

Deduce that the suspension of X, SX, is a simply-connected homology 4-manifold that is not a 4-manifold.

10. Let $|K|$ be an orientable homology n-manifold, and let (L, M, N) be a triple of subcomplexes of K. Consider the diagram

$$\begin{array}{ccccccccc}
\cdots \to & H^r(L,M) & \xrightarrow{j^*} & H^r(L,N) & \xrightarrow{i^*} & H^r(M,N) & \xrightarrow{\delta^*} & H^{r+1}(L,M) & \to \cdots \\
& \downarrow D & & \downarrow D & & \downarrow D & & \downarrow D & \\
\cdots \to & H_{n-r}(\overline{M},\overline{L}) & \xrightarrow[i_*]{} & H_{n-r}(\overline{N},\overline{L}) & \xrightarrow[j_*]{} & H_{n-r}(\overline{N},\overline{M}) & \xrightarrow[\partial_*]{} & H_{n-r-1}(\overline{M},\overline{L}) & \to \cdots
\end{array}$$

where the rows are exact sequences of triples, and D is the isomorphism of Theorem 5.3.13. Show that the diagram is commutative up to sign; more precisely, that

$$i_*D = Dj^*, \quad j_*D = Di^*, \quad \text{and} \quad \partial_*D = (-1)^{n-r}D\delta^*$$

(use Chapter 4, Exercise 8).

Establish a similar result for homology and cohomology with coefficients Z_2, if $|K|$ is not necessarily orientable.

11. Let X be a homology n-manifold with boundary, and suppose $H_1(X; Z_2) = 0$. Show that X is orientable.

12. Let $|K|$ be an orientable homology n-manifold with boundary, and let (L, M) be a pair of subcomplexes of K, such that $|L| \cap \partial|K| = \varnothing$. Prove that $H^r(L, M) \cong H_{n-r}(\overline{M}, L)$, for each r.

13. Let $|K|$ be a closed homology n-manifold, and let (L, M) be a pair of subcomplexes such that $|L|$ is a homology n-manifold with boundary, and $\partial|L| = |L \cap M|$, $|K| = |L \cup M|$. Show that $|M|$ is also a homology n-manifold with boundary, and $\partial|M| = \partial|L|$. (*Hint:* prove this by induction on n, using the Mayer–Vietoris sequence on the links of points in $\partial|L|$, and the fact that $H_n(X) = 0$ for coefficients Z or Z_p, if X is an orientable homology n-manifold with non-empty boundary.)

14. Let (K, L) be a triangulation of an orientable homology n-manifold with boundary, where $|L| = \partial|K|$. If $i: |\overline{L}| \to |K|$ is the inclusion map, show that $i_*: H_r(\overline{L}) \to H_r(K)$ is an isomorphism for all r.

15. Given two closed homology n-manifolds $|K|$ and $|L|$, the *connected sum* $|K| \# |L|$ is defined as follows (compare Chapter 3, Exercise 14). Choose n-simplexes σ, τ in K and L respectively, and in $|K - \sigma| \cup |L - \tau|$, identify points of $|\dot{\sigma}|$ with corresponding points of $|\dot{\tau}|$, under some simplicial homeomorphism of $|\dot{\sigma}|$ onto $|\dot{\tau}|$. (This definition can be made independent of everything except the homeomorphism classes of

$|K|$ and $|L|$.) Show that $|K| \# |L|$ is a closed homology n-manifold, and that if $|K|$ and $|L|$ are orientable, then

$$H_n(|K| \# |L|) \cong Z,$$
$$H_r(|K| \# |L|) \cong H_r(K) \oplus H_r(L) \qquad (0 < r < n).$$

Prove also that $|K| \# |L|$ cannot be orientable unless both $|K|$ and $|L|$ are.

16. Let $|K|$ be a closed orientable homology n-manifold ($n \geqslant 2$). Let σ and τ be disjoint n-simplexes of K, and let L be the simplicial complex obtained from $K - (\sigma \cup \tau)$ and $\dot{\sigma} \times I$ by identifying $\dot{\sigma} \times 0$ with $\dot{\sigma}$ and $\dot{\sigma} \times 1$ with $\dot{\tau}$ (using a simplicial homeomorphism). Show that this can be done in such a way that $|L|$ is an orientable homology n-manifold, and

$$H_r(L) \cong H_r(K), \qquad r \neq n-1, 1$$

$$\left.\begin{array}{r} H_{n-1}(L) \cong H_{n-1}(K) \oplus Z \\ H_1(L) \cong H_1(K) \oplus Z \end{array}\right\} \quad (n \geqslant 3)$$

$$H_1(L) \cong H_1(K) \oplus Z \oplus Z \qquad (n = 2).$$

(The construction of L from K is a special case of a construction known as *surgery*: in general this consists in replacing a subspace homeomorphic to $S^r \times E^{n-r}$ by $E^{r+1} \times S^{n-r-1}$, which has the same boundary $S^r \times S^{n-r-1}$. In the above example $r = 0$.)

17. Show that in Exercise 16 the homology manifolds $|K|$ and $|L|$ are cobordant. Deduce that each of the orientable triangulable 2-manifolds M_g is the boundary of a 3-manifold.

Use a similar method to prove that N_h is the boundary of a 3-manifold if h is even, and that N_1 and N_h are cobordant if h is odd. (*Hint:* use Chapter 3, Exercise 14 to show that N_2 is the space obtained from two copies of N_1 by performing the construction of Exercise 16.)

NOTES ON CHAPTER 5

Cohomology. Cohomology theory originated with the 'pseudocycles' of Lefschetz [89], Chapter 6, and was developed further by Alexander [10], Whitney [163] and Lefschetz [91], Chapter 3. It was Whitney who invented the word 'cohomology'.

Corresponding to the Čech homology groups, one can define Čech cohomology groups: see for example Eilenberg and Steenrod [56], Chapter 9.

The Alexander–Poincaré Duality Theorem. The original references are Poincaré [116, 117] and Alexander and Veblen [12] for Theorem 5.3.17, and Alexander [8] for Theorem 5.3.19. The idea of combining these two theorems as in Theorem 5.3.13 is due to Lefschetz [89], Chapter 3.

In fact the assumption of triangulability in Theorem 5.3.13 is not really necessary, and was made only in order to simplify the proof. The more general theorem states that, if M is any (orientable) n-manifold, and (A, B) is a pair of closed subspaces of M, then

$$H_c^r(A, B) \cong H_{n-r}(M - B, M - A),$$

where H_c^r denotes Čech cohomology 'with compact supports'. A proof will be found in Spanier [131], Chapter 6 (see also Greenberg [60], Section 27).

In the case of non-orientable manifolds, Theorem 5.3.13 can be improved to give an isomorphism between cohomology with Z coefficients and homology with 'twisted integer' coefficients. See, for example, Swan [141], Chapter 11.

For more results along the lines of Proposition 5.3.22, see Chapter 8, Exercise 17.

The Jordan Curve Theorem. The result that the complement in S^3 of a subspace homeomorphic to S^1 has two connected components was first stated by Jordan [82], although his proof contained some gaps. The first rigorous proof was given by Veblen [146].

The Lefschetz Duality Theorem. Theorem 5.4.13 is due to Lefschetz [87, 88, 89]. In fact it holds for arbitrary manifolds with boundary: see Greenberg [60], Section 28, or Spanier [131], Chapter 6.

Cobordism. The concept of cobordism is due to Thom [143], who gave necessary and sufficient conditions for two differentiable manifolds to be cobordant. This work was extended to 'orientation-preserving' cobordism of differentiable manifolds by Milnor [105] and Wall [150]. The position with regard to cobordism of non-differentiable manifolds is, however, less satisfactory. A certain amount is known about manifolds of low dimensions (see Wall [151]), and for combinatorial manifolds the problem has been reduced to an (as yet unsolved) problem in homotopy theory (Williamson [164], Armstrong and Zeeman [15], Rourke and Sanderson [123]; see also Browder, Liulevicius and Peterson [28]).

Surgery and differentiable manifolds. The technique of surgery is due to Milnor [106]. Both cobordism and surgery have proved extremely useful tools in the study of manifolds. The interested reader should consult the excellent survey article of Smale [128].

CHAPTER 6

GENERAL HOMOTOPY THEORY

6.1 Introduction

In the last two chapters we have investigated algebraic invariants defined for various spaces. Although the techniques were powerful enough to prove some quite important theorems, the situation is somewhat unsatisfactory, because the definition of homology and cohomology appeared to be almost entirely algebraic. Aesthetically, at least, it would be more satisfying to perform as many of the manipulations as possible with the spaces themselves, rather than with groups, and also to ensure as far as possible that any constructions involved are homotopy-invariant.

The first aim in this chapter, then, will be the definition of constructions for topological spaces analogous to the direct sum, tensor product and $\pitchfork$ constructions for groups. We shall see that these constructions have many of the properties of their algebraic counterparts, and that there is a form of duality between the analogues of $\otimes$ and $\pitchfork$.

We shall then go on to consider in some detail the set $[X, Y]$ of homotopy classes of maps from a space X to a space Y. In many cases this set can be given the structure of a group, and it is of course a homotopy-type invariant of both X and Y. In the following two chapters it will become apparent that the set $[X, Y]$ generalizes both the fundamental group of Chapter 3 and the cohomology groups of Chapter 5, and so is an appropriate concept for unifying previous techniques. The present chapter, however, is concerned with the basic properties of $[X, Y]$, and with general methods for calculation: in particular we shall establish results similar to the very useful exact sequence theorems of Chapters 4 and 5.

The geometric analogues of algebraic constructions will be discussed in Section 6.2, and the set $[X, Y]$ in Section 6.3. Section 6.4 is concerned with exact sequence theorems involving $[X, Y]$, and Section 6.5 with certain important special cases of these exact sequences.

6.2 Some geometric constructions

Throughout this section, and indeed throughout the rest of this chapter, we shall assume unless otherwise stated that all spaces have base points, and that all continuous maps and homotopies are base-point-preserving, that is, homotopies will always be relative to base points. This assumption will usually not be made explicit: thus for example a map $f: X \to Y$ will always be taken to mean a based continuous map between spaces with base point.

Examples 6.2.1

(a) The spaces considered in Section 1.4 are given 'standard' base points as follows. The base point of I, the unit interval $[0, 1]$, is 1, and the base point of J, the double unit interval $[-1, 1]$, is -1; E^n and S^{n-1} each have base point $(-1, 0, \ldots, 0)$. Thus the identity map from J to E^1, and the standard map $\theta: E^n \to S^n$, are based maps (but $l: I \to J$ is not).

(b) Given a collection of based spaces X_a ($a \in A$), where x_a is the base point of X_a, the product $\underset{A}{\times} X_a$ is always given the base point (x_a). Thus for example J^n has base point $(-1, -1, \ldots, -1)$, and so $\rho: E^n \to J^n$ is *not* a based map (if $n \geqslant 2$).

Note that, if $f_a: X_a \to Y_a$ are based maps ($a \in A$), then so is $\times f_a: \times X_a \to \times Y_a$. Moreover, if each X_a is a copy of a single space X, the *diagonal map* $\Delta_X: X \to \times X_a$, defined by $\Delta_X(x) = (x_a)$, where $x_a = x$ for each a, is a based map (it is continuous by Proposition 1.4.21(c)). ∎

It was mentioned in Section 6.1 that one of the aims of this chapter is to ensure as far as possible that all geometric constructions are homotopy-invariant. Having already introduced the *product*, we start by checking its homotopy properties; and inevitably this involves first investigating maps between spaces, and then spaces themselves.

Theorem 6.2.2 *Given collections of (based) spaces X_a, Y_a ($a \in A$), and maps $f_a \simeq g_a: X_a \to Y_a$, then $\times f_a \simeq \times g_a$.*

Proof. Let $F_a: X_a \times I \to Y_a$ be the homotopy between f_a and g_a. Then $F: (\times X_a) \times I \to \times Y_a$, defined by

$$F((x_a), t) = (F_a(x_a, t)) \qquad (t \in I)$$

is clearly continuous, and is a (based) homotopy between $\times f_a$ and $\times g_a$. ∎

Corollary 6.2.3 *If each f_a is a homotopy equivalence, then so is $\times f_a$.*

Proof. Let $g_a: Y_a \to X_a$ be a homotopy inverse to f_a, for each $a \in A$. Then

$$(\times g_a)(\times f_a) = \times(g_a f_a) \simeq \times(1_{X_a}) = 1_{\times X_a}.$$

Similarly $(\times f_a)(\times g_a) \simeq 1_{\times Y_a}$. ∎

That is to say, the homotopy type of $\bigtimes X_a$ depends only on that of each X_a (clearly a similar proof will show that $\times f_a$ is a homeomorphism if each f_a is).

The first two new constructions in this chapter, the geometric analogues of the direct sum and tensor product, both make use of identification maps in their definitions. Since we shall be particularly interested in the homotopy properties of these constructions, it is convenient first to investigate the homotopy properties of identification spaces. These depend on the result that if $p: X \to Y$ is an identification map, then $p \times 1: X \times I \to Y \times I$ is also an identification map. This in turn is a special case of the more general result in which I is replaced by an arbitrary space Z; but a difficulty arises here, since this result would not be true without some restriction on the spaces involved (see Exercise 1). The following theorem covers all the cases that we shall need.

Theorem 6.2.4

(a) *If $p: X \to Y$ is an identification map, and Z is a locally compact Hausdorff space, then $p \times 1: X \times Z \to Y \times Z$ is an identification map.*

(b) *If A is a compact subspace of a space X, and $p: X \to X/A$ is the identification map, then for any space Z, $p \times 1: X \times Z \to (X/A) \times Z$ is an identification map.*

(*Note.* In this theorem, maps are not assumed to be base-point-preserving.)

Proof.

(a) Certainly $p \times 1$ is onto, and it is continuous by Proposition 1.4.21. It remains, then, to show that if $U \subset Y \times Z$ is a set such that $(p \times 1)^{-1}(U)$ is open, then U is itself open.

Let (y, z) be a point of U, and choose a point $x \in X$ such that $p(x) = y$. Thus $(p \times 1)(x, z) = (y, z)$, and $(x, z) \in (p \times 1)^{-1}(U)$. Since this set is open, and Z is locally compact and Hausdorff, Proposition 1.4.9 shows that there is an open set V in Z, containing z, such that $(x, z') \in (p \times 1)^{-1}(U)$ for all $z' \in \bar{V}$, and $\bar{V}$ is compact: see Fig. 6.1.

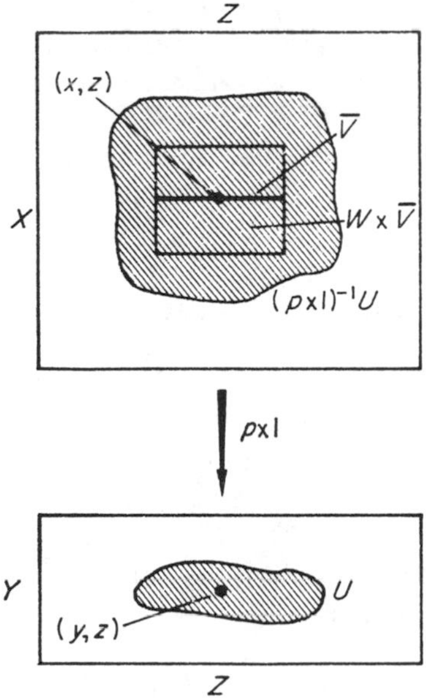

Fig. 6.1

Now each point $(x, z') \in x \times \bar{V}$ has an open neighbourhood of the form $A \times B$ contained in $(p \times 1)^{-1}(U)$, where A is open in X and B is open in Z. Since $\bar{V}$ is compact, a finite number of such B's will suffice to cover $\bar{V}$, and so if W is the intersection of the corresponding A's, W is open, $x \in W$, and $W \times \bar{V}$ is still contained in $(p \times 1)^{-1}(U)$. Notice also that $p^{-1}p(W) \times \bar{V} \subset (p \times 1)^{-1}(U)$, since $p(W) \times \bar{V} \subset U$.

Now consider all open sets W containing x, such that $W \times \bar{V} \subset (p \times 1)^{-1}(U)$. By taking their union, we might as well assume that W is the largest such set, in the sense that every such set is contained in W. In this case, $p^{-1}p(W) = W$: for certainly $W \subset p^{-1}p(W)$, and if x' is any point of $p^{-1}p(W)$, then $x' \times \bar{V} \subset (p \times 1)^{-1}(U)$; the same argument as before shows that there must be an open set W' containing x', such that $W' \times \bar{V} \subset (p \times 1)^{-1}(U)$, and so x' must be in W, for otherwise $W \cup W'$ would be strictly larger than W, although $(W \cup W') \times \bar{V} \subset (p \times 1)^{-1}(U)$; thus $p^{-1}p(W) \subset W$, and so $p^{-1}p(W) = W$.

Since p is an identification map, it follows that $p(W)$ is open in Y. But $(y, z) \in p(W) \times V \subset U$, so that U must be open.

(b) Again it is sufficient to show that if $U \subset (X/A) \times Z$ is a set such that $(p \times 1)^{-1}(U)$ is open, then U is open. As in case (a), let (y, z) be a point of U, and choose $x \in X$ such that $p(x) = y$.

If $x \in A$, then $A \times z \subset (p \times 1)^{-1}(U)$. Since A is compact, a similar argument to that used in case (a) shows that there exist open sets $V \subset X$, $W \subset Z$ such that $A \times z \subset V \times W \subset (p \times 1)^{-1}(U)$. But then $(y, z) \in p(V) \times W \subset U$; $p(V)$ is open since $p^{-1}p(V) = V$ (because $A \subset V$), and so $p(V) \times W$ is open.

If on the other hand $x \notin A$, there certainly exist open sets $V \subset X$, $W \subset Z$ such that $(x, z) \in V \times W \subset (p \times 1)^{-1}(U)$; and if $V \cap A = \varnothing$, then $p(V) \times W$ is open. However, if $V \cap A \neq \varnothing$, then $(p(A), z) \in U$, and we have already seen that we can then write

$$(p(A), z) \in p(V') \times W' \subset U.$$

But then $(x, z) \in p(V \cup V') \times (W \cap W') \subset U$; $p(V \cup V')$ is open, since $A \subset V'$, and so once again (x, z) is contained in an open subset of U. It follows that U is open, and so $p \times 1$ is an identification map. ∎

Theorem 6.2.4 has particularly useful corollaries on the homotopy properties of quotient spaces.

Proposition 6.2.5 *Given maps of pairs $f, g\colon (X, A) \to (Y, B)$, such that $f \simeq g$ as maps of pairs, then the induced maps $\bar{f}, \bar{g}\colon X/A \to Y/B$ are homotopic.*

(*Note.* The maps f and g need not be in any sense base-point-preserving; but if we take as base points of X/A, Y/B the points to which A, B respectively are identified, then, $\bar{f}, \bar{g}$ and the homotopy between them are all based.)

Proof. Let $F\colon (X \times I, A \times I) \to (Y, B)$ be the homotopy between f and g. Certainly F induces a function $\bar{F}\colon (X/A) \times I \to Y/B$ such that the diagram

$$\begin{array}{ccc} X \times I & \xrightarrow{F} & Y \\ {\scriptstyle p\times 1}\downarrow & & \downarrow{\scriptstyle q} \\ (X/A) \times I & \xrightarrow[\bar{F}]{} & Y/B \end{array}$$

is commutative, where p and q are the identification maps. But $\bar{F}(p \times 1) = qF$ is continuous, and hence $\bar{F}$ is continuous, since $p \times 1$ is an identification map (I is locally compact and Hausdorff). Thus $\bar{F}$ is a (based) homotopy between $\bar{f}$ and $\bar{g}$. ∎

Note that Proposition 6.2.5 remains true if $A = B = \varnothing$, provided that $X/\varnothing$, for example, is interpreted as the disjoint union of X with another point x_0, which is taken to be the base point of $X/\varnothing$.

Corollary 6.2.6 *If $f\colon (X, A) \to (Y, B)$ is a homotopy equivalence of pairs, then $\bar{f}\colon X/A \to Y/B$ is a (based) homotopy equivalence.* ∎

Corollary 6.2.7 *If (X, A) has the absolute homotopy extension property (see Theorem 2.4.1), and A is contractible, then the identification map $p\colon X \to X/A$ is a homotopy equivalence.*

Proof. Let $f\colon (X, x_0) \to (X, A)$ be the inclusion map, where x_0 is the base point of X, assumed to be in A. Since A is contractible, there exists a homotopy $F\colon A \times I \to A$ such that $F \mid (A \times 0)$ is the identity map and $F(A \times 1) = x_0$. This homotopy can be extended to a homotopy $F\colon X \times I \to X$, such that $F \mid (X \times 0)$ is the identity map; let $g\colon (X, A) \to (X, x_0)$ be $F \mid (X \times 1)$. Then fg is homotopic to the identity map (as a map of pairs) by the homotopy F, and the same is true of gf. Hence f is a homotopy equivalence, and therefore so is $\bar{f} = p$. ∎

In particular, therefore, $X \simeq X/A$ if (X, A) is a triangulable pair and A is contractible. See Exercise 2, however, for an example of a pair of spaces where this result does not hold.

Having investigated the homotopy properties of identification maps, we are now in a position to define the geometric analogue of the direct sum. It might be thought that the disjoint union would be suitable, but since in this chapter we assume that all spaces have base points, this is inappropriate, since there is no canonically defined base point in the disjoint union. This difficulty is easily overcome by identifying together the base points of each space.

Definition 6.2.8 Let X_a $(a \in A)$ be a collection of (disjoint) spaces, with base points $x_a \in X_a$. The *one-point union* (or *wedge*) $\bigvee_A X_a$ is defined to be the quotient space X/X_0, where X is the disjoint union of the spaces X_a, and X_0 is the subspace consisting of all the base points x_a; the base point of $\bigvee_A X_a$ is the point corresponding to X_0. In other words, $\bigvee_A X_a$ is the space obtained from X by identifying together the base points x_a.

As in the case of other constructions, if A is a finite set we shall often use the notation $X_1 \vee X_2 \vee \cdots$ instead of $\bigvee X_n$.

There is an analogue for the one-point union of the diagonal map. If each X_a is a copy of a single space X, the *folding map* $\nabla_X\colon \bigvee X_a \to X$ is induced by the map of the disjoint union to X that sends the point x in X_a to the point x in X, for each $a \in A$. ∇_X is continuous by Proposition 1.4.23(a).

To show that the one-point union has desirable homotopy properties, we prove a result similar to Theorem 6.2.2.

Theorem 6.2.9 *Given collections of spaces* X_a, Y_a $(a \in A)$, *and based maps* $f_a\colon X_a \to Y_a$, *there exists a map* $\bigvee f_a\colon \bigvee X_a \to \bigvee Y_a$, *with the following properties.*

(a) *If* $g_a\colon Y_a \to Z_a$ $(a \in A)$ *are further maps, then* $(\bigvee g_a)(\bigvee f_a) = \bigvee (g_a f_a)$.

(b) *If* $f_a \simeq g_a\colon X_a \to Y_a$ $(a \in A)$, *then* $\bigvee f_a \simeq \bigvee g_a$.

(c) *If each* $f_a\colon X_a \to Y_a$ *is a copy of a single map* $f\colon X \to Y$, *then* $f\nabla_X = \nabla_Y(\bigvee f_a)$.

Proof. $\bigvee f_a$ is the map induced by the obvious map of the disjoint unions. Properties (a) and (c) are clear from this definition, and property (b) follows from Proposition 6.2.5. ∎

Corollary 6.2.10 *If each* f_a *is a homotopy equivalence, so is* $\bigvee f_a$. ∎

Of course, a similar argument shows that $\bigvee f_a$ is a homeomorphism if each f_a is.

If A is a finite set, it is possible to regard $\bigvee X_a$ as a subspace of $\times X_a$, by means of the following result.

Proposition 6.2.11 *If* A *is a finite set, there is a homeomorphism of* $\bigvee X_a$ *onto the subspace* X *of* $\times X_a$ *consisting of all points with at most one co-ordinate different from the base point.*

Proof. There is an obvious map f of the disjoint union of the X_a to X, that sends the point x in X_a to the point of X whose 'a' co-ordinate is x and whose other co-ordinates are all base points. In fact, f is an identification map: it is certainly onto, and if U is a subset of X such that $f^{-1}(U)$ is open, then U is open. For $f^{-1}(U) = \bigcup_A (U \cap X_a)$ (where we identify X_a with its image under f); so if $f^{-1}(U)$ is open, each $U \cap X_a$ is open in X_a, and so U is open in X, since

$$U = \begin{cases} X \cap \bigcup_A \left((U \cap X_a) \times \bigtimes_{b \neq a} X_b\right), & \text{if } U \text{ does not contain the base point,} \\ X \cap \bigtimes_A (U \cap X_a), & \text{if } U \text{ contains the base point.} \end{cases}$$

Since f identifies together the base points of all the X_a, it follows that f induces a homeomorphism from $\bigvee X_a$ to X. ∎

The reader should notice where this proof breaks down if A is not a finite set: an arbitrary product of open sets is not necessarily open in a topological product.

The next construction is the geometric analogue of the tensor product. As in the algebraic situation, the definition is given only for a pair of spaces, and in fact the construction is not in general associative (though compare Theorem 6.2.23).

Definition 6.2.12 Given (based) spaces X and Y, the *reduced product* (or *smash product*) $X \wedge Y$ is defined to be the quotient space $(X \times Y)/(X \vee Y)$, where $X \vee Y$ is regarded as a subspace of $X \times Y$ as in Proposition 6.2.11. The base point of $X \wedge Y$ is of course the point corresponding to $X \vee Y$. Points of $X \wedge Y$ will be written in the form $x \wedge y$: this denotes the equivalence class of (x, y) in $X \times Y$.

The reduced product has mapping and homotopy properties that resemble those of the ordinary product and one-point union.

Theorem 6.2.13 *Given spaces X, Y, A, B, and based maps $f\colon X \to A$, $g\colon Y \to B$, there exists a map $f \wedge g\colon X \wedge Y \to A \wedge B$, with the following properties.*

(a) *If $h\colon A \to C$, $k\colon B \to D$ are further maps, then $(h \wedge k)(f \wedge g) = (hf) \wedge (kg)$.*

(b) *If $f \simeq f'\colon X \to A$ and $g \simeq g'\colon Y \to B$, then $f \wedge g \simeq f' \wedge g'$.*

Proof. The map $f \times g\colon X \times Y \to A \times B$ has the property that

$$f \times g(X \vee Y) \subset A \vee B;$$

hence $f \times g$ induces a map $f \wedge g\colon X \wedge Y \to A \wedge B$, and property (a) is obvious. As for (b), we note that the homotopy F between $f \times g$ and $f' \times g'$, constructed in the proof of Theorem 6.2.2, is in fact a homotopy of maps of pairs from $(X \times Y, X \vee Y)$ to $(A \times B, A \vee B)$, and so by Proposition 6.2.5 induces a homotopy between $f \wedge g$ and $f' \wedge g'$. ∎

Corollary 6.2.14 *If f and g are homotopy equivalences, so is $f \wedge g$.* ∎

Once again, of course, a similar proof shows that $f \wedge g$ is a homeomorphism if both f and g are.

The point in working with the reduced product, rather than the ordinary product, is that its properties are often more convenient when dealing with based spaces. For example, it is useful that each pair of points (x, y), in which either is a base point, becomes the base point of $X \wedge Y$. Moreover, the reduced product is particularly appropriate in any discussion of spheres, as the following proposition demonstrates.

Proposition 6.2.15 *For each $m, n \geqslant 0$, S^{m+n} is homeomorphic to $S^m \wedge S^n$.*

Proof. Consider the composite

$$E^m \times E^n \xrightarrow{p \times q} (E^m/S^{m-1}) \times (E^n/S^{n-1}) \xrightarrow{r} (E^m/S^{m-1}) \wedge (E^n/S^{n-1}),$$

where p, q and r are the obvious identification maps. Since $p \times q = (p \times 1)(1 \times q)$, and the composite of identification maps is again an identification map, Theorem 6.2.4 shows that this composite is an identification map. Morever its effect is to identify together points of $E^m \times S^{n-1} \cup S^{m-1} \times E^n$. Hence the standard homeomorphism

$$h_{m,n}\colon E^{m+n}, S^{m+n-1} \to E^m \times E^n, E^m \times S^{n-1} \cup S^{m-1} \times E^n$$

induces a homeomorphism, for which the same notation is used:

$$h_{m,n}\colon E^{m+n}/S^{m+n-1} \to (E^m/S^{m-1}) \wedge (E^n/S^{n-1})$$

(notice that this is a based map). But E^m/S^{m-1}, for example, is known to be homeomorphic to S^m, so that there is a (based) homeomorphism $h\colon S^{m+n} \to S^m \wedge S^n$, that makes the following diagram commutative:

$$\begin{array}{ccc} E^{m+n}/S^{m+n-1} & \xrightarrow{h_{m,n}} & (E^m/S^{m-1}) \wedge (E^n/S^{n-1}) \\ \downarrow{\scriptstyle\theta} & & \downarrow{\scriptstyle\theta\wedge\theta} \\ S^{m+n} & \xrightarrow[h]{} & S^m \wedge S^n. \end{array}$$ ∎

Although the homeomorphism h of Proposition 6.2.15 is easy to define, there are other more-or-less 'obvious' maps from S^{m+n} to $S^m \wedge S^n$, and for some purposes it is necessary to relate these. For example, one may regard S^{m+n} as a subspace of E^{m+n+1}, and consider the composite map

$$S^{m+n} \xrightarrow{h_{m+1,n}} S^m \times E^n \cup E^{m+1} \times S^{n-1} \xrightarrow{p} S^m \wedge (E^n/S^{n-1}) \xrightarrow{1\wedge\theta} S^m \wedge S^n,$$

where p is the identification map that identifies together points of $E^{m+1} \times S^{n-1}$ and $(-1, 0, \ldots, 0) \times E^n$. In fact this differs from the 'standard' homeomorphism of Proposition 6.2.15 only by a homotopy.

Proposition 6.2.16 *This map is homotopic to h.*

Proof. Consider the effect on a point $(x_1, \ldots, x_{m+n+1})$ of S^{m+n} of applying the composite map

$$S^{m+n} \xrightarrow{h_{m+1,n}} S^m \times E^n \cup E^{m+1} \times S^{n-1} \xrightarrow{p} S^m \wedge (E^n/S^{n-1})$$
$$\xrightarrow{1\wedge\theta} S^m \wedge S^n \xleftarrow{\theta\wedge\theta} (E^m/S^{m-1}) \wedge (E^n/S^{n-1})$$
$$\xleftarrow{h_{m,n}} E^{m+n}/S^{m+n-1} \xrightarrow{\theta} S^{m+n}.$$

Either $(x_1, \ldots, x_{m+n+1})$ is sent to the base point (for example, if $h_{m+1,n}(x_1, \ldots, x_{m+n+1}) \in E^{m+1} \times S^{n-1})$, or we can trace $(x_1, \ldots, x_{m+n+1})$ through the various maps as follows:

$$(x_1, \ldots, x_{m+n+1}) \to ((ax_1, \ldots, ax_{m+1}), (bx_{m+2}, \ldots, bx_{m+n+1})) \quad \text{under } h_{m+1,n}$$

$$\to (ax_1, \ldots, ax_{m+1}) \wedge (bx_{m+2}, \ldots, bx_{m+n+1}) \text{ under } p$$

$$\to (a_2x_2, \ldots, a_{m+1}x_{m+1}) \wedge (bx_{m+2}, \ldots, bx_{m+n+1}) \quad \text{under } (\theta \wedge \theta)^{-1}(1 \wedge \theta) = \theta^{-1} \wedge 1$$

$$\to (b_2x_2, \ldots, b_{m+1}x_{m+1}, \ldots, b_{m+n+1}x_{m+n+1}) \quad \text{under } h_{m,n}^{-1}$$

$$\to (x_0, c_2x_2, \ldots, c_{m+n+1}x_{m+n+1}) \text{ under } \theta,$$

where $a, b, a_2, \ldots, a_{m+1}, b_2, \ldots, b_{m+n+1}, c_2, \ldots, c_{m+n+1}$ are non-negative numbers. Thus if some $c_rx_r \neq 0$, $(x_1, \ldots, x_{m+n+1})$ is not sent to $(-x_1, \ldots, -x_{m+n+1})$; and if $c_2x_2 = \cdots = c_{m+n+1}x_{m+n+1} = 0$, then $(x_1, \ldots, x_{m+n+1})$ is sent to $(\pm 1, 0, \ldots, 0)$. But both $(1, 0, \ldots, 0)$ and $(-1, 0, \ldots, 0)$ are sent to themselves, so that in no case is $(x_1, \ldots, x_{m+n+1})$ sent to $(-x_1, \ldots, -x_{m+n+1})$. It follows from Corollary 2.2.4 that the composite map is homotopic to the identity map, so that $(1 \wedge \theta)ph_{m+1,n} \simeq (\theta \wedge \theta)h_{m,n}\theta^{-1} = h$. ∎

Another 'obvious' map from S^{m+n} to $S^m \wedge S^n$ is $(\theta \wedge 1)qh_{m,n+1}$, where $q\colon S^{m-1} \times E^{n+1} \cup E^m \times S^n \to (E^m/S^{m-1}) \wedge S^n$ is the obvious identification map. An argument similar to that of Proposition 6.2.16 shows that this map is homotopic to ϕh, where

$$\phi(x_1, \ldots, x_{m+n+1}) = (x_{m+1}, x_1, \ldots, x_m, x_{m+2}, \ldots, x_{m+n+1})$$

(so that, by Example 4.4.11, $\phi_*\colon H_{m+n}(S^{m+n}) \to H_{m+n}(S^{m+n})$ is multiplication by $(-1)^m$).

Apart from its applications to spheres, however, the reduced product is also useful in constructing an analogue for arbitrary based spaces of the suspension construction of Definition 4.4.8.

Definition 6.2.17 The *reduced suspension* of a space X, sX, is defined to be $X \wedge S^1$.

Thus for example the reduced suspension of S^n is homeomorphic to S^{n+1}; and it is immediate from Corollary 6.2.14 that $sX \simeq sY$ if $X \simeq Y$. The notation sX is used to prevent confusion with the suspension SX of a triangulated space X: if X is triangulable the two suspensions closely resemble each other, but are not quite identical, as the next proposition shows.

Proposition 6.2.18 *Let x_0 be the base point of a (not necessarily triangulable) space X. Then sX is homeomorphic to the quotient space $(X \times I)/(X \times 0 \cup x_0 \times I \cup X \times 1)$.*

(See Fig. 6.2, in which the thick line is supposed to be identified to a point.)

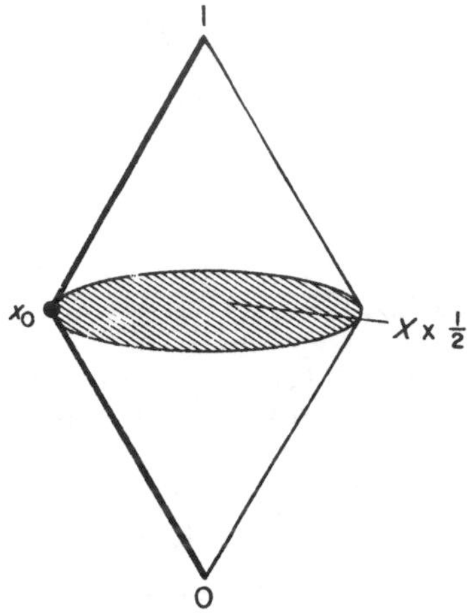

Fig. 6.2

Proof. This is rather similar to Proposition 6.2.15. Now the composite of standard maps

$$I \xrightarrow{l} J = E^1 \xrightarrow{\theta} S^1$$

is an identification map, and identifies together the points 0 and 1 (in fact $\theta l(t) = (\cos(2t - 1)\pi, \sin(2t - 1)\pi)$). Thus if $p: X \times S^1 \to X \wedge S^1$ is the identification map, Theorem 6.2.4(b) shows that the composite map

$$X \times I \xrightarrow{1 \times \theta l} X \times S^1 \xrightarrow{p} X \wedge S^1$$

is also an identification map, and its effect is to identify together points of $X \times 0 \cup x_0 \times I \cup X \times 1$. It follows that $p(1 \times \theta l)$ induces a homeomorphism

$$(X \times I)/(X \times 0 \cup x_0 \times I \cup X \times 1) \to X \wedge S^1 = sX. \blacksquare$$

Corollary 6.2.19 *If X is a polyhedron, and x_0 is a vertex, there is a homotopy equivalence $p: SX \to sX$, such that if $f: X \to Y$ is a continuous map of polyhedra, the diagram*

$$\begin{array}{ccc} SX & \xrightarrow{Sf} & SY \\ \downarrow{\scriptstyle p} & & \downarrow{\scriptstyle p} \\ sX & \xrightarrow[f \wedge 1]{} & sY \end{array}$$

is commutative.

Proof. It is easy to see that SX is homeomorphic to the space obtained from $X \times I$ by identifying $X \times 1$ and $X \times 0$ to points (these points correspond to a and b respectively in Definition 4.4.8). Thus there is an identification map $p: SX \to sX$, that identifies $x_0 \times I$ to a point. Now $x_0 \times I$ is contractible, and by Theorem 2.4.1 the pair $(X \times I, x_0 \times I)$ has the absolute homotopy extension property, since $X \times I$ is a polyhedron and $x_0 \times I$ is a subpolyhedron. Hence p is a homotopy equivalence, by Corollary 6.2.7.

That $p(Sf) = (f \wedge 1)p$ is an easy consequence of the definition of p and the fact that Sf is induced by $f \times 1: X \times I \to Y \times I$. ∎

Observe that if $X = S^n$, $p: S(S^n) \to sS^n$ gives yet another homotopy equivalence from S^{n+1} to $S^n \wedge S^1$, if $S(S^n)$ is identified with S^{n+1} as in Example 4.4.9. However, if $q: S^n \times E^1 \cup E^{n+1} \times S^0 \to S(S^n)$ is the map that identifies the two components of $E^{n+1} \times S^0$ to points and sends E^1 to I by l^{-1}, it is easy to see that the diagram

$$\begin{array}{ccc} S^{n+1} & \xrightarrow{h_{n+1,1}} & S^n \times E^1 \cup E^{n+1} \times S^0 \\ & \searrow \quad \swarrow{\scriptstyle q} & \\ & S(S^n) & \end{array}$$

is homotopy-commutative, in the sense that $qh_{n+1,1}$ is homotopic to the identity map $S^{n+1} \to S(S^n)$. It follows that $p: S^{n+1} \to s(S^n)$ is homotopic to the map considered in Proposition 6.2.16, and so is homotopic to the 'standard' homeomorphism h of Proposition 6.2.15.

It will be seen that the reduced suspension is in keeping with the spirit of this chapter, in which the usual policy is to identify to the base point anything that involves the base points of the original spaces. In the same spirit, the cone construction of Chapter 4, Exercise 5, can be adapted to arbitrary spaces as follows.

Definition 6.2.20 The *reduced cone* on a space X, cX, is defined to be $X \wedge I$ (recall that the base point of I is always taken to be 1).

The reduced cone bears the same relation to the join of a polyhedron and a point that the reduced suspension bears to the suspension of a polyhedron in the sense of Definition 4.4.8. We shall not give the details here, but merely note that the property of being contractible holds for any reduced cone.

Proposition 6.2.21 *For any space X, cX is contractible.*

Proof. Since I is contractible (to the point 1), Corollary 6.2.14 yields

$$cX = X \wedge I \simeq X \wedge 1,$$

which is clearly a single point. ▮

This discussion of the reduced product is concluded with results on the composition of the reduced product and one-point union constructions, and on the associativity of reduced products.

Theorem 6.2.22 *Given three spaces X, Y and Z, $(X \vee Y) \wedge Z$ is homeomorphic to $(X \wedge Z) \vee (Y \wedge Z)$.* (Compare Proposition 4.5.7(b).)

Proof. The map $f: X \times Y \times Z \to X \times Z \times Y \times Z$, defined by $f(x, y, z) = (x, z, y, z)$, is clearly continuous, since if $A \times B \times C \times D$ is an open set in $X \times Z \times Y \times Z$, $f^{-1}(A \times B \times C \times D) = A \times C \times (B \cap D)$. This composes with the product of identification maps to give a map

$$X \times Y \times Z \to (X \wedge Z) \times (Y \wedge Z),$$

and if $X \vee Y$, for example, is regarded as a subspace of $X \times Y$ by Proposition 6.2.11, this map sends $(X \vee Y) \times Z$ to $(X \wedge Z) \vee (Y \wedge Z)$. Moreover, $(X \vee Y) \vee Z$ is sent to the base point, so that f induces a map

$$g: (X \vee Y) \wedge Z \to (X \wedge Z) \vee (Y \wedge Z),$$

where $g((x, y_0) \wedge z) = x \wedge z$ in $X \wedge Z$ and $g((x_0, y) \wedge z) = y \wedge z$ in $Y \wedge Z$.

Conversely, define

$$h: (X \wedge Z) \vee (Y \wedge Z) \to (X \vee Y) \wedge Z$$

by $h = [(i_X \wedge 1_Z) \vee (i_Y \wedge 1_Z)]\nabla$, where $i_X: X \to X \vee Y$ is defined by $i_X(x) = (x, y_0)$, and i_Y is similarly defined. Then $h(x \wedge z) = (x, y_0) \wedge z$ and $h(y \wedge z) = (x_0, y) \wedge z$, so that both gh and hg are identity maps, and hence g is a homeomorphism. ▮

In particular, by taking $Z = S^1$, this proves that $s(X \vee Y)$ is homeomorphic to $sX \vee sY$.

Theorem 6.2.23 *If X and Y are compact, and X is Hausdorff, then $(X \wedge Y) \wedge Z$ is homeomorphic to $X \wedge (Y \wedge Z)$.*

Proof. Write p for the various identification maps of the form $X \times Y \to X \wedge Y$, and consider the diagram

$$\begin{array}{ccc} X \times Y \times Z & \xrightarrow{1} & X \times Y \times Z \\ \downarrow{\scriptstyle p \times 1} & & \downarrow{\scriptstyle 1 \times p} \\ (X \wedge Y) \times Z & & X \times (Y \wedge Z) \\ \downarrow{\scriptstyle p} & & \downarrow{\scriptstyle p} \\ (X \wedge Y) \wedge Z & & X \wedge (Y \wedge Z). \end{array}$$

Now $p \times 1$ is an identification map by Theorem 6.2.4(b), since $X \vee Y$ is compact if X and Y are compact. Also, $1 \times p$ is an identification map by Theorem 6.2.4(a), since X is locally compact and Hausdorff. Since both $p(p \times 1)$ and $p(1 \times p)$ identify to points those points of $X \times Y \times Z$ that have at least one co-ordinate equal to a base point, $1 : X \times Y \times Z \to X \times Y \times Z$ induces maps

$$f : (X \wedge Y) \wedge Z \to X \wedge (Y \wedge Z)$$

and

$$g : X \wedge (Y \wedge Z) \to (X \wedge Y) \wedge Z$$

that are clearly homeomorphisms. ∎

A similar proof works if Y and Z are compact and Z is Hausdorff, so that in particular $s(sX)$ is homeomorphic to $X \wedge S^2$ for any space X. See also Exercise 4, for another set of conditions on X, Y and Z that makes $(X \wedge Y) \wedge Z$ homeomorphic to $X \wedge (Y \wedge Z)$.

The next and last construction in this section is the geometric analogue of $\pitchfork$. Given spaces X and Y, it is reasonably obvious that we should consider a space whose points are the continuous maps from X to Y, but it is not immediately clear how to topologize this space. We shall use what is known as the 'compact-open' topology, for reasons that will become clear in the proof of Theorem 6.2.25 below.

Definition 6.2.24 Given spaces X and Y, with base points x_0 and y_0 respectively, the *mapping space* Y^X consists of all (based) maps from X to Y. The base point of Y^X is the 'constant map', that sends all of X to y_0, and Y^X is topologized by taking as a sub-base of open sets all subsets of Y^X of the form

$$W_{K,U} = \{f : X \to Y \mid f(K) \subset U\},$$

where K is a compact subspace of X and U is an open subspace of Y. This topology on Y^X is called the *compact-open topology*.

As with the other constructions in this section, the next step is to prove a theorem on maps and homotopies.

Theorem 6.2.25 *Given spaces X, Y, A and B, and based maps $f: A \to X$, $g: Y \to B$, there exists a map $g^f: Y^X \to B^A$, with the following properties.*

(a) *If $h: C \to A$ and $k: B \to D$ are further maps, then*

$$(k^h)(g^f) = (kg)^{fh}: Y^X \to D^C.$$

(b) *If $f \simeq f': A \to X$ and $g \simeq g': Y \to B$, then $g^f \simeq (g')^{f'}$.*

Proof. Given a point λ of Y^X, that is, a map $\lambda: X \to Y$, define

$$g^f(\lambda) = g\lambda f.$$

This is certainly a map from A to B, and if λ is the constant map from X to Y, then $g^f(\lambda)$ is the constant map from A to B. However, it is not obvious that g^f is continuous, and this must be proved next.

Take a sub-basic open set $W_{K,U}$ in B^A, where $K \subset A$ is compact and $U \subset B$ is open. Then

$$\begin{aligned}(g^f)^{-1}(W_{K,U}) &= \{\lambda: X \to Y \mid g\lambda f(K) \subset U\} \\ &= \{\lambda: X \to Y \mid \lambda f(K) \subset g^{-1}(U)\}.\end{aligned}$$

But $f(K)$ is a compact subspace of X, and $g^{-1}(U)$ is an open subspace of Y, so that $(g^f)^{-1}(W_{K,U}) = W_{f(K),g^{-1}(U)}$, a sub-basic open set of Y^X. Hence g^f is continuous. (The reader will see now why the compact-open topology is used: continuous images of compact sets are compact, and inverse images of open sets are open.)

Property (a) follows immediately from the definition of g^f. As for property (b), this is rather more complicated. Let $F: A \times I \to X$ and $G: Y \times I \to B$ be the homotopies between f, f' and g, g' respectively, and let $f_t: A \to X$, $g_t: Y \to B$ $(0 \leqslant t \leqslant 1)$ be the maps defined by $f_t(a) = F(a, t)$, $g_t(y) = G(y, t)$. Then certainly the set of maps $(g_t)^{f_t}: Y^X \to B^A$ starts with g^f and ends with $(g')^{f'}$, but we have to prove that this process defines a *continuous* map $Y^X \times I \to B^A$.

To do so, define functions (which will afterwards be proved to be continuous) $\theta: Y^X \times I \to (Y \times I)^X$ and $\phi: B^{A \times I} \times I \to B^A$ by the rules

$$[\theta(\lambda, t)](x) = (\lambda(x), t) \qquad (x \in X, \lambda: X \to Y, t \in I)$$

$$[\phi(\mu, t)](a) = \mu(a, t) \qquad (a \in A, \mu: A \times I \to B).$$

Now consider the composite map

$$Y^X \times I \xrightarrow{1 \times \Delta} Y^X \times I \times I \xrightarrow{\theta \times 1} (Y \times I)^X \times I$$
$$\xrightarrow{G^F \times 1} B^{A \times I} \times I \xrightarrow{\phi} B^A,$$

where $\Delta: I \to I \times I$ is the diagonal map. Under this composite, the pair (λ, t) $(\lambda: X \to Y, t \in I)$ is sent to the map that sends $a \in A$ to $g_t \lambda f_t(a) \in B$; that is, for a given t the composite is exactly $(g_t)^{f_t}$. So the composite is the homotopy we want, and is continuous provided θ and ϕ are continuous.

To deal with θ, consider the set $W_{K,U} \subset (Y \times I)^X$, where $K \subset X$ is compact and $U \subset Y \times I$ is open, and suppose that $\theta(\lambda, t) \in W_{K,U}$. Then $\lambda(K) \times t \subset U$. Now for each point $(y, t) \in \lambda(K) \times t$, there are open sets $V_y \subset Y$, $T_y \subset I$, such that

$$(y, t) \in V_y \times T_y \subset U.$$

The open sets V_y cover $\lambda(K)$, which is compact, so that a finite subcollection of them, say $V_{y_1}, \ldots, V_{y_n}$, will suffice to cover $\lambda(K)$. Thus if $V = \bigcup_{r=1}^{n} V_{y_r}$ and $T = \bigcap_{r=1}^{n} T_{y_r}$, V and T are open sets, and

$$\lambda(K) \times t \subset V \times T \subset U.$$

Now consider $W_{K,V} \times T \subset Y^X \times I$. Certainly (λ, t) is in this subset, and if (λ', t') is any other point in it, $\lambda'(K) \times t' \subset V \times T \subset U$, so that $\theta(\lambda', t') \in W_{K,U}$, and $(\lambda', t') \in \theta^{-1}(W_{K,U})$. It follows that $\theta^{-1}(W_{K,U})$ is open, so that θ is continuous.

The proof that ϕ is continuous is similar. This time consider $W_{K,U} \subset B^A$, where $K \subset A$ is compact and $U \subset B$ is open, and suppose that $\phi(\mu, t) \in W_{K,U}$. Then $\mu(K \times t) \subset U$, or $K \times t \subset \mu^{-1}(U)$, which is an open set in $A \times I$. The same argument as before shows that there exists an open set $T \subset I$ such that

$$K \times t \subset K \times T \subset \mu^{-1}(U),$$

and since I is locally compact and Hausdorff there is an open set $V \subset I$ such that $t \in V \subset \bar{V} \subset T$ and $\bar{V}$ is compact. Now consider $W_{K \times \bar{V}, U} \times V \subset B^{A \times I} \times I$. Again (μ, t) is in this subset, and if (μ', t') is another point in it, $\mu'(K \times t') \subset \mu'(K \times \bar{V}) \subset U$, so that $\phi(\mu', t') \in W_{K,U}$. Hence $\phi^{-1}(W_{K,U})$ is open and ϕ is continuous.

As has already been remarked, the continuity of θ and ϕ is sufficient to prove that $g^f \simeq (g')^{f'}$. ▮

Corollary 6.2.26 *If f and g are homotopy equivalences, so is g^f.* ▮

Naturally, also, g^f is a homeomorphism if f and g are.

For the next result, suppose that Z is a subspace of Y, so that a set $U \subset Z$ is open if and only if it is of the form $V \cap Z$, where V is open in Z. Certainly Z^X is a subset of Y^X, and in fact the compact-open topology on Z^X coincides with the topology as a subspace of Y^X.

Proposition 6.2.27 *If Z is a subspace of Y, then Z^X is a subspace of Y^X.*

Proof. We have to show that a set is open in Z^X if and only if it is the intersection with Z^X of a set that is open in Y^X. Now if $i: Z \to Y$ is the inclusion map, $i^1: Z^X \to Y^X$ is continuous, so that if $U \subset Y^X$ is open, $U \cap Z^X = (i^1)^{-1}(U)$ is open in Z^X. To prove the converse, it is sufficient to consider an open set in Z^X of the form $W_{K,U}$, where $K \subset X$ is compact and $U \subset Z$ is open. But $U = V \cap Z$, where V is open in Y; and

$$\begin{aligned} W_{K,V} \cap Z^X &= \{f: X \to Y \mid f(K) \subset V \text{ and } f(X) \subset Z\} \\ &= \{f: X \to Z \mid f(K) \subset V \cap Z = U\} \\ &= W_{K,U}. \end{aligned}$$

That is, an open set in Z^X is the intersection with Z^X of an open set in Y^X. ∎

As examples of mapping spaces, we can define spaces that are in a sense dual to the reduced suspension and reduced cone constructions.

Definition 6.2.28 Given a space X, the *path space* LX is defined to be X^I, and the *loop space* ΩX is X^{S^1}.

Thus the points of LX are the paths in X that end at the base point x_0, and it is easy to see that the points of ΩX may be regarded as loops in X based at x_0: more precisely, ΩX may be identified with the subspace $(1)^{\theta l}(\Omega X)$ of LX. The relationship of ΩX and LX to sX and cX will become clearer later, but it is worth noticing the following analogue of Proposition 6.2.21 here.

Proposition 6.2.29 *For any space X, LX is contractible.*

Proof. By Corollary 6.2.26, $LX = X^I \simeq X^1$. But X^1 is a single point. ∎

We end this section with some results on the composition of the mapping space construction with products, reduced products and one-point unions. Most of these amount to proving that certain maps are continuous, and the first of these is the 'evaluation map'.

Definition 6.2.30 Given spaces X and Y, define a function

$$f\colon Y^X \times X \to Y$$

by the rule $f(\lambda, x) = \lambda(x)$ ($\lambda\colon X \to Y, x \in X$). If λ is the constant map, or x is the base point x_0, then $f(\lambda, x) = y_0$, the base point of Y. That is, $f(Y^X \vee X) = y_0$, so that f induces a function $e\colon Y^X \wedge X \to Y$, called the *evaluation map*.

We do not claim that e is always continuous (see Exercise 6). However, it is if X is a reasonably well-behaved space.

Theorem 6.2.31 *If X is locally compact and Hausdorff, then $e\colon Y^X \wedge X \to Y$ is continuous.*

Proof. It is sufficient to show that $f\colon Y^X \times X \to Y$ is continuous, and the proof of this follows the pattern familiar from Theorem 6.2.25. Suppose then that $U \subset Y$ is open, and that $f(\lambda, x) \in U$. Then $\lambda(x) \in U$ and $x \in \lambda^{-1}(U)$, which is open in X. Since X is locally compact and Hausdorff, there exists an open set V in X, such that $x \in V \subset \bar{V} \subset \lambda^{-1}(U)$, and $\bar{V}$ is compact. Consider $W_{\bar{V},U} \times V \subset Y^X \times X$: this contains (λ, x), and if (λ', x') is another point in it, then

$$f(\lambda', x') = \lambda'(x') \subset \lambda'(\bar{V}) \subset U.$$

Thus $f^{-1}(U)$ is open, and f is continuous. ∎

The next few results show that mapping spaces obey rules similar to the index laws for real numbers, at least if the spaces involved are sufficiently well-behaved.

Theorem 6.2.32 *Given spaces X, Y and Z, where X and Y are Hausdorff, $Z^{X \vee Y}$ is homeomorphic to $Z^X \times Z^Y$.*

Proof. Let x_0 and y_0 be the base points of X and Y respectively, and define maps

$$i_X\colon X \to X \vee Y, \qquad i_Y\colon Y \to X \vee Y$$

by $i_X(x) = (x, y_0)$, $i_Y(y) = (x_0, y)$ (using Proposition 6.2.11 to identify $X \vee Y$ with the subspace $X \times y_0 \cup x_0 \times Y$ of $X \times Y$). Now define a function $\theta\colon Z^X \times Z^Y \to (Z \vee Z)^{X \vee Y}$ by $\theta(\lambda, \mu) = \lambda \vee \mu$, where $\lambda\colon X \to Z$ and $\mu\colon Y \to Z$, and consider the composite functions

$$\phi\colon Z^{X\vee Y} \xrightarrow{\Delta} Z^{X\vee Y} \times Z^{X\vee Y} \xrightarrow{1_{i_X} \times 1_{i_Y}} Z^X \times Z^Y$$

and

$$\psi\colon Z^X \times Z^Y \xrightarrow{\theta} (Z \vee Z)^{X\vee Y} \xrightarrow{\nabla^1} Z^{X\vee Y},$$

where Δ is the diagonal map, and $\nabla: Z \vee Z \to Z$ is the folding map. Given $\nu: X \vee Y \to Z$, $\phi(\nu) = (\nu i_X, \nu i_Y)$, and given $\lambda: X \to Z$ and $\mu: Y \to Z$, $\psi(\lambda, \mu) = \nabla(\lambda \vee \mu)$. Thus $\phi\psi$ and $\psi\phi$ are identity functions, and the only point that remains in showing that ϕ is a homeomorphism is to show that θ is continuous (it is certainly base-point-preserving).

To do so, consider the set $W_{K,U}$, where $K \subset X \vee Y$ is compact and $U \subset Z \vee Z$ is open. Now

$$\begin{aligned}\theta^{-1}(W_{K,U}) &= \{(\lambda, \mu) \mid (\lambda \vee \mu)(K) \subset U\} \\ &= \{(\lambda, \mu) \mid \lambda(K \cap X) \subset U \cap (Z \times z_0) \\ &\qquad\qquad \text{and } \mu(K \cap Y) \subset U \cap (z_0 \times Z)\},\end{aligned}$$

where z_0 is the base point of Z, and X and Y are identified with their images in $X \vee Y$. Certainly $U_1 = U \cap (Z \times z_0)$ and $U_2 = U \cap (z_0 \times Z)$ are open, since U is the intersection with $Z \vee Z$ of an open set in $Z \times Z$. But since X and Y are Hausdorff, so is $X \times Y$ and hence $X \vee Y$: thus K, X and Y are closed in $X \vee Y$, so that $K \cap X$ and $K \cap Y$ are closed and hence compact. That is, $\theta^{-1}(W_{K,U}) = W_{K\cap X, U_1} \times W_{K \cap Y, U_2}$, so that θ is continuous. Hence ϕ is a homeomorphism. ∎

There is a similar result involving $(Y \times Z)^X$ and $Y^X \times Z^X$, though it is a little more difficult this time to prove that the maps involved are continuous. We need the following lemma.

Lemma 6.2.33 *Let X be a Hausdorff space, and let $\mathscr{S}$ be a sub-base of open sets for a space Y. Then the sets of the form $W_{K,U}$, for $K \subset X$ compact and $U \in \mathscr{S}$, form a sub-base of open sets for Y^X.*

Proof. Let $K \subset X$ be compact, $V \subset Y$ be open, and let $\lambda \in W_{K,V}$. Now it is certainly true that $V = \bigcup V_\alpha$, where each V_α is a finite intersection of sets in $\mathscr{S}$. Then $K \subset \bigcup \lambda^{-1}(V_\alpha)$; hence, since K is compact, a finite collection of the sets $\lambda^{-1}(V_\alpha)$, say $\lambda^{-1}(V_1), \dots, \lambda^{-1}(V_n)$, suffice to cover K. Since K is a compact Hausdorff space, it is regular, and so given a point $x \in K$, which must be in some $\lambda^{-1}(V_r)$, there exists an open set A_x in K such that

$$x \in A_x \subset \bar{A}_x \subset K \cap \lambda^{-1}(V_r).$$

Again, a finite collection of the sets A_x will cover K, and their closures are each contained in just one set of the form $\lambda^{-1}(V_r)$. Thus by taking suitable unions of $\bar{A}_x$'s, we can write $K = \bigcup_{r=1}^{n} K_r$, where $K_r \subset \lambda^{-1}(V_r)$ and K_r is closed and so compact. It follows that $\lambda \in \bigcap_{r=1}^{n} W_{K_r, V_r} \subset W_{K,V}$, since if $\mu(K_r) \subset V_r$ for each r, then $\mu(K) \subset \bigcup_{r=1}^{n} V_r \subset V$. But if, say, $V_r = \bigcap_{s=1}^{m} U_s$, for $U_s \in \mathscr{S}$, then $W_{K_r,V_r} = \bigcap_{s=1}^{m} W_{K_r,U_s}$. Hence λ is contained in a finite intersection of sets of the form $W_{K_r,U}$, for $U \in \mathscr{S}$, and this intersection is contained in $W_{K,V}$. ∎

Theorem 6.2.34 *Given spaces X, Y and Z, where X is Hausdorff, $(Y \times Z)^X$ is homeomorphic to $Y^X \times Z^X$.*

Proof. This is now very similar to Theorem 6.2.32. Let $p_Y: Y \times Z \to Y$ and $p_Z: Y \times Z \to Z$ be the maps defined by $p_Y(y, z) = y$ and $p_Z(y, z) = z$, and define a function $\theta: Y^X \times Z^X \to (Y \times Z)^{X \times X}$ by $\theta(\lambda, \mu) = \lambda \times \mu$, where $\lambda: X \to Y$ and $\mu: X \to Z$. Consider the composites

$$\phi: (Y \times Z)^X \xrightarrow{\Delta} (Y \times Z)^X \times (Y \times Z)^X \xrightarrow{p_Y^1 \times p_Z^1} Y^X \times Z^X,$$

$$\psi: Y^X \times Z^X \xrightarrow{\theta} (Y \times Z)^{X \times X} \xrightarrow{1^\Delta} (Y \times Z)^X,$$

where each Δ is a diagonal map. If $\nu: X \to Y \times Z$, then $\phi(\nu) = (p_Y\nu, p_Z\nu)$, and if $\lambda: X \to Y$, $\mu: X \to Z$, then $\psi(\lambda, \mu) = (\lambda \times \mu)\Delta$. Thus $\phi\psi$ and $\psi\phi$ are identity functions, and it remains only to prove that θ is continuous.

Since X is Hausdorff, by Lemma 6.2.33 it is sufficient to consider sets of the form $W_{K,U \times V}$, where $K \subset X \times X$ is compact and $U \subset Y$, $V \subset Z$ are open. Then

$$\begin{aligned} \theta^{-1}(W_{K,U \times V}) &= \{(\lambda, \mu) \mid (\lambda \times \mu)(K) \subset U \times V\} \\ &= \{(\lambda, \mu) \mid K \subset \lambda^{-1}(U) \times \mu^{-1}(V)\}. \end{aligned}$$

But if $p_1, p_2: X \times X \to X$ are the maps defined like p_Y and p_Z, then $p_1(K)$ and $p_2(K)$ are compact, and $K \subset \lambda^{-1}(U) \times \mu^{-1}(V)$ if and only if $p_1(K) \times p_2(K) \subset \lambda^{-1}(U) \times \mu^{-1}(V)$. Hence

$$\theta^{-1}(W_{K,U \times V}) = W_{p_1(K),U} \times W_{p_2(K),V},$$

and so θ is continuous. ∎

At this point we possess rules for manipulating mapping spaces, analogous to the index laws $a^{b+c} = a^b.a^c$ and $(a.b)^c = a^c.b^c$ for real numbers, and it remains to investigate what rule, if any, corresponds to the index law $a^{b.c} = (a^b)^c$. To this end, we start by defining the 'association map'.

Definition 6.2.35 Given spaces X, Y and Z, the *association map* is the function $\alpha\colon Z^{X\wedge Y} \to (Z^Y)^X$ defined by

$$[\alpha\lambda(x)](y) = \lambda(x \wedge y) \qquad (x \in X, y \in Y, \lambda\colon X \wedge Y \to Z).$$

To justify this definition, we have to show that $\alpha(\lambda)$ really is an element of $(Z^Y)^X$, that is, is a continuous based map from X to Z^Y. Now for a fixed x, the function $\alpha\lambda(x)\colon Y \to Z$ is certainly a continuous based map, so that at least $\alpha(\lambda)$ is a function from X to Z^Y; and it is obviously base-point-preserving.

Proposition 6.2.36 *$\alpha(\lambda)\colon X \to Z^Y$ is continuous.*

Proof. Consider $W_{K,U}$, where $K \subset Y$ is compact and $U \subset Z$ is open. If $x \in X$ is a point such that $\alpha\lambda(x) \in W_{K,U}$, then $\lambda p(x \times K) \subset U$, or $x \times K \subset (\lambda p)^{-1}(U)$, where $p\colon X \times Y \to X \wedge Y$ is the identification map. As in the proof of Theorem 6.2.25, there exists an open set $V \subset X$ such that $x \times K \subset V \times K \subset (\lambda p)^{-1}(U)$. But for any point $x' \in V$, $\lambda p(x' \times K) \subset \lambda p(V \times K) \subset U$, so that $\alpha\lambda(x') \in W_{K,U}$. That is,

$$x \in V \subset (\alpha\lambda)^{-1}(W_{K,U}),$$

so that $(\alpha\lambda)^{-1}(W_{K,U})$ is open, and hence $\alpha(\lambda)$ is continuous. ∎

Thus at least $\alpha(\lambda)$ is an element of $(Z^Y)^X$, for each $\lambda \in Z^{X\wedge Y}$ Moreover, the function α is obviously base-point-preserving; but unfortunately α is not always continuous unless X is a Hausdorff space.

Proposition 6.2.37 *If X is Hausdorff, the association map $\alpha\colon Z^{X\wedge Y} \to (Z^Y)^X$ is continuous.*

Proof. By Lemma 6.2.33, it suffices to consider $\alpha^{-1}(W_{K,U})$, where $K \subset X$ is compact and $U \subset Z^Y$ is of the form $W_{L,V}$, for $L \subset Y$ compact and $V \subset Z$ open. Now

$$\begin{aligned}\alpha^{-1}(W_{K,U}) &= \{\lambda \mid (\alpha\lambda)(K) \subset W_{L,V}\}\\ &= \{\lambda \mid \lambda p(K \times L) \subset V\}\\ &= W_{p(K\times L),V}.\end{aligned}$$

But $p(K \times L)$ is a compact subset of $X \wedge Y$, so that α is continuous. ∎

Of course, we should like to be able to say that α is a homeomorphism, but this is not true without imposing more conditions on the spaces involved.

Theorem 6.2.38

(a) *For all spaces* X, Y *and* Z, *the function* $\alpha: Z^{X \wedge Y} \to (Z^Y)^X$ *is* (1-1).

(b) *If* Y *is locally compact and Hausdorff, then* α *is also onto.*

(c) *If both* X *and* Y *are compact and Hausdorff, then* α *is a homeomorphism.*

Proof.

(a) Let $\lambda, \mu: X \wedge Y \to Z$ be two maps such that $\alpha(\lambda) = \alpha(\mu)$. Then for all $x \in X$, $y \in Y$, we have

$$\begin{aligned} \lambda(x \wedge y) &= [\alpha\lambda(x)](y) \\ &= [\alpha\mu(x)](y) \\ &= \mu(x \wedge y), \end{aligned}$$

so that $\lambda = \mu$.

(b) Given a map $\lambda: X \to Z^Y$, let $\mu: X \wedge Y \to Z$ be the composite

$$X \wedge Y \xrightarrow{\lambda \wedge 1} Z^Y \wedge Y \xrightarrow{e} Z,$$

where e is the evaluation map. By Theorem 6.2.31 e, and hence μ, are continuous. But if $x \in X$ and $y \in Y$, we have

$$\begin{aligned} [\alpha\mu(x)](y) &= \mu(x \wedge y) \\ &= e(\lambda \wedge 1)(x \wedge y) \\ &= [\lambda(x)](y), \end{aligned}$$

so that $\alpha(\mu) = \lambda$, and hence α is onto.

(c) Certainly α is continuous, (1-1) and onto, so we have only to show that the inverse function to α is continuous.

Consider the map $\theta: (Z^Y)^X \times X \times Y \to Z$ given by composing the 'evaluation maps' $f: (Z^Y)^X \times Y \to Z^Y$ (or rather $f \times 1_Y$) and $f: Z^Y \times Y \to Z$; by Theorem 6.2.31 this is continuous. Now since X and Y are compact, if $p: X \times Y \to X \wedge Y$ is the identification map, then by Theorem 6.2.4(b)

$$1 \times p: (Z^Y)^X \times X \times Y \to (Z^Y)^X \times (X \wedge Y)$$

is also an identification map. And since θ maps $(Z^Y)^X \times Y \times x_0$ and $(Z^Y)^X \times y_0 \times X$ to z_0, it follows that θ induces a map

$$\phi: (Z^Y)^X \times (X \wedge Y) \to Z.$$

This in turn induces

$$\psi: (Z^Y)^X \wedge (X \wedge Y) \to Z;$$

but then $\alpha(\psi)$ is a map $(Z^Y)^X \to Z^{X \wedge Y}$, and it is easy to see that this map is the inverse of α. Hence α is a homeomorphism. ▮

6.3 Homotopy classes of maps

It has already been noted in Chapter 2 that, given (based) spaces X and Y, the relation between (based) maps $X \to Y$ of being homotopic (by a based homotopy) is an equivalence relation. It therefore makes sense to write $[X, Y]$ for the set of equivalence classes, under this equivalence relation.

Example 6.3.1 The set $[S^1, Y]$ is in (1-1)-correspondence with $\pi_1(Y, y_0)$. For $\pi_1(Y, y_0)$ is, as a set, the set of (pairwise) homotopy classes of maps of pairs $(I, 0 \cup 1) \to (Y, y_0)$, and this, by Proposition 6.2.5, is in (1-1)-correspondence with the set $[I/(0 \cup 1), Y]$. But $I/(0 \cup 1)$ is homeomorphic to S^1. ▮

So $[X, Y]$ is a generalization of at least the fundamental group. In fact, as we shall see later, by a suitable choice of either X or Y most of the standard algebraic invariants of topology, for example, homotopy and cohomology groups, can be obtained. However, the immediate task is to investigate two problems suggested by Example 6.3.1: in what way do maps of spaces give rise to functions on $[X, Y]$, and in what circumstances can $[X, Y]$ be given a group structure?

The first of these problems is quite simple, and the situation is entirely analogous to that of Theorem 3.2.8.

Theorem 6.3.2 *A (based) map $f\colon Y_0 \to Y_1$ gives rise to a function*

$$f_*\colon [X, Y_0] \to [X, Y_1],$$

with the following properties.

(a) *If $f'\colon Y_0 \to Y_1$ is another map, and $f \simeq f'$, then $f_* = f'_*$.*
(b) *If $1\colon Y \to Y$ is the identity map, then 1_* is the identity function.*
(c) *If $g\colon Y_1 \to Y_2$ is another map, then $(gf)_* = g_* f_*$.*

Proof. Write $[\lambda]$ for the equivalence class of a map $\lambda\colon X \to Y_0$ in the set $[X, Y_0]$. Define f_* by the rule $f_*[\lambda] = [f\lambda]$: this clearly depends only on the class of λ, and property (a) is obvious. Moreover properties (b) and (c) follow immediately from this definition. ▮

Corollary 6.3.3 *If $f\colon Y_0 \to Y_1$ is a homotopy equivalence, then f_* is a (1-1)-correspondence.* ▮

Of course, similar results hold about maps of X rather than Y.

Since the proofs are almost identical, we shall merely state these results.

Theorem 6.3.4 *A map $f: X_0 \to X_1$ gives rise to a function*

$$f^*: [X_1, Y] \to [X_0, Y],$$

with the following properties.

(a) *If $f': X_0 \to X_1$ is another map, and $f \simeq f'$, then $f^* = (f')^*$.*
(b) *If $1: X \to X$ is the identity map, then 1^* is the identity function.*
(c) *If $g: X_1 \to X_2$ is another map, then $(gf)^* = f^*g^*$.* ▮

Corollary 6.3.5 *If $f: X_0 \to X_1$ is a homotopy equivalence, then f^* is a (1-1)-correspondence.* ▮

Thus the set $[X, Y]$ depends only, up to (1-1)-correspondence, on the homotopy types of X and Y. Indeed, there is a sort of converse to Corollaries 6.3.3 and 6.3.5.

Theorem 6.3.6

(a) *If $f: Y_0 \to Y_1$ is a map such that $f_*: [X, Y_0] \to [X, Y_1]$ is a (1-1)-correspondence for all spaces X, then f is a homotopy equivalence.*

(b) *Similarly, if $g: X_0 \to X_1$ gives rise to a (1-1)-correspondence $g^*: [X_1, Y] \to [X_0, Y]$ for all spaces Y, then g is a homotopy equivalence.*

Proof.

(a) In particular, $f_*: [Y_1, Y_0] \to [Y_1, Y_1]$ is a (1-1)-correspondence, so that there exists a map $g: Y_1 \to Y_0$ such that $f_*[g] = [1_{Y_1}]$, or $fg \simeq 1_{Y_1}$. Thus for any X, f_*g_* is the identity function on $[X, Y_1]$, and so g_* is the inverse (1-1)-correspondence to f_*. In particular, $g_*[f] = [1_{Y_0}]$, so that we also have $gf \simeq 1_{Y_0}$. Hence f is a homotopy equivalence.

(b) is proved similarly. ▮

Of course, Theorem 6.3.6 is much too general to be of practical use in showing that a given map is a homotopy equivalence. However, for a large class of spaces (including all polyhedra), it is sufficient in (a) to consider only $X = S^n$, for all n: this is J. H. C. Whitehead's theorem, which will be proved in Chapter 7.

We turn now to the second problem: when is $[X, Y]$ a group? The answer is that it is if Y is 'group-like' in the sense of the next definition, or if X has corresponding 'dual' properties.

Definition 6.3.7 A space Y is called an *H-space* if there exists a map

$$m: Y \times Y \to Y,$$

such that $mi_1 \simeq mi_2 \simeq 1_Y$, where $i_1, i_2: Y \to Y \times Y$ are the maps defined by $i_1(y) = (y, y_0)$, $i_2(y) = (y_0, y)$ (y_0 is the base point of Y). An H-space Y is said to be *associative* if $m(m \times 1) \simeq m(1 \times m): Y \times Y \times Y \to Y$, and an *inverse* is a map $u: Y \to Y$ such that

$$m(u \times 1)\Delta_Y \simeq m(1 \times u)\Delta_Y \simeq e_Y,$$

where e_Y is the constant map that sends all of Y to y_0, and Δ_Y is the diagonal map.

For convenience, we shall say that Y is an AHI if it is an associative H-space with an inverse.

It will be seen that an AHI Y is 'group-like', in the sense that if we write $y_1 . y_2$ for $m(y_1, y_2)$ and y^{-1} for $u(y)$, we almost have the properties

$$y_0 . y = y . y_0 = y,$$

$$(y_1 . y_2) . y_3 = y_1 . (y_2 . y_3),$$

$$y . y^{-1} = y^{-1} . y = y_0,$$

except that all equalities are, as it were, replaced by homotopies. However, since the set $[X, Y]$ involves only homotopy classes of maps, it will be no surprise that $[X, Y]$ is a genuine group whenever Y is an AHI.

Theorem 6.3.8 *If X is any space and Y is an AHI, then $[X, Y]$ can be given the structure of a group.*

Proof. Given two maps $f, g: X \to Y$, let $f.g$ be $m(f \times g)\Delta_X$: this is certainly another continuous map from X to Y. Moreover, given further maps $f', g': X \to Y$, such that $f \simeq f'$ and $g \simeq g'$, then $f.g \simeq f'.g'$ by Theorem 6.2.2, and so a multiplication in $[X, Y]$ can be unambiguously defined by $[f].[g] = [f.g]$.

It remains to show that this multiplication satisfies the axioms for a group. First, given a third map $h: X \to Y$, we have

$$\begin{aligned}(f.g).h &= m(f.g \times h)\Delta \\ &= m[m(f \times g)\Delta \times h]\Delta \\ &= m(m \times 1)(f \times g \times h)\Delta.\end{aligned}$$

Similarly $f.(g.h) = m(1 \times m)(f \times g \times h)\Delta$, so that $(f.g).h \simeq f.(g.h)$, and $([f].[g]).[h] = [f].([g].[h])$.

Secondly, if $e: X \to Y$ is the constant map,

$$\begin{aligned} f.e &= m(f \times e)\Delta \\ &= mi_1 f \\ &\simeq f, \end{aligned}$$

and similarly $e.f \simeq f$, so that $[e]$ is a unit element for $[X, Y]$.

Lastly, we may define $[f]^{-1} = [uf]$, since

$$\begin{aligned} (uf).f &= m(uf \times f)\Delta \\ &= m(u \times 1)\Delta f \\ &\simeq e, \end{aligned}$$

and similarly $f.(uf) \simeq e$. ∎

Naturally also, maps of X give rise to homomorphisms, not just functions.

Proposition 6.3.9 *If $g: X_0 \to X_1$ is a map, and Y is an AHI, then $g^*: [X_1, Y] \to [X_0, Y]$ is a homomorphism. In particular, g^* is an isomorphism if g is a homotopy equivalence.*

Proof. Given maps $f_1, f_2: X_1 \to Y$, we have

$$\begin{aligned} (f_1 . f_2)g &= m(f_1 \times f_2)\Delta g \\ &= m(f_1 \times f_2)(g \times g)\Delta \\ &= m(f_1 g \times f_2 g)\Delta \\ &= (f_1 g).(f_2 g). \end{aligned}$$

Thus $g^*([f_1].[f_2]) = g^*[f_1].g^*[f_2]$. ∎

Before giving examples of AHI spaces, let us examine the 'dual' situation, in which $[X, Y]$ becomes a group because of properties possessed by X rather than Y.

Definition 6.3.10 A space X is called an *H'-space* if there exists a map

$$\mu: X \to X \vee X,$$

such that $p_1\mu \simeq p_2\mu \simeq 1_X$, where p_1 and p_2 are the restrictions to $X \vee X$ of the 'projection maps' $p_1, p_2: X \times X \to X$ defined by $p_1(x_1, x_2) = x_1$, $p_2(x_1, x_2) = x_2$ (as usual, we regard $X \vee X$ as a subspace of $X \times X$, by Proposition 6.2.11). An H'-space X is said to be *associative* if $(\mu \vee 1)\mu \simeq (1 \vee \mu)\mu: X \to X \vee X \vee X$, and an *inverse* is a map $v\colon X \to X$ such that $\nabla_X(v \vee 1)\mu \simeq \nabla_X(1 \vee v)\mu \simeq e_X$, where ∇_X is the folding map. Again, we shall say that X is an AH'I if it is an associative H'-space with an inverse.

Notice that the definition of an H'-space closely resembles that of an H-space: we merely turn all the maps round and use the one-point union instead of the product. For this reason, we shall not always prove in full both 'dual' versions of theorems involving H- and H'-spaces. The reader should have no difficulty, for example, in filling in the details in such theorems as the following.

Theorem 6.3.11 *If X is an AH'I and Y is any space, $[X, Y]$ can be given the structure of a group. Moreover, if $g\colon Y_0 \to Y_1$ is a map, $g_*\colon [X, Y_0] \to [X, Y_1]$ is a homomorphism, and so is an isomorphism if g is a homotopy equivalence.*

Proof. Given maps $f_1, f_2\colon X \to Y$, define $f_1 . f_2 = \nabla(f \vee g)\mu$, and proceed as in the proofs of Theorem 6.3.8 and Proposition 6.3.9. ∎

As a first example, we shall show that S^1 is an AH'I. In fact this example is really as general as we shall need, since it will be proved afterwards that, as a consequence, sX is an AH'I, and ΩX an AHI, for any space X whatsoever.

Proposition 6.3.12 *S^1 is an AH'I.*

Proof. This is very similar to Corollary 3.2.6; not unexpectedly, perhaps, in virtue of Example 6.3.1. Let $\theta l\colon I \to S^1$ be the composite of standard maps, and use it to denote points of S^1 by real numbers t such that $0 \leqslant t \leqslant 1$; that is, denote the point $\theta l(t)$ merely by t. Define a map

$$\nu\colon I \to S^1 \vee S^1$$

by

$$\nu(t) = \begin{cases} (2t, 0) & (0 \leqslant t \leqslant \frac{1}{2}) \\ (0, 2t - 1) & (\frac{1}{2} \leqslant t \leqslant 1). \end{cases}$$

This is certainly continuous, and since $\nu(0) = \nu(1)$, ν induces a (based) map $\mu\colon S^1 \to S^1 \vee S^1$.

To show that this map makes S^1 into an H'-space, consider the composite

$$I \xrightarrow{\nu} S^1 \vee S^1 \xrightarrow{p_1} S^1.$$

Now

$$p_1\nu(t) = \begin{cases} 2t & (0 \leqslant t \leqslant \frac{1}{2}) \\ 0 & (\frac{1}{2} \leqslant t \leqslant 1), \end{cases}$$

so that $p_1\nu$ is just the product loop of θl and the 'constant path', and hence $p_1\nu \simeq \theta l$, rel 0, 1, by Corollary 3.2.6(b). Hence by Proposition 6.2.5 $p_1\mu \simeq 1$; similarly, $p_2\mu \simeq 1$.

S^1 is associative by a similar argument using Corollary 3.2.6(a), and the map $v\colon S^1 \to S^1$ given by $v(t) = 1 - t$ is an inverse, by using Corollary 3.2.6(c). ▮

Corollary 6.3.13 $[S^1, Y]$ *and* $\pi_1(Y, y_0)$ *are isomorphic groups.*

Proof. The (1-1)-correspondence $\pi_1(Y, y_0) \to [S^1, Y]$ in Example 6.3.1 is clearly a homomorphism. ▮

We show next that, because S^1 is an AH'I, sX is an AH'I and ΩX is an AHI, for any space X. In fact, a slightly more general result is true.

Theorem 6.3.14 *Given spaces* X *and* Y,

(a) $X \wedge Y$ *is an AH'I if either* X *or* Y *is;*

(b) *if moreover* X *is Hausdorff,* Y^X *is an AHI if either* X *is an AH'I or* Y *is an AHI.*

Proof.

(a) Let X be an AH'I, with map $\mu\colon X \to X \vee X$, and inverse $v\colon X \to X$. Define $\bar{\mu}\colon X \wedge Y \to (X \wedge Y) \vee (X \wedge Y)$ to be the composite

$$X \wedge Y \xrightarrow{\mu \wedge 1} (X \vee X) \wedge Y \xrightarrow{g} (X \wedge Y) \vee (X \wedge Y),$$

where g is the homeomorphism of Theorem 6.2.22. Now there is an obvious commutative diagram

$$\begin{array}{ccc} (X \vee X) \wedge Y & \xrightarrow{g} & (X \wedge Y) \vee (X \wedge Y) \\ & {\scriptstyle p_1 \wedge 1} \searrow \quad \swarrow {\scriptstyle \bar{p}_1} & \\ & X \wedge Y & \end{array}$$

where p_1 and $\bar{p}_1$ are the 'projection maps'; thus

$$\bar{p}_1\bar{\mu} = (p_1 \wedge 1)(\mu \wedge 1) = p_1\mu \wedge 1 \simeq 1,$$

by Theorem 6.2.13. Similarly $\bar{p}_2\bar{\mu} \simeq 1$.

Similar arguments show that $X \wedge Y$ is associative, and that the map $\bar{v} = v \wedge 1\colon X \wedge Y \to X \wedge Y$ is an inverse for $X \wedge Y$. And of course the same proof works if Y rather than X is an AH'I.

(b) Suppose that X is Hausdorff and an AH'I. Let

$$\bar{m}\colon Y^X \times Y^X \to Y^X$$

be the composite

$$Y^X \times Y^X \xrightarrow{\psi} Y^{X \vee X} \xrightarrow{1^\mu} Y^X,$$

where ψ is the homeomorphism of Theorem 6.2.32. Now proceed as in (a), with the inverse $\bar{u} = 1^v \colon Y^X \to Y^X$.

Similarly, if X is Hausdorff and Y is an AHI, with map $m \colon Y \times Y \to Y$ and inverse $u \colon Y \to Y$, then Y^X becomes an AHI under $\bar{m}$, the composite

$$Y^X \times Y^X \xrightarrow{\psi} (Y \times Y)^X \xrightarrow{m^1} Y^X,$$

where this time ψ is the homeomorphism of Theorem 6.2.34. Of course, the inverse is $\bar{u} = u^1 \colon Y^X \to Y^X$. ∎

In particular, then, sX is an AH′I and ΩY is an AHI for any spaces X and Y, and so $[sX, Y]$ and $[X, \Omega Y]$ are groups for any X and Y.

Specializing further, since by Proposition 6.2.15 S^n is homeomorphic to sS^{n-1} for all $n \geqslant 1$, it is easy to see that S^n is an AH′I, and hence $[S^n, Y]$ is a group for all spaces Y.

Definition 6.3.15 For any (based) space Y, and $n \geqslant 1$, the group $[S^n, Y]$ is called the ***nth homotopy group*** of Y, and is usually written $\pi_n(Y)$.

Notice that this definition can even be extended to the case $n = 0$: for based maps $S^0 \to Y$ are in (1-1)-correspondence with points of Y, and their homotopies correspond to paths in Y; thus the set $[S^0, Y]$ is what we have previously referred to as $\pi_0(Y)$.

Example 6.3.16 $\pi_r(S^n) = 0$ if $r < n$. For by Corollary 2.2.4 a map $f \colon X \to S^n$ is homotopic to the constant map if it is not onto; but by the Simplicial Approximation Theorem a map $f \colon S^r \to S^n$ is homotopic to a simplicial map with some triangulations, and this cannot be onto if $r < n$. ∎

Definition 6.3.15 is somewhat unsatisfactory as it stands, since it appears to depend on the choice of a map $\mu \colon S^n \to S^n \vee S^n$ that makes S^n into an AH′I. However, this ambiguity is more apparent than real if $n > 1$, for we shall prove in Chapter 7 that all such maps are homotopic; and if $n = 1$ there are only two homotopy classes of such maps, which give rise to isomorphic group structures in π_1: see Exercise 9.

Ambiguities in the definition of the group structure also appear to arise in more general situations. For example, consider the set $[X_1 \wedge X_2, Y]$: if both X_1 and X_2 are AH′I's, then by Theorem

6.3.14 $X_1 \wedge X_2$ is an AH'I in two ways, and so there appear to be two different group structures in $[X_1 \wedge X_2, Y]$. Fortunately, however, these two group structures always coincide.

Theorem 6.3.17 *Let X_1 and X_2 be AH'I's, with corresponding maps $\mu_1: X_1 \to X_1 \vee X_1$ and $\mu_2: X_2 \to X_2 \vee X_2$, and let*

$$\bar{\mu}_1, \bar{\mu}_2: X_1 \wedge X_2 \to (X_1 \wedge X_2) \vee (X_1 \wedge X_2)$$

be the two maps, as in Theorem 6.3.14(a), that make $X_1 \wedge X_2$ into an AH'I. Then $\bar{\mu}_1 \simeq \bar{\mu}_2$.

Proof. Let $p_1, p_2: X_1 \vee X_1 \to X_1$ and $q_1, q_2: X_2 \vee X_2 \to X_2$ be the 'projection maps'. Now

$$\mu_1(x_1 \wedge x_2) = \begin{cases} (p_1\mu_1(x_1) \wedge x_2, \text{base point}) \text{ if } \mu_1(x_1) \in X_1 \times \text{base point} \\ (\text{base point}, p_2\mu_1(x_1) \wedge x_2) \text{ if } \mu_1(x_1) \in \text{base point} \times X_1. \end{cases}$$

Let $\alpha = [(1 \wedge q_1\mu_2) \vee (1 \wedge q_2\mu_2)]\bar{\mu}_1: X_1 \wedge X_2 \to (X_1 \wedge X_2) \vee (X_1 \wedge X_2)$; then certainly $\alpha \simeq \bar{\mu}_1$, and

$$\alpha(x_1 \wedge x_2) = \begin{cases} (p_1\mu_1(x_1) \wedge q_1\mu_2(x_2), \text{base point}) \text{ if } \mu_1(x_1) \in X_1 \times \text{base point} \\ (\text{base point}, p_2\mu_1(x_1) \wedge q_2\mu_2(x_2)) \text{ if } \mu_1(x_1) \in \text{base point} \times X_1. \end{cases}$$

But $q_1\mu_2(x_2) = \text{base point}$ unless $\mu_2(x_2) \in X_2 \times \text{base point}$, and $q_2\mu_2(x_2) = \text{base point}$ unless $\mu_2(x_2) \in \text{base point} \times X_2$. So the effect of α is given more precisely by the formulae

$$\alpha(x_1 \wedge x_2) = \begin{cases} (p_1\mu_1(x_1) \wedge q_1\mu_2(x_2), \text{base point}) \; (\mu_1(x_1) \in X_1 \times \text{base point}, \mu_2(x_2) \in X_2 \times \text{base point}) \\ (\text{base point}, p_2\mu_1(x_1) \wedge q_2\mu_2(x_2)) \; (\mu_1(x_1) \in \text{base point} \times X_1, \mu_2(x_2) \in \text{base point} \times X_2) \\ \text{base point} \quad (\text{otherwise}). \end{cases}$$

Similarly, by symmetry, $\alpha \simeq \bar{\mu}_2$, so that $\bar{\mu}_1 \simeq \bar{\mu}_2$. ∎

Corollary 6.3.18 *For any Y, the group structures in $[X_1 \wedge X_2, Y]$, defined by $\bar{\mu}_1$ and $\bar{\mu}_2$, are the same.* ∎

In fact the proof of Theorem 6.3.17 has a rather surprising consequence. Let us call an H'-space X (with map $\mu: X \to X \vee X$) *commutative* if $\mu \simeq \tau\mu$, where $\tau: X \vee X \to X \vee X$ is the restriction of the map $X \times X \to X \times X$ that sends (x_1, x_2) to (x_2, x_1).

Proposition 6.3.19 *With the same data as in Theorem 6.3.17, $X_1 \wedge X_2$ is a commutative AH'I.*

Proof. Let

$$\beta = [(1 \wedge q_2\mu_2) \vee (1 \wedge q_1\mu_2)]\bar{\mu}_1 \colon X_1 \wedge X_2 \to (X_1 \wedge X_2) \vee (X_1 \wedge X_2);$$

again $\beta \simeq \bar{\mu}_1$, and

$$p(x_1 \wedge x_2) = \begin{cases} (p_1\mu_1(x_1) \wedge q_2\mu_2(x_2), \text{ base point}) \ (\mu_1(x_1) \in X_1 \times \text{ base point}, \ \mu_2(x_2) \in \text{base point} \times X_2) \\ (\text{base point}, \ p_2\mu_1(x_1) \wedge q_1\mu_2(x_2)) \ (\mu_1(x_1) \in \text{base point} \times X_1, \ \mu_2(x_2) \in X_2 \times \text{ base point}) \\ \text{base point} \quad (\text{otherwise}). \end{cases}$$

Similarly, by symmetry, $\beta \simeq \tau\bar{\mu}_2$. Hence $\bar{\mu}_1 \simeq \tau\bar{\mu}_2 \simeq \tau\bar{\mu}_1$. ∎

The point of Proposition 6.3.19 is that commutative AH'I's give rise to abelian groups.

Proposition 6.3.20 *If X is a commutative AH'I, and Y is any space, then $[X, Y]$ is an abelian group.*

Proof. Given maps $f, g \colon X \to Y$, we have

$$\begin{aligned} f \cdot g &= \nabla(f \vee g)\mu \\ &\simeq \nabla(f \vee g)\tau\mu \\ &= \nabla(g \vee f)\mu \\ &= g \cdot f. \end{aligned}$$

Thus $[f]\cdot[g] = [g]\cdot[f]$. ∎

Corollary 6.3.21 *For any spaces X and Y, $[s(sX), Y]$ is an abelian group. In particular, $\pi_n(Y)$ is abelian for $n \geqslant 2$.* ∎

As usual, there are 'dual' results to 6.3.17–6.3.21, involving AHI's instead of AH'I's.

Theorem 6.3.22 *Let X be a Hausdorff space, and an AH'I, with map $\mu \colon X \to X \vee X$; let Y be an AHI, with map $m \colon Y \times Y \to Y$. Then if $\bar{m}_1, \bar{m}_2 \colon Y^X \times Y^X \to Y^X$ are the maps arising from μ and m respectively, in Theorem 6.3.14(b), we have $\bar{m}_1 \simeq \bar{m}_2$.*

Proof. By definition, $\bar{m}_1$ is the composite

$$Y^X \times Y^X \xrightarrow{\theta} (Y \vee Y)^{X \vee X} \xrightarrow{\Delta^1} Y^{X \vee X} \xrightarrow{1^\mu} Y^X,$$

where θ is the map defined in the proof of Theorem 6.2.32. Let $\alpha = 1^\mu \nabla^1 (mi_1 \vee mi_2)^1 \theta \colon Y^X \times Y^X \to Y^X$, where $i_1, i_2 \colon Y \to Y \times Y$

are the usual inclusions; then $\alpha \simeq \overline{m}_1$. But if $f, g \colon X \to Y$ are two maps, we have

$$\alpha(f, g) = \nabla(mi_1 \vee mi_2)(f \vee g)\mu.$$

That is, if $x \in X$,

$$[\alpha(f, g)](x) = \begin{cases} m(fp_1\mu(x), y_0) & (\mu(x) \in X \times x_0) \\ m(y_0, gp_2\mu(x)) & (\mu(x) \in x_0 \times X), \end{cases}$$

where $p_1, p_2 \colon X \vee X \to X$ are the 'projection maps'. But this is the same as the map that sends x to $m(fp_1\mu(x), gp_2\mu(x))$, so that

$$\alpha(f, g) = m(f \times g)(p_1\mu \times p_2\mu)\Delta.$$

It follows that we can also write

$$\alpha = m^1 1^\Delta 1^{(p_1\mu \times p_2\mu)}\theta,$$

where now $\theta \colon Y^X \times Y^X \to (Y \times Y)^{X \times X}$ is the map defined in Theorem 6.2.34. Hence $\alpha \simeq m^1 1^\Delta \theta = \overline{m}_2$. ▮

Corollary 6.3.23 *For any Z, the two group structures in $[Z, Y^X]$, defined by $\overline{m}_1$ and $\overline{m}_2$, are the same.* ▮

Let us call an H-space Y (with map $m \colon Y \times Y \to Y$) *commutative* if $m \simeq m\tau$, where $\tau \colon Y \times Y \to Y \times Y$ is defined by $\tau(y_1, y_2) = (y_2, y_1)$.

Proposition 6.3.24 *With the same data as in Theorem 6.3.22, Y^X is a commutative AHI.* ▮

Proposition 6.3.25 *If Y is a commutative AHI, and X is any space, then $[X, Y]$ is an abelian group.* ▮

Corollary 6.3.26 *For any spaces X and Y, $[X, \Omega(\Omega Y)]$ is an abelian group.* ▮

In fact there is hardly any need to prove Corollary 6.3.26 as a separate result, since it is true that for any spaces X and Y, $[sX, Y]$ and $[X, \Omega Y]$ are isomorphic groups: thus Corollary 6.3.26 follows from Corollary 6.3.21. The theorem that $[sX, Y] \cong [X, \Omega Y]$ is a special case of a more general result: recall from Theorem 6.2.38 that, if Y is locally compact and Hausdorff, the association map gives a (1-1)-correspondence between maps $X \wedge Y \to Z$ and $X \to Z^Y$. Indeed, homotopies correspond as well, so that we have

Theorem 6.3.27 *If Y is locally compact and Hausdorff, α induces a (1-1)-correspondence $\bar{\alpha} \colon [X \wedge Y, Z] \to [X, Z^Y]$.*

Proof. Let $F: (X \wedge Y) \times I \to Z$ be a homotopy between maps $f, g: X \wedge Y \to Z$. If $p: X \times Y \to X \wedge Y$ is the identification map, $F(p \times 1): X \times Y \times I \to Z$ is a map that sends $X \times y_0 \times I$ and $x_0 \times Y \times I$ to z_0, and so induces a map $F': (X \times I) \wedge Y \to Z$. Then $\alpha(F'): X \times I \to Z^Y$ sends $x_0 \times I$ to the base point, and is clearly a homotopy between $\alpha(f)$ and $\alpha(g)$.

Conversely, if $\alpha(f) \simeq \alpha(g)$, then since α is a (1-1)-correspondence we may assume that the homotopy between them is of the form $\alpha(F)$, where $F: (X \times I) \wedge Y \to Z$ is a map such that $F[(x, t) \wedge y_0] = z_0$ for all $(x, t) \in X \times I$. So if $q: X \times I \times Y \to (X \times I) \wedge Y$ is the identification map, Fq may be regarded as a map $X \times Y \times I \to Z$ that sends $X \times y_0 \times I$ and $x_0 \times Y \times I$ to z_0. Since, by Theorem 6.2.4, $(p \times 1)$ is an identification map, it follows that Fq induces a map $F': (X \wedge Y) \times I \to Z$, which is clearly a homotopy between f and g. ∎

In particular, there is a (1-1)-correspondence between $[sX, Y]$ and $[X, \Omega Y]$, for all X and Y. The next step is to prove that this (1-1)-correspondence preserves the multiplication, and so is an isomorphism. In fact a more general result is true: the (1-1)-correspondence of Theorem 6.3.27 is an isomorphism whenever the sets concerned are groups.

Theorem 6.3.28 *$\bar{\alpha}$ is an isomorphism in any of the following cases.*

(a) *X an AH'I; group structures defined by the AH'I spaces $X \wedge Y$, X respectively.*

(b) *Z an AHI; groups defined by the AHI's Z and Z^Y.*

(c) *Y an AH'I; groups defined by the AH'I $X \wedge Y$ and the AHI Z^Y.*

Proof.

(a) Let $\mu: X \to X \vee X$ be the map that makes X an AH'I, and consider two maps $f, g: X \wedge Y \to Z$. Now for $x \in X$ and $y \in Y$, we have

$$(f \cdot g)(x \wedge y) = \begin{cases} f(p_1\mu(x) \wedge y) & (\mu(x) \in X \times x_0) \\ g(p_2\mu(x) \wedge y) & (\mu(x) \in x_0 \times X), \end{cases}$$

where $p_1, p_2: X \vee X \to X$ are the 'projection maps'. But

$$[(\alpha f)(x)](y) = f(x \wedge y),$$

and

$$(\alpha f . \alpha g)(x) = \begin{cases} (\alpha f)(p_1\mu(x)) & (\mu(x) \in X \times x_0) \\ (\alpha g)(p_2\mu(x)) & (\mu(x) \in x_0 \times X), \end{cases}$$

so that clearly $\alpha f.\alpha g = \alpha(f.g)$, and $\bar{\alpha}$ is a homomorphism and hence an isomorphism.

(b) Let $m\colon Z \times Z \to Z$ be the map that makes Z an AHI. This time

$$\begin{aligned}[\alpha(f.g)(x)](y) &= (f.g)(x \wedge y)\\ &= m(f(x \wedge y), g(x \wedge y))\\ &= m([(\alpha f)(x)](y), [(\alpha g)(x)](y))\\ &= [(\alpha f.\alpha g)(x)](y),\end{aligned}$$

so that $\alpha f.\alpha g = \alpha(f.g)$.

(c) Again, let $\mu\colon Y \to Y \vee Y$ make Y into an AH′I. Then

$$(f.g)(x \wedge y) = \begin{cases} f(x \wedge p_1\mu(y)) & (\mu(y) \in Y \times y_0)\\ g(x \wedge p_2\mu(y)) & (\mu(y) \in y_0 \times Y).\end{cases}$$

But $(\alpha f.\alpha g)(x) = \overline{m}(\alpha f(x), \alpha g(x))$, where $\overline{m}\colon Z^Y \times Z^Y \to Z^Y$ is induced by μ. Thus

$$\begin{aligned}[(\alpha f.\alpha g)(x)](y) &= [\overline{m}(\alpha f(x), \alpha g(x))](y)\\ &= [\nabla\{\alpha f(x) \vee \alpha g(x)\}\mu](y)\\ &= \begin{cases} f(x \wedge p_1\mu(y)) & (\mu(y) \in Y \times y_0)\\ g(x \wedge p_2\mu(y)) & (\mu(y) \in y_0 \times Y),\end{cases}\end{aligned}$$

so that once again $\alpha f.\alpha g = \alpha(f.g)$. ∎

Corollary 6.3.29 *For any spaces X and Y, $\bar{\alpha}\colon [sX, Y] \to [X, \Omega Y]$ is an isomorphism.* ∎

Corollary 6.3.30 *For any space Y, and $n > 1$, $\pi_n(Y) \cong \pi_{n-1}(\Omega Y)$.*

Proof. By definition, $\pi_n(Y) = [sS^{n-1}, Y]$. By Corollary 6.3.18 we may assume that the group structure is defined by the AH′I structure of S^{n-1}; thus by Theorem 6.3.28(a), $\pi_n(Y) \cong [S^{n-1}, \Omega Y] = \pi_{n-1}(\Omega Y)$. ∎

Notice that, by Corollary 6.3.23, the group structure in $\pi_{n-1}(\Omega Y)$ may equally well be taken to be that defined by the AHI ΩY. Thus even $\pi_0(\Omega Y)$ is a group, which by Corollary 6.3.29 is isomorphic to $\pi_1(Y)$.

6.4 Exact sequences

From the discussion of the sets $[X, Y]$, it is clear that the problem of classifying spaces up to homotopy equivalence is intimately bound up with the calculation of $[X, Y]$. It is particularly important to be able to identify these sets in the cases where they are groups; and the clue to how to proceed is provided by Chapters 4 and 5, where the most useful tools were the exact homology and cohomology sequences of pairs and triples.

We wish, then, to establish results analogous to these exact sequences, for the sets $[X, Y]$. Now in general the set $[X, Y]$ is not a group, although it has a 'distinguished element', namely the class of the constant map from X to Y (this, of course, is the unit element of $[X, Y]$ if it happens to be a group). Let us call a set with a distinguished element a *based set*: we must first define the notion of an exact sequence of based sets.

Definition 6.4.1 Given a function $f\colon A \to B$ between based sets (with distinguished elements a_0 and b_0), write $\operatorname{Im} f = \{f(a) \mid a \in A\}$ and $\operatorname{Ker} f = \{a \mid f(a) = b_0\}$. A sequence of based sets and functions

$$\cdots \longrightarrow A_i \xrightarrow{f_i} A_{i+1} \xrightarrow{f_{i+1}} \cdots$$

is called an *exact sequence* if, for each i, $\operatorname{Im} f_i = \operatorname{Ker} f_{i+1}$. (Note that this coincides with the usual definition if the sets are groups, the functions are homomorphisms, and the distinguished element of each group is its unit element.)

The aim in this section is to show that a map $f\colon A \to B$ gives rise to an exact sequence involving the sets $[A, Y]$ and $[B, Y]$, for any space Y; there is also a 'dual' result involving the sets $[X, A]$ and $[X, B]$. If A is a subspace of B, and f is the inclusion map, this sequence ought to resemble the exact cohomology sequence. It is to be expected, therefore, that the sequence will also involve something like the relative cohomology of a pair. Now we have seen, in Chapter 4, Exercise 5, that if (K, L) is a simplicial pair, $H_*(K, L) \cong \tilde{H}_*(K \cup CL)$, where $K \cup CL$ denotes K with a 'cone' attached to L. This result suggests how to define the 'relative set' that appears as the third object in the exact sequence involving $[A, Y]$ and $[B, Y]$.

Definition 6.4.2 Given a map $f\colon A \to B$, the *mapping cone* C_f is defined to be the space obtained from B and cA (the reduced cone), by identifying, for each $a \in A$, the points $a \wedge 0 \in cA$ and $f(a)$ in B. The base point of C_f is, of course, the point to which $a_0 \wedge t$ and b_0 are identified, for all $t \in I$, where a_0 and b_0 are the base points of A and B respectively. See Fig. 6.3, in which the thick line is supposed to be identified to a point.

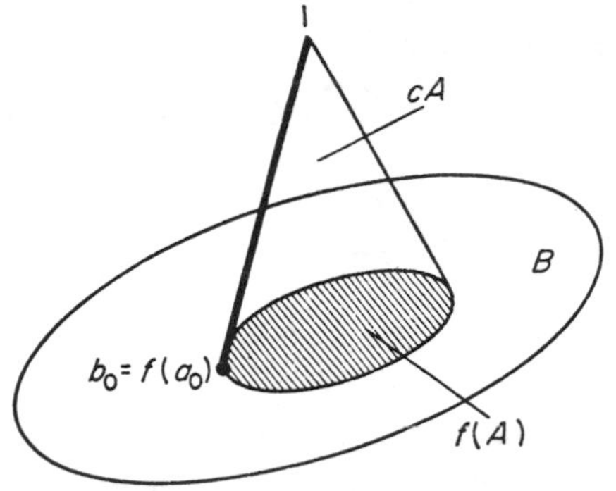

Fig. 6.3

Write f' for the 'inclusion map' of B in C_f; more precisely, f' is the inclusion of B in the disjoint union of B and cA, composed with the identification map onto C_f.

Theorem 6.4.3 *For any space* Y, *the sequence of based sets and functions*

$$[C_f, Y] \xrightarrow{(f')^*} [B, Y] \xrightarrow{f^*} [A, Y]$$

is exact.

Proof. The map $f'f: A \to C_f$ is the same as the composite

$$A \xrightarrow{i} cA \to C_f,$$

where i is defined by $i(a) = a \wedge 0$, and the second map is the 'inclusion map', defined similarly to f'. Now by Proposition 6.2.21 cA is contractible, so that $f'f \simeq e$, the constant map from A to C_f. Thus $f^*f'^* = e^*$, and $\operatorname{Im} f'^* \subset \operatorname{Ker} f^*$.

Conversely, let $g: B \to Y$ be a map such that $f^*[g]$ is the distinguished element of $[A, Y]$, so that $gf \simeq e$, the constant map from A to Y. Let $F: A \times I \to Y$ be the homotopy; since

$$F(a_0, t) = F(a, 1) = y_0, \qquad \text{all } a \in A, t \in I,$$

F induces a map $\bar{F}: cA \to Y$. And since $\bar{F}(a \wedge 0) = F(a, 0) = gf(a)$, the maps g and $\bar{F}$ together induce a map $G: C_f \to Y$, where clearly $f'G = g$. That is, $[g] \in \operatorname{Im} f'^*$, so that $\operatorname{Ker} f^* \subset \operatorname{Im} f'^*$, and hence $\operatorname{Ker} f^* = \operatorname{Im} f'^*$. ∎

Of course, the exact sequence is an exact sequence of groups and homomorphisms, if Y happens to be an AHI.

The construction of C_f can be iterated, and we then obtain a long sequence of spaces and maps

$$A \xrightarrow{f} B \xrightarrow{f'} C_f \xrightarrow{f''} C_{f'} \xrightarrow{f^{(3)}} C_{f''} \longrightarrow \cdots.$$

Corollary 6.4.4 *For any space Y, the sequence of based sets*

$$\cdots \longrightarrow [C_{f'}, Y] \xrightarrow{(f'')^*} [C_f, Y] \xrightarrow{(f')^*} [B, Y] \xrightarrow{f^*} [A, Y]$$

is exact (and if Y is an AHI, it is an exact sequence of groups and homomorphisms). ▮

Thus we have a long exact sequence, but the resemblance to the exact cohomology sequence is no longer clear. However, let us examine $C_{f'}$ more closely. $C_{f'}$ is the space obtained from C_f by 'attaching' cB, and since cB already includes a copy of B, it is easy to see that $C_{f'}$ is the space obtained from cA and cB by identifying, for each $a \in A$, the points $a \wedge 0$ and $f(a) \wedge 0$: see Fig. 6.4.

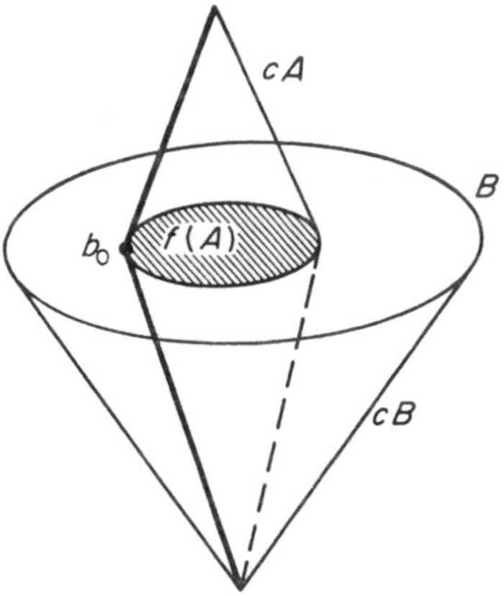

Fig. 6.4

It will be noticed that cB contains a copy of $c(fA)$, and Fig. 6.4 suggests that it might be possible to 'shrink away' $cB - c(fA)$, so as to leave something like sA. This is indeed the case, as the next theorem shows.

Theorem 6.4.5. $C_{f'} \simeq sA$.

Proof. As usual, denote points of S^1 by numbers t, with $0 \leqslant t \leqslant 1$, where 0 and 1 both represent the base point. Define $\theta: sA \to C_{f'}$ by

$$\theta(a \wedge t) = \begin{cases} f(a) \wedge (1 - 2t) \text{ in } cB & (0 \leqslant t \leqslant \frac{1}{2}) \\ a \wedge (2t - 1) \text{ in } cA & (\frac{1}{2} \leqslant t \leqslant 1). \end{cases}$$

(Or rather, θ is this map into the disjoint union of cB and cA, followed by the identification map onto $C_{f'}$.) This is well-defined, since $f(a) \wedge 1$ and $a \wedge 1$ both represent the base point of $C_{f'}$, and (for $t = \frac{1}{2}$) $f(a) \wedge 0$ and $a \wedge 0$ represent the same point of $C_{f'}$. Moreover θ is continuous by Proposition 1.4.15(d).

Also define $\phi: C_{f'} \to sA$, by

$$\begin{cases} \phi(a \wedge t) = a \wedge t \in sA, & \text{for points of } cA \\ \phi(b \wedge t) = \text{base point}, & \text{for points of } cB. \end{cases}$$

(Strictly speaking, ϕ is the map induced by this map from the disjoint union of cA and cB. This map does induce ϕ, since $\phi(a \wedge 0) = a \wedge 0 \in sA = \text{base point} = \phi(f(a) \wedge 0)$.)

It remains to prove that $\phi\theta$ and $\theta\phi$ are homotopic to the respective identity maps. Now $\phi\theta: sA \to sA$ is given by the formulae

$$\phi\theta(a \wedge t) = \begin{cases} \text{base point} & (0 \leqslant t \leqslant \frac{1}{2}) \\ a \wedge (2t - 1) & (\frac{1}{2} \leqslant t \leqslant 1). \end{cases}$$

But this is the same as the composite

$$sA \xrightarrow{\bar{\mu}} sA \vee sA \xrightarrow{p_2} sA,$$

where $\bar{\mu}$ is defined as in Theorem 6.3.14(a), using the map $\mu: S^1 \to S^1 \vee S^1$ of Proposition 6.3.12. Hence $\phi\theta \simeq 1$.

On the other hand $\theta\phi$ is given by

$$\begin{cases} \theta\phi(a \wedge t) = \begin{cases} f(a) \wedge (1 - 2t) & (0 \leqslant t \leqslant \frac{1}{2}) \\ a \wedge (2t - 1) & (\frac{1}{2} \leqslant t \leqslant 1) \end{cases} \\ \theta\phi(a \wedge t) = \text{base point}. \end{cases}$$

To construct a homotopy $F: C_{f'} \times I \to C_{f'}$, between $\theta\phi$ and 1, define $F_A: cA \times I \to C_{f'}$ by

$$F_A(a \wedge t, s) = \begin{cases} f(a) \wedge (1 - 2t - s(1 - t)) & (0 \leqslant t \leqslant (1 - s)/(2 - s)) \\ a \wedge (2t - 1 + s(1 - t)) & ((1 - s)/(2 - s) \leqslant t \leqslant 1) \end{cases}$$

and $F_B: cB \times I \to C_{f'}$ by

$$F_B(b \wedge t, s) = b \wedge (1 - s(1 - t)) \qquad (0 \leqslant s \leqslant 1).$$

Now F_A is continuous, because it is induced by a continuous map of $(A \times I) \times I \to C_{f'}$ (using Proposition 1.4.15: the definitions of F_A coincide when $t = (1 - s)/(2 - s)$), and $p \times 1: (A \times I) \times I \to cA \times I$ is an identification map, where $p: A \times I \to cA$ is the standard identification. Similarly F_B is continuous; and since

$$F_A(a \wedge 0, s) = f(a) \wedge (1 - s) = F_B(f(a) \wedge 0, s),$$

F_A and F_B together induce a homotopy $F: C_{f'} \times I \to C_{f'}$, which is continuous by an argument similar to that used for F_A and F_B. Moreover for $t = 1$, $F_A(a \wedge 1, s) = F_B(b \wedge 1, s) =$ base point, so that F is a *based* homotopy; and clearly F is a homotopy between $\theta\phi$ and 1. Hence θ and ϕ are homotopy equivalences. ▌

It follows, of course, that $C_{f''} \simeq sB$, $C_{f^{(3)}} \simeq s(C_f)$, and so on; in fact each space in the sequence

$$A \xrightarrow{f} B \xrightarrow{f'} C_f \xrightarrow{f''} C_{f''} \longrightarrow \cdots$$

can be identified, up to homotopy equivalence, with an iterated suspension of A, B or C_f.

In particular $f^{(3)}$ is more-or-less a map from sA to sB, and it would be very convenient if this map were $f \wedge 1$. This is not quite true, since instead of the identity map of S^1, we must use the inverse map $v: S^1 \to S^1$ of Proposition 6.3.12.

Proposition 6.4.6 *The diagram*

$$\begin{array}{ccc} C_{f'} & \xrightarrow{f^{(3)}} & C_{f''} \\ {\scriptstyle\theta}\uparrow & & \downarrow{\scriptstyle\phi} \\ sA & \xrightarrow[f \wedge v]{} & sB \end{array}$$

is homotopy-commutative, where θ and ϕ are homotopy equivalences defined as in Theorem 6.4.5 (that is, $\phi f^{(3)}\theta \simeq f \wedge v$).

Proof. $\phi f^{(3)}$ maps points of C_f to the base point, and points of cB to sB by the rule $(b \wedge t) \to (b \wedge t)$ in sB. Thus

$$\phi f^{(3)}\theta(a \wedge t) = \begin{cases} f(a) \wedge (1 - 2t) & (0 \leqslant t \leqslant \frac{1}{2}) \\ \text{base point} & (\frac{1}{2} \leqslant t \leqslant 1), \end{cases}$$

and so $\phi f^{(3)}\theta = f \wedge \bar{v}$, where $\bar{v}: S^1 \to S^1$ is defined by

$$\bar{v}(t) = \begin{cases} 1 - 2t & (0 \leqslant t \leqslant \frac{1}{2}) \\ 0 & (\frac{1}{2} \leqslant t \leqslant 1). \end{cases}$$

But $v \simeq \bar{v}$ by an obvious homotopy, so that $f \wedge v \simeq f \wedge v$. ▌

To sum up, we have (almost) proved

Theorem 6.4.7 *A map $f: A \to B$ gives rise to a sequence of spaces and maps*

$$A \xrightarrow{f} B \xrightarrow{f_1} C_f \xrightarrow{f_2} sA \xrightarrow{f \wedge 1} sB \xrightarrow{f_1 \wedge 1} sC_f \longrightarrow \cdots,$$

such that, for any space Y, the sequence

$$\cdots \longrightarrow [sB, Y] \xrightarrow{(f \wedge 1)^*} [sA, Y] \xrightarrow{f_2^*} [C_f, Y] \xrightarrow{f_1^*} [B, Y] \xrightarrow{f^*} [A, Y]$$

is an exact sequence of sets. If Y is an AHI, this is an exact sequence of groups; in any case it is an exact sequence of groups as far as $[sA, Y]$, and an exact sequence of abelian groups as far as $[s(sA), Y]$.

Proof. Let $f_1 = f' : B \to C_f$, and $f_2 = \phi f'' : C_f \to sA$. Consider the diagram

$$\begin{array}{ccccccccc}
\cdots \longrightarrow [sB, Y] & \xrightarrow{(f\wedge v)^*} & [sA, Y] & \xrightarrow{(f_2)^*} & [C_f, Y] & \xrightarrow{(f_1)^*} & [B, Y] & \xrightarrow{f^*} & [A, Y] \\
\downarrow{\scriptstyle \phi^*} & & \downarrow{\scriptstyle \phi^*} & \nearrow{\scriptstyle (f'')^*} & & & & & \\
\cdots \longrightarrow [C_{f''}, Y] & \xrightarrow[(f^{(3)})^*]{} & [C_{f'}, Y] & & & & & &
\end{array}$$

By Proposition 6.4.6 this diagram is commutative; and each ϕ^* is a (1-1)-correspondence, which sends distinguished elements to distinguished elements. Thus the upper row is an exact sequence. But $f \wedge v = (f \wedge 1)(1 \wedge v) : sA \to sB$, and $(1 \wedge v)^* : [sA, Y] \to [sA, Y]$ is the function that sends each element into its inverse ($[sA, Y]$ is a group, since sA is an AH'I). Since the image of $(f \wedge 1)^* : [sB, Y] \to [sA, Y]$ is a subgroup, this means that $\operatorname{Im}(f \wedge 1)^* = \operatorname{Im}(f \wedge v)^*$; also, of course, $\operatorname{Ker}(f \wedge 1)^* = \operatorname{Ker}(f \wedge v)^*$. Thus $(f \wedge v)^*$ can be replaced by $(f \wedge 1)^*$ without sacrificing exactness.

That the sets are (abelian) groups (and the functions homomorphisms) in the stated circumstances is an immediate corollary of Theorem 6.3.8, Theorem 6.3.11 and Corollary 6.3.21. ∎

As has already been pointed out, the exact sequence of Theorem 6.4.7 resembles the exact cohomology sequence of a pair. In Chapters 4 and 5 a useful property of such sequences was that a map of pairs gave rise to a commutative diagram of exact sequences; and this property holds also for the exact sequences of Theorem 6.4.7.

Proposition 6.4.8 *A commutative diagram of spaces and maps*

$$\begin{array}{ccc}
A & \xrightarrow{f} & B \\
\downarrow{\scriptstyle \lambda} & & \downarrow{\scriptstyle \mu} \\
A' & \xrightarrow[f']{} & B'
\end{array}$$

gives rise to a commutative diagram

$$\begin{array}{ccccccccc}
A & \xrightarrow{f} & B & \xrightarrow{f_1} & C_f & \xrightarrow{f_2} & sA & \longrightarrow & \cdots \\
\downarrow{\scriptstyle \lambda} & & \downarrow{\scriptstyle \mu} & & \downarrow{\scriptstyle \nu} & & \downarrow{\scriptstyle \lambda \wedge 1} & & \\
A' & \xrightarrow[f']{} & B' & \xrightarrow[f_1']{} & C_{f'} & \xrightarrow[f_2']{} & sA' & \longrightarrow & \cdots
\end{array}$$

Proof. Let ν be the map induced by the map from the disjoint union of B and cA to $C_{f'}$ given by combining μ and $\lambda \wedge 1$ (certainly $\mu f(a) = (\lambda \wedge 1)(a \wedge 0)$ in $C_{f'}$). Then obviously $\nu f_1 = f'_1\mu$ and $(\lambda \wedge 1)f_2 = f'_2\nu$. ∎

For any space Y, we therefore obtain a commutative diagram involving the two exact sequences of sets of homotopy classes of maps into Y. In fact a similar result holds if we merely have $\mu f \simeq f'\lambda$ instead of $\mu f = f'\lambda$, but this is a little more difficult to prove: see Exercise 11.

As usual, there is a 'dual' result to Theorem 6.4.7, which gives an exact sequence of sets of homotopy classes of maps *from* a space X, rather than *to* a space Y. The method of proof is very similar to that of Theorem 6.4.7, so that we shall not give all the details in full.

The 'dual' to the mapping cone is the *mapping path-space.*

Definition 6.4.9 Given a map $f: A \to B$, the *mapping path-space* L_f is the subspace of $A \times LB$ of pairs (a, λ) such that $f(a) = \lambda(0)$. The base point of L_f is (a_0, e), where a_0 is the base point of A and $e: I \to B$ is the constant map.

Let $f': L_f \to A$ be the map defined by $f'(a, \lambda) = a$.

Proposition 6.4.10 *For any space X, the sequence*

$$[X, L_f] \xrightarrow{f'_*} [X, A] \xrightarrow{f_*} [X, B]$$

is an exact sequence of based sets.

Proof. Since $ff'(a, \lambda) = f(a) = \lambda(0)$, the map ff' is the same as the composite

$$L_f \to LB \xrightarrow{p} B,$$

where the first map is defined similarly to f', and p is defined by $p(\lambda) = \lambda(0)$. It is easy to see that p is continuous; and LB is contractible by Proposition 6.2.29: hence ff' is homotopic to the constant map, and $\operatorname{Im} f'_* \subset \operatorname{Ker} f_*$.

Conversely, given a map $g: X \to A$ such that fg is homotopic to the constant map from X to B, let $F: X \times I \to B$ be this homotopy. F induces a map $F': cX \to B$, and hence $\alpha(F'): X \to LB$, where α is the association map. Then $G = (g \times \alpha F')\Delta: X \to A \times LB$ is a map into L_f, and $f'G = g$. Hence $\operatorname{Ker} f_* \subset \operatorname{Im} f'_*$. ∎

By iterating the definition of L_f, we obtain the sequence of spaces maps

$$\cdots \longrightarrow L_{f''} \xrightarrow{f^{(3)}} L_{f'} \xrightarrow{f''} L_f \xrightarrow{f'} A \xrightarrow{f} B.$$

Corollary 6.4.11 *For any space X, the sequence*

$$\cdots \longrightarrow [X, L_{f'}] \xrightarrow{f''_*} [X, L_f] \xrightarrow{f'_*} [X, A] \xrightarrow{f_*} [X, B]$$

is an exact sequence of based sets (an exact sequence of groups if X is an $AH'I$). ∎

Theorem 6.4.12 $L_{f'} \simeq \Omega B$.

Proof. $L_{f'}$ is the subspace of $A \times LB \times LA$ consisting of points (a, λ, μ) such that $f(a) = \lambda(0)$ and $a = f'(a, \lambda) = \mu(0)$: see Fig. 6.5.

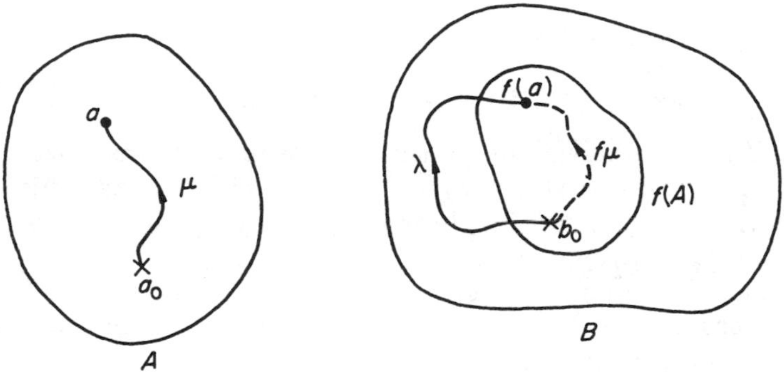

Fig. 6.5

Now $f\mu$ is a path in B from b_0 to $f(a)$, so that we can define a map $\theta: L_{f'} \to \Omega B$ by 'sticking together' the two paths λ and $f\mu$ in B. More precisely, represent points of S^1 as usual by numbers t, $0 \leqslant t \leqslant 1$, and define

$$[\theta(a, \lambda, \mu)](t) = \begin{cases} f\mu(1 - 2t) & (0 \leqslant t \leqslant \frac{1}{2}) \\ \lambda(2t - 1) & (\frac{1}{2} \leqslant t \leqslant 1). \end{cases}$$

Certainly θ is continuous, since it corresponds under the association map to an obviously continuous map $L_{f'} \wedge I \to B$. Also define $\phi: \Omega B \to L_{f'}$ by $\phi(\lambda) = (a_0, \lambda, e)$, where $\lambda: S^1 \to B$, $e: I \to A$ is the constant map, and on the right-hand side λ is regarded as a map $I \to B$. Then ϕ is continuous as a map into $A \times LB \times LA$, and its image is contained in $L_{f'}$, since $f(a_0) = \lambda(0)$ and $a_0 = e(0)$.

Now $\theta\phi: \Omega B \to \Omega B$ is given by

$$[\theta\phi(\lambda)](t) = \begin{cases} b_0 & (0 \leqslant t \leqslant \frac{1}{2}) \\ \lambda(2t - 1) & (\frac{1}{2} \leqslant t \leqslant 1). \end{cases}$$

So $\theta\phi \simeq 1$, since it is the composite

$$\Omega B \xrightarrow{i_2} \Omega B \times \Omega B \xrightarrow{\bar{m}} \Omega B,$$

where $\bar{m}$ is induced by $\mu: S^1 \to S^1 \vee S^1$.

On the other hand, $\phi\theta(a, \lambda, \mu) = (a_0, \nu, e)$, where

$$\nu(t) = \begin{cases} f\mu(1 - 2t) & (0 \leqslant t \leqslant \frac{1}{2}) \\ \lambda(2t - 1) & (\frac{1}{2} \leqslant t \leqslant 1). \end{cases}$$

To construct a homotopy between $\phi\theta$ and 1, define $F_B: L_{f'} \times I \to LB$ by

$$[F_B(a, \lambda, \mu, s)](t) = \begin{cases} f\mu(1 - 2t - s(1 - t)) & (0 \leqslant t \leqslant (1 - s)/(2 - s)) \\ \lambda(2t - 1 + s(1 - t)) & ((1 - s)/(2 - s) \leqslant t \leqslant 1), \end{cases}$$

and $F_A: L_{f'} \times I \to LA$ by

$$[F_A(a, \lambda, \mu, s)](t) = \mu(1 - s(1 - t)) \qquad (0 \leqslant s \leqslant 1).$$

Now F_B is continuous, since it corresponds under the association map to a continuous map $(L_{f'} \times I) \wedge I \to B$ (which in turn is induced by a continuous map $L_{f'} \times I \times I \to B$). Similarly F_A is continuous; and since $[F_B(a, \lambda, \mu, s)](0) = f\mu(1 - s)$ and $[F_A(a, \lambda, \mu, s)](0) = \mu(1 - s)$, F_B and F_A combine to give a homotopy $F: L_{f'} \times I \to L_{f'}$ (the 'A-co-ordinate' of $F(a, \lambda, \mu, s)$ is $\mu(1 - s)$). And clearly F is a homotopy between $\phi\theta$ and 1. ▌

Proposition 6.4.13 *The diagram*

$$\begin{array}{ccc} L_{f''} & \xrightarrow{f^{(3)}} & L_{f'} \\ {\scriptstyle\phi}\uparrow & & \downarrow{\scriptstyle\theta} \\ \Omega A & \xrightarrow[f^{v}]{} & \Omega B \end{array}$$

is homotopy-commutative, where $v: S^1 \to S^1$ is the inverse map, and ϕ and θ are as in Theorem 6.4.12.

Proof. $L_{f''}$ is a subspace of $L_f \times LA \times L(L_f)$; and if $\lambda \in \Omega A$, then $\phi(\lambda) = (l_0, \lambda, e)$, where l_0 is the base point of L_f and $e: I \to L_f$ is the constant map. Thus $f^{(3)}\phi(\lambda) = (l_0, \lambda)$. But if $L_{f'}$ is regarded as a subspace of $A \times LB \times LA$, (l_0, λ) becomes (a_0, e, λ), where $e: I \to B$ is the constant map. It follows that

$$[\theta f^{(3)}\phi(\lambda)](t) = \begin{cases} f\lambda(1 - 2t) & (0 \leqslant t \leqslant \frac{1}{2}) \\ b_0 & (\frac{1}{2} \leqslant t \leqslant 1). \end{cases}$$

Hence $\theta f^{(3)}\phi = f^{\bar{v}}$, where $\bar{v}$ is as in Proposition 6.4.6. Since $\bar{v} \simeq v$, this means that $\theta f^{(3)}\phi \simeq f^{v}$. ▌

Theorem 6.4.14 *A map $f: A \to B$ gives rise to a sequence of spaces and maps*

$$\cdots \longrightarrow \Omega L_f \xrightarrow{f_1{}^1} \Omega A \xrightarrow{f^1} \Omega B \xrightarrow{f_2} L_f \xrightarrow{f_1} A \xrightarrow{f} B,$$

such that, for any space X, the sequence

$$\cdots \longrightarrow [X, \Omega L_f] \xrightarrow{(f_1{}^1)_*} [X, \Omega A] \xrightarrow{(f^1)_*} [X, \Omega B] \xrightarrow{(f_2)_*}$$
$$[X, L_f] \xrightarrow{(f_1)_*} [X, A] \xrightarrow{f_*} [X, B]$$

is an exact sequence of based sets. If X is an $AH'I$, this is an exact sequence of groups; in any case it is an exact sequence of groups as far as $[X, \Omega B]$, and an exact sequence of abelian groups as far as $[X, \Omega(\Omega B)]$.

Proof. Let $f_1 = f': L_f \to A$, and $f_2 = f''\phi: \Omega B \to L_f$. The proof now proceeds as in Theorem 6.4.7, since $f^v = 1^v f^1$, and $(1^v)_*: [X, \Omega B] \to [X, \Omega B]$ sends each element to its inverse. ∎

Proposition 6.4.15 *A commutative diagram*

$$\begin{array}{ccc} A & \xrightarrow{f} & B \\ {\scriptstyle\mu}\downarrow & & \downarrow{\scriptstyle\lambda} \\ A' & \xrightarrow[f']{} & B' \end{array}$$

gives rise to a commutative diagram

$$\begin{array}{ccccccccc} \cdots \longrightarrow & \Omega B & \xrightarrow{f_2} & L_f & \xrightarrow{f_1} & A & \xrightarrow{f} & B \\ & {\scriptstyle\lambda^1}\downarrow & & \downarrow{\scriptstyle\nu} & & \downarrow{\scriptstyle\mu} & & \downarrow{\scriptstyle\lambda} \\ \cdots \longrightarrow & \Omega B' & \xrightarrow[f_2']{} & L_{f'} & \xrightarrow[f_1']{} & A' & \xrightarrow[f']{} & B' \end{array}$$

Proof. Let ν be the restriction to L_f of $\mu \times \lambda^1: A \times LB \to A' \times LB'$. ∎

Thus for any space X, we obtain a commutative diagram involving the two exact sequences of homotopy classes of maps from X.

6.5 Fibre and cofibre maps

This section is concerned with a further investigation of the spaces C_f and L_f. If the map $f: A \to B$ satisfies certain conditions, it is possible to identify C_f and L_f, up to homotopy equivalence, as a quotient space of B and a subspace of A, respectively: for example, if (B, A) is a polyhedral pair, and $i: A \to B$ is the inclusion map, then $C_i \simeq B/A$.

As usual, there are two sets of 'dual' results, involving C_f and L_f respectively. We start with the results on C_f.

Definition 6.5.1 A map $f: A \to B$ is called a *cofibre map* if, whenever we are given a space X, a map $g: B \to X$ and a homotopy $H: A \times I \to X$, starting with gf, there exists a homotopy $G: B \times I \to X$ that starts with g, and satisfies $H = G(f \times 1)$.

Thus if A is a subspace of B, the inclusion map $i: A \to B$ is a cofibre map if the pair (B, A) has the absolute homotopy extension property (see Section 2.4) (the converse is not true in general, since the definition of cofibre maps refers to *based* maps and homotopies, but the absolute homotopy extension property refers to maps and homotopies that are not necessarily based). In particular, by Theorem 2.4.1, the inclusion of a subpolyhedron in a polyhedron is always a cofibre map.

Theorem 6.5.2 *If $f: A \to B$ is a cofibre map, then $C_f \simeq B/f(A)$.*

Proof. Let $\lambda: C_f \to B/f(A)$ be the map induced by the identification map $B \to B/f(A)$ and the constant map $cA \to B/f(A)$. We show that λ is a homotopy equivalence by constructing a homotopy inverse, and this is where we need to know that f is a cofibre map.

Now $f_1 f: A \to C_f$ is homotopic to the constant map, by the homotopy $H: A \times I \to C_f$ given by $H(a, t) = a \wedge t$ $(\in cA)$. Since f is a cofibre map, there exists a homotopy $G: B \times I \to C_f$ that starts with f_1 and satisfies $H = G(f \times 1)$. Let $g_1: B \to C_f$ be the final map of G; then $g_1 f(A) =$ base point, so that g_1 induces a map $\mu: B/f(A) \to C_f$.

To show that $\mu\lambda \simeq 1: C_f \to C_f$, note that we already have a homotopy $G: B \times I \to C_f$. Define also $J: cA \times I \to C_f$ by

$$J(a \wedge t, s) = a \wedge (s + t(1 - s)) \qquad (a \in A, \quad s, t \in I).$$

J is continuous by Theorem 6.2.4; and since

$$J(a \wedge 0, s) = a \wedge s = G(f(a), s),$$

J and G combine to induce a homotopy $C_f \times I \to C_f$, between the identity map and $\mu\lambda$.

Finally, to show that $\lambda\mu \simeq 1: B/f(A) \to B/f(A)$, we have only to remark that $\lambda G: B \times I \to B/f(A)$ sends $f(A) \times I$ to the base point, and so induces a homotopy $B/f(A) \times I \to B/f(A)$ between the identity and $\lambda\mu$. ▮

Corollary 6.5.3 *If (B, A) has the absolute homotopy extension property, and $i: A \to B$ is the inclusion map, then $C_i \simeq B/A$. Thus for any space Y, the sequence*

$$\cdots \to [sB, Y] \xrightarrow{(i \wedge 1)^*} [sA, Y] \xrightarrow{(i_2\mu)^*} [B/A, Y] \xrightarrow{p^*} [B, Y] \xrightarrow{i^*} [A, Y]$$

is exact, where $p: B \to B/A$ is the identification map. Moreover, if (B', A') also has the absolute homotopy extension property, and $f: (B, A) \to (B', A')$ is a map of pairs, there is a commutative diagram

$$\begin{array}{ccccccccc}
\cdots \longrightarrow & [sA, Y] & \xrightarrow{(i_2\mu)^*} & [B/A, Y] & \xrightarrow{p^*} & [B, Y] & \xrightarrow{i_*} & [A, Y] \\
& \uparrow{\scriptstyle (f\wedge 1)^*} & & \uparrow{\scriptstyle \bar{f}^*} & & \uparrow{\scriptstyle f^*} & & \uparrow{\scriptstyle f^*} \\
\cdots \longrightarrow & [sA', Y] & \xrightarrow[(i_2'\mu')^*]{} & [B'/A', Y] & \xrightarrow[(p')^*]{} & [B', Y] & \xrightarrow[(i')^*]{} & [A', Y].
\end{array}$$

Proof. By Theorems 6.4.7 and 6.5.2, the sequence would certainly be exact if we wrote λi_1 instead of p; but clearly $\lambda i_1 = p$. And by Proposition 6.4.8 a map g gives rise to a commutative diagram of exact sequences, since $\mu^* = \lambda^{*-1}$, and λ commutes with maps induced by f. ∎

Theorem 6.5.2 also provides another more-or-less standard map from E^n/S^{n-1} to S^n. For the pair (E^n, S^{n-1}), being triangulable, has the absolute homotopy extension property, and so there is a homotopy equivalence $\mu: E^n/S^{n-1} \to C_i$, where $i: S^{n-1} \to E^n$ is the inclusion map. This may be composed with $i_2: C_i \to sS^{n-1}$, and the inverse of $h: S^n \to S^{n-1} \wedge S^1$ (the homeomorphism of Proposition 6.2.15) to yield

$$h^{-1}i_2\mu: E^n/S^{n-1} \to S^n.$$

Proposition 6.5.4 *$h^{-1}i_2\mu \simeq \phi\theta$, where $\theta: E^n/S^{n-1} \to S^n$ is the standard homeomorphism, and $\phi(x_1, \ldots, x_{n+1}) = (-x_{n+1}, x_1, \ldots, x_n)$.*

Proof. In defining $\mu: E^n/S^{n-1} \to C_i$, we have to construct a homotopy $G: E^n \times I \to C_i = E^n \cup cS^{n-1}$, and take the final map $g_1: E^n \to C_i$, which induces μ. Now it is easy to see that a suitable map g_1 is given by

$$g_1(x_1, \ldots, x_n) = \begin{cases} (2x_1, \ldots, 2x_n) \in E^n & (r \leqslant \frac{1}{2}) \\ (x_1/r, \ldots, x_n/r) \wedge (2r - 1) \in cS^{n-1} & (r \geqslant \frac{1}{2}), \end{cases}$$

where $r = \|x\|$. Hence $i_2\mu: E^n/S^{n-1} \to S^{n-1} \wedge S^1$ is given by

$$i_2\mu(x_1, \ldots, x_n) = \begin{cases} \text{base point} & (r \leqslant \frac{1}{2}) \\ (x_1/r, \ldots, x_n/r) \wedge (2r - 1) & (r \geqslant \frac{1}{2}). \end{cases}$$

But $\phi\theta(x_1, \ldots, x_n) = ((x_1/r) \sin \pi r, \ldots, (x_n/r) \sin \pi r, -\cos \pi r)$, and it is easy to see that $i_2\mu$ is homotopic to the composite of $\phi\theta$ with the map $p: S^n \to S^{n-1} \wedge S^1$ discussed after Corollary 6.2.19. Since $p \simeq h$, it follows that $h^{-1}i_2\mu \simeq \phi\theta$. ∎

So far Theorem 6.5.2 will seem rather special, since the only maps known to be cofibre maps are the inclusions of subpolyhedra in polyhedra. However, cofibre maps are much more common than this state of affairs suggests; indeed, every map is, to within homotopy equivalence, a cofibre map.

Theorem 6.5.5 *Any map $f: A \to B$ is the composite of a cofibre map and a homotopy equivalence.*

Proof. Let the *mapping cylinder* of f, M_f, be the space obtained from B and $(A \times I)/(a_0 \times I)$ by identifying, for each $a \in A$, the points $(a, 1)$ and $f(a)$: see Fig. 6.6, in which the thick line is supposed to be identified to a point (the base point of M_f).

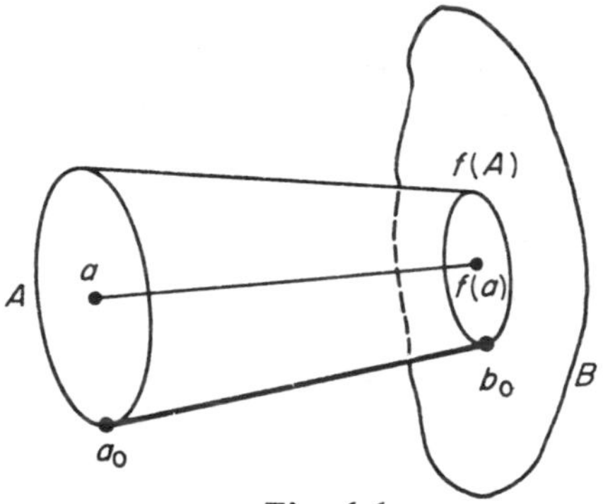

Fig. 6.6

Let $g: A \to M_f$ be the inclusion of A in $A \times I$ (as $A \times 0$), followed by the identification map, and let $h: M_f \to B$ be the map induced by the identity map of B and the map from $A \times I$ to B that sends each (a, t) to $f(a)$ (so that h, as it were, shrinks $A \times I$ down the 'strings' joining a and $f(a)$).

Clearly $f = hg$, so that it remains to prove that g is a cofibre map and that h is a homotopy equivalence. We deal with g first.

Suppose, then, that we have $k: M_f \to X$, and a homotopy $H: A \times I \to X$ starting with kg. To construct the corresponding homotopy $G: M_f \times I \to X$, define $G_B: B \times I \to X$ by

$$G_B(b, s) = k(b) \quad (0 \leqslant s \leqslant 1),$$

and $G_A: (A \times I) \times I \to X$ by

$$G_A(a, t, s) = \begin{cases} k(a, (2t - s)/(2 - s)) & (0 \leqslant s \leqslant 2t) \\ H(a, s - 2t) & (2t \leqslant s \leqslant 1). \end{cases}$$

Now G_A is continuous, since if $s = 2t$, $k(a, 0) = kg(a) = H(a, 0)$. Moreover $G_A(a, 1, s) = k(a, 1) = kf(a) = G_B(f(a), s)$, so that G_A and G_B together induce (using Theorem 6.2.4) a homotopy $G: M_f \times I \to X$. Clearly G starts with k, and

$$G(g \times 1)(a, s) = G(a, 0, s) = H(a, s),$$

so that $G(g \times 1) = H$. Hence g is a cofibre map.

To show that h is a homotopy equivalence, define $j: B \to M_f$ to be (the restriction of) the identification map onto M_f. Then $hj = 1_B$, and $jh: M_f \to M_f$ is given by

$$\begin{cases} jh(b) = b \\ jh(a, t) = f(a). \end{cases}$$

A homotopy $H: M_f \times I \to M_f$ between 1 and jh can be defined by 'sliding down the strings from a to $f(a)$'; more precisely, H is defined by

$$\begin{cases} H(b, s) = b \\ H(a, t, s) = (a, t + s(1 - t)). \end{cases}$$

As usual, Theorem 6.2.4 shows that this is continuous. ∎

The 'dual' results involve the space L_f, and certain maps known as *fibre maps*: this is historically the older concept, and explains the use of the term 'cofibre map' in Definition 6.5.1.

Definition 6.5.6 A map $f: A \to B$ is a *fibre map* if, whenever we are given a space X, a map $g: X \to A$ and a homotopy $H: X \times I \to B$ that starts with fg, there exists a homotopy $G: X \times I \to A$ that starts with g and satisfies $fG = H$.

If f is a fibre map, the *fibre* of f, F, is defined by $F = f^{-1}(b_0)$, where b_0 is the base point of B. It is a subspace of A.

Theorem 6.5.7 *If $f: A \to B$ is a fibre map, then $L_f \simeq F$.*

Proof. Recall that L_f is the subspace of $A \times LB$ of pairs (a, ν) such that $f(a) = \nu(0)$. Thus we can define $\lambda: F \to L_f$ by $\lambda(a) = (a, e)$, where $e: I \to B$ is the constant map: certainly $(a, e) \in L_f$ if $f(a) = b_0$.

Now consider $ff_1: L_f \to B$. This is homotopic to the constant map, by a homotopy $H: L_f \times I \to B$, where $H(a, \nu, t) = \nu(t)$ $(0 \leqslant t \leqslant 1)$. Since f is a fibre map, there exists a homotopy $G: L_f \times I \to A$ that starts with f_1 and satisfies $fG = H$. In particular, if $g_1\colon L_f \to A$ is the final map of G, $fg_1(L_f) = b_0$, so that $g_1(L_f) \subset F$; denote g_1, regarded as a map into F, by μ.

Now $G(\lambda \times 1)\colon F \times I \to A$ is a map into F, since $fG(a, e, t) = H(a, e, t) = e(t) = b_0$; thus $G(\lambda \times 1)$ provides a homotopy between 1_F and $\mu\lambda$. On the other hand, a homotopy between the identity map of L_f and $\lambda\mu$ can be constructed by combining $G\colon L_f \times I \to A$ with $J\colon L_f \times I \to LB$, where J is defined by

$$[J(a, \nu, s)](t) = \nu(s + t(1 - s))$$

(this is continuous, since it corresponds under the association map to a continuous map from $(L_f \times I) \wedge I$ to B). G and J combine to give a homotopy $L_f \times I \to L_f$, between 1 and $\lambda\mu$, since

$$fG(a, \nu, s) = H(a, \nu, s) = \nu(s) = [J(a, \nu, s)](0). \qquad \blacksquare$$

Corollary 6.5.8 *If $f\colon A \to B$ is a fibre map, with fibre F, then for any space X, the sequence*

$$\cdots \longrightarrow [X, \Omega B] \xrightarrow{(\mu f_2)_*} [X, F] \xrightarrow{i_*} [X, A] \xrightarrow{f_*} [X, B]$$

is exact, where $i\colon F \to A$ is the inclusion map. Moreover, if $f'\colon A' \to B'$ is a fibre map, with fibre F', and

$$\begin{array}{ccc} A & \xrightarrow{f} & B \\ {\scriptstyle g}\downarrow & & \downarrow{\scriptstyle h} \\ A' & \xrightarrow[f']{} & B' \end{array}$$

is a commutative diagram of spaces and maps, then $g(F) \subset F'$, and there is a commutative diagram

$$\begin{array}{ccccccc} \cdots \longrightarrow [X, \Omega B] & \xrightarrow{(\mu f_2)_*} & [X, F] & \xrightarrow{i_*} & [X, A] & \xrightarrow{f_*} & [X, B] \\ {\scriptstyle (h^1)_*}\downarrow & & \downarrow{\scriptstyle g_*} & & \downarrow{\scriptstyle g_*} & & \downarrow{\scriptstyle h_*} \\ \cdots \longrightarrow [X, \Omega B'] & \xrightarrow[(\mu' f_2')_*]{} & [X, F'] & \xrightarrow[i_*']{} & [X, A'] & \xrightarrow[f_*']{} & [X, B']. \end{array}$$

Proof. Since $f_1\lambda = i\colon F \to A$, Theorems 6.4.14 and 6.5.7 show that the sequence is exact. And the commutative diagram follows from Proposition 6.4.14. $\blacksquare$

Corollary 6.5.9 *If $f\colon A \to B$ is a fibre map, with fibre F, there is an exact sequence*

$$\cdots \longrightarrow \pi_n(F) \xrightarrow{i_*} \pi_n(A) \xrightarrow{f_*} \pi_n(B) \longrightarrow \pi_{n-1}(F) \longrightarrow \cdots$$
$$\longrightarrow \pi_0(F) \xrightarrow{i_*} \pi_0(A) \xrightarrow{f_*} \pi_0(B)$$

called the 'exact homotopy sequence' of the fibre map f. It is an exact sequence of groups and homomorphisms as far as $\pi_1(B)$.

Proof. The diagram

$$\begin{array}{ccc} {[X, \Omega A]} & \xrightarrow{(f^1)_*} & {[X, \Omega B]} \\ {\scriptstyle\bar{\alpha}}\uparrow & & \uparrow{\scriptstyle\bar{\alpha}} \\ {[sX, A]} & \xrightarrow[f_*]{} & {[sX, B]} \end{array}$$

is clearly commutative, where $\bar{\alpha}$ is the isomorphism of Corollary 6.3.29. Now apply Corollary 6.5.8 with $X = S^0$, and identify the groups and homomorphisms up to $[S^0, \Omega B]$. ∎

Lastly, the 'dual' of Theorem 6.5.5 is true.

Theorem 6.5.10 *Any map $f: A \to B$ is the composite of a homotopy equivalence and a fibre map.*

Proof. Let I^+ be the disjoint union of I and a point p, where p is taken to be the base point of I^+. Thus for any (based) space B, B^{I^+} may be regarded as the set of maps from I to B that are *not* necessarily base-point-preserving. This allows us to define the 'dual' of the mapping cylinder: we let P_f be the subspace of $A \times B^{I^+}$ consisting of pairs (a, λ) such that $f(a) = \lambda(1)$, and we take (a_0, e) to be the base point of P_f, where e is the constant map.

Now define $g: P_f \to B$ by $g(a, \lambda) = \lambda(0)$, and $h: A \to P_f$ by $h(a) = (a, e_{f(a)})$, where $e_{f(a)}: I^+ \to B$ is the map that sends all of I to $f(a)$. Then g is continuous, since it is easy to see that the map $B^{I^+} \to B$ that sends λ to $\lambda(0)$ is; also h is continuous, since the map $B \to B^{I^+}$ that sends b to e_b corresponds under the association map to the map $B \wedge I^+ \to B$ that sends $b \wedge t$ to b for all $t \in I$ (notice that $B \wedge I^+$ may be identified with $(B \times I)/(b_0 \times I)$). Moreover $gh(a) = e_{f(a)}(0) = f(a)$, so that $f = gh$, and it remains only to prove that g is a fibre map and h is a homotopy equivalence. As in Theorem 6.5.5, we deal with g first.

Suppose, then, that we have $k: X \to P_f$, and a homotopy $H: X \times I \to B$ that starts with gk. Now $k: X \to P_f \subset A \times B^{I^+}$ has two components: a map $k_1: X \to A$ and a map $k_2: X \to B^{I^+}$, which under the association map corresponds to a map $X \wedge I^+ \to B$. Since $X \wedge I^+ = (X \times I)/(x_0 \times I)$, the latter map can be composed with the identification map to yield $k': X \times I \to B$, where $k'(x, t) = [k_2(x)](t)$. Now

$$H(x, 0) = gk(x) = [k_2(x)](0) = k'(x, 0),$$

so that we can define $G'_B: (X \times I) \times I \to B$ by

$$G'_B(x, s, t) = \begin{cases} k'(x, (2t - s)/(2 - s)) & (0 \leqslant s \leqslant 2t) \\ H(x, s - 2t) & (2t \leqslant s \leqslant 1). \end{cases}$$

This is the base point if $x = x_0$, so that G'_B induces a map $(X \times I) \wedge I^+ \to B$, and hence a homotopy $G_B: X \times I \to B^{I^+}$. Thus we can define a homotopy $G: X \times I \to A \times B^{I^+}$ by

$$G(x, s) = (k_1(x), G_B(x, s)).$$

In fact this is a map into P_f, since

$$fk_1(x) = [k_2(x)](1) = k'(x, 1) = G'_B(x, s, 1) = [G_B(x, s)](1).$$

Moreover $G(x, 0) = (k_1(x), k_2(x)) = k(x)$, so that G starts with k; and $gG(x, s) = [G_B(x, s)](0) = G'_B(x, s, 0) = H(x, s)$, so that $gG = H$. Hence g is a fibre map.

To show that h is a homotopy equivalence, define $j: P_f \to A$ by $j(a, \lambda) = a$. Then $jh = 1_A$, and $hj(a, \lambda) = (a, e_{f(a)})$. A homotopy $H: P_f \times I \to P_f$ between 1 and hj is given by 'contracting the paths λ'; more precisely, by defining

$$H(a, \lambda, s) = (a, \lambda_s),$$

where λ_s is the map from I^+ to B defined by $\lambda_s(t) = \lambda(t + s(1 - t))$. Now H is continuous, since the map $B^{I^+} \times I \to B^{I^+}$ that sends (λ, s) to λ_s is just the homotopy induced as in Theorem 6.2.25(b) by the homotopy $I^+ \times I \to I^+$ that sends (t, s) to $(t + s(1 - t))$. And since H is obviously a map into P_f, H is a homotopy between 1 and hj. ∎

EXERCISES

1. Show that the results of Theorem 6.2.4 are not true without some restriction on the spaces involved, by means of the following example. For each integer $n \geqslant 1$, let I_n be a copy of the unit interval I. Let X be the disjoint union of the I_n, let $Y = \bigvee I_n$, and let $p: X \to Y$ be the identification map. Then if Q denotes the rationals, topologized as a subset of the real line, $p \times 1: X \times Q \to Y \times Q$ is not an identification map. (*Hint:* for each $n \geqslant 1$, enumerate the rationals in $[-1/n, 1/n]$ as $q_1, q_2, q_3, \ldots$, and let $U_{n,r} = (q_r - 1/(n.2^r), q_r + 1/(n.2^r)) \cap Q$. Let V_n be the subset of $I_n \times Q$ of the form

$$I_n \times \{(-\infty, -1/n) \cap Q\} \cup \bigcup_{r=1}^{\infty} (1 - 1/2^r, 1] \times U_{n,r} \cup I_n \times \{(1/n, \infty) \cap Q\}.$$

Finally, let V be the subset of $Y \times Q$ of the form $(p \times 1)(\bigcup_{n=1}^{\infty} V_n)$. Then $(p \times 1)^{-1}(V) = \bigcup V_n$, and so is open; but V is not open, since no open neighbourhood of $(y_0, 0)$ can be contained in V.)

2. Let $X = S^1$, and $A = S^1 - p$, where p is any point of S^1 other than the base point. Show that A is contractible, but $X \not\simeq X/A$.

3. Given spaces X and Y, the *join* $X * Y$ is defined to be the space obtained from $X \times Y \times I$ by identifying $(x, y, 0)$ with $(x, y_0, 0)$ and $(x, y, 1)$ with $(x_0, y, 1)$ for each $x \in X$, $y \in Y$. Show that if X and Y are polyhedra, this definition coincides, up to homeomorphism, with that of Definition 2.3.17.

   Now define the ***reduced join*** $X \mathbin{\dot{*}} Y$ to be the space obtained from $X * Y$ by identifying to a point the subspace $(x_0 \times Y \times I) \cup (X \times y_0 \times I)$. Prove that $X \mathbin{\dot{*}} Y$ and $s(X \wedge Y)$ are homeomorphic, and deduce that $X * Y \simeq s(X \wedge Y)$ if X and Y are polyhedra.

4. Show that $(X \wedge Y) \wedge Z$ and $X \wedge (Y \wedge Z)$ are homeomorphic if both X and Z are locally compact and Hausdorff.

5. Define functions

$$\theta: Y^X \times Z \to (Y \times Z)^X, \qquad \phi: Y^{X \times Z} \times Z \to Y^X$$

   by the rules

$$[\theta(\lambda, z)](x) = (\lambda(x), z) \qquad (x \in X, z \in Z, \lambda: X \to Y),$$

$$[\phi(\mu, z)](x) = \mu(x, z) \qquad (x \in X, z \in Z, \mu: X \times Z \to Y).$$

   Show that θ is always continuous, and that ϕ is continuous if Z is locally compact and Hausdorff.

6. Show that the evaluation map $e: Y^X \wedge X \to Y$ is not always continuous, by taking $X = Q$ and $Y = R^1$, both with base point 0. (*Hint:* no compact subset of Q can contain all the rationals in an interval.)

7. Deduce from Exercise 6 that the association map $\alpha: Z^{X \wedge Y} \to (Z^Y)^X$ is not always onto.

8. Given spaces X and Y, such that $X \simeq Y$, show that Y is an AHI if and only if X is. Similarly, show that Y is an AH'I if and only if X is.

9. Let $\mu: S^1 \to S^1 \vee S^1$ be a map that makes S^1 into an AH'I. Show that there are only two homotopy classes of such maps, and that if $\pi_1(Y)_1$ and $\pi_1(Y)_2$ denote the corresponding group structures in $[S^1, Y]$, then $\pi_1(Y)_1 \cong \pi_1(Y)_2$. (*Hint:* use Theorem 3.3.18 to calculate $\pi_1(S^1 \vee S^1)$, and show that there are only two elements that give an associative H'-structure in S^1.)

10. Show that $\pi_1(Y)$ is abelian if Y is an H-space (not necessarily associative or with inverse).

11. Given a homotopy-commutative diagram

$$\begin{array}{ccc} A & \xrightarrow{f} & B \\ {\scriptstyle\lambda}\downarrow & & \downarrow{\scriptstyle\mu} \\ A' & \xrightarrow[f']{} & B', \end{array}$$

show that there exists a map $\nu: C_f \to C_{f'}$, such that the diagram

$$\begin{array}{ccccccccc} A & \xrightarrow{f} & B & \xrightarrow{f_1} & C_f & \xrightarrow{f_2} & sA & \longrightarrow & \cdots \\ {\scriptstyle\lambda}\downarrow & & \downarrow{\scriptstyle\mu} & & \downarrow{\scriptstyle\nu} & & \downarrow{\scriptstyle\lambda\wedge 1} & & \\ A' & \xrightarrow[f']{} & B' & \xrightarrow[f'_1]{} & C_{f'} & \xrightarrow[f'_2]{} & sA' & \longrightarrow & \cdots \end{array}$$

is also homotopy-commutative. Prove also that ν is a homotopy equivalence if both λ and μ are. (*Hint:* let F be the homotopy between μf and $f'\lambda$, and define ν by

$$\begin{cases} \nu(b) = \mu(b) \quad (b \in B) \\ \nu(a \wedge t) = \begin{cases} F(a, 2t) & (0 \leqslant t \leqslant \frac{1}{2}) \\ \lambda(a) \wedge (2t - 1) & (\frac{1}{2} \leqslant t \leqslant 1). \end{cases} \end{cases}$$

In order to prove that ν is a homotopy equivalence if λ and μ are, write $\nu = \nu(\lambda, \mu, F)$, and let $\bar{\nu} = \nu(\bar{\lambda}, \bar{\mu}, \bar{F})$, where $\bar{\lambda}$ and $\bar{\mu}$ are homotopy inverses to λ and μ, and $\bar{F}$ is a homotopy between $\bar{\mu}f'$ and $f\bar{\lambda}$. Show that $\bar{\nu}\nu \simeq \nu(1_A, 1_B, F')$, where F' is some homotopy between f and itself, and deduce that $\bar{\nu}\nu$ is a homotopy equivalence. A similar argument shows that $\nu\bar{\nu}$ is a homotopy equivalence, and it is easy to conclude that therefore ν is a homotopy equivalence (although $\bar{\nu}$ is *not* necessarily a homotopy inverse!).)

12. Establish the 'dual' results to those of Exercise 11.

13. Given a map $f: S^n \to X$, let Y be the adjunction space $X \cup_f E^{n+1}$. Show that $Y \simeq C_f$, and deduce that $S^m \times S^n \simeq C_\theta$, where $\theta: S^{m+n-1} \to S^m \vee S^n$ is a certain map. (*Hint:* regard S^{m+n-1} as the subspace $E^m \times S^{n-1} \cup S^{m-1} \times E^n$ of $E^m \times E^n$.)

14. Given a space X, and elements $\alpha \in \pi_m(X)$, $\beta \in \pi_n(X)$, the *Whitehead product* $[\alpha, \beta] \in \pi_{m+n-1}(X)$ is defined as follows. Represent α and β by maps $f: S^m \to X$, $g: S^n \to X$ respectively, and let $[\alpha, \beta]$ be the element represented by the composite

$$S^{m+n-1} \xrightarrow{\theta} S^m \vee S^n \xrightarrow{f \vee g} X \vee X \xrightarrow{\nabla} X,$$

where θ is the map considered in Exercise 13. Show that if X is an H-space, $[\alpha, \beta] = 0$ for all α and β. Conversely, prove that S^n is an H-space if $[\iota_n, \iota_n] = 0$, where $\iota_n \in \pi_n(S^n)$ is the element represented by the identity map.

15. Given two spaces X and Y, let $i: X \vee Y \to X \times Y$ be the inclusion map. By considering the map $f: c(X \times Y) \to sX \vee sY$ defined by

$$f((x,y) \wedge t) = \begin{cases} (x \wedge 2t, \text{ base point}) & (0 \leqslant t \leqslant \frac{1}{2}) \\ (\text{base point}, y \wedge (2t-1)) & (\frac{1}{2} \leqslant t \leqslant 1), \end{cases}$$

show that $i_2: C_i \to s(X \vee Y)$ is homotopic to the constant map. Deduce the following results.
(a) If $s: \pi_n(X) \to \pi_{n+1}(sX)$ is the homomorphism that sends the element represented by $f: S^n \to X$ to the element represented by $f \wedge 1: S^n \wedge S^1 \to X \wedge S^1$, then $s[\alpha, \beta] = 0$ for all α, β.
(b) If i is a cofibre map, then $s(X \times Y) \simeq sX \vee sY \vee s(X \wedge Y)$.

16. Given any map $f: A \to B$, prove that $f_1: B \to C_f$ is a cofibre map and that $f_1: L_f \to A$ is a fibre map.

17. Show that, if (X, A) has the absolute homotopy extension property, and $f \simeq g: A \to Y$, then the adjunction spaces $Y \cup_f X$ and $Y \cup_g X$ are homotopy-equivalent. (*Hint:* let Z_f be the space obtained from $X \times 0 \cup (A \times I)/(x_0 \times I) \cup Y$ by identifying $(a, 1)$ with $f(a)$, for each $a \in A$; prove that $Y \cup_f X \simeq Z_f$ and that $Z_f \simeq Z_g$.)

18. A map $f: A \to B$ is called a *Serre fibre map* if it satisfies Definition 6.5.6 for all polyhedra X (rather than for all spaces X); and the *fibre* F is once again $f^{-1}(b_0)$. Show that there is a map $g: F \to L_f$ with the property that, if X is any polyhedron, $g_*: [X, F] \to [X, L_f]$ is a (1-1)-correspondence. Deduce that, if $i: F \to A$ is the inclusion map, there is an exact sequence

$$\cdots \longrightarrow \pi_n(F) \xrightarrow{i_*} \pi_n(A) \xrightarrow{f_*} \pi_n(B) \longrightarrow \pi_{n-1}(F) \longrightarrow \cdots.$$

19. A map $f: A \to B$ is called a *local product*, with fibre F, if for each point $b \in B$ there exists an open neighbourhood U of b and a (not necessarily based) homeomorphism $h_U: U \times F \to f^{-1}(U)$, such that $fh_U(b' \times F) = b'$ for all $b' \in U$. Prove that a local product is a Serre fibre map. (*Hint:* let K be a simplicial complex, and suppose given a map $g: |K| \to A$ and a homotopy $H: |K| \times I \to B$, such that H starts with fg. Triangulate I by a simplicial complex L with vertices $0 = t_0 < t_1 < \cdots < t_n = 1$, and choose $r \geqslant 0$, such that for each simplex σ of $K^{(r)}$ and each i, $\sigma \times [t_i, t_{i+1}]$ is mapped by H into one of the open neighbourhoods U. Now construct a homotopy $G: |K^{(r)}| \times I \to A$ that starts with g and satisfies $fG = H$, by induction on the skeletons of $K^{(r)}$.)

20. Prove that the map $f: R^1 \to S^1$, defined by $f(x) = (\cos 2\pi x, \sin 2\pi x)$, is a local product. Deduce that $\pi_n(S^1) = 0$ for $n > 1$.

21. Let $f: S^n \to RP^n$ be the identification map provided by Proposition 1.4.40. Show that f is a local product, and deduce that $\pi_r(RP^n) \cong \pi_r(S^n)$ for $r > 1$. (*Hint:* consider the open sets U_r in RP^n, where U_r is the set of points $[x_1, \ldots, x_{n+1}]$ such that $x_r \neq 0$.)

22. A map $f: A \to B$ is called a *covering map* if it is a local product, with a discrete space as fibre. Show that $f_*: \pi_n(A) \to \pi_n(B)$ is isomorphic for $n > 1$ and (1-1) for $n = 1$. Prove also that if $g: I \to B$ is a (based) path in B, there exists a unique (based) path $h: I \to A$ such that $fh = g$.

23. Let $f: A \to B$ be a covering map, and let X be a path-connected and locally path-connected space (X is said to be *locally path-connected* if, for each point $x \in X$ and open set U containing x, there exists an open set V, such that $x \in V \subset U$ and any two points of V can be connected by a path in U). Prove that, if $g: X \to B$ is a map such that $g_*\pi_1(X) \subset f_*\pi_1(A)$, there exists a unique map $h: X \to A$ such that $fh = g$. (*Hint:* use Exercise 22 to define the *function h*, for each point of X, and then show that h is continuous and unique.)

24. Let $f: A \to B$ and $f': A' \to B$ be two covering maps such that $f_*\pi_1(A) = f'_*\pi_1(A')$, and suppose that both A and A' are path-connected and locally path-connected. Prove that A and A' are homeomorphic.

25. Let B be a path-connected, locally path-connected space, that is also 'weakly locally simply-connected' (that is, for each point $x \in B$, and open set U containing x, there exists an open set V, such that $x \in V \subset U$, and every loop in V based at x is contractible in B). Let G be any subgroup of $\pi_1(B)$. Prove that there exists a space A and a covering map $f: A \to B$, such that $f_*\pi_1(A) = G$. (*Hint:* define an equivalence relation R in B^I, by $uRv \Leftrightarrow [u^{-1}.v] \in G$, and let $A = B^I/R$.) Show also that A is determined up to homeomorphism by B and G: thus in particular there is essentially only one such space A if $G = 0$; in this case $\pi_1(A) = 0$ and A is called the *universal cover* of B.

NOTES ON CHAPTER 6

Identification maps. Theorem 6.2.4(a) is due to Cohen [41] and (b) to Puppe [119]. It is possible to remove the restrictions on the spaces by retopologizing the product: see R. Brown [32].

Associativity of the reduced product. It can be shown that $(X \wedge Y) \wedge Z$ and $X \wedge (Y \wedge Z)$ are always homotopy-equivalent (though not necessarily by a *based* homotopy equivalence): see Puppe [119].

Mapping spaces. The compact-open topology (due to Fox [58]) is not the only possible topology for Y^X; for example, we could take as a sub-base all sets of the form $W_{x,U}$, where x is a point of X and U is an open subset of Y: this is the topology of *pointwise convergence*. For a discussion of these and other topologies, see Fox [58] or Kelley [85], Chapter 7.

Some attempts have been made to circumvent the difficulties caused by the fact that in general the evaluation map is not continuous, and the association map is not a homeomorphism. Spanier's method [130] is to weaken the definition of continuity, but perhaps the more satisfactory method is that of R. Brown [33], who shows that if one of the 'extraordinary products' of [32] is used instead of the ordinary topological product, the evaluation map is continuous and the association map is a homeomorphism.

H-spaces. These were first introduced by Hopf [71]. It should be noted that Definition 6.3.7 is not absolutely standard, since some authors require that mi_1 and mi_2 should both coincide with the identity map, instead of merely being homotopic to it; similar remarks apply to the definitions of associativity and commutativity.

In certain circumstances an associative H-space Y will automatically have an inverse: for example, if Y is path-connected, and a CW-complex in the sense of Chapter 7. See James [79] or Sibson [126].

It is possible for Y^X to be an H-space, even though Y is not an H-space and X is not an H'-space: see R. Brown [34] for an example.

H-spaces have been popular objects of study by topologists for some time. For a variety of results, see Browder [26, 27], Browder and Thomas [29], James [80], Stasheff [133, 134] and Dold and Lashof [45], in addition to the papers already mentioned.

The 'dual' notion of an H'-space is due to Eckmann and Hilton [49].

Homotopy groups are due originally to Čech [38], but we follow the notation of Hurewicz [74], which has since become standard.

Duality. For more details of the 'duality' exhibited in Chapter 6, see Hilton [63].

Exact sequences. Theorem 6.4.7 is essentially due to Barratt [19], but we follow the exposition of Puppe [119]. The 'dual' Theorem 6.4.14 was first proved (in a less general form) by Peterson [115].

Cofibre maps. The result that a polyhedral pair has the absolute homotopy extension property can be generalized, and in Chapter 7 we shall prove that the same result holds with 'polyhedral' replaced by 'CW'. For another set of conditions under which a pair of spaces has the absolute homotopy extension property, see Hu [72], or [73], p. 31, Ex. O.

The *mapping cylinder* was first defined by J. H. C. Whitehead [159].

Fibre maps. Definition 6.5.6 is but one of many definitions of 'fibre map' or 'fibre space'. Our definition is that of Hurewicz [76], and the weaker version given in Exercise 18 is due to Serre [125]. That a Serre fibre map need not be a Hurewicz fibre map is shown by an example of R. Brown [34]. A rather different method of weakening Definition 6.5.6 will be found in Dold [44], and possibly the weakest definition of all, in which a map $f: A \to B$ is called a *quasifibration* if the result of Corollary 6.5.9 holds, is due to Dold and Thom [46].

A quite different approach is exemplified by the *local product* of Exercise 19, which with a little extra structure becomes the *fibre bundle* of Whitney

[162] (see also Steenrod [137]). The theory of fibre bundles, particularly those with a vector space as fibre, has been greatly developed in recent years, and has led to the construction of powerful new topological invariants. For an outline of this theory, see Atiyah and Hirzebruch [18] or Atiyah [17].

Finally, fibre bundles have been generalized by Milnor [107] and Rourke and Sanderson [123].

The Whitehead product. Whitehead's original definition will be found in [157], and generalizations in Hilton [63]. Adams [2] has proved that $[\iota_n, \iota_n] = 0$ only if $n = 1, 3$ or 7, so that S^n is an H-space only for these values of n.

Exercise 15(a) is capable of considerable generalization, at least if X is a sphere: the 'suspension homomorphism' s and a homomorphism defined by a certain Whitehead product can be fitted into an exact sequence, the *EHP-sequence* of G. W. Whitehead [154] (see also James [78]).

Covering spaces. For details of Exercises 22–25, see Hu [73], Chapter 3, or Hilton and Wylie [64], Chapter 6. In Exercise 25, if B is a polyhedron, then A may be taken to be a polyhedron as well: for a proof see, for example, Seifert and Threlfall [124], Chapter 8.

CHAPTER 7

HOMOTOPY GROUPS AND CW-COMPLEXES

7.1 Introduction

We have already, in Chapter 6, defined the homotopy groups of a (based) space Y, and established some of their properties: those, at least, that are shared by the more general sets and groups of the form $[X, Y]$. The first object of this chapter is to continue this investigation, but with special reference now to the groups $\pi_n(Y)$. Most of these results will be true only for the homotopy groups, and not for the more general situation: for example, we shall calculate the groups $\pi_n(S^n)$ $(n \geqslant 1)$, and prove an important theorem concerning the homotopy groups of CW-complexes. On the other hand we shall also investigate the effect on $\pi_n(Y)$ of changing the base point (compare Theorem 3.2.16), and establish calculation theorems for $\pi_n(X \times Y)$, $\pi_n(X \vee Y)$; and although these results will be given only in terms of homotopy groups, they are capable of generalization to sets of the form $[X, Y]$: see Exercises 1, 6 and 7.

The rest of the chapter will be concerned with CW-complexes. These are at once generalizations and simplifications of the notion of a simplicial complex. A simplicial complex—or rather its polyhedron—may be thought of as a space built up by successively attaching simplexes along their boundaries. Now a simplex and its boundary form a triangulation of (E^n, S^{n-1}) for some n, so that in fact a polyhedron is formed by successively attaching *cells* by maps of their boundaries. However, the cells have to be triangulated, and the 'attaching maps' involved have to be simplicial homeomorphisms onto their images. A *CW-complex*, on the other hand, is built up by successively attaching cells by any continuous maps of their boundaries (not necessarily homeomorphisms onto their images), and the number of cells is not restricted to be finite. This has many advantages: for example, a polyhedron can often be regarded as a CW-complex with far fewer cells than there were simplexes originally (for instance, S^n is a CW-complex with only two cells), and the product of two polyhedra is a CW-complex in a natural way, since the product of two simplexes is a cell, but not a simplex in general. CW-complexes are also the natural setting for the theorem of J. H. C. Whitehead: given a map $f: X \to Y$ of path-connected CW-complexes, such that $f_*: \pi_n(X) \to \pi_n(Y)$ is isomorphic for all n, then f is a homotopy equivalence.

Section 7.2 contains the standard results on homotopy groups, and Section 7.3 the definition and elementary properties of CW-complexes. Theorems on the calculation of homotopy groups of CW-complexes (in particular $\pi_n(S^n)$) are proved in Section 7.4, and the theorem of J. H. C. Whitehead in Section 7.5.

7.2 Homotopy groups

In Chapters 3 and 6 we have already defined $\pi_n(Y)$ for any based space Y, seen that a based map $f: X \to Y$ gives rise to homomorphisms $f_*: \pi_n(X) \to \pi_n(Y)$ for $n \geqslant 1$, and proved that $\pi_n(Y)$ is abelian for $n \geqslant 2$. Moreover, a fibre map $f: A \to B$ gives rise to an exact sequence of homotopy groups. Our first task in this section is to extend to $\pi_n(Y)$ the result that the definitions of $\pi_1(Y)$, given in Chapter 3 and as $[S^1, Y]$ in Chapter 6, give isomorphic groups. The point is that the definition analogous to that of Chapter 3 is often easier to work with than the definition as $[S^n, Y]$.

Let I^n be the product of n copies of the unit interval I, and ∂I^n its 'boundary'; thus I^n is the subset of R^n of points $(x_1, \ldots, x_n)$ such that $0 \leqslant x_r \leqslant 1$ for $1 \leqslant r \leqslant n$, and ∂I^n is the subset of points with at least one co-ordinate equal to 0 or 1. Now the composite of standard maps

$$I^n \xrightarrow{\iota^n} J^n \xrightarrow{\rho^n} E^n \xrightarrow{\theta} S^n$$

sends ∂I^n to the base point $(-1, 0, \ldots, 0)$, and so induces a (based) homeomorphism $I^n/\partial I^n \to S^n$. Thus by composing with this homeomorphism, a based map $S^n \to Y$ may be regarded as a map $I^n/\partial I^n \to Y$, or alternatively as a map of pairs $(I^n, \partial I^n) \to (Y, y_0)$, where y_0 is the base point. Moreover, by Proposition 6.2.5, a based homotopy between maps $S^n \to Y$ corresponds to a homotopy of maps of pairs between the corresponding maps $(I^n, \partial I^n) \to (Y, y_0)$. Hence we have proved

Proposition 7.2.1 *The elements of $\pi_n(Y)$ are in* (1-1)-*correspondence with homotopy classes,* rel ∂I^n, *of maps $(I^n, \partial I^n) \to (Y, y_0)$. Moreover, if $f: Y \to Z$ is a based map, the image under f_* of the element of $\pi_n(Y)$ represented by a map $g: (I^n, \partial I^n) \to (Y, y_0)$ is just the homotopy class of the composite $fg: (I^n, \partial I^n) \to (Z, z_0)$.* ∎

Indeed, it is easy to complete this interpretation of π_n, by specifying the group structure in terms of maps of $(I^n, \partial I^n)$.

Proposition 7.2.2 *Given two maps $f, g\colon (I^n, \partial I^n) \to (Y, y_0)$, define $f \circ g\colon (I^n, \partial I^n) \to (Y, y_0)$ by*

$$f \circ g(x_1, \ldots, x_n) = \begin{cases} f(x_1, \ldots, x_{n-1}, 2x_n) & (0 \leqslant x_n \leqslant \frac{1}{2}) \\ g(x_1, \ldots, x_{n-1}, 2x_n - 1) & (\frac{1}{2} \leqslant x_n \leqslant 1). \end{cases}$$

Then the definition of $f \circ g$ extends to homotopy classes rel ∂I^n *of such maps, and gives a definition of multiplication in $\pi_n(Y)$ that coincides with the original one. Moreover, the same definition results if f and g are 'composed' using any other co-ordinate instead of x_n.*

Proof. It may be assumed that f and g are composites

$$(I^n, \partial I^n) \xrightarrow{\phi} (S^n, s_0) \xrightarrow{f'} (Y, y_0),$$
$$(I^n, \partial I^n) \xrightarrow{\phi} (S^n, s_0) \xrightarrow{g'} (Y, y_0),$$

where ϕ is the above composite of standard maps and s_0 is the base point $(-1, 0, \ldots, 0)$ of S^n. Now $f'.g' = \nabla(f' \vee g')\bar{\mu}$, where $\bar{\mu}\colon S^n \to S^n \vee S^n$ is a map defined by the standard homeomorphism $S^n = S^{n-1} \wedge S^1$ and the map $\mu\colon S^1 \to S^1 \vee S^1$ of Proposition 6.3.12. But the proof of Proposition 6.3.12 shows that the diagram

$$\begin{array}{ccc} I^n/\partial I^n & \xrightarrow{\phi} & S^n = S^{n-1} \wedge S^1 \\ \downarrow{\scriptstyle \nu} & & \downarrow{\scriptstyle \bar{\mu}} \\ (I^n/\partial I^n) \vee (I^n/\partial I^n) & \xrightarrow{\phi \vee \phi} & S^n \vee S^n = (S^{n-1} \wedge S^1) \vee (S^{n-1} \wedge S^1) \end{array}$$

is commutative, where ν is defined by

$$\nu(x_1, \ldots, x_n) = \begin{cases} ((x_1, \ldots, x_{n-1}, 2x_n), \text{ base point}) & (0 \leqslant x_n \leqslant \frac{1}{2}) \\ (\text{base point}, (x_1, \ldots, x_{n-1}, 2x_n - 1)) & (\frac{1}{2} \leqslant x_n \leqslant 1). \end{cases}$$

Hence $(f'.g')\phi = f \circ g$, so that, since the definition of $f \circ g$ obviously extends to homotopy classes, the two definitions of multiplication in $\pi_n(Y)$ coincide.

Moreover, Corollary 6.3.18 shows that, if we write

$$S^n = (S^{p-1} \wedge S^1) \wedge (S^{n-p-1} \wedge S^1)$$

and use the H'-structure of S^n determined by that of $S^{p-1} \wedge S^1$ to define a multiplication in $\pi_n(Y)$, this multiplication is the same as the previous one. But this multiplication corresponds as above to that given by defining

$$f \circ g(x_1, \ldots, x_n) = \begin{cases} f(x_1, \ldots, 2x_p, \ldots, x_n) & (0 \leqslant x_p \leqslant \frac{1}{2}) \\ g(x_1, \ldots, 2x_p - 1, \ldots, x_n) & (\frac{1}{2} \leqslant x_p \leqslant 1), \end{cases}$$

so that it is immaterial which co-ordinate we use to 'compose' f and g. ∎

The next point to consider is the effect on $\pi_n(Y)$ of changing the base point y_0. We already know what happens if $n = 1$ (see Theorem 3.2.16), and the result for π_n is the obvious generalization. In order to state the theorem, write $\pi_n(Y, y_0)$ instead of $\pi_n(Y)$ for the nth homotopy group of Y, with the base point y_0.

Theorem 7.2.3 *Let y_0 and y_1 be two base points lying in the same path-component of a space Y. A path u in Y from y_0 to y_1 gives rise to an isomorphism $u_\#: \pi_n(Y, y_0) \to \pi_n(Y, y_1)$ $(n \geqslant 1)$, with the following properties.*

(a) *If* $u \simeq v$ rel 0, 1, *then* $u_\# = v_\#$.

(b) $(e_{y_0})_\#$ *is the identity isomorphism.*

(c) *If w is a path from y_1 to y_2, then* $(u.w)_\# = w_\# u_\#$.

(d) *If λ: $Y \to Z$ is a map such that $\lambda(y_0) = z_0$ and $\lambda(y_1) = z_1$, then* $\lambda_* u_\# = (\lambda u)_\# \lambda_*: \pi_n(Y, y_0) \to \pi_n(Z, z_1)$.

Proof. We use the interpretation of π_n given by Propositions 7.2.1 and 7.2.2. Suppose given, then, an element of $\pi_n(Y, y_0)$, represented by a map $f: (I^n, \partial I^n) \to (Y, y_0)$. To define $u_\#[f]$, let $f': (I^n, \partial I^n) \to (Y, y_1)$ be any map that is homotopic to f by a homotopy $F: I^n \times I \to Y$ such that $F(x, t) = u(t)$ for all $x \in \partial I^n$, $t \in I$; put $u_\#[f] = [f'] \in \pi_n(Y, y_1)$. To justify this definition, we have to show that such maps f' always exist, and that $u_\#[f]$ does not depend on the particular choice of f'.

To show that such maps f' exist, note that, since the pair $(I^n, \partial I^n)$ is clearly triangulable, it has the absolute homotopy extension property. The path u may be regarded as a homotopy of ∂I^n, which extends to a homotopy F of I^n that starts with f, and whose final map is a suitable f'.

We can show that $u_\#[f]$ does not depend on the choice of f', and at the same time prove (a), as follows. Let v be another path from y_0 to y_1, and let $G: I^n \times I \to Y$ be a homotopy that starts with f, and satisfies $G(x, t) = v(t)$ for all $x \in \partial I^n$, $t \in I$; write g for the final map of G. Of course, $f' \simeq g$ by the homotopy H formed by composing the reverse of F with G; but unfortunately H is not in general a homotopy relative to ∂I^n. This difficulty can be overcome by using the absolute homotopy extension property again: $H(x, t) = u^{-1}.v(t)$ if $x \in \partial I^n$, and since $u \simeq v$ rel 0, 1 we have $u^{-1}.v \simeq e_{y_1}$ rel 0, 1. By combining this homotopy with the constant homotopy of f' and g, we obtain a homotopy starting with the restriction of H to $I^n \times 0 \cup \partial I^n \times I \cup I^n \times 1$, and since this subspace can be triangulated as a subpolyhedron of $I^n \times I$, this homotopy can be extended to a homotopy of $I^n \times I$ that starts with H. The final map of this homotopy is again a homotopy between f' and g, but by construction it is a homotopy relative to ∂I^n. It follows that $[f'] = [g] \in \pi_n(Y, y_1)$, so that (by taking $u = v$) $u_{\#}[f]$ is independent of the choice of f', and $u_{\#}[f] = v_{\#}[f]$ if $u \simeq v$ rel 0, 1.

Properties (b)–(d) are clear from the definition of $u_{\#}$, and in particular (a)–(c) show that $u_{\#}$ is a (1-1)-correspondence. It remains, then, to show that $u_{\#}$ is a homomorphism. Let $f, g: (I^n, \partial I^n) \to (Y, y_0)$ be two maps, let $F, G: I^n \times I \to Y$ be homotopies starting with f, g respectively, such that $F(x, t) = G(x, t) = u(t)$ for all $x \in \partial I^n$, $t \in I$, and let f', g' be the final maps of F, G respectively. Define $F \circ G: I^n \times I \to Y$ by

$$F \circ G(x_1, \ldots, x_n, t) = \begin{cases} F(x_1, \ldots, x_{n-1}, 2x_n, t) & (0 \leqslant x_n \leqslant \frac{1}{2}) \\ G(x_1, \ldots, x_{n-1}, 2x_n - 1, t) & (\frac{1}{2} \leqslant x_n \leqslant 1); \end{cases}$$

then $F \circ G$ is a homotopy between $f \circ g$ and $f' \circ g'$, and $F \circ G(x, t) = u(t)$ for all $x \in \partial I^n$, $t \in I$. Hence

$$u_{\#}([f][g]) = u_{\#}[f \circ g] = [f' \circ g'] = [f'][g'] = u_{\#}[f].u_{\#}[g],$$

so that $u_{\#}$ is a homomorphism, and therefore an isomorphism. ∎

Corollary 7.2.4 *Let $f: X \to Y$ be a homotopy equivalence. Then if x_0 is any base point of X, and $y_0 = f(x_0)$,*

$$f_*: \pi_n(X, x_0) \to \pi_n(Y, y_0)$$

is an isomorphism for all $n \geqslant 1$.

Proof. Let g be a homotopy inverse to f, and let F be the homotopy between gf and 1_X; let $x_1 = g(y_0)$. Now if $\alpha: I^n \to X$ represents an element of $\pi_n(X, x_0)$, $F(\alpha \times 1)$ is a homotopy between α and $gf\alpha$. This homotopy is not in general relative to ∂I^n, but its restriction to $\partial I^n \times I$ defines a path u, say, from x_0 to x_1. Thus

$$g_* f_* = u_{\#}: \pi_n(X, x_0) \to \pi_n(X, x_1),$$

so that $g_* f_*$ is an isomorphism. Similarly, $f_* g_*$ is an isomorphism, so that f_* and g_* are themselves isomorphic. ∎

It follows from Theorem 7.2.3 that each loop u based at y_0 gives rise to an isomorphism $u_{\#}: \pi_n(Y, y_0) \to \pi_n(Y, y_0)$ ($n \geqslant 1$) that depends only on the class of u in $\pi_1(Y, y_0)$. Thus $\pi_1(Y, y_0)$ acts as a 'group of automorphisms' of $\pi_n(Y, y_0)$, which, as we saw in Chapter 3, are actually inner automorphisms if $n = 1$. In certain circumstances these automorphisms all reduce to the identity automorphism.

Definition 7.2.5 A space Y is *n-simple* if, for each point $y_0 \in Y$, and each loop u based at y_0, $u_{\#}: \pi_n(Y, y_0) \to \pi_n(Y, y_0)$ is the identity isomorphism.

It is easy to see, using Theorem 7.2.3, that if Y is n-simple, the isomorphism $u_{\#}$ determined by a path u from y_0 to y_1 depends only on y_0 and y_1, and not on the particular path u.

Theorem 7.2.6 *Let Y be a path-connected space. Then*

(a) *Y is n-simple if and only if the condition of Definition 6.2.5 holds for just one choice of base point y_0;*

(b) *if Y is simply-connected, Y is n-simple for all n;*

(c) *if Y is an H-space (not necessarily associative or with inverse), Y is n-simple for all n;*

(d) *if Y is n-simple and $X \simeq Y$, then X is n-simple.*

Proof.

(a) Suppose the condition of Definition 7.2.5 holds for y_0; let u be a path from y_0 to another point y_1. Now if v is a loop based at y_1, $u.v.u^{-1}$ is a loop based at y_0, and $(u.v.u^{-1})_{\#}$ is the identity isomorphism. But $(u.v.u^{-1})_{\#} = (u_{\#})^{-1}v_{\#}u_{\#}$, so that $v_{\#} = u_{\#}(u_{\#})^{-1}$, which is the identity isomorphism.

(b) This is obvious.

(c) Let y_0 be the base point, and let $m: Y \times Y \to Y$ be the 'H-space map'. Define $m': Y \to Y$ by $m'(y) = m(y, y_0)$, and let $f: I^n \to Y$ be a map representing an element of $\pi_n(Y, y_0)$. Then $[f] = [m'f]$, since $m' \simeq 1_Y$ rel y_0; and the composite

$$Y \times I \xrightarrow{1 \times u} Y \times Y \xrightarrow{m} Y$$

(where u is a loop based at y_0) is a homotopy between m' and itself, whose restriction to $y_0 \times I$ is a loop v which is homotopic, rel 0, 1, to u. Hence, by composing this homotopy with $f \times 1$,

$$u_{\#}[f] = v_{\#}[f] = v_{\#}[m'f] = [m'f] = [f],$$

so that $u_{\#}$ is the identity isomorphism.

(d) Let $f: X \to Y$ be the homotopy equivalence; let x_0 be a base point for X, and let $y_0 = f(x_0)$. For each $[u] \in \pi_1(X, x_0)$, $x \in \pi_n(X, x_0)$,

$$f_*(x) = (fu_\#)f_*(x) = f_*u_\#(x),$$

by Theorem 7.2.3(d). Since $f_\#$ is an isomorphism by Corollary 7.2.4, it follows that $u_\#(x) = x$. ∎

Thus if Y is an n-simple space, any consideration of base points is irrelevant when working with $\pi_n(Y)$. This is true not only in the sense of Definition 7.2.5, but also in that the elements of $\pi_n(Y)$ may be regarded as homotopy classes of maps $S^n \to Y$, that are not necessarily base-point-preserving.

Proposition 7.2.7 *If Y is a path-connected n-simple space, the elements of $\pi_n(Y)$ are in* (1-1)*-correspondence with the homotopy classes of maps $S^n \to Y$.*

Proof. Let $f: S^n \to Y$ be a map, and suppose that $f(s_0) = y_1$, where s_0 is the base point $(-1, 0, \ldots, 0)$ of S^n. Then $[f]$, the *based* homotopy class of f, is an element of $\pi_n(Y, y_1)$, and if u is any path from y_1 to y_0, $u_\#[f] \in \pi_n(Y, y_0)$.

Now let $g: S^n \to Y$ be another map, with $g(s_0) = y_2$. If v is a path from y_2 to y_0, then $v_\#[g] \in \pi_n(Y, y_0)$; and if $f \simeq g$ by a homotopy F, where $F \mid (s_0 \times I)$ defines a path w from y_1 to y_2, then $w_\#[f] = [g]$ in $\pi_n(Y, y_2)$. Hence

$$v_\#[g] = (w \,.\, v)_\#[f] = u_\#[f],$$

since Y is n-simple. Thus the homotopy class of f defines a unique element of $\pi_n(Y, y_0)$; but conversely an element of $\pi_n(Y, y_0)$ is a based homotopy class of based maps, which is contained in a unique (unbased) homotopy class. ∎

Corollary 7.2.8 *If X and Y are path-connected n-simple spaces, and $f: X \to Y$ is a map, $f_*: \pi_n(X) \to \pi_n(Y)$ is given by $f_*[g] = [fg]$, where $g: S^n \to X$, and $[\quad]$ now denotes unbased homotopy classes.* ∎

The next topic in this section is *relative* homotopy groups. These bear much the same relation to ordinary homotopy groups as relative homology and cohomology groups do to those of a single space, and once again are most often used, via an exact sequence, for calculation purposes.

For the definition, let X be a space with base point x_0, and let Y be a subspace containing x_0. Let $i: Y \to X$ be the inclusion map, and let L_i be the mapping path-space of i.

Definition 7.2.9 For $n \geqslant 2$, the nth *relative homotopy group* of (X, Y), $\pi_n(X, Y)$, is defined by $\pi_n(X, Y) = \pi_{n-1}(L_i)$. We can also define $\pi_1(X, Y) = \pi_0(L_i)$, though in general this is just a set, not a group.

Sometimes we write $\pi_n(X, Y, x_0)$, instead of $\pi_n(X, Y)$, if we wish to draw attention to the particular base point. Notice that $\pi_n(X, Y)$ is abelian for $n \geqslant 3$.

The definition of $\pi_n(X, Y)$ may seem a little obscure, but with the aid of Theorem 6.3.27 we can give an alternative definition on the lines of Propositions 7.2.1 and 7.2.2.

Proposition 7.2.10 *The elements of $\pi_n(X, Y, x_0)$ are in (1-1)-correspondence with homotopy classes of maps (of triples) $(I^n, \partial I^n, D^{n-1}) \to (X, Y, x_0)$, where D^{n-1} is the closure of $\partial I^n - I^{n-1} \times 0$. Given two such maps, f and g, define $f \circ g$ by*

$$f \circ g(x_1, \ldots, x_n) = \begin{cases} f(2x_1, x_2, \ldots, x_n) & (0 \leqslant x_1 \leqslant \frac{1}{2}) \\ g(2x_1 - 1, x_2, \ldots, x_n) & (\frac{1}{2} \leqslant x_1 \leqslant 1); \end{cases}$$

this definition extends to homotopy classes and gives a definition of multiplication in $\pi_n(X, Y)$ $(n \geqslant 2)$ that coincides with that of Definition 7.2.9. Moreover the same definition results if we 'compose' f and g using any other co-ordinate except x_n.

Proof. By definition, L_i is the subspace of $Y \times X^I$ consisting of pairs (y, λ) such that $\lambda(0) = y$. This is homeomorphic to the subspace L of X^I consisting of (based) maps λ such that $\lambda(0) \in Y$: we merely let (y, λ) correspond to λ, and note that the map sending λ to $(\lambda(0), \lambda)$ is continuous by Theorem 6.2.31. Now by an obvious modification of Theorem 6.3.27, the elements of $\pi_n(X, Y)$, which may be taken to be homotopy classes of maps $(I^{n-1}, \partial I^{n-1}) \to (L, l_0)$, where l_0 is the base point of L, are in (1-1)-correspondence with homotopy classes of maps $(I^{n-1} \wedge I, \partial I^{n-1} \wedge I) \to (X, x_0)$ that send $I^{n-1} \wedge 0$ to Y. By Proposition 6.2.5, these in turn may be regarded as homotopy classes of maps

$$(I^{n-1} \times I, I^{n-1} \times 0 \cup \partial I^{n-1} \times I \cup I^{n-1} \times 1, I^{n-1} \times 1 \cup \partial I^{n-1} \times I) \to (X, Y, x_0);$$

that is, as classes of maps

$$(I^n, \partial I^n, D^{n-1}) \to (X, Y, x_0).$$

(If $n = 1$, I^0 is to be interpreted as the pair of points 0 and 1, with base point 1, and ∂I^0 and D^0 as the point 1.)

Finally, it is clear from Proposition 7.2.2 that if we define $f \circ g$ by 'composing' along any co-ordinate of I^n except the last, the resulting multiplication in $\pi_{n-1}(L)$, and hence in $\pi_n(X, Y)$, is the correct one. ∎

Notice that a map $(I^n, \partial I^n, D^{n-1}) \to (X, Y, x_0)$ represents the identity element of $\pi_n(X, Y)$ if and only if it is homotopic, as a map of triples, to a map that sends I^n to Y. For such a map corresponds to a map $(I^{n-1}, \partial I^{n-1}) \to (L, l_0)$ whose image is contained in Y^I; but Y^I is contractible.

An obvious corollary of Proposition 7.2.10 is that if the subspace Y happens to be just x_0, then $\pi_n(X, Y, x_0) \simeq \pi_n(X, x_0)$ (at least if $n \geqslant 2$). Thus there is no ambiguity in the notation $\pi_n(X, x_0)$: it may equally well be interpreted as the nth homotopy group of X, with base point x_0, or as the nth relative homotopy group of the pair (X, x_0).

Proposition 7.2.11 *A based map of pairs $\lambda: (X, Y) \to (A, B)$ gives rise to a homomorphism $\lambda_*: \pi_n(X, Y) \to \pi_n(A, B)$ $(n \geqslant 2)$, with the following properties.*

(a) *If $\lambda \simeq \mu$ (as based maps of pairs), then $\lambda_* = \mu_*$.*
(b) *The identity map gives rise to the identity isomorphism.*
(c) *If $\mu: (A, B) \to (C, D)$ is another based map of pairs, then*

$$(\mu\lambda)_* = \mu_*\lambda_*.$$

Proof. By Proposition 6.4.15, the commutative square

$$\begin{array}{ccc} Y & \xrightarrow{i} & X \\ \lambda\downarrow & & \downarrow\lambda \\ B & \xrightarrow[i']{} & A \end{array}$$

(where i and i' are the inclusion maps) gives rise to a commutative diagram

$$\begin{array}{ccccccc} \cdots & \longrightarrow & L_i & \longrightarrow & Y & \longrightarrow & X \\ & & \bar{\lambda}\downarrow & & \downarrow\lambda & & \downarrow\lambda \\ \cdots & \longrightarrow & L_{i'} & \longrightarrow & B & \longrightarrow & A. \end{array}$$

Define $\lambda_*: \pi_n(X, Y) \to \pi_n(A, B)$ to be $\bar{\lambda}_*: \pi_{n-1}(L_i) \to \pi_{n-1}(L_{i'})$; then if an element of $\pi_n(X, Y)$ is represented by a map $f: (I^n, \partial I^n, D^{n-1}) \to (X, Y, x_0)$, it is easy to see that $\lambda_*[f] = [\lambda f]$, so that properties (a)–(c) are clear. ∎

It is also possible to extend Theorem 7.2.3 to relative homotopy groups.

Theorem 7.2.12 *Let (X, Y) be a pair of spaces, and let x_0, x_1 be two base points in the same path-component of Y. A path u in Y from x_0 to x_1 gives rise to an isomorphism $u_{\#}: \pi_n(X, Y, x_0) \to \pi_n(X, Y, x_1)$ $(n \geqslant 2)$, with the following properties.*

(a) *If $u \simeq v$ rel 0, 1 (as paths in Y), then $u_{\#} = v_{\#}$.*
(b) *$(e_{x_0})_{\#}$ is the identity isomorphism.*
(c) *If w is a path in Y from x_1 to x_2, then $(u.w)_{\#} = w_{\#}u_{\#}$.*
(d) *If $\lambda: (X, Y) \to (A, B)$ is a map such that $\lambda(x_0) = a_0$ and $\lambda(x_1) = a_1$, then $\lambda_* u_* = (\lambda u)_{\#}\lambda_*: \pi_n(X, Y, x_0) \to \pi_n(A, B, a_1)$.*

Proof. Given a map $f: (I^n, \partial I^n, D^{n-1}) \to (X, Y, x_0)$, let $f': (I^n, \partial I^n, D^{n-1}) \to (X, Y, x_1)$ be any map that is homotopic to f by a homotopy $F: I^n \times I \to X$ that sends $\partial I^n \times I$ to Y and satisfies $F(d, t) = u(t)$ for all $d \in D^{n-1}$, $t \in I$; define $u_{\#}[f] = [f'] \in \pi_n(X, Y, x_1)$. As in the proof of Theorem 7.2.3, this defines an isomorphism $u_{\#}$ that depends only on the homotopy class of f, and satisfies properties (a)–(d). ∎

We shall say that the pair (X, Y) is *(relatively) n-simple*, if, for each point x_0 in Y and each loop u in Y based at x_0, $u_{\#}: \pi_n(X, Y, x_0) \to \pi_n(X, Y, x_0)$ is the identity isomorphism. In this case, if u is a path in Y from x_0 to x_1, the isomorphism $u_{\#}$ depends only on x_0 and x_1, and not on the path u itself.

Theorem 7.2.13 *Let Y be a path-connected space. Then*

(a) *(X, Y) is n-simple if the definition holds for just one point x_0 of Y*
(b) *if Y is simply-connected, (X, Y) is n-simple for all $n \geqslant 2$;*
(c) *if (X, Y) is n-simple and $(A, B) \simeq (X, Y)$ (as pairs), then (A, B) is n-simple.*

Proof. This is an obvious modification of Theorem 7.2.6, and the proof is left to the reader. ∎

Example 7.2.14 The pair (E^n, S^{n-1}) is (relatively) n-simple, for all $n \geqslant 2$. This is obvious from Theorem 7.2.13(b) unless $n = 2$. To deal with the case $n = 2$, consider a representative map $f: (I^2, \partial I^2, D^1) \to (E^2, S^1, s_0)$, and a loop u in S^1 based at s_0. A representative map for $u_{\#}[f]$ is $f': (I^2, \partial I^2, D^1) \to (E^2, S^1, s_0)$, where $f' \mid I^1$ is the product loop $u^{-1}.(f \mid I^1).u$; but since $\pi_1(S^1)$ is abelian, this product loop is homotopic, rel 0, 1, to the loop $f|I^1$. In other words, by extending this homotopy, first to the constant homotopy on D^1, and then to I^2, we may assume that $f'|\partial I^2 = f|\partial I^2$. But since E^2 is convex, it follows that $f' \simeq f$ rel ∂I^2, by a linear homotopy; hence $u_{\#}[f] = [f'] = [f]$. ∎

If (X, Y) is relatively n-simple, the elements of $\pi_n(X, Y)$ may be defined without reference to base points.

Proposition 7.2.15 *Let Y be path-connected and (X, Y) be n-simple. The elements of $\pi_n(X, Y)$ are in (1-1)-correspondence with the homotopy classes of maps $(I^n, \partial I^n) \to (X, Y)$. Moreover, given two such maps f and g, such that $f(1, x_2, \ldots, x_n) = g(0, x_2, \ldots, x_n)$, the product $[f][g]$ is the homotopy class of the map $f \circ g$, where*

$$f \circ g(x_1, \ldots, x_n) = \begin{cases} f(2x_1, x_2, \ldots, x_n) & (0 \leqslant x_1 \leqslant \frac{1}{2}) \\ g(2x_1 - 1, x_2, \ldots, x_n) & (\frac{1}{2} \leqslant x_1 \leqslant 1). \end{cases}$$

Similar remarks apply if any other co-ordinate is used instead of x_1 (even if x_n is used).

Proof. Let $f: (I^n, \partial I^n) \to (X, Y)$ be a map. Now D^{n-1} is clearly contractible, so that $f|D^{n-1}$ is homotopic (as a map into Y) to a map to a single point x_1, say. This homotopy may be extended to ∂I^n and then to I^n, to give a final map $f': (I^n, \partial I^n, D^{n-1}) \to (X, Y, x_1)$, such that $f' \simeq f$ as maps of the pair $(I^n, \partial I^n)$. We now have $[f'] \in \pi_n(X, Y, x_1)$, and if u is any path in Y from x_1 to x_0, $u_\#[f'] \in \pi_n(X, Y, x_0)$.
Now let $g: (I^n, \partial I^n) \to (X, Y)$ be another map, homotopic as a map of pairs to $g': (I^n, \partial I^n, D^{n-1}) \to (X, Y, x_2)$, and let v be a path in Y from x_2 to x_0. If $f \simeq g$ as maps of pairs, then $f' \simeq g'$ as maps of pairs, by a homotopy F, say. But $F \mid (D^{n-1} \times I)$ is homotopic, rel $D^{n-1} \times 0 \cup D^{n-1} \times 1$, to a map that sends each $D^{n-1} \times t$ to a single point (because $D^{n-1} \times I$ may be contracted to I by a deformation retraction that sends each $D^{n-1} \times t$ to t). By extending *this* homotopy to $(I^n \times I) \times I$, we may assume that the homotopy F between f' and g' sends each $D^{n-1} \times t$ to a point, and so defines a path w from x_1 to x_2. It follows that $w_\#[f'] = [g'] \in \pi_n(X, Y, x_2)$, so that

$$v_\#[g'] = (w.v)_\#[f'] = u_\#[f'],.$$

since (X, Y) is n-simple. That is to say, the element of $\pi_n(X, Y, x_0)$ determined by f depends only on the homotopy class of f as a map of pairs, and not on the choice of f' or u.

Now suppose that f and g are maps such that $f(1, x_2, \ldots, x_n) = g(0, x_2, \ldots, x_n)$. Choose $f' \simeq f$ as above, the homotopy being F, and let $G: D^{n-1} \times I \to Y$ be the homotopy defined by $G(x_1, \ldots, x_n, t) = F(1 - x_1, \ldots, x_n, t)$; extend G to the whole of I^n, to give a final map $g' \simeq g$. The effect of this is to ensure that

$$F(1, x_2, \ldots, x_n, t) = G(0, x_2, \ldots, x_n, t),$$

so that $f' \circ g'$, as defined in Proposition 7.2.10, is homotopic to $f \circ g$, as defined in the present proposition. It follows that $f \circ g$ represents the element $[f'][g']$ of $\pi_n(X, Y)$; and a similar argument applies if we use any other co-ordinate instead of x_1, except x_n.

To prove that even x_n may be used, suppose that f and g are maps $(I^n, \partial I^n) \to (X, Y)$ such that $f(x_1, \ldots, x_{n-1}, 1) = g(x_1, \ldots, x_{n-1}, 0)$. Define $\bar{f}$ by $\bar{f}(x_1, \ldots, x_n) = f(x_1, \ldots, x_{n-2}, 1 - x_n, x_{n-1})$. By extending the standard homeomorphism between $(I^2, \partial I^2)$ and (E^2, S^1) to a homeomorphism between $(I^n, \partial I^n)$ and $(I^{n-2} \times E^2, I^{n-2} \times S^1 \cup \partial I^{n-2} \times E^2)$, and rotating E^2 through an angle $\pi/2$, we can see that $f \simeq \bar{f}$ as maps of pairs: see Fig. 7.1.

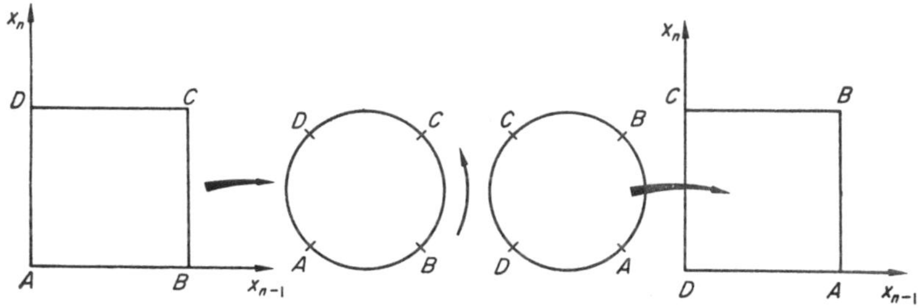

Fig. 7.1

Hence f and $\bar{f}$ represent the same element of $\pi_n(X, Y)$. But if $f \circ g$ is defined by

$$f \circ g(x_1, \ldots, x_n) = \begin{cases} f(x_1, \ldots, x_{n-1}, 2x_n) & (0 \leqslant x_n \leqslant \frac{1}{2}) \\ g(x_1, \ldots, x_{n-1}, 2x_n - 1) & (\frac{1}{2} \leqslant x_n \leqslant 1), \end{cases}$$

then

$$(\overline{f \circ g})(x_1, \ldots, x_n) = \begin{cases} f(x_1, \ldots, 1 - x_n, 2x_{n-1}) & (0 \leqslant x_{n-1} \leqslant \frac{1}{2}) \\ g(x_1, \ldots, 1 - x_n, 2x_{n-1} - 1) & (\frac{1}{2} \leqslant x_{n-1} \leqslant 1), \end{cases}$$

which is the map obtained by 'composing' $\bar{f}$ and $\bar{g}$ along the x_{n-1}-co-ordinate. Hence $f \circ g$ represents the element $[\bar{f}][\bar{g}] = [f][g]$ in $\pi_n(X, Y)$. ∎

A corollary of the proof of Proposition 7.2.15 is the following.

Corollary 7.2.16 *If (X, Y) is 2-simple, $\pi_2(X, Y)$ is abelian.*

Proof. Choose a base point $x_0 \in Y$, and let f, g be two maps $(I^2, \partial I^2, D^1) \to (X, Y, x_0)$. Then $f \circ g$ is defined by

$$f \circ g(x_1, x_2) = \begin{cases} f(2x_1, x_2) & (0 \leqslant x_1 \leqslant \frac{1}{2}) \\ g(2x_1 - 1, x_2) & (\frac{1}{2} \leqslant x_1 \leqslant 1). \end{cases}$$

Hence

$$(\overline{\overline{f \circ g}})(x_1, x_2) = \begin{cases} g(1 - 2x_1, 1 - x_2) & (0 \leqslant x_1 \leqslant \frac{1}{2}) \\ f(2 - 2x_1, 1 - x_2) & (\frac{1}{2} \leqslant x_1 \leqslant 1), \end{cases}$$

so that $(\overline{\overline{f \circ g}}) = \bar{\bar{g}} \circ \bar{\bar{f}}$, and hence $[f][g] = [g][f] \in \pi_2(X, Y)$. ▮

Example 7.2.17 Let (X, Y) be 2-simple, and suppose given a map $f\colon (I^2, \partial I^2) \to (X, Y)$. Now we can divide I^2 into four small squares, by cutting each unit interval in half at the point $\frac{1}{2}$; and then four maps $f_{\epsilon_1,\epsilon_2}\colon (I^2, \partial I^2) \to (X, Y)$ ($\epsilon_1, \epsilon_2 = 0$ or 1) can be defined by restricting f to each of the four squares: more precisely, define

$$f_{\epsilon_1,\epsilon_2}(x_1, x_2) = f(2x_1 - \epsilon_1, 2x_2 - \epsilon_2):$$

see Fig. 7.2.

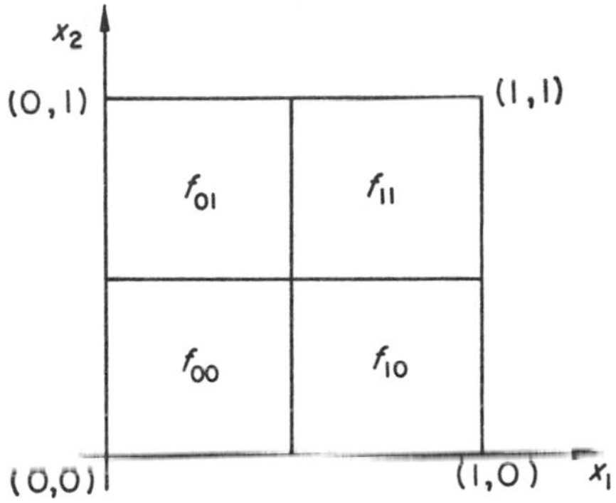

Fig. 7.2

Then if each map $f_{\epsilon_1,\epsilon_2}$ sends ∂I^2 to Y, we have

$$[f] = [f_{00}] + [f_{01}] + [f_{10}] + [f_{11}]$$

in $\pi_2(X, Y)$. For, by Proposition 7.2.15, the right-hand side is represented by the map $(f_{00} \circ f_{01}) \circ (f_{10} \circ f_{11})$, which coincides with f if composition inside the brackets refers to the x_2-co-ordinate, and composition between the brackets refers to the x_1-co-ordinate.

This result can clearly be extended to $\pi_n(X, Y)$, if (X, Y) is n-simple: if $f\colon (I^n, \partial I^n) \to (X, Y)$ is a map, then by halving each unit interval I^n is subdivided into 2^n hypercubes, and so we obtain 2^n maps of I^n to X; if each of these sends ∂I^n to Y, then we obtain 2^n elements of $\pi_n(X, Y)$ whose sum is $[f]$. ▮

The most important property of the relative homotopy groups, however, is that they can be fitted into an exact sequence, and hence used for calculation purposes. As in the case of homology groups, there are exact sequences of a pair and of a triple.

Theorem 7.2.18 *Let (X, Y) be a pair of spaces, with base point $x_0 \in Y$. There is an exact sequence*

$$\cdots \longrightarrow \pi_n(Y) \xrightarrow{i_*} \pi_n(X) \xrightarrow{j_*} \pi_n(X, Y) \xrightarrow{\partial_*} \pi_{n-1}(Y) \longrightarrow \cdots$$
$$\longrightarrow \pi_1(X, Y) \xrightarrow{\partial_*} \pi_0(Y) \xrightarrow{i_*} \pi_0(X),$$

called the 'exact homotopy sequence of the pair (X, Y)'. Moreover, a (based) map of pairs $f: (X, Y) \to (A, B)$ gives rise to a commutative diagram involving the exact homotopy sequences of (X, Y) and (A, B).

Similarly, if Z is a subspace of Y containing x_0, there is an exact sequence

$$\cdots \longrightarrow \pi_n(Y, Z) \xrightarrow{i_*} \pi_n(X, Z) \xrightarrow{j_*} \pi_n(X, Y) \xrightarrow{\partial_*} \pi_{n-1}(Y, Z) \longrightarrow \cdots$$
$$\longrightarrow \pi_1(Y, Z) \xrightarrow{i_*} \pi_1(X, Z) \xrightarrow{j_*} \pi_1(X, Y),$$

called the 'exact homotopy sequence of the triple (X, Y, Z)'. Again, a (based) map of triples $f: (X, Y, Z) \to (A, B, C)$ gives rise to a commutative diagram of exact homotopy sequences.

In the exact sequence of a triple, i_* and j_* are induced by the inclusion maps $i: (Y, Z) \to (X, Z)$, $j: (X, Z) \to (X, Y)$, and ∂_* is given by restricting a map $(I^n, \partial I^n, D^{n-1}) \to (X, Y, x_0)$ to a map $(I^{n-1}, \partial I^{n-1}) \to (Y, x_0) \subset (Y, Z)$. The homomorphisms in the exact sequence of a pair may be similarly interpreted.

Proof. Since (apart from the last few terms) the exact sequence of the pair (X, Y) is obtained from the exact sequence of the triple (X, Y, Z) by putting $Z = x_0$, we shall prove the theorem only for the exact sequence of the triple, and leave to the reader the modifications necessary to deal with the end of the exact sequence of the pair.

Let α be the inclusion map of Z in Y, and let L_α, as usual, be the mapping path-space of α. As in Proposition 7.2.10, L_α may be identified with the subspace L of Y^I consisting of maps λ such that $\lambda(0) \in Z$; and then the standard map $\alpha_1: L \to Z$ is interpreted as a map $L \to Z$ given by $\alpha_1(\lambda) = \lambda(0)$. Let M and N be the spaces similarly obtained from the inclusions $Z \subset X$, $Y \subset X$ respectively, and let $\phi: M \to Z$, $\theta: N \to Y$ be the obvious maps (defined by taking the initial points of paths). Now M and N are both subspaces of X^I, and in fact M is a subspace of N: let $\beta: M \to N$ be the inclusion map. Finally, let L' be the space obtained from β, so that L' is the subspace of N^I of paths starting in M; let $\beta_1: L' \to M$ be the obvious map, and $\psi: L' \to L$ the restriction of $\theta^1: N^I \to Y^I$, so that we have the following commutative diagram, in which the rows are sequences of spaces and maps as in Theorem 6.4.14.

$$\begin{array}{ccccccc} \cdots \longrightarrow \Omega N & \xrightarrow{\beta_2} & L' & \xrightarrow{\beta_1} & M & \xrightarrow{\beta} & N \\ \downarrow{\scriptstyle \theta^1} & & \downarrow{\scriptstyle \psi} & & \downarrow{\scriptstyle \phi} & & \downarrow{\scriptstyle \theta} \\ \cdots \longrightarrow \Omega Y & \xrightarrow[\alpha_2]{} & L & \xrightarrow[\alpha_1]{} & Z & \xrightarrow[\alpha]{} & Y. \end{array}$$

We claim that ψ is a homotopy equivalence. For, by Proposition 6.2.27 and Theorem 6.2.38(c), we may identify L' with the subspace of $X^{I \wedge I}$ consisting of (based) maps $\lambda: I \wedge I \to X$ such that $\lambda(I \wedge 0) \subset Y$ and $\lambda(0 \wedge 0) \in Z$; and ψ, β_1 are given by restricting such a map to $I \wedge 0$, $0 \wedge I$ respectively. But consider the map $f: I \wedge I \to I$ given by $f(x_1 \wedge x_2) = \min\,[x_1 + x_2, 1]$: this induces a map $\chi: L \to L'$, and clearly $\psi\chi = 1_L$. Morever if $g: I \to I \wedge I$ is the inclusion map as $I \wedge 0$, we have $gf \simeq 1 \text{ rel } I \wedge 0$, so that $\chi\psi \simeq 1_L$, by Theorem 6.2.25. In other words, ψ is a homotopy equivalence. Notice also that $\beta_1\chi$ is just the inclusion map $L \to M$ given by the inclusion of Y^I in X^I.

Hence, as in Corollary 6.5.9, we obtain an exact sequence

$$\cdots \longrightarrow \pi_n(L) \xrightarrow{(\beta_1\chi)_*} \pi_n(M) \xrightarrow{\beta_*} \pi_n(N) \xrightarrow{\gamma} \pi_{n-1}(L) \longrightarrow \cdots,$$

where γ is the composite

$$\pi_n(N) \xrightarrow{\cong} \pi_{n-1}(\Omega N) \xrightarrow{\theta^1_*} \pi_{n-1}(\Omega Y) \xrightarrow{(\alpha_2)_*} \pi_{n-1}(L).$$

Certainly the homotopy groups of L, M and N are the relative homotopy groups of the pairs (Y, Z), (X, Z) and (X, Y) respectively, so that it remains only to interpret the maps. Since $\beta_1\chi$ and β are the obvious inclusion maps, it is easy to see from the proof of Proposition 7.2.11 that $(\beta_1\chi)_*$ and β_* may be identified with

$$i_*: \pi_{n+1}(Y, Z) \to \pi_{n+1}(X, Z), \qquad j_*: \pi_{n+1}(X, Z) \to \pi_{n+1}(X, Y)$$

respectively. Moreover, since there is a commutative diagram

$$\begin{array}{ccc} S^n & \longrightarrow & S^{n-1} \wedge S^1 \\ \uparrow & & \uparrow \\ I^n/\partial I^n & \xrightarrow{\mu} & (I^{n-1}/\partial I^{n-1}) \wedge (I/\partial I), \end{array}$$

where the maps are standard homeomorphisms (μ is induced by the obvious map $I^n \to I^{n-1} \times I$), the isomorphism $\pi_n(N) \to \pi_{n-1}(\Omega N)$ is given by sending a map $g: I^n/\partial I^n \to N$ to the map $I^{n-1}/\partial I^{n-1} \to$

$N^{I/\partial I}$ that corresponds to $g\mu^{-1}$ under the association map (of course, $I/\partial I$ is identified with S^1). Since $\theta: N \to Y$ is given by evaluation at 0, it follows as in the proof of Proposition 7.2.10 that γ corresponds to the map $\partial_*: \pi_{n+1}(X, Y) \to \pi_n(Y, Z)$ given by restricting a map $(I^{n+1}, \partial I^{n+1}, D^n) \to (X, Y, x_0)$ to a map from $(I^n \times 0, \partial I^n \times 0)$.

Finally, a map of triples $f: (X, Y, Z) \to (A, B, C)$ gives rise as in Proposition 6.4.15 to commutative diagrams involving the spaces L, L', M, N and the corresponding spaces formed from A, B, C, and hence gives rise to a commutative diagram involving the exact homotopy sequences of the triples (X, Y, Z) and (A, B, C). ∎

As an example of the use of the exact homotopy sequence, we shall establish a useful formula for the groups $\pi_n(X \vee Y)$ $(n \geqslant 2)$. However, this depends also on a knowledge of $\pi_n(X \times Y)$, so that we first need

Theorem 7.2.19 *Let X and Y be based spaces. Then*

$$\pi_n(X \times Y) \cong \pi_n(X) \oplus \pi_n(Y) \qquad (n \geqslant 1).$$

Proof. By Theorem 6.2.34, the spaces $(X \times Y)^{S^n}$ and $X^{S^n} \times Y^{S^n}$ are homeomorphic, so that there is a (1-1)-correspondence between based maps $S^n \to X \times Y$ and pairs of maps $S^n \to X$, $S^n \to Y$, where a map $f: S^n \to X \times Y$ corresponds to the pair $(p_X f, p_Y f)$ $(p_X, p_Y$ are the projection maps of $X \times Y$ onto X and $Y)$. Since the same result is true with S^n replaced by $S^n \times I$, this (1-1)-correspondence extends to homotopy classes of maps, that is, to a (1-1)-correspondence

$$\theta: \pi_n(X \times Y) \to \pi_n(X) \oplus \pi_n(Y).$$

It remains to show that θ is a homomorphism; but this is trivial, since $\theta(x) = (p_X)_* x \oplus (p_Y)_* x$. ∎

Notice that if $i_X, i_Y: X, Y \to X \times Y$ are the inclusions as $X \times y_0$, $x_0 \times Y$ respectively, then the homomorphism $\phi: \pi_n(X) \oplus \pi_n(Y) \to \pi_n(X \times Y)$ defined by $\phi(x \oplus y) = (i_X)_* x + (i_Y)_* y$ is the inverse isomorphism to θ. For

$$\begin{aligned} \theta\phi(x \oplus y) &= [(p_X i_X)_* x + (p_X i_Y)_* y] \oplus [(p_Y i_X)_* x + (p_Y i_Y)_* y] \\ &= x \oplus y, \end{aligned}$$

since $p_X i_X = 1_X$, $p_Y i_Y = 1_Y$, and $p_X i_Y$, $p_Y i_X$ are constant maps.

Theorem 7.2.20 *Let X and Y be based spaces. Then, for* $n \geqslant 2$,

$$\pi_n(X \vee Y) \cong \pi_n(X) \oplus \pi_n(Y) \oplus \pi_{n+1}(X \times Y, X \vee Y).$$

Proof. Since $n \geqslant 2$, there is an exact sequence of abelian groups

$$\cdots \longrightarrow \pi_{n+1}(X \times Y) \xrightarrow{j_*} \pi_{n+1}(X \times Y, X \vee Y) \xrightarrow{\partial_*} \pi_n(X \vee Y) \xrightarrow{i_*} \pi_n(X \times Y).$$

On the other hand, a homomorphism $\psi: \pi_n(X \times Y) \to \pi_n(X \vee Y)$ can be defined by $\psi = (i_X p_X)_* + (i_Y p_Y)_*$. Then $\psi i_* = (i i_X p_X)_* + (i i_Y p_Y)_*$, which is the identity isomorphism of $\pi_n(X \times Y)$, by the remark after Theorem 7.2.19. Hence i_* is onto, j_* is the zero map, and the exact sequence splits, so that by Proposition 1.3.26 we have

$$\begin{aligned}\pi_n(X \vee Y) &\cong \pi_n(X \times Y) \oplus \pi_{n+1}(X \times Y, X \vee Y)\\ &\cong \pi_n(X) \oplus \pi_n(Y) \oplus \pi_{n+1}(X \times Y, X \vee Y). \blacksquare\end{aligned}$$

At first sight this theorem is not very helpful, since we are unlikely to know $\pi_{n+1}(X \times Y, X \vee Y)$ if we do not know the homotopy groups of $X \vee Y$. However, in many cases it is possible to prove, by some other method, that $\pi_{n+1}(X \times Y, X \vee Y) = 0$, so that $\pi_n(X \vee Y)$ is just the direct sum of $\pi_n(X)$ and $\pi_n(Y)$; we shall examine this point in detail in Section 7.4. In fact the general problem of calculating homotopy groups is very difficult, but is reasonably manageable provided that we confine our attention to fairly 'well-behaved' spaces such as CW-complexes. The next section contains the definition and elementary properties of CW-complexes, and in Section 7.4 we shall return to the problem of calculating their homotopy groups.

7.3 CW-complexes

As has already been suggested, we wish to generalize and simplify the notion of simplicial complexes, by building up spaces by successively attaching cells to, say, a discrete set of points. This will generalize the idea of a polyhedron, because the cells are attached by arbitrary continuous maps, and at the same time greater generality will be obtained by allowing more than a finite number of cells.

It would be possible to give the definition of a CW-complex directly in terms of attaching cells. However, it is usually more convenient in practice to have a somewhat different definition, which will afterwards be proved to be equivalent to this intuitive idea: see Theorem 7.3.12.

Definition 7.3.1 A *CW-complex* is a Hausdorff space K, together with an indexing set A_n for each integer $n \geqslant 0$, and maps

$$\phi_\alpha: E^n \to K \quad (\text{all } n \geqslant 0, \alpha \in A_n),$$

such that the following properties are satisfied, where

$$e^n = \{x \in R^n \mid d(x, 0) < 1\} \qquad (n \geqslant 1).$$

(a) $K = \bigcup \phi^n_\alpha(e^n)$, for all $n \geqslant 0$ and $\alpha \in A_n$ (we interpret e^0 and E^0 as a single point).

(b) $\phi^n_\alpha(e^n) \cap \phi^m_\beta(e^m)$ is empty unless $n = m$ and $\alpha = \beta$; and $\phi^n_\alpha|e^n$ is (1-1) for all $n \geqslant 0$ and $\alpha \in A_n$.

(c) Let $K^n = \bigcup \phi^m_\alpha(e^m)$, for all $0 \leqslant m \leqslant n$ and all $\alpha \in A_m$. Then $\phi_\alpha(S^{n-1}) \subset K^{n-1}$ for each $n \geqslant 1$ and $\alpha \in A_n$.

(d) A subset X of K is closed if and only if $(\phi^n_\alpha)^{-1}(X)$ is closed in E^n, for each $n \geqslant 0$ and $\alpha \in A_n$.

(e) For each $n \geqslant 0$ and $\alpha \in A_n$, $\phi^n_\alpha(E^n)$ is contained in the union of a finite number of sets of the form $\phi^m_\beta(e^m)$.

The maps ϕ^n_α are called the *characteristic maps* for K, and the subspaces $\phi^n_\alpha(E^n)$ are the *n-cells* of K. K^n is called the *n-skeleton* of K, and if $K^n = K$ for some n, the smallest such n is called the *dimension* of K (if no such n exists, K is said to be *infinite-dimensional*). Notice that, unlike a simplicial complex, which is merely a set of simplexes, a CW-complex is itself a topological space: there is thus no need for the notation $|K|$.

Property (d) is sometimes expressed by saying that K has the *weak topology*, and property (e) by saying that K is *closure-finite*. Hence the initials 'CW', which stand for 'closure-finite with the weak topology'.

As a first example, we show that every polyhedron is a CW-complex.

Proposition 7.3.2 *Let K be a simplicial complex. Then $|K|$ is a CW-complex.*

Proof. Certainly $|K|$ is Hausdorff, since it is a subspace of some Euclidean space. For each n-simplex σ of K, let $\phi^n_\sigma: (E^n, S^{n-1}) \to (\sigma, |\dot\sigma|)$ be a homeomorphism: for example, that given in Example 2.3.13. Then if A_n denotes the set of all n-simplexes of K, the characteristic maps ϕ^n_σ make $|K|$ into a CW-complex, since properties (a)–(e) are satisfied: (a) and (b) follow from Proposition 2.3.6, (d) follows from Proposition 2.3.8, and (c) and (e) are obvious. ∎

Examples 7.3.3 It follows, for example, that S^n, the torus T, and real projective n-space RP^n are all CW-complexes, since obviously any space homeomorphic to a CW-complex is itself a CW-complex (for the proof that RP^n is triangulable, see Chapter 3, Exercise 7). However, one of the advantages of CW-complexes is that, because of their greater generality, it is usually possible to express a given polyhedron as a CW-complex with fewer cells than the original number of simplexes.

(a) Consider the standard map $\theta\colon (E^n, S^{n-1}) \to (S^n, s_0)$, where s_0 is the point $(-1, 0, \ldots, 0)$; $\theta|e^n$ is a homeomorphism onto its image. Since there is also an obvious map $\phi\colon E^0 \to s_0$, it follows that S^n is a CW-complex with one 0-cell and one n-cell, and characteristic maps ϕ, θ.

(b) Consider the torus T, formed from the square $ABCD$ by identifying the edges AB, DC, and AD, BC: see Fig. 7.3.

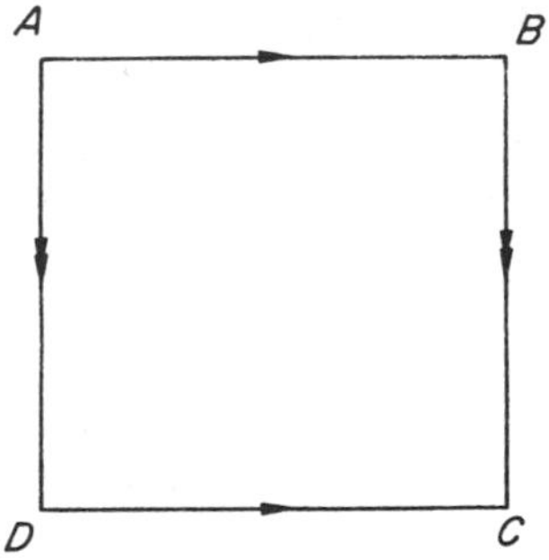

Fig. 7.3

Define maps $\phi^0\colon E^0 \to T$, $\phi_1^1, \phi_2^1\colon E^1 \to T$, and $\phi^2\colon E^2 \to T$ by sending E^0 to the point to which the four vertices A, B, C and D are identified, E^1 to AB, AD, respectively (so that ± 1 go to A, B and A, D respectively), and by mapping E^2 homeomorphically onto the square $ABCD$, and composing this map with the identification map onto T. It is easy to see that these characteristic maps make T into a CW-complex with one 0-cell, two 1-cells, and one 2-cell.

(c) By Proposition 1.4.40, RP^n may be regarded as the space obtained from E^n by identifying antipodal points of S^{n-1}. Since this identification turns S^{n-1} into RP^{n-1}, RP^n is the adjunction space $RP^{n-1} \cup_f E^n$, where $f\colon S^{n-1} \to RP^{n-1}$ is the identification map. In turn, RP^{n-1} is $RP^{n-2} \cup_f E^{n-1}$, and so on; in other words, RP^n is obtained from RP^0 (a single point) by successively attaching one cell of each dimension $1, 2, \ldots, n$. Let $\phi^r\colon E^r \to RP^n$ $(0 \leqslant r \leqslant n)$ be the composite of the identification map onto RP^r and the inclusion of RP^r in RP^n: it is now easy to see that these characteristic maps make RP^n into a CW-complex with one cell of each dimension $0, 1, \ldots, n$ (properties (a)–(c) and (e) are obvious, and (d) is fairly easy; in any case it follows from Theorem 7.3.12 below). ▌

On the other hand, not every space is a CW-complex, since non-Hausdorff spaces exist (see also Example 7.3.10).

We next establish the standard elementary properties of CW-complexes.

Proposition 7.3.4 *Let K be a CW-complex, and let X be any space. A function $f\colon K \to X$ is continuous if and only if each $f\phi^n_\alpha$ is continuous, for each $n \geqslant 0$ and $\alpha \in A_n$.*

Proof. Certainly each $f\phi^n_\alpha$ is continuous if f is. Conversely, let A be a closed subset of X. Then each $(\phi^n_\alpha)^{-1}(f^{-1}A)$ is closed in E^n, so that $f^{-1}A$ is closed in K by property (d). Hence f is continuous. ∎

Definition 7.3.5 Given a CW-complex K, a subspace L is called a *subcomplex* if, for each $n \geqslant 0$, there exists a subset B_n of A_n such that

(a) $L = \bigcup \phi^n_\alpha(e^n)$, for all $n \geqslant 0$ and $\alpha \in B_n$;
(b) $\phi^n_\alpha(E^n) \subset L$ for all $n \geqslant 0$ and $\alpha \in B_n$.

L is called a *finite* subcomplex if it has only a finite number of cells.

Notice that arbitrary unions and intersections of subcomplexes are again subcomplexes.

Proposition 7.3.6 *Let K be a CW-complex. For each $n \geqslant 0$ and $\alpha \in A_n$, $\phi^n_\alpha(E^n)$ is contained in a finite subcomplex of K.*

Proof. By property (e), $\phi^n_\alpha(E^n)$ is contained in the union L of a finite number of sets of the form $\phi^m_\beta(e^m)$. However, L may not be a subcomplex, since it may not satisfy (b) of Definition 7.3.5. But if $\phi^m_\beta(e^m)$ is a set of L such that $\phi^m_\beta(E^m)$ is not contained in L, then by properties (c) and (e) we can always add a finite number of sets $\phi^n_\gamma(e^p)$ (with $p < m$), so as to include $\phi^m_\beta(S^{m-1})$. Thus, by working down in dimensions, we can add a finite number of sets $\phi^p_\gamma(e^p)$ to L until L becomes a (finite) subcomplex. ∎

Proposition 7.3.7 *If L is a subcomplex of a CW-complex K, then L is a CW-complex and is a closed subspace of K.*

Proof. Certainly L is Hausdorff, and satisfies properties (a)–(c) and (e) of Definition 7.3.1, with A_n replaced by B_n. Moreover the maps $\phi^n_\alpha\colon E^n \to L$ $(\alpha \in B_n)$ are continuous, so that certainly $(\phi^n_\alpha)^{-1}X$ is closed in E^n whenever X is closed in L.

We can complete the proof of (d), and at the same time show that L is a closed subspace of K, by showing that, if X is a subspace of L

such that $(\phi^n_\alpha)^{-1}X$ is closed in E^n for all $n \geqslant 0$ and $\alpha \in B_n$, then X is closed in K: for then X is closed in L since $X = X \cap L$, and L is closed in K since we may take $X = L$.

Suppose then that $X \subset L$ and $(\phi^n_\alpha)^{-1}X$ is closed in E^n for all n and $\alpha \in B_n$. Then each $(\phi^n_\alpha)^{-1}X$ is compact, since E^n is, and so each $X \cap \phi^n_\alpha(E^n)$ is compact, since the maps ϕ^n_α are continuous. Since the union of a finite number of compact sets is again compact, this implies that $X \cap M$ is compact for any finite subcomplex M contained in L; and hence $X \cap M$ is compact for any finite subcomplex M whatever (because $M \cap L$ is a subcomplex, and $X \subset L$). Thus $X \cap M$ is closed in M, since K (and hence M) is Hausdorff. It follows from Proposition 7.3.6 that $X \cap \phi^n_\alpha(E^n)$ is closed in $\phi^n_\alpha(E^n)$ for all n and $\alpha \in A_n$, that is, $(\phi^n_\alpha)^{-1}X$ is closed in E^n for all n and $\alpha \in A_n$. Hence, by property (d), X is closed in K. ▮

Proposition 7.3.8 *If K is a CW-complex, the path components of K are subcomplexes. And if K is connected, it is path-connected.*

Proof. Since each $\phi^n_\alpha(e^n)$ and $\phi^n_\alpha(E^n)$ is path-connected, the path-components are certainly subcomplexes, for if X is a path component, $X = \bigcup \phi^n_\alpha(e^n)$, for all n and α such that $X \cap \phi^n_\alpha(e^n) \neq \varnothing$. To prove the second remark, suppose that K is connected but not path-connected. Then the path components form a family of disjoint subcomplexes, whose union is K. By selecting one and taking the union of the others, K can be expressed as the union of two disjoint subcomplexes, each of which is a closed subspace of K by Proposition 7.3.7. Hence K is disconnected, contrary to hypothesis. ▮

Proposition 7.3.9 *If X is a compact subspace of a CW-complex K, it is contained in a finite subcomplex.*

Proof. Choose a point x_α in each non-empty set $X \cap \phi^n_\alpha(e^n)$, and let P be the set of all these points. If Q is any subset of P, each set $Q \cap \phi^n_\alpha(E^n)$ is finite, by property (e), and hence closed, since K is Hausdorff. Hence each $(\phi^n_\alpha)^{-1}Q$ is closed, and so Q is closed in K. Thus P is a discrete subspace of K, and hence of X. Since X is compact, it follows that P must be finite: hence X meets only a finite number of sets of the form $\phi^n_\alpha(e^n)$, and their union is contained in a finite subcomplex as in the proof of Proposition 7.3.6. ▮

Example 7.3.10 Let X be the subspace of R^1 consisting of the points 0 and $1/n$, for all integers $n \geqslant 1$. Now the path components of X are just the single points (since each point $1/n$ is both open and closed); so if X were homotopy-equivalent to a CW-complex K, K would have to have an infinite number of path components. But if $f: X \to K$ were a homotopy equivalence, $f(X)$ would be compact, since X is, and so would have to be contained in a finite subcomplex of K. Thus $f(X)$ would be contained in the union of a finite number of path components, and this contradicts the assumption that f is a homotopy equivalence. Hence X is not homotopy-equivalent to a CW-complex. ∎

In some contexts theorems valid for CW-complexes are also valid for any space having the homotopy type of a CW-complex. The above example shows that not every space is of this type.

We wish now to reconcile the intuitive idea of a space built up by attaching cells with the formal definition of a CW-complex. For this purpose, we must first be quite precise about what is meant by a 'space built up by attaching cells'.

Definition 7.3.11 A *cellular space* is a topological space K, with a sequence of subspaces

$$K^0 \subset K^1 \subset K^2 \subset \cdots \subset K,$$

such that $K = \bigcup_{n=0}^{\infty} K^n$, and the following properties hold.

(a) K^0 is a discrete space.

(b) For each $n > 0$, there exists an indexing set A_n, and continuous maps $\phi^n_\alpha: S^{n-1} \to K^{n-1}$ for each $\alpha \in A_n$. Moreover, K^n is the space obtained from K^{n-1} and (disjoint) copies E^n_α of E^n (one for each $\alpha \in A_n$) by identifying the points x and $\phi^n_\alpha(x)$ for each $x \in S^{n-1}_\alpha$ and each $\alpha \in A_n$.

(c) A subset X of K is closed if and only if $X \cap K^n$ is closed in K^n, for each $n \geqslant 0$.

Note that property (c) is automatically satisfied if K is 'finite-dimensional', that is, all sets A_n are empty for sufficiently large n.

Theorem 7.3.12 *Every CW-complex is a cellular space, and every cellular space is a CW-complex.*

Proof. Suppose first that K is a CW-complex. Then the n-skeletons form a sequence of subspaces $K^0 \subset K^1 \subset K^2 \subset \cdots \subset K$. K^0 is discrete, since it is a CW-complex and each point is a subcomplex: thus each subset is a subcomplex and hence closed in K^0. Moreover the characteristic maps $\phi^\alpha_n: E^n \to K$ ($n \geqslant 0$, $\alpha \in A_n$) restrict to maps of S^{n-1} to K^{n-1} for each $n > 0$. Now a subset X of K^n is closed in K^n if and only if $(\phi^m_\alpha)^{-1}X$ is closed in E^m, for each $m \leqslant n$ and $\alpha \in A_m$; that is, if and only if $X \cap K^{n-1}$ is closed in K^{n-1} and each $(\phi^n_\alpha)^{-1}X$ is closed in E^n. In other words, the topology of K^n is exactly the identification topology produced from the disjoint union of K^{n-1} and copies E^n_α of E^n, by identifying x with $\phi^n_\alpha(x)$ for each $x \in S^{n-1}_\alpha$ and $\alpha \in A_n$. Hence property (b) of Definition 7.3.11 is satisfied. Finally, $X \subset K$ is closed if and only if $(\phi^n_\alpha)^{-1}X$ is closed for all $n \geqslant 0$ and $\alpha \in A_n$; that is, if and only if $X \cap K^n$ is closed in K^n for each $n \geqslant 0$. So K is a cellular space.

Conversely, suppose given a cellular space K, as in Definition 7.3.11. In order to show that K is a CW-complex, it is first necessary to show that K is Hausdorff, and in fact this is the most difficult part of the proof.

Suppose then that we have two distinct points x and y in K. Choose the smallest n such that x and y are both in K^n, and suppose that x, say, is in e^n_α for some $\alpha \in A_n$ (points of K^{n-1} or E^n_α are identified with their images in K^n). Now, even if y is also in e^n_α, there exists a number $\epsilon > 0$ such that $\|x\| < 1 - 2\epsilon$ and $y \notin \{z \in e^n \mid \|z - x\| < 2\epsilon\}$. Let

$$U_n = \{z \in e^n_\alpha \mid \|z - x\| < \epsilon\},$$

$$V_n = K^n - \{z \in e^n_\alpha \mid \|z - x\| \leqslant \epsilon\};$$

then U_n and V_n are open sets in K^n, containing x and y respectively, such that $U_n \cap V_n = \varnothing$.

What we now have to do is to 'thicken' U_n and V_n to disjoint open sets in K. So suppose, as an inductive hypothesis, that U_m, V_m are disjoint open sets in K^m $(m \geqslant n)$, such that $U_m \cap K^n = U_n$ and $V_m \cap K^n = V_n$. The sets $X_\alpha = (\phi^{m+1}_\alpha)^{-1}U_m$ and $Y_\alpha = (\phi^{m+1}_\alpha)^{-1}V_m$ are then disjoint open sets in S^m, for each $\alpha \in A_{m+1}$. Define

$$X'_\alpha = \{z \in e^{m+1}_\alpha \mid \|z\| > \tfrac{1}{2} \text{ and } z/\|z\| \in X_\alpha\}$$

$$Y'_\alpha = \{z \in e^{m+1}_\alpha \mid \|z\| > \tfrac{1}{2} \text{ and } z/\|z\| \in Y_\alpha\};$$

see Fig. 7.4 overleaf.

Now let $U_{m+1} = U_m \cup (\bigcup_\alpha X'_\alpha)$ and $V_{m+1} = V_m \cup (\bigcup_\alpha Y'_\alpha)$; then U_{m+1} and V_{m+1} are disjoint open sets in K^{m+1}, and $U_{m+1} \cap K^m = U_m$, $V_{m+1} \cap K^m = V_m$, so that the inductive step is complete. Finally, let $U = \bigcup_{m \geqslant n} U_m$ and $V = \bigcup_{m \geqslant n} V_m$: then U and V are disjoint open sets in K (by property (c) of Definition 7.3.11), $x \in U$ and $y \in V$. Hence K is Hausdorff.

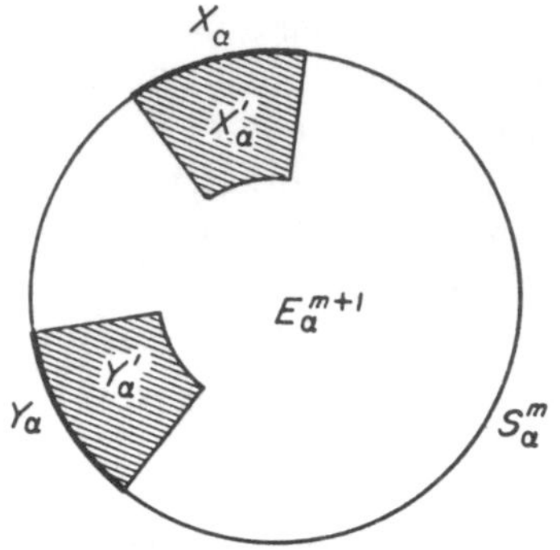

Fig. 7.4

To complete the proof, extend the maps $\phi_\alpha^n\colon S_\alpha^{n-1} \to K^{n-1}$ to maps $\phi_\alpha^n\colon E_\alpha^n \to K^n \subset K$ ($n \geqslant 1$) by using the inclusion maps of each e_α^n, and suppose as an inductive hypothesis that with these characteristic maps K^n is a CW-complex (certainly K^0 is a CW-complex). Then K^{n+1} automatically satisfies properties (a)–(c) of Definition 7.3.1; also (d) is true, since

$$\begin{aligned} X \subset K^{n+1} \text{ closed} &\Leftrightarrow X \cap K^n \text{ closed in } K^n \text{ and } (\phi_\alpha^{n+1})^{-1}X \text{ closed} \\ &\quad\ \text{in } E_\alpha^{n+1} \text{ for all } \alpha \in A_{n+1} \\ &\Leftrightarrow (\phi_\alpha^m)^{-1}X \text{ closed for all } 0 \leqslant m \leqslant n+1 \text{ and} \\ &\quad\ \alpha \in A_m, \end{aligned}$$

by the inductive hypothesis. And (e) is satisfied since, by Proposition 7.3.9, $\phi_\alpha^{n+1}(S_\alpha^n)$ is contained in a finite subcomplex of K^n; hence $\phi_\alpha^{n+1}(E_\alpha^{n+1})$ is contained in the union of this subcomplex and $\phi_\alpha^{n+1}(e_\alpha^{n+1})$.

Hence each K^n is a CW-complex. It follows that K is also a CW-complex, for the only non-trivial thing to check is property (d), and this follows from property (c) of Definition 7.3.11. ∎

We next investigate to what extent the constructions of Section 6.2 can be applied to CW-complexes. In order to deal with the one-point union and reduced product, we must first consider quotient spaces.

Theorem 7.3.13 *Let (K, L) be a CW-pair; that is, K is a CW-complex and L is a subcomplex. Then K/L is a CW-complex.*

Proof. It is first necessary to show that K/L is Hausdorff. This will follow from the fact that K is Hausdorff, provided that, given a point $x \in K - L$, there exist disjoint open sets U, V in K, with $x \in U$ and $L \subset V$. But this can obviously be established by the argument in the proof of Theorem 7.3.12: if $x \in \phi_\alpha^n(e^n)$, enclose x and L^n in disjoint open sets in K^n ($\phi_\alpha^n(e^n) \cap L^n$ is empty), and then 'thicken' these open sets to make open sets in K.

If now A_n and B_n are the indexing sets for the cells of K, L respectively, let $C_n = A_n - B_n$ if $n > 0$, and let $C_0 = (A_0 - B_0) \cup \{\alpha\}$, where $\alpha \in B_0$ indexes one particular 0-cell of L. Let $p: K \to K/L$ be the identification map; we shall show that $p\phi_\alpha^n$ ($\alpha \in C_n$, $n \geqslant 0$) are characteristic maps for K/L. To do so, we have to check properties (a)–(e) of Definition 7.3.1.

(a) Each $\phi_\alpha^n(e^n)$ is in L or $K - L$. Since the points of K/L are those of $K - L$, together with one extra, representing L, (a) follows.

(b) This is true for the same reason.

(c) For each $\alpha \in C_n$, $p\phi_\alpha^n(S^{n-1}) \subset p(K^{n-1}) = (K/L)^{n-1}$.

(d) $X \subset K/L$ is closed if and only if $p^{-1}X$ is closed, that is, $(p\phi_\alpha^n)^{-1}X$ is closed in E^n for each $\alpha \in A_n$. But if $\alpha \in B_n$, $(p\phi_\alpha^n)^{-1}X$ is either E^n or empty (according as X meets L or not), and so is closed in any case.

(e) For each $\alpha \in C_n$, $\phi_\alpha^n(E^n)$ is contained in a finite union of sets of the form $\phi_\beta^m(e^m)$ ($\beta \in A_m$). Hence $p\phi_\alpha^n(E^n)$ is contained in the union of the corresponding sets $p\phi_\beta^m(e^m)$; and $p\phi_\beta^m(e^m)$ is the point representing L if $\beta \notin C_m$. ∎

Since the disjoint union of a collection of CW-complexes is obviously another CW-complex, we also have

Corollary 7.3.14 *If K_a ($a \in A$) are a collection of CW-complexes, then $\bigvee_A K_a$ is a CW-complex (assuming that the base points are all 0-cells).*

Proof. Clearly K_0, the union of the base points, is a subcomplex of the disjoint union K of the complexes K_a. Hence $\bigvee K_a = K/K_0$ is a CW-complex. ∎

Example 7.3.15 Let K be a CW-complex. By Theorem 7.3.13, K^n/K^{n-1} is a CW-complex, for all $n \geqslant 0$ (if $n = 0$, K^{n-1} is empty, and we interpret K^n/K^{n-1} as $(K^n)^+$, the disjoint union of K^n with an extra point). Moreover, even if $n = 0$, the characteristic maps

$$\phi_\alpha^n: (E_\alpha^n, S_\alpha^{n-1}) \to (K^n, K^{n-1}) \qquad (\alpha \in A_n)$$

induce a map

$$\phi: \bigvee_{A_n} (E^n/S^{n-1})_\alpha \to K^n/K^{n-1}.$$

Now ϕ is continuous, (1-1) and onto, and, by Proposition 7.3.4, ϕ^{-1} is continuous. Hence ϕ is a homeomorphism; and since E^n/S^{n-1} is homeomorphic to S^n we have thus proved that K^n/K^{n-1} is homeomorphic to a one-point union of a collection of S^n's, one for each n-cell of K. ▌

The situation with regard to products is rather more complicated, because in general the product of two CW-complexes is not a CW-complex: the product topology may fail to be that defined by Definition 7.3.1(d) (see the notes at the end of the chapter). However, in two important special cases this difficulty does not arise.

Theorem 7.3.16 *If K and L are CW-complexes, so is $K \times L$, provided that*

(a) *one of K, L is locally compact; or*
(b) *both K and L have a countable number of cells.*

Proof. Certainly $K \times L$ is Hausdorff. If K has indexing sets A_n and characteristic maps ϕ^n_α, and L has indexing sets B_n and characteristic maps ψ^n_β, we wish to show that $K \times L$ is a CW-complex with characteristic maps $\phi^n_\alpha \times \psi^m_\beta$, for all $\alpha \in A_n$, $\beta \in B_m$ (E^{n+m} is identified with $E^n \times E^m$ by the standard homeomorphism $h_{n,m}$). It is easy to see that properties (a)–(c) and (e) of Definition 7.3.1 are satisfied; but as we have already said, there is no guarantee that (d) will be true in general.

Let us write $K \bar{\times} L$ for the space $K \times L$, retopologized so as to be a CW-complex; that is, retopologized so that $X \subset K \bar{\times} L$ is closed if and only if $(\phi^n_\alpha \times \psi^m_\beta)^{-1}X$ is closed in $E^n \times E^m$, for all n, m, α, β. Now the (pointwise) identity function $i: K \bar{\times} L \to K \times L$ is continuous, (1-1) and onto. Thus, in order to complete the proof of the theorem, it is sufficient to show that the identity function $j: K \times L \to K \bar{\times} L$ is also continuous; for then $K \times L$ and $K \bar{\times} L$ will be homeomorphic and $K \times L$ will have the correct topology as a CW-complex. The proof that j is continuous differs in the two cases.

(a) Suppose that K is locally compact. Now for each n, m, α, β, the map $j(\phi^n_\alpha \times \psi^m_\beta): E^n \times E^m \to K \bar{\times} L$ is continuous. As usual, let X^+ denote the disjoint union of X with an extra point, which is taken to be the base point of X^+; thus $X^+ \wedge Y^+ = (X \times Y)^+$. Moreover $j(\phi^n_\alpha \times \psi^m_\beta)$ may be regarded as a *based* map $(E^n)^+ \wedge (E^m)^+ \to K \bar{\times} L$, and so we may apply the association map to obtain a map $(E^n)^+ \to (K \bar{\times} L)^{(E^m)^+}$, one for each α and β. Since this is continuous for each α, and $(E^m)^+$ is locally compact and Hausdorff, Proposition

7.3.4 and Theorem 6.2.38 show that the map $K^+ \wedge (E^m)^+ \to K \bar{\times} L$ that corresponds to $j(1 \times \psi_\beta^m)\colon K \times E^m \to K \bar{\times} L$ is also continuous. And then a similar argument with K^+ (which is also locally compact and Hausdorff) shows that $j\colon K \times L \to K \bar{\times} L$ is continuous.

(b) Let $X \subset K \bar{\times} L$ be an open set, and let $Y = j^{-1}X \subset K \times L$. Let (k, l) be a point of Y, and enumerate the cells of K and L so that k, l are in the first cells of K, L respectively. Let K_r, L_r denote the unions of the first r cells of K, L respectively. Now, by definition of the topology of $K \bar{\times} L$, if CX denotes the complement of X, then each $(\phi_\alpha^n \times \psi_\beta^m)^{-1}j^{-1}CX = (\phi_\alpha^n \times \psi_\beta^m)^{-1}CY$ is closed in $E^n \times E^m$, and so compact; hence $CY \cap [\phi_\alpha^n(E^n) \times \psi_\beta^m(E^m)]$ is compact and so closed. Since K_r and L_r are finite unions of cells, it follows that $CY \cap (K_r \times L_r)$ is closed, and so $Y \cap (K_r \times L_r)$ is open in $K_r \times L_r$, for each r.

Now suppose, as an inductive hypothesis, that we have sets U_r, V_r, open in K_r, L_r respectively, such that $k \in U_1 \subset \cdots \subset U_r$ and $l \in V_1 \subset \cdots \subset V_r$; suppose also that $\bar{U}_r \times \bar{V}_r \subset Y \cap (K_r \times L_r)$. This is certainly true if $r = 1$, since K_1 and L_1 are compact Hausdorff. Since K_{r+1} and L_{r+1} are also compact Hausdorff, there exist sets U_{r+1}, V_{r+1}, open in K_{r+1}, L_{r+1} respectively, such that $\bar{U}_r \times \bar{V}_r \subset U_{r+1} \times V_{r+1} \subset \bar{U}_{r+1} \times \bar{V}_{r+1} \subset Y \cap (K_{r+1} \times L_{r+1})$, and this is sufficient to complete the inductive step.

Finally, let $U = \bigcup_{r=1}^{\infty} U_r$ and $V = \bigcup_{r=1}^{\infty} V_r$; then $(k, l) \in U \times V \subset Y$. Moreover each U_r is open in K_r, and hence $U_r \cap K_s$ is open in K_s for all $s \leqslant r$; hence $U \cap K_s = \bigcup_{r \geqslant s} (U_r \cap K_s)$ is open in K_s, for each s. It follows that each $U \cap \phi_\alpha^n(E^n)$ is open in $\phi_\alpha^n(E^n)$, so that $(\phi_\alpha^n)^{-1}U$ is open in E^n and hence U is open in K. Similarly V is open in L, so that $U \times V$, and hence Y, is open in $K \times L$. Thus j is continuous. ▌

Corollary 7.3.17 *If K and L are CW-complexes, so is $K \wedge L$, if either*

(a) *one of K, L is locally compact; or*
(b) *both K and L have a countable number of cells.* ▌

Thus, in particular, cK and sK are CW-complexes if K is (so also is SK, the 'unreduced' suspension of K).

Theorem 7.3.16 is often used in constructing homotopies of CW-complexes. For I is a CW-complex in an obvious way (it has one 0-cell at each end and a single 1-cell); also I is locally compact, so that $K \times I$ is a CW-complex whenever K is.

Example 7.3.18 Let K be a CW-complex of dimension n, and for each $\alpha \in A_n$ let V_α be the subspace $\phi^n_\alpha\{x \in E^n \mid \|x\| \leqslant \frac{1}{2}\}$; let $V = \bigcup_\alpha V_\alpha$. Then K^{n-1} is a strong deformation retract of $K - V$.

To prove this, it is sufficient to construct a homotopy $F: K \times I \to K$, starting with the identity map, such that $F((K - V) \times I) \subset K - V$, $F((K - V) \times 1) = K^{n-1}$ and F is constant on K^{n-1}. This can be done by taking F to be the identity homotopy on K^{n-1}, and to be radial projection from the origins in the n-cells; more precisely,

$$F(x, t) = x \quad \text{if} \quad x \in K^{n-1},$$

$$F(\phi^n_\alpha(y), t) = \begin{cases} \phi^n_\alpha((1 + t)y), y \in E^n, \|y\| \leqslant 1/(1 + t) \\ \phi^n_\alpha(y/\|y\|), y \in E^n, \|y\| \geqslant 1/(1 + t), \alpha \in A_n. \end{cases}$$

Now this is certainly continuous on each cell of $K \times I$, and the definitions coincide for points of $\phi^n_\alpha(S^{n-1}) \times I$. Hence F is continuous, by Proposition 7.3.4. And clearly F has the required properties. ∎

The same idea is used to prove what is perhaps the most important result about CW-complexes, namely that a CW-pair always has the absolute homotopy extension property.

Theorem 7.3.19 *Let (K, L) be a CW-pair. Then (K, L) has the absolute homotopy extension property.*

Proof. What we must show is that, given a map $f: (K \times 0) \cup (L \times I) \to Y$, f can be extended to a map $K \times I \to Y$. This is done by extending f inductively to $M^n \times I$, where $M^n = K^n \cup L$.

First, then, extend f to $M^0 \times I$ by defining $f(x, t) = f(x)$, for any 0-cell x of $K - L$. This is continuous, since it is continuous on each cell of the CW-complex $(K \times 0) \cup (M^0 \times I)$.

Next suppose that f has been extended to a map $f: (K \times 0) \cup (M^{n-1} \times I) \to Y$. For each n-cell $\phi^n_\alpha(E^n)$ of $K - L$. consider the composite map

$$(E^n \times 0) \cup (S^{n-1} \times I) \xrightarrow{\phi^n_\alpha \times 1} (K \times 0) \cup (M^{n-1} \times I) \xrightarrow{f} Y.$$

Now a retraction $\rho: E^n \times I \to (E^n \times 0) \cup (S^{n-1} \times I)$ (regarded as subspaces of $R^n \times R^1$) can be defined by radial projection from the point $(0, \ldots, 0, 2)$: see Fig. 7.5.
This combines with the above composite map to extend it to a map of $E^n \times I$ to Y; and since each ϕ^n_α is (1-1) on e^n, these maps combine to give an extension of f to a function $f: (K \times 0) \cup (M^n \times I) \to Y$. Moreover this extension is continuous: for $(K \times 0) \cup (M^n \times I)$ is a

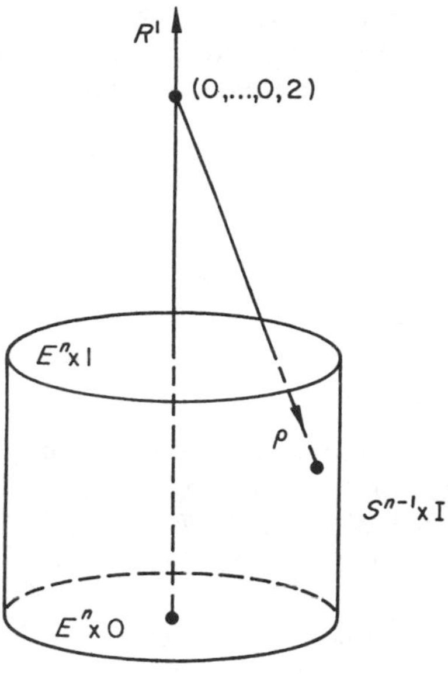

Fig. 7.5

CW-complex, and the composite of each of its characteristic maps with f is continuous; hence f is continuous by Proposition 7.3.4.

The inductive step is now complete, so that f can be extended to each $(K \times 0) \cup (M^n \times I)$. Hence f can be extended to a function $f: K \times I \to Y$, which once again is continuous by Proposition 7.3.4, since $K \times I$ is a CW-complex. ∎

Corollary 7.3.20 *If (K, L) is a CW-pair, the inclusion map $i: L \to K$ is a cofibre map.* ∎

In particular, the sequence of Corollary 6.5.3 is exact for any CW-complex and subcomplex.

We end this section with a further consideration of the situation revealed by Theorem 7.3.16. The method of proof was to show that $K \mathbin{\bar{\times}} L$ was always a CW-complex, and then to show that $K \mathbin{\bar{\times}} L$ coincided with $K \times L$ in certain circumstances. It follows, then, that when dealing with CW-complexes it is usually more convenient to topologize the product of K and L as $K \mathbin{\bar{\times}} L$, rather than to use the standard product topology. There is also a corresponding version of the reduced product, defined by $K \mathbin{\bar{\wedge}} L = (K \mathbin{\bar{\times}} L)/(K \vee L)$, where $K \vee L$ is regarded as the subspace $K \mathbin{\bar{\times}} l_0 \cup k_0 \mathbin{\bar{\times}} L$ of $K \mathbin{\bar{\times}} L$: note that $K \mathbin{\bar{\times}} l_0$, for example, is homeomorphic to $K \times l_0$, and hence to

K, since l_0 is certainly locally compact. Again, $K \overline{\wedge} L$ is always a CW-complex, and coincides with $K \wedge L$ if either K or L is locally compact, or if K and L each have a countable number of cells.

A further advantage of using the products $\overline{\times}$ and $\overline{\wedge}$ is that they are both strictly associative, and behave well with respect to identification maps.

Proposition 7.3.21 *The products $\overline{\times}$ and $\overline{\wedge}$ are associative, for CW-complexes. Moreover, if (K, L) and (M, N) are CW-pairs, $(K/L) \overline{\wedge} (M/N)$ is homeomorphic to $(K \overline{\times} M)/(L \overline{\times} M \cup K \overline{\times} N)$.*

Proof. Clearly $K \overline{\times} (L \overline{\times} M)$ and $(K \overline{\times} L) \overline{\times} M$ have exactly the same cells and characteristic maps, so that a homeomorphism between them can easily be constructed by using Proposition 7.3.4. A similar argument works for $K \overline{\wedge} (L \overline{\wedge} M)$ and $(K \overline{\wedge} L) \overline{\wedge} M$, and also shows that $(K/L) \overline{\wedge} (M/N)$ and $(K \overline{\times} L)/(L \overline{\times} M \cup K \overline{\times} N)$ are homeomorphic. ∎

Because of Proposition 7.3.21 brackets can be omitted from such expressions as $K \overline{\wedge} L \overline{\wedge} M$, without causing ambiguity.

We next show that the topology of $K \overline{\times} L$ does not in fact depend on the structure of K and L as CW-complexes; indeed, $\overline{\times}$ can be defined for arbitrary topological spaces.

Proposition 7.3.22 *Given a space X, let $k(X)$ be X, retopologized so that a subset A of X is closed in $k(X)$ if and only if $A \cap C$ is closed in X, for all closed compact subsets C of X. Then if K and L are CW-complexes, we have*

(a) $k(K) = K$;
(b) $k(K \times L) = K \overline{\times} L$.

Proof. First note that the above description of $k(X)$ does define a topology, since $(A \cup B) \cap C = (A \cap C) \cup (B \cap C)$, and if $\{A_\alpha\}$ is any collection of subsets, then $(\bigcap A_\alpha) \cap C = \bigcap (A_\alpha \cap C)$.

To prove (a), notice that if A is a subset of K such that $A \subset C$ is closed for all closed compact sets C, then in particular $A \cap \phi_\alpha^n(E^n)$ is closed for all n and α. Hence each $(\phi_\alpha^n)^{-1}A$ is closed, and so A is itself closed. Conversely, it is obvious that each $A \subset C$ is closed if A is closed.

The proof of (b) is similar: certainly every set that is closed in $k(K \times L)$ is closed in $K \overline{\times} L$. Conversely, if A is closed in $K \overline{\times} L$, then each $(\phi_\alpha^n \times \psi_\beta^m)^{-1}A$ is closed in $E^n \times E^m$, and hence is compact. Thus each $A \cap (\phi_\alpha^n \times \psi_\beta^m)(E^n \times E^m)$ is compact, and hence closed in

$K \times L$. It follows that the intersection of A with each compact subset of $K \times L$ is closed, since any such compact subset is contained in the product of its projections onto K and L, and hence is contained in a finite union of products of cells. Hence A is closed in $k(K \times L)$. ∎

Moreover, continuous maps of CW-complexes induce continuous maps of their $\overline{\times}$ product.

Proposition 7.3.23 *If $f: X \to Y$ is a continuous map of spaces, and Y is Hausdorff, then the corresponding map $f: k(X) \to k(Y)$ is also continuous.*

Proof. Suppose that A is a closed subset of $k(Y)$. For any closed compact set C in X, we have

$$f^{-1}(A) \cap C = f^{-1}(A \cap f(C)) \cap C;$$

since $f(C)$ is compact and hence closed, $A \cap f(C)$ is closed in Y, and so $f^{-1}(A \cap f(C)) \cap C$ is closed in X. Hence $f^{-1}(A)$ is closed in $k(X)$, and so $f: k(X) \to k(Y)$ is continuous. ∎

Corollary 7.3.24 *Let K, L, M and N be CW-complexes. Continuous maps $f: K \to M$, $g: L \to N$ induce continuous maps $f \mathbin{\overline{\times}} g: K \mathbin{\overline{\times}} L \to M \mathbin{\overline{\times}} N$ and $f \mathbin{\overline{\wedge}} g: K \mathbin{\overline{\wedge}} L \to M \mathbin{\overline{\wedge}} N$, with properties similar to those of $f \times g$ and $f \wedge g$. Moreover the diagonal maps $\Delta: K \to K \times K$, $\Delta: K \to K \mathbin{\overline{\wedge}} K$, defined by $\Delta(x) = (x, x)$, are continuous.* ∎

7.4 Homotopy groups of CW-complexes

Section 7.2 was concerned with general results on homotopy groups, including the exact sequences of pairs and triples, and theorems on the homotopy groups of products and one-point unions. In this section we shall pursue these ideas further, so as to obtain more precise results when the spaces involved are CW-complexes.

It is not possible to get very far without knowing the groups $\pi_r(S^n)$, at least for $r \leqslant n$. We have already seen in Example 6.3.16 that $\pi_r(S^n) = 0$ for $r < n$, so that our first task is to calculate $\pi_n(S^n)$. Now we already know that $\pi_1(S^1) \cong Z$; what we shall do is to prove inductively that $\pi_n(S^n) \cong Z$ for all $n \geqslant 1$.

The method of proof is to construct a homomorphism $d: \pi_n(S^n) \to Z$, and to show by induction that d is onto and (1-1). The homomorphism is defined by attaching to each map $f: S^n \to S^n$ an integer, called its *degree*.

Definition 7.4.1 For each $n \geqslant 0$, let σ_n be a generator of $\tilde{H}_n(S^n) \cong Z$. Given a map $f: S^n \to S^n$, the *degree* of f, $d(f)$, is defined by $f_*(\sigma_n) = d(f)\sigma_n$.

Clearly $d(f)$ does not depend on the choice of the generator σ_n; and given two maps $f, g: S^n \to S^n$, we have $d(fg) = d(f)d(g)$. Moreover, homotopic maps have the same degree, so that d may be regarded as a function from $\pi_n(S^n)$ to Z.

Proposition 7.4.2 *For $n \geqslant 1$, $d: \pi_n(S^n) \to Z$ is a homomorphism.*

Proof. Let $f, g: S^n \to S^n$ be two based maps, and consider $f.g = \nabla(f \vee g)\mu$, where $\mu: S^n \to S^n \vee S^n$ is the 'H'-space map'. Now it is clear from a triangulation that $\tilde{H}_n(S^n \vee S^n) \cong \tilde{H}_n(S^n) \oplus \tilde{H}_n(S^n)$, and that $(f \vee g)_*(x \oplus y) = f_*(x) \oplus g_*(y)$, $\nabla_*(x \oplus y) = x + y$. Also $\mu_*(x) = x \oplus x$, since $p_1\mu \simeq p_2\mu \simeq 1$, where $p_1, p_2: S^n \vee S^n \to S^n$ are the projection maps.

Hence

$$\begin{aligned}(f.g)_*(\sigma_n) &= \nabla_*(f \vee g)_*\mu_*(\sigma_n)\\ &= \nabla_*(f \vee g)_*(\sigma_n \oplus \sigma_n)\\ &= \nabla_*(f_*\sigma_n \oplus g_*\sigma_n)\\ &= f_*(\sigma_n) + g_*(\sigma_n),\end{aligned}$$

so that $d([f][g]) = d[f] + d[g]$. ∎

Corollary 7.4.3 *For $n \geqslant 1$, d is onto.*

Proof. The identity map of S^n has degree 1. ∎

It remains to prove that d is also (1-1), and for this two lemmas are necessary.

Lemma 7.4.4 *Let $f, g: X \to S^n$ be two maps, and suppose that there exists a non-empty open set $U \subset S^n$, such that the sets $f^{-1}(s)$ and $g^{-1}(s)$ coincide for all $s \in U$. Then $f \simeq g$.*

Proof. Let V be a non-empty open set such that $\bar{V} \subset U$, and let $W = f^{-1}(V) = g^{-1}(V)$. Since $S^n - V \subset S^n -$ point, which is homeomorphic to e^n, it follows that $f|(X - W) \simeq g|(X - W)$, by a homotopy that corresponds under the homeomorphism to a linear homotopy. In particular the homotopy is constant on $(X - W) \cap f^{-1}(U)$, and so can be fitted together with the constant homotopy on $\bar{W} \subset f^{-1}(U)$ to yield the required homotopy. ∎

For the second lemma, let x and y be two points in R^n, and let L be the straight-line segment joining them. Choose $\epsilon > 0$, and write $M = \{z \in R^n \mid d(z, L) \leqslant \epsilon\}$, $N = \{z \in R^n \mid d(z, L) = \epsilon\}$: see Fig. 7.6.

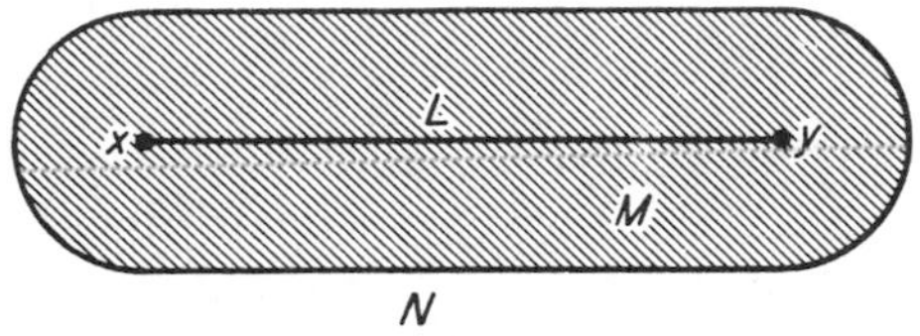

Fig. 7.6

Lemma 7.4.5 *There exists a homeomorphism $h: M \to M$ such that $h(x) = y$ and $h(z) = z$ for all $z \in N$.*

Proof. Consider a line segment l starting at x. It is clear that l meets N at a unique point, z say, and that all points of l between x and z lie in M. Thus points of M can be expressed uniquely in the form $\lambda x + (1 - \lambda)z$, where $0 \leqslant \lambda \leqslant 1$ and $z \in N$. Similarly, points can be uniquely expressed in the form $\lambda y + (1 - \lambda)z$; define h by

$$h(\lambda x + (1 - \lambda)z) = \lambda y + (1 - \lambda)z.$$

Then h is (1-1) and onto, and maps x to y, leaving fixed points of N; the proof that h and its inverse are continuous is left as an exercise to the reader. ∎

Theorem 7.4.6 *For $n \geqslant 1$, d is* (1-1).

Proof. This is proved by induction on n. The induction starts, since $d: \pi_1(S^1) \to Z$ is an isomorphism: this is because $\pi_1(S^1)$ is known to be isomorphic to Z, and d is onto. Suppose, then, that the theorem is true for $n - 1$, and consider a based map $f: S^n \to S^n$ of degree 0. Since $n \geqslant 2$, S^n is n-simple, so that it is sufficient, by Proposition 7.2.7, to show that f is homotopic to the constant map as an *unbased* map. This is done by constructing a homotopy between f and a map of the form Sg, where $g: S^{n-1} \to S^{n-1}$ is another map of degree 0 (we identify $S(S^{n-1})$ with S^n as in Example 4.4.9).

Let $N = (0, \ldots, 0, 1)$ and $S = (0, \ldots, 0, -1)$ be the 'north and south poles' of S^n, respectively, and let S^n_+ and S^n_- be the (open) 'north and south hemispheres', defined by $x_{n+1} > 0$ and $x_{n+1} < 0$ respectively. Triangulate S^n so that N and S are in the interiors of n-simplexes. By the Simplicial Approximation Theorem, we may assume that f is a simplicial map from some subdivision to this triangulation, in which case $f^{-1}(N)$ and $f^{-1}(S)$ are just finite sets of points, say

$$f^{-1}(N) = p_1, \ldots, p_j,$$

$$f^{-1}(S) = q_1, \ldots, q_k.$$

The first step is to alter f by a homotopy so that all the p's are in S^n_+ and all the q's are in S^n_-. Now the standard map $\theta: (E^n, S^{n-1}) \to (S^n, (-1, 0, \ldots, 0))$ can be modified in an obvious way to give a homeomorphism $\phi: S^n - S \to e^n$. In e^n, each point $\phi(p_i)$ not in $\phi(S^n_+)$ (and with $p_i \neq S$) can be joined by a straight-line segment to a point r_i in $\phi(S^n_+)$, and similarly the points $\phi(q_i)$ not in $\phi(S^n_-)$ may be joined to points s_i in $\phi(S^n_-)$. Moreover we may choose the points r_i, s_i so that the line segments are all disjoint: for since only a finite number of points is involved, there is a point x in $\phi(S^n_+)$ such that each straight line through x meets at most one of the points $\phi(p_i)$, $\phi(q_i)$; and the line segments may then be chosen to be segments of lines through x: see Fig. 7.7.

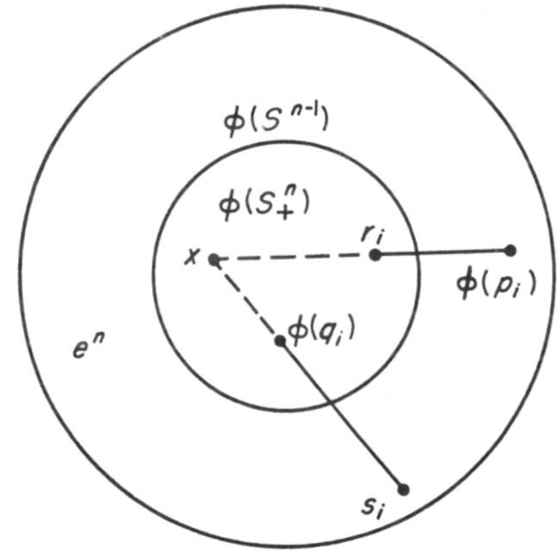

Fig. 7.7

Since the line segments are compact, there exists $\epsilon > 0$ such that the 'closed ϵ-neighbourhoods' of the line segments (in the sense of Lemma 7.4.5) are also disjoint, and are still in e^n. Hence, by Lemma 7.4.5, there exists a homeomorphism of e^n, fixed outside these ϵ-neighbourhoods, sending each $\phi(p_i)$ to r_i and each $\phi(q_i)$ to s_i. This homeomorphism may be transferred back to S^n to give a homeomorphism $h: S^n \to S^n$ that pushes each p_i into S^n_+ and each q_i into S^n_- (at least if no p_i is S: but otherwise the same technique can be used first to push p_i away from S). Moreover, Lemma 7.4.4 shows that $h \simeq 1$, so that by replacing f by fh^{-1} if necessary, we may assume that each p_i is in S^n_+ and each q_i in S^n_-.

The next step is to 'straighten out' f on S^n_+. More precisely, define $f_1: S^n \to S^n$ by

$$f_1 \mid (S^n_- \cup S^{n-1}) = f \mid (S^n_- \cup S^{n-1}),$$

$$f_1(\lambda N + (1 - \lambda)x) = \lambda N + (1 - \lambda)f(x) \qquad (x \in S^{n-1}, 0 \leqslant \lambda \leqslant 1).$$

(For the second line of the definition, we identify points of $S^n - S$ with their images under ϕ in e^n; certainly $f(S^{n-1}) \subset S^n - S$.) Since $S^n_+ \cup S^{n-1}$ is compact, its image under f is a closed set that does not contain S; hence there is an open set U containing S with $f^{-1}(U) \subset S^n_-$. It follows that $f^{-1}(x) = f_1^{-1}(x)$ for all $x \in U$, so that by Lemma 7.4.4 we have $f \simeq f_1$.

Next we straighten out f on S^n_- as well, by defining $f_2: S^n \to S^n$ by

$$f_2 \mid (S^n_+ \cup S^{n-1}) = f_1 \mid (S^n_+ \cup S^{n-1}),$$

$$f_2(\lambda S + (1 - \lambda)x) = \lambda S + (1 - \lambda)f_1(x) \qquad (x \in S^{n-1}, 0 \leqslant \lambda \leqslant 1).$$

As before, $f_2 \simeq f_1 \simeq f$, and $f_2(N) = N$, $f_2(S) = S$, $f_2(S^{n-1}) \subset S^n - (N \cup S)$. f_2 is not quite a suspended map, but we can make it so by moving the image of $f_2(S^{n-1})$ up or down meridians of S^n until it lies in S^{n-1}. This gives a new map $f_3: S^n \to S^n$, which is homotopic to f_2 by Corollary 2.2.4, and which is a suspension of a map $g: S^{n-1} \to S^{n-1}$.

Since $d(f) = 0$, it follows from Theorem 4.4.10 that $d(g) = 0$ as well. By the inductive hypothesis, this means that g is homotopic to the constant map, and hence, by Corollary 6.2.19, f is homotopic to the suspension of the constant map, which in turn is homotopic to the constant map since it is not onto. ∎

To sum up, $d: \pi_n(S^n) \to Z$ is an isomorphism for all $n \geqslant 1$. This result, apart from being important for calculation purposes, has many useful applications to the homotopy theory of spheres. Most of these depend on the following result, which gives the degrees of some standard homeomorphisms.

Proposition 7.4.7 *Given a permutation ρ of $1, 2, \ldots, n$, let $f: S^{n-1} \to S^{n-1}$ be the homeomorphism defined by $f(x_1, \ldots, x_n) = (x_{\rho(1)}, \ldots, x_{\rho(n)})$. Then $d(f)$ is $+1$ or -1 according as ρ is even or odd. Similarly, if $g: S^{n-1} \to S^{n-1}$ is defined by*

$$g(x_1, \ldots, x_n) = (x_1, \ldots, -x_r, x_{r+1}, \ldots, x_n),$$

then $d(g) = -1$.

Proof. It is clearly sufficient to consider $\bar{S}^{n-1}$ instead of S^{n-1} when the result is immediate from Example 4.4.11. ▮

Corollary 7.4.8 *Every homotopy equivalence of S^n is homotopic to a homeomorphism.* ▮

Example 7.4.9 Let $\tau: S^m \wedge S^n \to S^n \wedge S^m$ be the map that exchanges the two factors. Then, if $S^m \wedge S^n$ and $S^n \wedge S^m$ are identified with S^{m+n} as in Proposition 6.2.15, τ has degree $(-1)^{mn}$ For clearly τ corresponds to the homeomorphism of S^{m+n} that sends $(x_1, \ldots, x_{m+n+1})$ to $(x_1, x_{m+2}, \ldots, x_{m+n+1}, x_2, \ldots, x_{m+1})$. ▮

Example 7.4.10 If $f: S^m \to S^m$ is a based map, and $f \wedge 1: S^m \wedge S^n \to S^m \wedge S^n$ is regarded as a map of S^{m+n} to itself, then $d(f \wedge 1) = d(f)$. Because of the associativity of the reduced products of spheres, it is sufficient to prove this in the case $n = 1$; but by Corollary 6.2.19 and the following remark, there is a homotopy-commutative diagram

$$\begin{array}{ccc} S^{m+1} & \xrightarrow{Sf} & S^{m+1} \\ \downarrow p & & \downarrow p \\ S^m \wedge S^1 & \xrightarrow[f \wedge 1]{} & S^m \wedge S^1 \end{array} \qquad (h: S^{m+1} \searrow S^m \wedge S^1,\ h: S^{m+1} \swarrow S^m \wedge S^1)$$

Thus $d(f \wedge 1)$, by which we really mean $d(h^{-1}(f \wedge 1)h)$, is the same as $d(Sf) = d(f)$. ▮

This technique also allows us to give a useful alternative description of the homomorphism ∂_* in the exact homotopy sequence of a pair.

Proposition 7.4.11 *Let (X, Y) be a pair of spaces, with base point $x_0 \in Y$. Define a function $\bar{\partial}: \pi_n(X, Y) \to \pi_{n-1}(Y)$ by representing an element of $\pi_n(X, Y)$ by a map $f: (I^n, \partial I^n, D^{n-1}) \to (X, Y, x_0)$, restricting f to ∂I^n, and regarding this, via standard homeomorphisms, as a map $S^{n-1} \to Y$. Then $\bar{\partial} = (-1)^n \partial_*$.*

Proof. Via the standard homeomorphism $I^n \xrightarrow{l^n} J^n \xrightarrow{\rho} E^n$, we may regard f as a based map $(E^n, S^{n-1}) \to (X, Y)$; and then $\bar{\partial}[f] = [f | Y]$. On the other hand, $\partial_*[f]$ is obtained by using $h_{n-1,1}: S^{n-1} \to S^{n-2} \times E^1 \cup E^{n-1} \times S^0$ to restrict f further to $E^{n-1} \times (-1)$, and then taking the induced map $E^{n-1}/S^{n-2} \to Y$, composed with $\theta^{-1}: S^{n-1} \to E^{n-1}/S^{n-2}$. That is to say, $\bar{\partial}[f]$ is $\partial_*[f]$, composed with the homotopy class of

$$S^{n-1} \xrightarrow{h_{n-1,1}} S^{n-2} \times E^1 \cup E^{n-1} \times S^0 \xrightarrow{p} E^{n-1}/S^{n-2} \xrightarrow{\theta} S^{n-1},$$

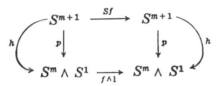

where p is the map that identifies $S^{n-2} \times E^1 \cup E^{n-1} \times (+1)$ to a point. Now it was noted after Proposition 6.2.16 that this composite would have degree $(-1)^{n-1}$ if p had identified $E^{n-1} \times (-1)$ instead of $E^{n-1} \times (+1)$ to a point; so this composite has degree $(-1)^n$ since multiplication by -1 of the last co-ordinate in S^{n-1} is a map of degree -1. Hence $\bar{\partial}[f]$ is $\partial_*[f]$, composed with a map of S^{n-1} of degree $(-1)^n$, and so $\bar{\partial} = (-1)^n \partial_*$. ▮

It is useful also to have a relative version of the results on degrees of maps of S^n.

Definition 7.4.12 For $n \geqslant 0$, let $\bar{\sigma}_n$ be a generator of $H_n(E^n, S^{n-1}) \cong Z$. Given a map $f: (E^n, S^{n-1}) \to (E^n, S^{n-1})$, the *degree* of f, $d(f)$, is defined by $f_*(\bar{\sigma}_n) = d(f).\bar{\sigma}_n$.

If (E^n, S^{n-1}) is identified with $(I^n, \partial I^n)$ via standard homeomorphisms, the degree defines a function $d: \pi_n(E^n, S^{n-1}) \to Z$.

Proposition 7.4.13 *For $n \geqslant 2$, d is an isomorphism. Moreover, the diagram*

$$\begin{array}{ccc} \pi_n(E^n, S^{n-1}) & \xrightarrow{\bar{\partial}} & \pi_{n-1}(S^{n-1}) \\ & {\scriptstyle d}\searrow \quad \swarrow{\scriptstyle d} & \\ & Z & \end{array}$$

is commutative.

Proof. Since E^n is contractible, $\pi_n(E^n) = \pi_{n-1}(E^n) = 0$, so that $\bar{\partial}$ is an isomorphism by Theorem 7.2.18. And since $\bar{\partial}$ is defined by restricting a map of $(I^n, \partial I^n, D^{n-1})$ to a map of ∂I^n, it follows from Theorem 4.4.3 that the diagram is commutative; hence $d: \pi_n(E^n, S^{n-1}) \to Z$ is also an isomorphism. ▮

An obvious argument with the exact homotopy sequence shows that in fact $\pi_r(E^n, S^{n-1}) \cong \pi_{r-1}(S^{n-1})$ for all r, so that we know $\pi_r(E^n, S^{n-1})$ for all $r \leqslant n$. The reader is warned, however, that $\pi_r(E^n, S^{n-1})$ and $\pi_r(S^n)$ are not necessarily zero for $r > n$: for example $\pi_3(S^2) \cong Z$ (see Exercise 19). Indeed, the calculation of $\pi_r(S^n)$ is one of the most difficult problems of homotopy theory, and remains unsolved for general r and n (see the notes at the end of the chapter).

Example 7.4.14 The result on $\pi_n(S^n)$ can be combined with Theorems 7.2.19 and 7.2.20 to give

$$\pi_n(S^n \times S^n) \cong Z \oplus Z,$$

$$\pi_n(S^n \vee S^n) \cong Z \oplus Z \oplus \pi_{n+1}(S^n \times S^n, S^n \vee S^n) \qquad (n \geqslant 2). ▮$$

The latter result is still somewhat unsatisfactory, since we do not know $\pi_{n+1}(S^n \times S^n, S^n \vee S^n)$ (in fact it is zero). As was suggested at the end of Section 7.2, we need a general theorem to the effect that $\pi_n(X, Y) = 0$ in certain circumstances, at least if X and Y are CW-complexes. This is the next theorem; and the method of proof will also yield information on a certain non-vanishing relative homotopy group as well.

Theorem 7.4.15 *Let K be an n-dimensional CW-complex ($n \geqslant 2$), and let L be a subcomplex that contains K^{n-1}. Then*

(a) $\pi_r(K, L) = 0$, $1 \leqslant r < n$.

(b) *Let the indexing sets for K and L be A_r, B_r respectively, and let ϕ^r_α be the characteristic maps. Then if (K, L) is relatively n-simple, the homomorphism*

$$\phi^n_*: \bigoplus_{A_n - B_n} \pi_n(E^n_\alpha, S^{n-1}_\alpha) \to \pi_n(K, L),$$

defined to be $(\phi^n_\alpha)_*$ on each $\pi_n(E^n_\alpha, S^{n-1}_\alpha)$, is onto.

Note. Since (E^n, S^{n-1}) is relatively n-simple (by Example 7.2.14), $(\phi^n)_*$ is defined even though ϕ^n_α need not be a based map. In fact we shall prove in Chapter 8 that ϕ^n_* is an isomorphism, so that $\pi_n(K, L)$ is a free abelian group with one generator for each $\alpha \in A_n - B_n$.

Proof. For each $\alpha \in A_n - B_n$, let U_α be the open subspace $\phi^n_\alpha\{x \in E^n \mid \|x\| < \frac{2}{3}\}$ of K, and let V be the closed subspace $\bigcup_{A_n - B_n} \phi^n_\alpha\{x \in E^n \mid \|x\| \leqslant \frac{1}{3}\}$: thus $K - V$ is open. Also, write W_α for $(K - V) \cap U_\alpha$: see Fig. 7.8.

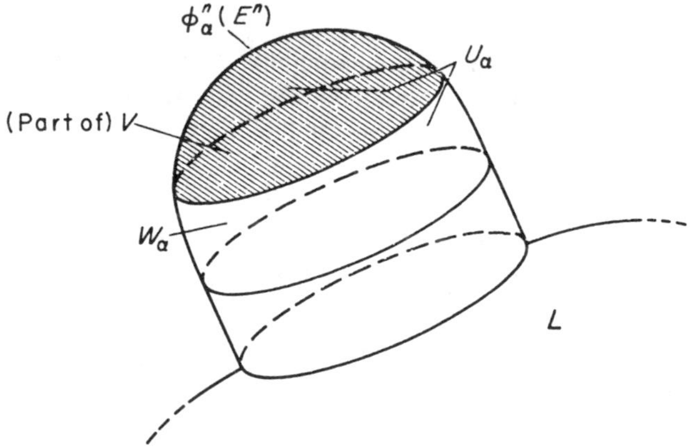

Fig. 7.8

We shall show that any map $f: (I^r, \partial I^r) \to (K, L)$ $(r \leqslant n)$ can be 'pushed off' V, and hence pushed into L. This will prove (a), and an extension of this method will prove (b).

Now I^r can be regarded as the product of r copies of I. Since I is a CW-complex with one 1-cell and two 0-cells, Theorem 7.3.16 yields a CW decomposition of I^r, in which there is just one r-cell. Indeed, if I is 'subdivided' by introducing a new 0-cell at $\frac{1}{2}$, this has the effect of subdividing I^r into 2^r hypercubes each of side $\frac{1}{2}$, and the corresponding CW decomposition has 2^r r-cells: see Fig. 7.9 for the case $r = 2$.

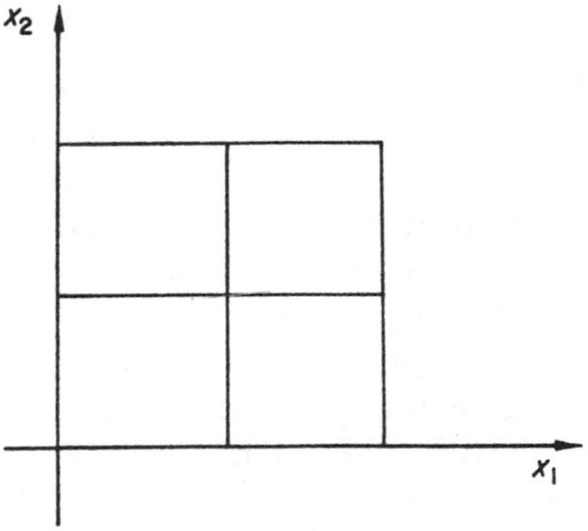

Fig. 7.9

This process can be iterated: at the next stage we obtain a CW-decomposition with 2^{2r} r-cells consisting of hypercubes of side $\frac{1}{4}$, and so on. Now we use an argument similar to that in the proof of the Simplicial Approximation Theorem: given a map $f: (I^r, \partial I^r) \to (K, L)$ $(r \leqslant n)$, the sets $f^{-1}(K - V)$, $f^{-1}(U_\alpha)$ form an open covering of I^r, so that by Theorem 1.4.35 we can iterate the subdivision process until I^r is subdivided into a CW-complex M, say, in which each r-cell (hypercube) is mapped by f into $K - V$ or into one of the sets U_α. Notice also that ∂I^r is a subcomplex of M.

The next step is to construct a map $g: M \to K$ such that

(a) for each m-cell $\psi^m_\beta(E^m)$ of M $(m < n)$, $f\psi^m_\beta(E^m) \subset K - V \Rightarrow g|\psi^m_\beta(E^m) = f|\psi^m_\beta(E^m)$; otherwise $f\psi^m_\beta(E^m) \subset U_\alpha \Rightarrow g\psi^m_\beta(E^m) \subset W_\alpha$;

(b) $f \simeq g$ rel ∂I^r, and points of M that are mapped by f into U_α remain in U_α throughout the homotopy.

This is done by induction on the skeletons of M, in the manner of Theorem 7.3.19. Suppose then that g has been defined on M^{m-1} $(m < n)$, so as to satisfy (a) and (b) (it is easy to define g on M^0, since each 0-cell that is mapped by f into U_α can be joined by a straight line to a point of W_α). Now consider an m-cell $\psi_\beta^m(E^m)$ of M such that $f\psi_\beta^m(E^m) \subset U_\alpha$; then $f\psi_\beta^m(S^{m-1}) \subset U_\alpha$ and $g\psi_\beta^m(S^{m-1}) \subset W_\alpha$. Since each characteristic map of M is actually a homeomorphism, $g|\psi_\beta^m(S^{m-1})$ represents an element of $\pi_{m-1}(W_\alpha) = 0$, since $W_\alpha \simeq S^{n-1}$ and $m < n$. Thus $g|\psi_\beta^m(S^{m-1})$ is homotopic to a constant map, and hence can be extended to a map $g: \psi_\beta^m(E^m) \to W$. Moreover the original homotopy between f and g on $\psi_\beta^m(S^{m-1})$ can be extended to a homotopy of $\psi_\beta^m(E^m)$ in U_α that starts with f and whose final map is g on $\psi_\beta^m(S^{m-1})$; and this final map is homotopic to g, rel $\psi_\beta^m(S^{m-1})$, by a linear homotopy. It follows that we can extend g to M^m so as still to satisfy (a) and (b), by using this construction on m-cells mapped into some U_α, and by defining $g = f$ (with the constant homotopy) on m-cells mapped into $K - V$, the resulting g (and homotopy) being continuous by Proposition 7.3.4. By induction, therefore, g can be extended to M^{n-1}, the extension to M^n (if $r = n$) being possible since (M^n, M^{n-1}) has the absolute homotopy extension property.

Since $f \simeq g$ rel ∂I^r, if f maps D^{r-1} to the base point, then $[f] = [g]$ in $\pi_r(K, L)$. If $r < n$, $[g]$ is the image under the inclusion map of an element of $\pi_r(K - V, L)$; but, as in Example 7.3.18, L is a strong deformation retract of $K - V$, so that $\pi_r(K - V, L) = 0$ and hence $[f] = [g] = 0$. It follows that $\pi_r(K, L) = 0$ for $r < n$, so that at this point the proof of (a) is complete.

To prove (b), note that we have proved that each element of $\pi_n(K, L)$ can be represented by a map $g: (I^n, \partial I^n) \to (K, L)$, that maps M^{n-1} to $K - V$ and each n-cell of M to $K - V$ or to one U_α, where M is the CW-decomposition of I^n obtained above. Now $(K, K - V) \simeq (K, L)$, so that, by Theorem 7.2.13(c), $(K, K - V)$ is n-simple as well as (K, L). Hence, by iteration of the construction in Example 7.2.17, $[g] \in \pi_n(K, K - V)$ is a sum of elements, each of which lies in the image of the homomorphism induced by an inclusion map $(U_\alpha, W_\alpha) \to (K, K - V)$. Since the deformation of $K - V$ onto L sends W_α onto $\phi_\alpha^n(S^{n-1})$, this proves that $[g] \in \pi_n(K, L)$ is a sum of elements, each of which lies in the image of some $(\phi_\alpha^n)_*: \pi_n(E_\alpha^n, S_\alpha^{n-1}) \to \pi_n(K, L)$. Hence $[g]$ is in the image of ϕ_*^n, and ϕ^n is onto. ∎

Example 7.4.16 By Example 7.4.14, for $n \geqslant 2$ we have

$$\pi_n(S^n \vee S^n) \cong Z \oplus Z \oplus \pi_{n+1}(S^n \times S^n, S^n \vee S^n).$$

Now S^n is a CW-complex with one 0-cell and one n-cell, so that $S^n \times S^n$ has one 0-cell, two n-cells and one $2n$-cell; moreover the n-skeleton (and indeed the $(2n - 1)$-skeleton) is $S^n \vee S^n$. It follows from Theorem 7.4.15 that $\pi_{n+1}(S^n \times S^n, S^n \vee S^n) = 0$, and $\pi_n(S^n \vee S^n) \cong Z \oplus Z$. A similar argument shows that

$$\pi_n(S^p \vee S^q) \cong \pi_n(S^p) \oplus \pi_n(S^q) \qquad (2 \leqslant n < p + q - 1). \blacksquare$$

It follows that, for $n \geqslant 2$, there is only one homotopy class of maps $\mu: S^n \to S^n \vee S^n$ that make S^n into an AH'I. For since $p_1\mu \simeq p_2\mu \simeq 1$, $[\mu]$ must be $1 \oplus 1$ in $Z \oplus Z \cong \pi_n(S^n \vee S^n)$. It follows that there is only one possible way of defining a group structure in $\pi_n(X)$, at least if $n \geqslant 2$ (for the case $n = 1$, see Chapter 6, Exercise 9).

Various general results on homotopy groups of CW-complexes can be deduced from Theorem 7.4.15.

Theorem 7.4.17 *Let (K, L) be a CW-pair, and let $i: K^n \cup L \to K$ be the inclusion map ($n \geqslant 0$). Then*

(a) *$i_*: \pi_r(K^n \cup L) \to \pi_r(K)$ is onto for $0 \leqslant r \leqslant n$ and (1-1) for $0 \leqslant r < n$; similarly for $i_*: \pi_r(K^n \cup L, L) \to \pi_r(K, L)$;*

(b) *$\pi_r(K, K^n \cup L) = 0$ for $1 \leqslant r \leqslant n$;*

Proof. Consider the exact homotopy sequence of the pair (K^{m+1}, K^m) $(m \geqslant 0)$:

$$\cdots \longrightarrow \pi_{r+1}(K^{m+1}, K^m) \longrightarrow \pi_r(K^m) \xrightarrow{i_*} \pi_r(K^{m+1}) \longrightarrow \pi_r(K^{m+1}, K^m) \longrightarrow \cdots,$$

where $i: K^m \to K^{m+1}$ once again denotes the inclusion map. Now by Theorem 7.4.15 $\pi_r(K^{m+1}, K^m) = 0$ for $1 \leqslant r \leqslant m$, so that $i_*: \pi_r(K^m) \to \pi_r(K^{m+1})$ is onto for $1 \leqslant r \leqslant m$ and (1-1) for $1 \leqslant r < m$. Moreover, since attaching cells clearly cannot increase the number of path-components, $i_*: \pi_0(K^m) \to \pi_0(K^{m+1})$ is always onto, and is (1-1) if $m > 0$.

Hence $i_*: \pi_r(K^n) \to \pi_r(K^m)$ is isomorphic for $r < n$ and onto for $r = n$, for all $m > n$. But elements of $\pi_r(K)$ are represented by maps of S^r to K, and since S^r is compact the images must be contained in finite skeletons. A similar argument applies to homotopies of S^r in K, so that $i_*: \pi_r(K^n) \to \pi_r(K)$ is isomorphic for $r < n$ and onto for $r = n$. To deduce the first part of (a), observe that $\pi_r(K^n \cup L) \leftarrow \pi_r(K^n \cup L^{n+1})$ is an isomorphism for all $r \leqslant n$, and $\pi_n(K^n \cup L^{n+1}) \to \pi_n(K^{n+1})$ is isomorphic if $r < n$, onto if $r = n$, by another application of Theorem 7.4.15.

The exact sequence of the pair $(K, K^n \cup L)$ now gives (b), and then the second part of (a) follows from the exact sequence of the triple $(K, K^n \cup L, L)$. $\blacksquare$

Theorem 7.4.17 may be expressed by the statement that $\pi_n(K, L)$ depends only on the $(n + 1)$-skeleton of K, and so this result extends Corollary 3.3.10, which was the case $n = 1$, K a polyhedron, and $L = \varnothing$.

7.5 The theorem of J. H. C. Whitehead and the Cellular Approximation Theorem

The main theorem in this section is the theorem of J. H. C. Whitehead, that states that if $f: K \to L$ is a map of CW-complexes that induces isomorphisms $f_*: \pi_r(K) \to \pi_r(L)$ for all $r \geqslant 0$, then f is a homotopy equivalence. It is convenient to have a special name for maps that induce isomorphisms of homotopy groups.

Definition 7.5.1 If X and Y are any spaces, a map $f: X \to Y$ is called a *weak homotopy equivalence* if $f_*: \pi_0(X) \to \pi_0(Y)$ is a (1-1)-correspondence, and $f_*: \pi_r(X, x_0) \to \pi_r(Y, f(x_0))$ is an isomorphism for all $r \geqslant 1$ and all points $x_0 \in X$.

Of course, if X and Y are path-connected, it is sufficient that $f_*: \pi_r(X, x_0) \to \pi_r(Y, f(x_0))$ should be an isomorphism for all $r \geqslant 1$ and just *one* point $x_0 \in X$.

Clearly every homotopy equivalence is a weak homotopy equivalence, and Whitehead's theorem states that the converse is true, provided X and Y are CW-complexes. The method of proof is to investigate first the special case in which f is an inclusion map, and then to deduce the general result by using the mapping cylinder. We start by proving a general result about inclusion maps that are weak homotopy equivalences.

Theorem 7.5.2 *Let (X, Y) be a pair of spaces, such that the inclusion map $i: Y \to X$ is a weak homotopy equivalence. Let K be a CW-complex, with a 0-cell as base point. Then for any choice of base point in Y, $i_*: [K, Y] \to [K, X]$ is a (1-1)-correspondence.*

Proof. We show first that i_* is onto. Suppose, then, that we have a based map $f: K \to X$; we shall show by induction on the skeletons of K that f can be deformed into Y. The argument is similar to that of Theorem 7.3.19 (indeed, it is a generalization of that argument): f is regarded as a map of $K \times 0$ to X, and is extended to a map $f: K \times I \to X$, such that $f(K \times 1) \subset Y$, and if L is any subcomplex of K that is mapped by f into Y, then $f(L \times I) \subset Y$: thus in particular the homotopy is a based homotopy.

Given such a subcomplex L, write $M^n = K^n \cup L$, and extend f as the constant homotopy to $(K \times 0) \cup (L \times I)$. If x is any 0-cell of $K - L$, there is a path $u: I \to X$ such that $u(0) = f(x)$ and $u(1) \in Y$; thus we can extend f to $M^0 \times I$ by setting $f(x, t) = u(t)$, $0 \leqslant t \leqslant 1$. This serves to start the induction; so we may now assume that f has been extended to a map $f: (K \times 0) \cup (M^{n-1} \times I) \to X$, such that $f(M^{n-1} \times 1) \subset Y$. For each n-cell $\phi_\alpha^n(E^n)$ of $K - L$, consider the composite

$$(E^n \times 0) \cup (S^{n-1} \times I) \xrightarrow{\phi_\alpha^n \times 1} (K \times 0) \cup (M^{n-1} \times I) \xrightarrow{f} X,$$

which sends $S^{n-1} \times 1$ to Y. Define a homeomorphism h of $E^n \times I$ to itself by

$$\begin{aligned}
h(x, 0) &= (x/2, 0) && (x \in E^n),\\
h(x, t) &= (\tfrac{1}{2}(1 + t)x, 0) && (x \in S^{n-1}, 0 \leqslant t \leqslant 1),\\
h(x, 1) &= (x/\|x\|, 2 - 2\|x\|) && (x \in E^n, \|x\| \geqslant \tfrac{1}{2}),\\
h(x, 1) &= (2x, 1) && (x \in E^n, \|x\| \leqslant \tfrac{1}{2}),
\end{aligned}$$

extending the definition inside $E^n \times I$ by regarding the inside as the join of $(E^n \times 0) \cup (S^{n-1} \times I) \cup (E^n \times 1)$ to $(0, \frac{1}{2})$ (we are, as it were, pulling $S^{n-1} \times I$ down into $E^n \times 0$: see Fig. 7.10).

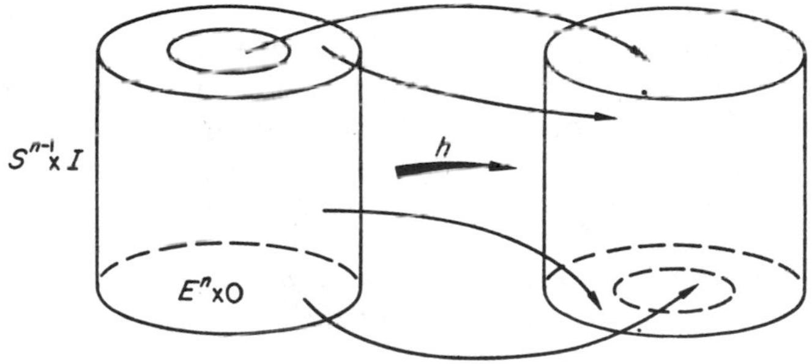

Fig. 7.10

The point of this definition is that $f(\phi_\alpha^n \times 1)h^{-1}$ is a map of (E^n, S^{n-1}) to (X, Y), which therefore represents an element of $\pi_n(X, Y)$, with some base point. But by the exact homotopy sequence $\pi_n(X, Y) = 0$; thus $f(\phi_\alpha^n \times 1)h^{-1}$ can be extended to a map of $E^n \times I$ that sends $E^n \times 1$ and $S^{n-1} \times I$ to Y. Hence, by applying h again, $f(\phi_\alpha^n \times 1)$ can be extended to a map of $E^n \times I$ that sends $E^n \times 1$ to Y. As in Theorem 7.3.19, this process defines a continuous extension

$f: (K \times 0) \cup (M^n \times I) \to X$ such that $f(M^n \times 1) \subset Y$; and hence a continuous extension $f: K \times I \to X$ such that $f(K \times 1) \subset Y$. It follows that $i_*: [K, Y] \to [K, X]$ is onto.

It is easy to deduce that i_* is also (1-1). For suppose $f, g: K \to Y$ are based maps such that $if \simeq ig$ by a based homotopy $F: K \times I \to X$. Since $K \times I$ is a CW-complex and $(K \times 0) \cup (k_0 \times I) \cup (K \times I)$ is a subcomplex, F can be deformed to a map $G: K \times I \to Y$ such that G coincides with F on $(K \times 0) \cup (k_0 \times I) \cup (K \times 1)$. That is, G is a based homotopy between f and g. ∎

The above *is* a generalization of Theorem 7.3.19, for we could apply it to the inclusion map $i: (K \times 0) \cup (L \times I) \to K \times I$ to obtain a retraction $K \times I \to (K \times 0) \cup (L \times I)$.

It is easy to extend Theorem 7.5.2 to an arbitrary weak homotopy equivalence, by using the mapping cylinder.

Corollary 7.5.3 *Given a weak homotopy equivalence $f: Y \to X$, and a CW-complex K, $f_*: [K, Y] \to [K, X]$ is a (1-1)-correspondence (where K has a 0-cell as base point, and Y, X have any base points that correspond under f).*

Proof. By Theorem 6.5.5, f is the composite

$$Y \xrightarrow{g} M_f \xrightarrow{h} X,$$

where M_f is the mapping cylinder, g is an inclusion map, and h is a homotopy equivalence. Since both f and h are weak homotopy equivalences, so is g; hence $g_*: [K, Y] \to [K, M_f]$ is a (1-1)-correspondence. But h_* is obviously a (1-1)-correspondence, and hence so is $f_* = h_* g_*$. ∎

Whitehead's theorem follows immediately.

Theorem 7.5.4 *If $f: K \to L$ is a weak homotopy equivalence of CW-complexes, f is a homotopy equivalence.*

Proof. By Corollary 7.5.3, $f_*: [L, K] \to [L, L]$ is a (1-1)-correspondence, so that there exists a map $g: L \to K$ such that $fg \simeq 1_L$. Then g is also a weak homotopy equivalence, so by a similar argument there exists $f': K \to L$ such that $gf' \simeq 1_K$. But then

$$f' \simeq (fg)f' \simeq f(gf') \simeq f,$$

so that $gf \simeq 1_K$ as well, and so g is a homotopy inverse to f. ∎

The reader should not be tempted to think that *every* weak homotopy equivalence is a homotopy equivalence: the assumption that K and L are CW-complexes is essential in Theorem 7.5.4.

Example 7.5.5 Let X be the subspace of R^2 consisting of straight line segments joining $(0, 1)$ to the points $(0, 0)$ and $(1/n, 0)$, for all positive integers n, and $(0, -1)$ to all the points $(0, 0)$ and $(-1/n, 0)$: see Fig. 7.11.

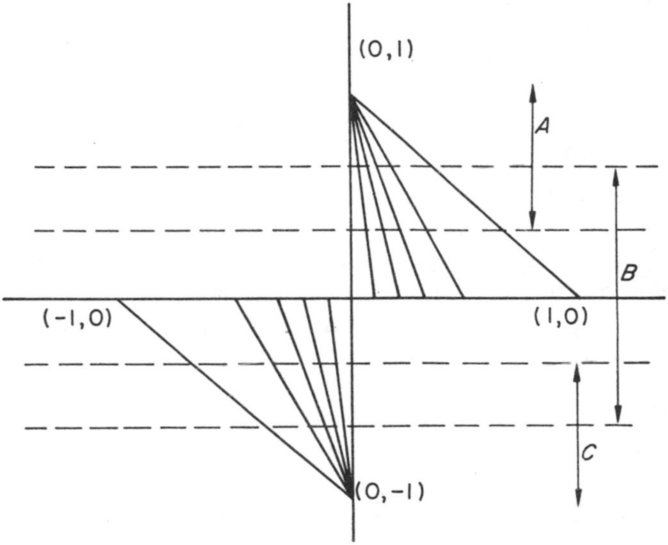

Fig. 7.11

We shall see that $\pi_n(X) = 0$ for all $n \geq 0$, but that X is not contractible. Thus the map that sends all of X to $(0, 0)$ is a weak homotopy equivalence that is not a homotopy equivalence. To prove the first assertion, take an open covering of X by three open sets A, B, C, defined by $x_2 > \frac{1}{3}$, $\frac{2}{3} > x_2 > -\frac{2}{3}$, $-\frac{1}{3} > x_2$ respectively. Then if $f: S^n \to X$ is any map, the sets $f^{-1}(A)$, $f^{-1}(B)$, $f^{-1}(C)$ form an open covering of S^n, with Lebesgue number δ, say. If S^n is triangulated so that the mesh is less than δ, only a finite number of simplexes are mapped into B, and since the image of each is path-connected, it follows that $f(S^n) \cap B$ is contained in a finite number of 'rays' from $(0, 1)$ or $(0, -1)$. That is, $f(S^n)$ is contained in Y, the union of A and C with a finite number of rays. Since it is easy to see that Y is contractible, this means that f is homotopic to the constant map in Y, so certainly in X. Hence $\pi_n(X) = 0$.

On the other hand X is not contractible For if it were, there would be a map $f: X \times I \to X$ starting with the identity map and ending with the constant map to some point $x_0 \in X$. Since I is compact, the continuity of f implies that, given $x \in X$ and $\epsilon > 0$, there exists δ

such that $d(x, y) < \delta \Rightarrow d(f(x, t), f(y, t)) < \epsilon$ for all $t \in I$ But for each integer $n > 0$, the homotopy f defines paths u^+ and u^- from $(1/n, 0)$, $(-1/n, 0)$ to x_0 respectively. Subdivide I (considered as a 1-simplex) so that each simplex of the subdivision is mapped by each of u^+ and u^- into just one of the sets A, B or C. Since $(1/n, 0)$ and $(-1/n, 0)$ are in different path components of B, there is a first vertex t such that either $u^+(t) \in A$ or $u^-(t) \in C$; if say, $u^+(t) \in A$, then $u^-(t)$ lies in the region $x_2 \leqslant 0$. Hence $d(u^+(t), u^-(t)) > \frac{1}{3}$, which contradicts the continuity of f, since if we take $x = (0, 0)$ and $\epsilon = \frac{1}{3}$, there is always an n such that $2/n < \delta$, for any δ. ▌

The last important theorem in this chapter is the Cellular Approximation Theorem, which in a sense is the analogue for CW-complexes of the Simplicial Approximation Theorem for simplicial complexes. The theorem states that, if $f: K \to L$ is a map between CW-complexes, then f is homotopic to a map that sends the n-skeleton of K into the n-skeleton of L, for each n.

Definition 7.5.6 If K and L are CW-complexes, a map $f: K \to L$ such that $f(K^n) \subset L^n$ for each $n \geqslant 0$ is called a *cellular map*.

Theorem 7.5.7 *If K and L are CW-complexes, and $f: K \to L$ is a map such that $f|M$ is cellular for some subcomplex M of K (possibly empty), then there exists a cellular map $g: K \to L$ such that $g|M = f|M$ and $g \simeq f$* rel *M.*

Proof. This is very similar to Theorem 7.5.2: by induction on the skeletons of K, we define a homotopy $F: K \times I \to L$ that starts with f, ends with a cellular map, and is the constant homotopy on $M \times I$. Since, for each 0-cell x of $K - M$, there is a path in L from $f(x)$ to a point of L^0, we can certainly define F on $K^0 \times I \cup M \times I$. Suppose, then, that F has been extended to $K^{n-1} \times I$, and that $F(K^{n-1} \times 1) \subset L^{n-1}$. Just as in Theorem 7.5.2, F can be extended to each n-cell of $K - M$, since $\pi_n(L, L^n) = 0$ by Theorem 7.4.17; and the result is a continuous extension such that $F(K^n \times 1) \subset L^n$. This completes the inductive step, and so gives the required homotopy $F: K \times I \to L$. ▌

The Cellular Approximation Theorem is particularly useful in view of the fact that the space obtained by attaching cells by cellular maps to a CW-complex is another CW-complex (this follows easily from Theorem 7.3.12). It is thus possible to make alterations in the homotopy groups of CW-complexes: an element of $\pi_n(K)$ can be represented by a cellular map $f: S^n \to K$, and this map can be used to attach an $(n + 1)$-cell to K, to form a new CW-complex K' in which $[f]$ is 'killed off'. This idea is formalized in the last two theorems of this chapter.

Theorem 7.5.8 *Given a CW-complex K and an integer $n \geqslant 0$, there exists a CW-complex L, having K as a subcomplex, such that, if $i: K \to L$ is the inclusion map,*

(a) *$i_*: \pi_r(K) \to \pi_r(L)$ is isomorphic for $r < n$;*
(b) *$\pi_n(L) = 0$.*

Proof. Let A be a set of generators for the group $\pi_n(K)$ (for example, the set of all elements of $\pi_n(K)$). For each $\alpha \in A$, take a representative (based) map $\phi_\alpha^n: S^n \to K$, which by Theorem 7.5.7 may be assumed to be cellular. Let L be the space obtained from K by attaching cells E_α^{n+1} by the maps ϕ_α^n, one for each $\alpha \in A$.

Then L is a CW-complex: for by Theorem 7.3.12 K is a cellular space, and hence so is L, since the maps ϕ_α^n send S^n into K^n. Also K is obviously a subcomplex of L. Moreover by Theorem 7.4.17(a) $i_*: \pi_r(K) = \pi_r(L^n \cup K) \to \pi_r(L)$ is isomorphic for $r < n$, and onto for $r = n$. But for each $\alpha \in A$, $i_*(\alpha) \in \pi_n(L)$ is represented by the map $i\phi_\alpha^n: S^n \to L$; and this is clearly homotopic to the constant map, since L has an $(n + 1)$-cell attached by ϕ_α^n. Hence $\pi_n(L) = 0$. ∎

This process can be iterated, so as to 'kill off' $\pi_r(K)$ for all $r \geqslant n$.

Theorem 7.5.9 *Given a CW-complex K and an integer $n \geqslant 0$, there exists a CW-complex L, having K as a subcomplex, such that, if $i: K \to L$ is the inclusion map,*

(a) *$i_*: \pi_r(K) \to \pi_r(L)$ is isomorphic for $r < n$;*
(b) *$\pi_r(L) = 0$ for $r \geqslant n$.*

Proof. By repeated applications of Theorem 7.5.8, there is a sequence of CW-complexes $K \subset L_1 \subset L_2 \subset \cdots$, each a subcomplex of the next, such that for each $m \geqslant 1$, if $i: K \to L_m$ is the inclusion map,

(a) $i_*: \pi_r(K) \to \pi_r(L_m)$ is isomorphic for $r < n$, and
(b) $\pi_r(L_m) = 0$ for $n \leqslant r < n + m$.

Let $L = \bigcup_{m=1}^{\infty} L_m$ (as a point set), topologized by the rule: $X \subset L$ is closed if and only if $X \cap L_m$ is closed in L_m, for each $m \geqslant 1$. This certainly is a topology, and L is a CW-complex by Theorem 7.3.12. Moreover each L_m, and K, is a subcomplex of L.

To prove (a) and (b), note that, given any r, $i_*: \pi_r(L^{r+1}) \to \pi_r(L)$

is an isomorphism. But L^{r+1} is the $(r + 1)$-skeleton of each L_m for which $n + m > r$, so that $i_*: \pi_r(L^{r+1}) \to \pi_r(L_m)$ is also an isomorphism for such m. Hence $i_*: \pi_r(L_m) \to \pi_r(L)$ is an isomorphism, and (a) and (b) are now immediate.

Example 7.5.10 We have already proved that

$$\pi_r(S^n) \cong \begin{cases} 0, & r < n \\ Z, & r = n. \end{cases}$$

It follows from Theorem 7.5.9 that there exists a CW-complex K such that $\pi_r(K) = 0$ for $r \neq n$, and $\pi_n(K) \cong Z$. Such a CW-complex is called an *Eilenberg–MacLane space* $K(Z, n)$: we shall see in Chapter 8 that these spaces are important in the cohomology theory of CW-complexes. ∎

EXERCISES

1. Let X and Y be path-connected spaces with base points x_0, y_0, and suppose that (X, x_0) has the absolute homotopy extension property. Show that a path u in Y from y_0 to y_1 gives rise to a (1-1)-correspondence $u_\#: [X, Y]_0 \to [X, Y]_1$ (where $[X, Y]_i$ denotes $[X, Y]$ with base points x_0, y_i respectively), with the following properties.
   (a) If $u \simeq v$ rel 0, 1, then $u_\# = v_\#$.
   (b) $(e_{y_0})_\#$ is the identity function.
   (c) If w is a path from y_1 to y_2, then $(u.w)_\# = w_\# u_\#$.
   (d) If $f: Y \to Z$ is a map such that $f(y_0) = z_0$ and $f(y_1) = z_1$, then $f_* u_\# = (fu)_\# f_*: [X, Y]_0 \to [X, Z]_1$.
   (e) If X is an AH'I, $u_\#$ is an isomorphism.

   Deduce that, if f in (d) is a homotopy equivalence (as an unbased map), then $f_*: [X, Y]_0 \to [X, Z]_0$ is a (1-1)-correspondence.
2. Let $\Omega_n(X)$ be the subgroup of $\pi_n(X)$ generated by all elements of the form $x - u_\# x$, where $[u] \in \pi_1(X)$ (if $n = 1$, this is to be interpreted as $x.(u_\# x)^{-1}$). Show that $\Omega_n(X)$ is a normal subgroup of $\pi_n(X)$. If X is path-connected, and $\pi_n^*(X)$ denotes $\pi_n(X)/\Omega_n(X)$, show that a homotopy class of (unbased) maps $f: S^n \to X$ defines a unique element of $\pi_n^*(X)$, and that a map $g: X \to Y$ between path-connected spaces gives rise to a homomorphism $g_*: \pi_n^*(X) \to \pi_n^*(Y)$.
3. Given elements $x, y \in \pi_2(X, Y)$, prove that $[\partial_* x]_\# y = x^{-1} y x$. (*Hint:* represent x and y by based maps $f, g: (E^2, S^1) \to (X, Y)$ such that f is the constant map on $x_2 \geqslant 0$ and g is the constant map on $x_2 \leqslant 0$; consider the effect of rotating E^2 through an angle π.) Deduce that (X, Y) is relatively 2-simple if $\pi_2(X, Y)$ is abelian and $i_* \pi_1(Y) = 0$.

4. Let $\Omega_n(X, Y)$ be the subgroup of $\pi_n(X, Y)$ generated by elements of the form $x - u_{\#}x$, where $[u] \in \pi_1(Y)$. Show that $\Omega_n(X, Y)$ is a normal subgroup of $\pi_n(X, Y)$. If Y is path-connected, and $\pi_n^*(X, Y)$ denotes $\pi_n(X, Y)/\Omega_n(X, Y)$, show that a homotopy class of maps $f: (I^n, \partial I^n) \to (X, Y)$ defines a unique element of $\pi_n^*(X, Y)$, and that the product of two such elements may be obtained as in Proposition 7.2.15.

Show that a map $g: (X, Y) \to (Z, W)$ gives rise to a homomorphism $g_*: \pi_n^*(X, Y) \to \pi_n^*(Z, W)$ ($n \geqslant 2$), and that there are homomorphisms $j_*: \pi_n^*(X) \to \pi_n^*(X, Y)$ (if $\pi_1(X, Y) = 0$), $\partial_*: \pi_n^*(X, Y) \to \pi_{n-1}^*(Y)$, such that $i_*j_* = 0$, $\partial_*j_* = 0$ and $i_*\partial_* = 0$.

5. Let (X, Y) be a pair of spaces. Deduce the following results from the exact homotopy sequence of (X, Y).

(a) If Y is a retract of X, then $\pi_n(X) \cong \pi_n(Y) \oplus \pi_n(X, Y)$ ($n \geqslant 2$).

(b) If 1_X is homotopic to a map of X into Y, then

$$\pi_n(Y) \cong \pi_n(X) \oplus \pi_{n+1}(X, Y) \qquad (n \geqslant 2).$$

(c) If $i: Y \to X$ is homotopic to the constant map, then

$$\pi_n(X, Y) \cong \pi_n(X) \oplus \pi_{n-1}(Y) \qquad (n \geqslant 3).$$

(All maps and homotopies are assumed to be based.)

6. Let X be an AH'I and Y, Z be any spaces. Prove that

$$[X, Y \times Z] \cong [X, Y] \oplus [X, Z].$$

7. Let X be a commutative AH'I. Prove that

$$[X, Y \vee Z] \cong [X, Y] \oplus [X, Z] \oplus [X, L_i],$$

where L_i is the mapping path-space of the inclusion map $Y \vee Z \to Y \times Z$. Show also that $[X, L_i]$ is in (1-1)-correspondence with homotopy classes of based maps $(X \wedge I, X \wedge 0) \to (Y \times Z, Y \vee Z)$.

8. If F denotes the real numbers R, the complex numbers C or the quaternions H, *F-projective space* of dimension n, FP^n, is defined to be the space $(F^{n+1} - 0)/S$, where S is the equivalence relation given by $xSy \Leftrightarrow x = fy$, for some $f \in F$. FP^n is given the identification topology, and the equivalence class of $(f_1, \ldots, f_{n+1})$ is written $[f_1, \ldots, f_{n+1}]$.

By writing points of E^{2n} in the form $(z_1, \ldots, z_n, r)$, where $0 \leqslant r \leqslant 1$ and $z_1, \ldots, z_n$ are complex numbers such that $|z_1|^2 + \cdots + |z_n|^2 = 1 - r^2$, prove that CP^n is homeomorphic to the space obtained from E^{2n} by identifying points of S^{2n-1} that are mapped to the same point under $p: S^{2n-1} \to CP^{n-1}$, where p is defined by $p(z_1, \ldots, z_n) = [z_1, \ldots, z_n]$. Deduce that CP^n is a CW-complex with one cell in each dimension $0, 2, \ldots, 2n$.

Similarly, show that HP^n is a CW-complex with one cell in each dimension $0, 4, \ldots, 4n$.

9. Prove that a CW-complex is normal.

10. Let K be a CW-complex with a finite number of cells. Use Theorem 7.3.12 and the Simplicial Approximation Theorem to show that K has the homotopy type of a polyhedron. Deduce that $\pi_1(K \vee L) \cong \pi_1(K) * \pi_1(L)$ for any CW-complexes K and L.

11. A Hausdorff space K is said to be (*infinitely*) *triangulated* if, for each $n \geqslant 0$, there exists an indexing set A_n and an n-simplex σ_n, and maps $\phi_\alpha^n: \sigma_n \to K$ for each $\alpha \in A_n$, with the following properties.

    (a) $K = \bigcup \phi_\alpha^n(\sigma_n)$, for all $n \geqslant 0$ and $\alpha \in A_n$.
    (b) Each ϕ_α^n is (1-1).
    (c) Given a face τ_m of σ_n, there exists a simplicial homeomorphism $\psi_\tau: \sigma_m \to \tau_m$ such that, given $\alpha \in A_n$, there exists $\beta \in A_m$ with $\phi_\beta^m = \phi_\alpha^n\psi_\tau$.
    (d) $\phi_\alpha^n(\sigma_n) \cap \phi_\beta^m(\sigma_m)$ is either empty, or is $\phi_\gamma^p(\sigma_p)$ for some p and $\gamma \in A_p$. In the latter case, there exist faces τ of σ_n, μ of σ_m, such that $\phi_\gamma^p = \phi_\alpha^n\psi_\tau = \phi_\beta^m\psi_\mu$.
    (e) A subset X of K is closed if and only if $(\phi_\alpha^n)^{-1}X$ is closed in σ_n, for each n and $\alpha \in A_n$.

    Prove that any space homeomorphic to a polyhedron is a triangulated space, and that any triangulated space is a CW-complex.

12. Prove the analogue of the Simplicial Approximation Theorem for maps of a (compact) polyhedron into a triangulated space. Deduce that any CW-complex is homotopy-equivalent to a triangulated space. Show also that a CW-complex with a countable number of cells has countable homotopy groups.

13. Let U be an open set in R^n. Show that U is a triangulated space, and hence is a CW-complex. (*Hint:* divide R^n into hypercubes of unit side, and triangulate each. For each $m \geqslant 0$, pick the simplexes of the mth derived complex that are contained in U, and observe that, since the mesh tends to zero as $m \to \infty$, each point of U is contained in at least one simplex. The resulting collection of simplexes is not a triangulation, but may be made so by subdivision.)

14. By using an argument similar to that in Theorem 7.3.12, show that a CW-complex is *locally contractible*, that is, given a point x and an open set U containing x, there exists a contractible open set V such that $x \in V \subset U$. Deduce that a CW-complex is locally path-connected and weakly locally simply-connected, in the sense of Chapter 6, Exercises 23 and 25.

15. Let K be a CW-complex, and let $\tilde{K}$ be its universal cover, with covering map $f: \tilde{K} \to K$ (see Chapter 6, Exercise 25). Prove that $\tilde{K}$ is also a CW-complex. (*Hint:* given a characteristic map $\phi_\alpha^n: E^n \to K$, a point $x \in \phi_\alpha^n(e^n)$ and a point $\tilde{x}$ such that $f(\tilde{x}) = x$, there is a unique map $\tilde{\phi}_\alpha^n: E^n \to \tilde{K}$ such that $\tilde{x} \in \tilde{\phi}_\alpha^n(e^n)$ and $f\tilde{\phi}_\alpha^n = \phi_\alpha^n$. Show that the set of all such $\tilde{\phi}_\alpha^n$ is a set of characteristic maps for a CW-decomposition of $\tilde{K}$.)

16. Let K be an n-dimensional CW-complex, and let L be a subcomplex that contains K^{n-1} ($n \geqslant 2$). Let the indexing sets for K and L be A_r, B_r respectively, and let ϕ_α^r be the characteristic maps. Prove that $\phi_*^n: \bigoplus_{A_n - B_n} \pi_n(E_\alpha^n, S_\alpha^{n-1}) \to \pi_n^*(K, L)$ is onto.

17. Let $f: X \to Y$ be a map such that $f_*: \pi_0(X) \to \pi_0(Y)$ is a (1-1)-correspondence, and $f_*: \pi_r(X, x_0) \to \pi_r(Y, f(y_0))$ is an isomorphism for $r < n$ and is onto for $r = n$, for all points $x_0 \in X$. Show that, for any CW-complex K, $f_*: [K, X] \to [K, Y]$ is a (1-1)-correspondence if dim $K < n$, and is onto if dim $K = n$.

18. Let K be a CW-complex, and let n be any positive integer. Show that there exists a space X and a map $f: X \to K$ such that
(a) $\pi_r(X) = 0$ for $r < n$;
(b) $f_*: \pi_r(X) \to \pi_r(K)$ is isomorphic for $r \geqslant n$.

19. Define $f: S^n \to RP^n$ by $f(x_1, \ldots, x_{n+1}) = [x_1, \ldots, x_{n+1}]$; show that f is a local product, with fibre S^0. Deduce that RP^n has the same homotopy groups as S^n, except for $\pi_1(RP^n) \cong Z_2$ ($n > 1$).

Similarly, show that there are local products $S^{2n+1} \to CP^n$, with fibre S^1, and $S^{4n+3} \to HP^n$, with fibre S^3. Deduce that $\pi_r(CP^n) \cong \pi_r(S^{2n+1})$ (except that $\pi_2(CP^n) \cong Z$), and that $\pi_r(HP^n) \cong \pi_r(S^{4n+3}) \oplus \pi_{r-1}(S^3)$.

Hence prove that $\pi_r(S^3) \cong \pi_r(S^2)$, $r \geqslant 3$, and that $\pi_r(S^4) \cong \pi_r(S^7) \oplus \pi_{r-1}(S^3)$.

20. Let G be a topological group (see Chapter 3, Exercise 10), and consider the map $p: G \times G \times I \to G \times I$ defined by $p(g, h, t) = (gh, t)$. Show that p induces a local product $q: G * G \to SG$ with fibre G, provided G is locally compact and Hausdorff. (*Hint:* consider the open sets in SG corresponding to $G \times [0, 1)$ and $G \times (0, 1]$.) By considering S^1 and S^3 as complex numbers and quaternions of unit modulus, respectively, show that S^1 and S^3 are topological groups. Hence, once again, deduce the existence of local products $S^3 \to S^2$, with fibre S^1, and $S^7 \to S^4$, with fibre S^3.

21. Let G be a topological group, except that the associative law is weakened to: $(gh)h^{-1} = g$, for all $g, h \in G$. Show that, provided G is locally compact and Hausdorff, $q: G * G \to SG$ is still a local product, with fibre G.

The *Cayley numbers* are, as an additive group, the direct sum $H \oplus H$ of two copies of the quaternions, and multiplication is defined by $(h_1, h_2).(k_1, k_2) = (h_1k_1 - \bar{k}_2h_2, k_2h_1 + h_2\bar{k}_1)$. The *conjugate* of (h_1, h_2) is $\overline{(h_1, h_2)} = (\bar{h}_1, -h_2)$: show that $(h_1, h_2).\overline{(h_1, h_2)} = (|h_1|^2 + |h_2|^2, 0)$ and is $(0, 0)$ if and only if $(h_1, h_2) = (0, 0)$. Hence define the *modulus* $|(h_1, h_2)| = (|h_1|^2 + |h_2|^2)^{1/2}$, and prove that the Cayley numbers of unit modulus form a group under multiplication, except that the associative law is weakened as above.

By identifying S^7 with this 'group', show that there is a local product $S^{15} \to S^8$, with fibre S^7. Deduce that

$$\pi_r(S^8) \cong \pi_r(S^{15}) \oplus \pi_{r-1}(S^7).$$

22. Let G be a topological group, and let $K \subset H$ be closed subgroups (that is, subgroups that are closed subspaces). Write G/H for the set of left cosets gH, topologized so that the quotient function $p: G \to G/H$ is an identification map.

The map p is said to have a *local cross-section* if there exists an open neighbourhood U of the point (H) in G/H and a map $f: U \to G$ such that $pf = 1_U$. Prove that, if p has a local cross-section, then the identification map $q: G/K \to G/H$ is a local product, with fibre H/K. (*Hint:* consider the open covering of G/H by open sets gU, for all $g \in G$, and define $\phi: gU \times H/K \to q^{-1}(gU)$ by $\phi(x, y) = g.f(g^{-1}x).y$.)

23. The *orthogonal group* $O(n)$ is the group of real $(n \times n)$ matrices A such that $AA' = I$, topologized as a subspace of R^{n^2}. Show that $O(n)$ is a topological group, and that if $O(n - 1)$ is regarded as the subgroup of matrices (a_{ij}) such that $a_{nn} = 1$, $a_{in} = a_{nj} = 0$ otherwise, then $O(n - 1)$ is a closed subgroup.

By identifying left cosets of $O(n - 1)$ with the last column of a representative matrix in $O(n)$, show that $O(n)/O(n - 1)$ is homeomorphic to S^{n-1}. Show also that the identification map $p: O(n) \to S^{n-1}$ has a local cross-section, by the following method. Given $(x_1, \dots, x_n) \in S^{n-1}$, with $x_n \neq 1$, regard $(x_1, \dots, x_n)$ as a column vector x; let e_r be the column vector with 1 in the rth place and 0's elsewhere, and define

$$f_r = e_r - (x_r/1 + x_n)(x + e_n) \qquad (1 \leqslant r \leqslant n - 1)$$
$$f_n = x;$$

now prove that $f(x) = (f_1, \dots, f_n)$ defines a local cross-section.

Deduce that the following identification maps are local products.

(a) $p: O(n) \to S^{n-1}$, with fibre $O(n - 1)$.
(b) $p: SO(n) \to S^{n-1}$, with fibre $SO(n - 1)$, where $SO(n)$ is the subgroup of $O(n)$ of matrices with determinant 1.
(c) $p: V_{n,k} \to S^{n-1}$, with fibre $V_{n-1,k-1}$, where $V_{n,k} = O(n)/O(n - k)$ (this space is called a *Stiefel manifold*).

Finally, show that $\pi_r(O(n)) \cong \pi_r(O(n + 1))$ for $r < n - 1$, with similar isomorphisms involving $SO(n)$ and $V_{n,k}$.

24. Consider S^3 as the topological group of quaternions of unit modulus, and regard S^2 as the subspace consisting of quaternions of the form $(0, b, c, d)$. Show that, for each $x \in S^3$, the map

$$y \to x.y.x^{-1}$$

is a linear map that sends S^2 into itself. Deduce that there is a map $h: S^3 \to SO(3)$, that induces a homeomorphism between RP^3 and $SO(3)$. Hence prove that $\pi_1(SO(n)) \cong Z_2$, $\pi_2(SO(n)) = 0$, $n \geqslant 3$.

25. The *unitary group* $U(n)$ is the group of complex $(n \times n)$ matrices A such that $A\bar{A}' = I$, and the *symplectic group* $Sp(n)$ is similarly defined, but using the quaternions. Show that $U(n)$ and $Sp(n)$ are topological groups, and that there are local products $U(n) \to S^{2n-1}$, with fibre $U(n-1)$, and $Sp(n) \to S^{4n-1}$, with fibre $Sp(n-1)$. If $SU(n)$ is the subgroup of $U(n)$ of matrices with determinant 1, show also that there is a local product $SU(n) \to S^{2n-1}$, with fibre $SU(n-1)$.

    Deduce that $\pi_r(U(n)) \cong \pi_r(U(n+1))$ and $\pi_r(SU(n)) \cong \pi_r(SU(n+1))$ for $r < 2n$, and that $\pi_r(Sp(n)) \cong \pi_r(Sp(n+1))$ for $r < 4n+2$. Show also that $\pi_1(U(n)) \cong Z$, $\pi_2(U(n)) = 0$ and $\pi_3(U(n)) \cong Z$, for $n \geqslant 2$, and that $\pi_1(Sp(n)) = \pi_2(Sp(n)) = 0$ and $\pi_3(Sp(n)) \cong Z$, for $n \geqslant 1$.

26. Given elements $\alpha \in \pi_{n+r}(S^n)$, $\beta \in \pi_{n+s}(S^n)$, show that the Whitehead product $[\alpha, \beta]$ is zero if either α or β is in the image of $p_*: \pi_*(SO(n+1)) \to \pi_*(S^n)$, where p is as in Exercise 23 (in this question, $(0, \ldots, 0, 1)$ is taken as the base point of S^n). Deduce that if the inclusion map $i: SO(n) \to SO(n+1)$ induces a monomorphism $i_*: \pi_{n+r-1}(SO(n)) \to \pi_{n+r-1}(SO(n+1))$, then $[\alpha, \beta] = [\beta, \alpha] = 0$ for all $\alpha \in \pi_{n+r}(S^n)$ and all β whatever.

    Establish similar results using $SU(n)$ and $Sp(n)$ in place of $SO(n)$, and also deduce that $[\alpha, \beta] = 0$ for all $\alpha, \beta \in \pi_*(S^2)$, except when α and β are both in $\pi_2(S^2)$.

NOTES ON CHAPTER 7

CW-complexes. The original definition, and most of the theorems of Section 7.3 (also Exercises 9 and 15) are the work of J. H. C. Whitehead [160]. In particular, Whitehead first proved Theorem 7.3.16(a), although (b) is due to Milnor [104]; for an example of two CW-complexes whose product is *not* a CW-complex, see Dowker [48]. The product $K \bar{\times} L$ was first considered by Spanier [129], though see also Kelley [85], Chapter 7, and R. Brown [32].

Various other constructions can be performed with CW-complexes to yield spaces that have at least the homotopy type of CW-complexes. For example, Milnor [104] proves that $K \times L$ and K^C have the homotopy type of CW-complexes for all CW-complexes K, L and all compact Hausdorff spaces C, and Stasheff [133] proves that, if $f: E \to B$ is a fibre map and B is a CW complex, then E is homotopy-equivalent to a CW-complex if and only if the same is true of the fibre. The special case of this result in which f is a covering map was established earlier by Whitehead [160]: indeed, in this case E actually *is* a CW-complex (*cf.* Exercise 15).

Although clearly not every space is a CW-complex, it is sufficient for many purposes to consider only CW-complexes rather than arbitrary topological spaces. For, by a theorem of J. H. C. Whitehead [161], given any space X, there exists a CW-complex K and a weak homotopy equivalence $f: K \to X$.

Calculation theorems. The result that $\pi_n(S^n) \cong Z$ is due to Brouwer [25] and Hopf [66], and Theorem 7.4.15 to J. H. C. Whitehead [156].

Section 7.5. Like much of the rest of this chapter, this is largely the work of J. H. C. Whitehead: Theorem 7.5.4 and 7.5.7 first appeared in [160], and Theorem 7.5.8 in [156].

The Hopf fibrings. The local products $S^3 \to S^2$, with fibre S^1, $S^7 \to S^4$, with fibre S^3, and $S^{15} \to S^8$, with fibre S^7, discussed in Exercises 20 and 21, were first discovered by Hopf [69, 70]. It might be supposed that these were but the first of a series of local products $S^{2^{n+1}-1} \to S^{2^n}$, with fibre S^{2^n-1}; but the result of Adams [2], mentioned in the notes on Chapter 6, shows that such local products exist only in the cases $n = 1$, 2 and 3.

Topological groups, local cross-sections, and homotopy groups. (Exercises 22–25.) For more details of these topics, see Steenrod [137]. The map $p: G \to G/H$, considered in Exercise 22, nearly always *has* a local cross-section: this is proved by Chevalley [40], Chapter 4, in the case where G is a *Lie group*, and in a more general situation by Mostert [108].

Many of the homotopy groups of the topological groups considered in Exercises 23–25 are known. For the groups $\pi_r(O(n))$ $(r < n - 1)$, $\pi_r(U(n))$ $(r < 2n)$ and $\pi_r(Sp(n))$ $(r < 4n + 2)$, see Bott [22]; many of the groups outside these ranges of dimensions have been calculated by Barratt and Mahowald [20]. The closely related homotopy groups of Stiefel manifolds have been investigated by Paechter [113].

Whitehead products. Exercise 26 is due to S. Thomeier.

Suggestions for further reading. One of the most important (and as yet unsolved) problems of homotopy theory is the calculation of the groups $\pi_r(S^n)$. There are two main lines of attack: the first is based on the EHP sequence of G. W. Whitehead [154], and has been exploited most fully by Toda [145]; the second attempts to calculate $\pi_r(S^n)$ only for $r < 2n - 1$ (when, by a theorem of Freudenthal [59], the groups depend only on $r - n$), and uses an algebraic machine known as the Adams spectral sequence (see Adams [1, 5]). Much work has been done on the latter method: see for example May [101, 102], Maunder [97, 98], Mahowald [93], and Mahowald and Tangora [95]. An attempt has been made to extend the Adams spectral sequence outside the range of dimensions $r < 2n - 1$ (see [23, 121]); this method also generalizes the EHP sequence.

A related problem is the determination of the image of the *J-homomorphism* $J: \pi_r(SO(n)) \to \pi_{r+n}(S^n)$ (for the definition, see G. W. Whitehead [152]). This problem has been almost completely solved by Adams [4] (see also Mahowald [94]).

CHAPTER 8

HOMOLOGY AND COHOMOLOGY OF CW-COMPLEXES

8.1 Introduction

This final chapter is concerned with various topics in the homology and cohomology theory of CW-complexes. We start by showing, in Section 8.2, that the homology and cohomology groups can be calculated directly from the cellular structure, using the cells in the same way that the simplexes are used in the simplicial homology groups of a polyhedron. This is the important basic result of this chapter, and in particular we shall see in Section 8.3 that it leads to a straightforward proof (for CW-complexes) of the theorem of Hurewicz that relates homotopy and homology groups.

In Section 8.4 we shall see how cohomology theory fits into the general scheme of Chapter 6. This will be done by showing that the cohomology groups of a CW-complex can be identified with the groups of homotopy classes of maps into Eilenberg–MacLane spaces; thus cohomology groups are 'dual' to homotopy groups, at least for CW-complexes. We shall also investigate more general 'cohomology theories', obtained by replacing the Eilenberg–MacLane spaces by other spaces.

Finally, in Section 8.5 cohomology theory will be sharpened by introducing a ring structure. As in the case of the Hurewicz theorem, it is possible to carry out the work for arbitrary spaces, but we shall confine our attention to CW-complexes, since the results of Section 8.2 will then greatly simplify the proofs.

8.2 The Excision Theorem and cellular homology

The chief aim of this section is to generalize to CW-complexes the result of Chapter 4 that $H_*(|K|) \cong H(C(K))$ for a simplicial complex K. It would be tempting to try to do so by the method of Section 4.3, that is, by showing first that the homology of a CW-complex K can be calculated from a subchain complex $\Delta(K)$ of $S(K)$, generated by the *cellular* maps from Δ_n to K, and secondly by taking a quotient of $\Delta(K)$ whose generators are in (1-1)-correspondence with the cells of K.

Now it is fairly easy to carry out the first of these steps, by using the Cellular Approximation Theorem in place of the Simplicial Approximation Theorem in the arguments of Section 4.3; but the second step would be much more difficult, because there is no coherent way of identifying the characteristic maps $\phi_\alpha^n\colon E^n \to K$ with cellular maps of Δ_n.

Because of these difficulties, we shall not pursue this line of attack. However, it is possible to establish the result we want by a different approach, based on the proof of Theorem 4.4.14: it will be proved that the homology groups of a CW-complex K can be calculated from a chain complex C in which $C_n = H_n(K^n, K^{n-1})$, and that C_n is a free abelian group whose generators are in (1-1)-correspondence with the n-cells of K. To make this argument work, it will be necessary to know $H_*(K^n, K^{n-1})$, and to this end we shall prove that $H_*(K, L) \cong \tilde{H}_*(K/L)$ for any CW-pair (K, L): since by Example 7.3.15 K^n/K^{n-1} is a one-point union of S^n's, this will suffice to calculate $H_*(K^n, K^{n-1})$. In turn, the result that $H_*(K, L) \cong \tilde{H}_*(K/L)$ is a corollary of the Excision Theorem, which is the analogue for arbitrary spaces of Example 4.3.6 for polyhedra.

Theorem 8.2.1 (The Excision Theorem.) *Let A and B be subspaces of a space X, and suppose that there exist open sets U, V, in X, such that $U \subset A$, $V \subset B$, and $X = U \cup V$. Then*

$$i_*\colon H_n(B, A \cap B) \to H_n(X, A)$$

is an isomorphism for all n, where i is the inclusion map.

Proof. We show first that every element of $H_*(X, A)$ can be represented by a linear combination of singular simplexes that are maps into either U or V. This is done by using a modification of the subdivision chain map ϕ of Definition 4.3.2.

Given any space X, define a homomorphism $\psi\colon S_n(X) \to S_n(X)$ (for each n) by the formula $\psi(\lambda) = \lambda_.\phi(1_n)$, where 1_n denotes the identity map of Δ_n, and $\phi\colon \Delta_n(K(\Delta_n)) \to \Delta_n((K(\Delta_n))')$ is the subdivision chain map. Now ψ is a chain map, because

$$\begin{aligned} \partial\psi(\lambda) &= \lambda_.\phi\partial(1_n) \\ &= \sum(-1)^r\lambda_.\phi F^r_.(1_{n-1}) \\ &= \sum(-1)^r(\lambda F^r)_.\phi(1_{n-1}) \\ &= \psi\partial(\lambda). \end{aligned}$$

(It is easy to see that $\phi F^r_.(1_{n-1}) = F^r_.\phi(1_{n-1})$, since F^r is (1-1) and simplicial.)

Moreover ψ is chain-homotopic to the identity chain isomorphism. The proof of this is exactly like that of Proposition 4.3.4: we construct suitable elements $y_{n+1} \in \Delta_{n+1}(M_n)$, where M_n is a triangulation of $\Delta_n \times I$ that has $K(\Delta_n)$ at the '0 end' and $(K(\Delta_n))'$ at the '1 end'. The details are left to the reader.

Clearly ψ can be extended to a chain map $\psi: S_n(X, A) \to S_n(X, A)$, which is chain-homotopic to the identity; also, ψ can be iterated. Now if λ is a singular n-simplex in X, the sets $\lambda^{-1}(U)$, $\lambda^{-1}(V)$ form an open covering of Δ_n; hence by Theorem 1.4.35 and Proposition 2.5.15 there exists an integer r such that $\psi^r(\lambda)$ is a linear combination of singular n-simplexes that map into either U or V. Clearly also the integer r can be chosen to have this property for a finite number of λ's simultaneously.

Now consider $x \in H_n(X, A)$, represented by $z \in Z_n(X, A)$. Since ψ is chain-homotopic to the identity, z and $\psi^r(z)$ differ by a boundary for each r, and so $\psi^r(z)$ also represents x. By choosing r large enough we can ensure that $\psi^r(z)$ is a linear combination of singular simplexes that each map into either U or V, and hence into either A or B; but then $\psi^r(z) \in Z_n(B, A \cap B)$. This proves that $i_*: H_n(B, A \cap B) \to H_n(X, A)$ is onto.

To show that i_* is also (1-1), consider $z \in Z_n(B, A \cap B)$ such that $i(z) \in B_n(X, A)$. Then z, regarded as a linear combination of singular simplexes in X, is of form $\partial x + y$, where $x \in S_{n+1}(X)$ and $y \in S_n(A)$. Choose r such that every singular simplex in $\psi^r(x)$ maps into U or V; thus $\psi^r(x) = a + b$, where $a \in S_{n+1}(A)$ and $b \in S_{n+1}(B)$. Hence $\psi^r(z) = \psi^r \partial x + \psi^r y$ and $\psi^r(z) - \partial b = \partial a + \psi^r(y)$. But $\psi^r(z) - \partial b \in S_n(B)$ and $\partial a + \psi^r(y) \in S_n(A)$, so that in fact both are in $S_n(A \cap B)$. It follows that $\psi^r(z) = \partial b + [\partial a + \psi^r(y)] \in B_n(B, A \cap B)$, so that $\psi^r(z)$, and hence z, represents the zero element of $H_n(B, A \cap B)$. Thus i_* is (1-1). ∎

An obvious modification of the above argument shows that, for any coefficient group G, $i_*: H_n(B, A \cap B; G) \to H_n(X, A; G)$ is an isomorphism. However, the corresponding result in cohomology does not follow quite so easily, since a representative cycle for an element of, say, $H^n(X, A; G)$ may be non-zero on an infinite number of singular n-simplexes, and so the argument involving ψ^r may not work. On the other hand the cohomology version can be deduced directly from the homology version by means of the following proposition.

Proposition 8.2.2 *Let D be a chain complex in which each D_n is a free abelian group, and is zero for $n < 0$. Let C be a subchain complex such that each D_n/C_n is also free abelian, and the inclusion* *chain map $f: C \to D$ induces an isomorphism $f_*: H(C) \to H(D)$. Then for any abelian group G, $(f \pitchfork 1)_*: H(D \pitchfork G) \to H(C \pitchfork G)$ is also an isomorphism.*

Proof. Write $E = \bigoplus E_n$, where $E_n = D_n/C_n$. By Theorem 4.4.2 we have $H(E) = 0$, and since E is free abelian, Proposition 5.2.8 and Theorem 4.4.2 again will show that $(f \pitchfork 1)_*$ is an isomorphism, provided that we can deduce that $H(E \pitchfork G) = 0$.

This is done by constructing a chain homotopy $h: E_n \to E_{n+1}$, between the identity and the zero chain map, rather as in the proof of Theorem 4.3.9. Suppose as an inductive hypothesis that we have constructed $h: E_r \to E_{r+1}$ for all $r < n$ (take h to be the zero homomorphism if $r < 0$). Then if $x \in E_n$, $\partial(x - h\partial x) = \partial x - \partial h \partial x = 0$, so that, since $Z(E) = B(E)$, there exists $hx \in E_{n+1}$ such that $x = \partial hx + h\partial x$: thus the inductive step is complete. But now $h \pitchfork 1$: $(E \pitchfork G)_{-n-1} \to (E \pitchfork G)_{-n}$ is a chain homotopy between the identity and zero; so we immediately have $H(E \pitchfork G) = 0$. ∎

Corollary 8.2.3 *With the data of Theorem 8.2.1,*

$$i^*: H^n(X, A; G) \to H^n(B, A \cap B; G)$$

is an isomorphism for all n, where G is any abelian group.

Proof. Put $D = S(X, A)$ and $C = S(B, A \cap B)$ in Proposition 8.2.2. ∎

Before deducing results about CW-complexes, we give a generalization of the suspension isomorphism of Theorem 4.4.10. For an arbitrary space X, this involves the 'unreduced suspension' SX, defined as in Corollary 6.2.19 to be the space obtained from $X \times I$ by identifying $X \times 1$ and $X \times 0$ to points (and given a map $f: X \to Y$, the corresponding map $Sf: SX \to SY$ is induced by $f \times 1: X \times I \to Y \times I$).

Theorem 8.2.4 *For each n, there is an isomorphism $s_*: \tilde{H}_n(X) \to \tilde{H}_{n+1}(SX)$, such that, if $f: X \to Y$ is any map, $s_* f_* = (Sf)_* s_*$.*

Proof. If K is a simplicial complex, it is easy to see, as in the proof of Thoerem 4.4.10, that s_* is $(-1)^{n+1}$ times the composite

$$\tilde{H}_n(K) \xleftarrow{\partial_*} H_{n+1}(K * a, K) \xrightarrow{i_*} H_{n+1}(SK, K * b) \xleftarrow{j_*} \tilde{H}_{n+1}(SK),$$

where i_* is an excision isomorphism. For an arbitrary space X, therefore, let s_* be $(-1)^{n+1}$ times the composite

$$\tilde{H}_n(X) \xleftarrow{\partial_*} H_{n+1}(C_1X, X) \xrightarrow{i_*} H_{n+1}(SX, C_0X) \xleftarrow{j_*} \tilde{H}_{n+1}(SX),$$

where C_0X, C_1X are the subspaces of SX corresponding to $X \times [0, \frac{1}{2}]$, $X \times [\frac{1}{2}, 1]$ respectively, and X is identified with $X \times \frac{1}{2}$.

To show that s_* is an isomorphism, observe that ∂_* and j_* are certainly isomorphic since C_1X and C_0X are clearly contractible, so that it is sufficient to prove that i_* is an isomorphism. Unfortunately this is not quite an immediate corollary of Theorem 8.2.1, since the open sets U, V do not exist. However, if we write $\bar{C}X$ for the subspace of SX corresponding to $X \times [\frac{1}{4}, 1]$, there is a commutative diagram

$$\begin{array}{ccc} H_{n+1}(\bar{C}X, X \times [\frac{1}{4}, \frac{1}{2}]) & \xrightarrow{i_*} & H_{n+1}(SX, C_0X) \\ {\scriptstyle i_*}\nwarrow & & \nearrow{\scriptstyle i_*} \\ & H_{n+1}(C_1X, X). & \end{array}$$

The top i_* is an isomorphism by Theorem 8.2.1, and the left-hand i_* is isomorphic since it is induced by an obvious homotopy equivalence; hence the right-hand i_* is isomorphic as well.

That $s_*f_* = (Sf)_* s_*$ is an immediate consequence of Theorem 4.4.3. ∎

Of course, any coefficient group may be used in Theorem 8.2.1, and there is a corresponding isomorphism $s^*: \tilde{H}^{n+1}(SX, G) \to \tilde{H}^n(X; G)$.

If K is a (based) CW-complex, the proof of Corollary 6.2.19 shows that the identification map $p: SK \to sK$ is a homotopy equivalence, so that, by composing with p_*, we may if we wish regard s_* as an isomorphism $s_*: \tilde{H}_n(K) \to \tilde{H}_{n+1}(sK)$ (and if $f: K \to L$ is a (based) map of CW-complexes, $s_*f_* = (f \wedge 1)_* s_*$). This version of s_* may be interpreted directly in terms of reduced cones and suspensions of CW-complexes.

Proposition 8.2.5 *Let K be a* CW-*complex. Then* $s_*: \tilde{H}_n(K) \to \tilde{H}_{n+1}(sK)$ *is* $(-1)^{n+1}$ *times the composite*

$$\tilde{H}_n(K) \xleftarrow{\partial_*} H_{n+1}(cK, K) \xrightarrow{q_*} H_{n+1}(sK, \text{point}) \xleftarrow{j_*} \tilde{H}_{n+1}(sK),$$

where $q: cK \to sK$ is the obvious identification map.

Proof. Consider the map $\phi: SK \to sK$ defined by

$$\phi(x, t) = \begin{cases} \text{base point}, & 0 \leqslant t \leqslant \frac{1}{2} \\ x \wedge (2t-1), & \frac{1}{2} \leqslant t \leqslant 1. \end{cases}$$

Clearly ϕ is homotopic to p, and maps C_1K onto cK and C_0K to the base point; our result therefore follows immediately from the commutative diagram

$$\begin{array}{ccccccc} & \overset{\partial_*}{\swarrow} & H_{n+1}(C_1K, K) & \xrightarrow{i_*} & H_{n+1}(SK, C_0K) & \xleftarrow{j_*} & \tilde{H}_{n+1}(SK) \\ \tilde{H}_n(K) & & \downarrow \phi_* & & \downarrow \phi_* & & \downarrow p_* \\ & \underset{\partial_*}{\nwarrow} & H_{n+1}(cK, K) & \xrightarrow[q_*]{} & H_{n+1}(sK, \text{point}) & \xleftarrow[j_*]{} & \tilde{H}_{n+1}(sK). \end{array}$$ ∎

We show next that if (K, L) is a (based) CW-pair, the sequence of spaces and maps

$$L \xrightarrow{i} K \xrightarrow{i_1} C_i \xrightarrow{i_2} sK \longrightarrow \dots$$

of Theorem 6.4.7 induces an exact sequence of homology groups, which can be identified with the exact homology sequence of (K, L).

Theorem 8.2.6 *There is an isomorphism* $\alpha: H_n(K, L) \to \tilde{H}_n(C_i)$, *such that, in the diagram*

$$\begin{array}{ccccc} \tilde{H}_n(K) & \xrightarrow{j_*} & H_n(K, L) & \xrightarrow{\partial_*} & \tilde{H}_{n-1}(L) \\ = \downarrow & & \downarrow \alpha & & \downarrow s_* \\ \tilde{H}_n(K) & \xrightarrow[(i_1)_*]{} & \tilde{H}_n(C_i) & \xrightarrow[(i_2)_*]{} & \tilde{H}_n(sL), \end{array}$$

we have $\alpha j_* = (i_1)_*$ *and* $s_* \partial_* = (-1)^{n+1}(i_2)_* \alpha$.

Proof. Let α be the composite

$$H_n(K, L) \xrightarrow{(i_1)_*} H_n(C_i, cL) \xleftarrow{j_*} \tilde{H}_n(C_i).$$

Certainly j_* is an isomorphism, and so also is $(i_1)_*$: this would be true as in the proof of Theorem 8.2.4 if the subspace cL of C_i were replaced by the 'unreduced cone' $(L \times I)/(L \times 1)$; but the identification of (base point) $\times I$ to a point makes no difference, by Corollary 6.2.7. That $\alpha j_* = (i_1)_*$ follows immediately from Theorem 4.4.3.

To prove that $s_* \partial_* = (-1)^{n+1}(i_2)_* \alpha$, consider an element $x \in H_n(K, L)$, represented by a cycle z. Regarded as an element of $\tilde{S}_n(K)$, $\partial z = y$, where $y \in \tilde{S}_{n-1}(L)$. Since cL is contractible, $y = \partial w$ for some $w \in \tilde{S}_n(cL)$; and then $z - w \in \tilde{S}_n(C_i)$ will do as a representative cycle for $\alpha(x)$. Now i_2 shrinks K to the base point, and is $q: cL \to sL$ on cL, so that $(i_2)_* \alpha(x) = [-q(w)]$, at least if we identify $\tilde{H}_n(sK)$ with $H_n(sK, \text{point})$.

On the other hand $\partial_*(x)$ is represented by y, and hence by Proposition 8.2.5 $s_* \partial_*(x) = (-1)^n [q(w)] = (-1)^{n+1}(i_2)_* \alpha(x)$.

Naturally there are corresponding results involving homology or cohomology groups with any coefficients.

Corollary 8.2.7 *Let (K, L) be a CW-pair, and let $\mu: K/L \to C_i$ be the homotopy equivalence of Theorem 6.5.2. Then if $p: K \to K/L$ is the identification map, $p_*: H_n(K, L) \to \tilde{H}_n(K/L)$ is an isomorphism, and in the diagram*

$$\begin{array}{ccccc} \tilde{H}_n(K) & \xrightarrow{j_*} & H_n(K, L) & \xrightarrow{\partial_*} & \tilde{H}_{n-1}(L) \\ {\scriptstyle =}\downarrow & & \downarrow{\scriptstyle p_*} & & \downarrow{\scriptstyle s_*} \\ \tilde{H}_n(K) & \xrightarrow[p_*]{} & \tilde{H}_n(K/L) & \xrightarrow[(i_2\mu)_*]{} & \tilde{H}_n(sL), \end{array}$$

we have $p_*j_* = p_*$ and $s_*\partial_* = (-1)^{n+1}(i_2\mu)_*p_*$.

Proof. It is sufficient to show that $\mu_*p_* = \alpha: H_n(K, L) \to \tilde{H}_n(C_i)$, or equivalently that $p_* = \lambda_*\alpha$, where $\lambda: C_i \to K/L$ is the homotopy inverse to μ in Theorem 6.5.2. But this follows from the commutative diagram

$$\begin{array}{ccccc} H_n(K, L) & \xrightarrow{i_*} & H_n(C_i, cL) & \xleftarrow{j_*} & H_n(C_i, \text{point}) \\ & {\scriptstyle p_*}\searrow & \downarrow{\scriptstyle \lambda_*} & \swarrow{\scriptstyle \lambda_*} & \\ & & H_n(K/L, \text{point}). & & \end{array}$$

∎

Observe that $p_*: H_n(K, L) \to \tilde{H}_n(K/L)$ is an isomorphism even if $L = \varnothing$, provided $K/\varnothing$ is interpreted as K^+, the disjoint union of K with another point.

Example 8.2.8 As remarked after Corollary 6.5.3,

$$\theta_* = (-1)^{n+1}(i_2\mu)_*: \tilde{H}_n(E^n/S^{n-1}) \to \tilde{H}_n(S^n),$$

where θ is the standard homeomorphism. It follows from Corollary 8.2.7 that $\theta_* = s_*\partial_*: H_n(E^n, S^{n-1}) \to \tilde{H}_n(S^n)$, and in particular that $\theta_*(\bar{\sigma}_n) = \sigma_n$, where σ_n and $\bar{\sigma}_n$ are the 'standard generators' of Example 4.4.11. ∎

Now that we have Corollary 8.2.7, we are in a position to prove the main theorem on the homology groups of a CW-complex K. The first step is to calculate $H_*(K^n, K^{n-1})$.

Proposition 8.2.9 *Let (K, L) be a CW-pair, with indexing sets A_n and B_n, and characteristic maps ϕ^n_α. Write $M^n = K^n \cup L$. Then the homomorphism $\phi^n_*: \bigoplus_{A_n - B_n} H_*(E^n_\alpha, S^{n-1}_\alpha) \to H_*(M^n, M^{n-1})$, defined to be $(\phi^n_\alpha)_*$ on each $H_*(E^n_\alpha, S^{n-1}_\alpha)$, is an isomorphism. That is,*

$H_r(M^n, M^{n-1}) = 0$ unless $r = n$, when it is a free abelian group with generators in (1-1)-correspondence with the n-cells of $K - L$.

Proof. Let X be the disjoint union $\bigcup_{A_n - B_n} E^n_\alpha$, and let Y be the corresponding union of the S^{n-1}_α. It is easy to see that

$$\bigoplus_{A_n - B_n} H_*(E^n_\alpha, S^{n-1}_\alpha) \cong H_*(X, Y),$$

where the isomorphism is induced by the inclusion maps of each E^n_α. On the other hand the obvious map $\phi^n: (X, Y) \to (M^n, M^{n-1})$ induces $\phi^n: X/Y \to M^n/M^{n-1}$, which by Example 7.3.15 is a homeomorphism. Hence by Corollary 8.2.7 $\phi^n_*: H_*(X, Y) \to H_*(M^n, M^{n-1})$ is an isomorphism. ∎

Thus we may define a chain complex $C(K, L) = \bigoplus C_n(K, L)$ by setting $C_n(K, L) = H_n(M^n, M^{n-1})$ (where M^n is interpreted as L if $n < 0$), and by taking as boundary homomorphism $\partial: C_n(K, L) \to C_{n-1}(K, L)$ the composite

$$H_n(M^n, M^{n-1}) \xrightarrow{\partial} H_{n-1}(M^{n-1}, L) \xrightarrow{j_*} H_{n-1}(M^{n-1}, M^{n-2})$$

(which is the same as $\partial_*: H_n(M^n, M^{n-1}) \to H_{n-1}(M^{n-1}, M^{n-2})$), or alternatively the composite

$$H_n(M^n, M^{n-1}) \xrightarrow{\partial_*} \tilde{H}_{n-1}(M^{n-1}) \xrightarrow{j_*} H_{n-1}(M^{n-1}, M^{n-2})).$$

Theorem 8.2.10 *For each n, $H_n(K, L) \cong H_n(C(K, L))$.*

Proof. This is almost identical with Theorem 4.4.14. Indeed, we can repeat the proof word-for-word, as far as the statement that

$$H_n(M^p, L) \cong H_n(C(K, L)),$$

where p is any integer greater than n, and it remains only to show that $H_n(K, L) \cong H_n(M^p, L)$ (there was no difficulty in Theorem 4.4.14, since we were dealing only with finite-dimensional complexes).

Now an element of $H_n(K, L)$ is represented by a cycle z, which is a (finite) linear combination of singular n-simplexes in K. Each of these singular simplexes is a map from Δ_n to K, whose image is compact and so contained in a finite subcomplex. Thus z is in fact a cycle of $S_n(M^p, L)$ for some p. Since $i_*: H_n(M^{n+1}, L) \to H_n(M^p, L)$ is an isomorphism for all $p > n$, this means that $H_n(M^{n+1}, L) \to H_n(K, L)$ is onto. But a similar argument shows that if $x \in H_n(M^{n+1}, L)$ is sent to zero in $H_n(K, L)$, then it must be sent to zero in some $H_n(M^p, L)$, and so $x = 0$; thus $H_n(M^{n+1}, L) \to H_n(K, L)$ is (1-1) as well. ∎

Corollary 8.2.11 *If $f: (K, L) \to (P, Q)$ is a cellular map of CW-pairs, $f_*: H_n(K, L) \to H_n(P, Q)$ is the homomorphism induced by the chain map $f_*: H_n(M^n, M^{n-1}) \to H_n(R^n, R^{n-1})$, where $R^n = P^n \cup Q$. In particular, if A and B are subcomplexes of K, such that $A \cup B = K$, then $i_*: H_n(B, A \cap B) \to H_n(K, A)$ is an isomorphism for all n.*

Proof. Since all the homomorphisms in the proof of Theorem 8.2.10 are homomorphisms in the exact sequences of triples, this follows immediately from the remark after Theorem 4.4.5 (which also shows that f_* *is* a chain map). ∎

For many purposes, it is convenient to have a more geometrical interpretation of the boundary homomorphism $\partial: C_n(K, L) \to C_{n-1}(K, L)$ and the chain map $f_*: C(K, L) \to C(P, Q)$. Now if $\bar{\sigma}_n$ is the standard generator of $H_n(E^n, S^{n-1})$, Proposition 8.2.9 shows that $C_n(K, L)$ may be identified with the free abelian group with generators the elements of $A_n - B_n$, by letting $\alpha \in A_n - B_n$ correspond to $(\phi_\alpha^n)_* \bar{\sigma}_n \in H_n(M^n, M^{n-1})$. Given $\alpha \in A_n - B_n$ and $\beta \in A_{n-1} - B_{n-1}$, let $d_{\alpha\beta}$ be the composite map

$$S^{n-1} \xrightarrow{\phi_\alpha^n} M^{n-1} \xrightarrow{p} M^{n-1}/M^{n-2} \xleftarrow{\phi^n} \bigvee (E^{n-1}/S^{n-2}) \xrightarrow{q_\beta} E^{n-1}/S^{n-2} \xrightarrow{\theta} S^{n-1},$$

where ϕ^n is the homeomorphism of Example 7.3.15 and q_β is the projection map corresponding to β.

Proposition 8.2.12 $\partial(\alpha) = \sum_{A_{n-1} - B_{n-1}} \partial_{\alpha\beta} \cdot \beta$, *where $\partial_{\alpha\beta}$ is the degree of $d_{\alpha\beta}$.*

Proof. The commutative diagram

$$\begin{array}{ccccc} H_n(M^n, M^{n-1}) & \xrightarrow{\partial_*} & \tilde{H}_{n-1}(M^{n-1}) & \xrightarrow{j_*} & H_{n-1}(M^{n-1}, M^{n-2}) \\ \uparrow{\scriptstyle (\phi_\alpha^n)_*} & & \uparrow{\scriptstyle (\phi_\alpha^n)_*} & & \\ H_n(E^n, S^{n-1}) & \xrightarrow[\partial_*]{} & \tilde{H}_{n-1}(S^{n-1}) & & \end{array}$$

shows that $\partial(\phi_\alpha^n)_* \bar{\sigma}_n = j_*(\phi_\alpha^n)_* \sigma_{n-1}$. But by Corollary 8.2.7 j_* may be identified with $p_*: \tilde{H}_{n-1}(M^{n-1}) \to \tilde{H}_{n-1}(M^{n-1}/M^{n-2})$, and by Proposition 8.2.9, if $x \in \tilde{H}_{n-1}(M^{n-1}/M^{n-2})$, then

$$x = \sum (\phi_\beta^{n-1})_*(q_\beta)_*(\phi_*^{n-1})^{-1}(x).$$

Hence

$$\begin{aligned} p_*(\phi^n_\alpha)_*\sigma_{n-1} &= \sum (\phi^{n-1}_\beta)_*\theta_*^{-1}\theta_*(q_\beta)_*(\phi^{n-1}_*)^{-1}p_*(\phi^n_\alpha)_*\sigma_{n-1} \\ &= \sum (\phi^{n-1}_\beta)_*\theta_*^{-1}\partial_{\alpha\beta}\sigma_{n-1} \\ &= \sum \partial_{\alpha\beta}(\phi^{n-1}_\beta)_*\bar{\sigma}_{n-1}, \end{aligned}$$

using Example 8.2.8. That is, $\partial(\alpha) = \sum \partial_{\alpha\beta}.\beta$. ∎

Similarly, let $f\colon (K, L) \to (P, Q)$ be a cellular map, let the indexing sets for P, Q be C_n, D_n, and the characteristic maps for P be ψ^n_β. Given $\alpha \in A_n - B_n$ and $\beta \in C_n - D_n$, let $f_{\alpha\beta}$ be the composite

$$S^n \xleftarrow{\theta} E^n/S^{n-1} \xrightarrow{\phi^n_\alpha} M^n/M^{n-1} \xrightarrow{f} R^n/R^{n-1} \xleftarrow{\psi^n} \bigvee (E^n/S^{n-1}) \xrightarrow{q_\beta} E^n/S^{n-1} \xrightarrow{\theta} S^n,$$

where $R^n = P^n \cup Q$.

Proposition 8.2.13 *f_* is induced by $f_.$, where $f_.(\alpha) = \sum d(f_{\alpha\beta})\beta$, $d(f_{\alpha\beta})$ being the degree of $f_{\alpha\beta}$.*

Proof. This is proved in a similar way to Proposition 8.2.12, using Corollary 8.2.11. We omit the details. ∎

In this discussion of CW-complexes, we have not so far mentioned homology with other coefficient groups, or cohomology. What we should like to know, of course, is that if G is any coefficient group, then

$$H_*(K, L; G) \cong H(C(K, L) \otimes G)$$

and

$$H^*(K, L; G) \cong H(C(K, L) \pitchfork G).$$

The next two theorems establish these isomorphisms.

Theorem 8.2.14 *Let (K, L) be a CW-pair, and G be an abelian group. Then for each n, $H_n(K, L; G) \cong H_n(C(K, L) \otimes G)$.*

Proof. Write C for $C(K, L)$ and $C(G)$ for the chain complex similarly defined by $C(G)_n = H_n(M^n, M^{n-1}; G)$. Now the proof of Theorem 8.2.10 clearly adapts to show that $H_n(K, L; G) \cong H_n(C(G))$, so that it remains only to produce a chain isomorphism $\alpha\colon C \otimes G \to C(G)$ (compare Chapter 4, Exercise 13).

Indeed, for any chain complex D whatever, we can define a homomorphism $\alpha\colon H(D) \otimes G \to H(D \otimes G)$ by $\alpha([z] \otimes g) \to [z \otimes g]$, where $z \in Z(D)$ and $g \in G$. Moreover if $f\colon D \to E$ is a chain map, then $(f \otimes 1)_*\alpha = \alpha(f_* \otimes 1)$, and there is a corresponding result for the homomorphism ∂_* in the exact sequence of Theorem 4.4.2. Thus by taking $D = S(M^n, M^{n-1})$ for various n, this gives rise to a chain map $\alpha\colon C \otimes G \to C(G)$, and we have only to show that

$$\alpha\colon H_*(M^n, M^{n-1}) \otimes G \to H_*(M^n, M^{n-1}; G)$$

is an isomorphism. By Proposition 8.2.9 it is enough to consider $(M^n, M^{n-1}) = (E^n, S^{n-1})$; and by Theorem 4.3.9 we may use the simplicial chain complex of $(K(\sigma), \dot{\sigma})$ instead of $S(E^n, S^{n-1})$, where σ is an n-simplex. But the result is now trivial, because this simplicial chain complex has only one generator. Hence $\alpha\colon C \otimes G \to C(G)$ is a chain isomorphism. ▮

Theorem 8.2.15 *For each n,* $H^n(K, L; G) \cong H_{-n}(C(K, L) \pitchfork G)$.

Proof. For any chain complex D, define $\alpha\colon H(D \pitchfork G) \to H(D) \pitchfork G$ by $(\alpha(x))y = \langle x, y\rangle$, where $x \in H(D \pitchfork G)$, $y \in H(D)$, and $\langle\ ,\ \rangle$ is the Kronecker product. By an argument similar to that in Theorem 8.2.14, α is an isomorphism if $D = S(X, Y)$, where X is a disjoint union of E^n's and Y is the corresponding union of S^{n-1}'s, and hence as in Proposition 8.2.9 α is an isomorphism if $D = S(M^n, M^{n-1})$. That is, $\alpha\colon C(G) \to C \pitchfork G$ is a chain isomorphism, where $C(G)_{-n} = H^n(M^n, M^{n-1}; G)$.

It remains to prove that $H^n(K, L; G) \cong H_{-n}(C(G))$. Now the argument of Theorem 8.2.10 will show that $H^n(M^p, L; G) \cong H_{-n}(C(G))$ for any integer $p > n$, but we cannot use the rest of that argument, since a representative cycle for an element of $H^n(K, L; G)$ may well be non-zero on an infinite number of singular n-simplexes. However, since $H_n(M^{n+1}, L) \cong H_n(K, L)$, we can immediately conclude that $H^n(K, L; G) \cong H^n(M^{n+1}, L; G)$, by Proposition 8.2.2. ▮

We end this section by calculating the homology and cohomology groups of some CW-complexes.

Examples 8.2.16

(a) The homology and cohomology of real projective spaces has already been calculated, in Examples 4.4.25 and 5.2.18. It is even easier to deal with complex and quaternionic projective spaces.

By Chapter 7, Exercise 8, CP^n is a CW-complex with one cell in each dimension $0, 2, \ldots, 2n$. Since $C_r(CP^n)$ is zero in alternate dimensions, $\partial = 0$, and Theorem 8.2.10 immediately yields

$$H_r(CP^n) \cong \begin{cases} Z, & r = 0, 2, \ldots, 2n \\ 0, & \text{otherwise.} \end{cases}$$

Similarly

$$H_r(HP^n) \cong \begin{cases} Z, & r = 0, 4, \ldots, 4n \\ 0, & \text{otherwise;} \end{cases}$$

and homology and cohomology with coefficients G are given by replacing all Z's by G's.

(b) Consider $S^p \times S^q$. Now S^p and S^q are CW-complexes with one 0-cell each, and one cell of dimension p, q respectively; hence

$$C_0(S^p \times S^q) \cong C_p(S^p \times S^q) \cong C_q(S^p \times S^q) \cong C_{p+q}(S^p \times S^q) \cong Z,$$

the other groups C_r being zero (if $p = q$, $C_p(S^p \times S^p) \cong Z \oplus Z$). Proposition 8.2.12 shows easily that all boundary homomorphisms are zero (if say $q = p + 1$, we need Chapter 6, Exercise 15 as well to show that $\partial: C_{p+1} \to C_p$ is zero). It follows that $H_r(S^p \times S^q) \cong C_r(S^p \times S^q)$ for all r, and also

$$H_r(S^p \times S^q; G) \cong C_r(S^p \times S^q) \otimes G,$$
$$H^r(S^p \times S^q; G) \cong C_r(S^p \times S^q) \pitchfork G. \quad \blacksquare$$

It will be noticed that $S^2 \times S^4$ and CP^3 have the same homology and cohomology groups, so that it is possible that they are homotopy-equivalent. In fact they are not, but the cohomology ring structure is necessary to prove this: see Example 8.5.12.

8.3 The Hurewicz Theorem

The Hurewicz Theorem states that, if X is a path-connected space and $\pi_r(X) = 0$ for $r < n$, then $\pi_n(X) \cong H_n(X)$ $(n \geqslant 2)$; there is also a corresponding version for relative homotopy and homology groups. Apart from the analogous theorem relating $\pi_1(X)$ and $H_1(X)$, we shall give a proof only for CW-complexes, since the proof for arbitrary spaces is more complicated (see the notes at the end of the chapter).

The method of proof is somewhat similar to that used in Section 7.4 to calculate $\pi_n(S^n)$: we first define homomorphisms $h_n: \pi_n(X) \to H_n(X)$, $h_n: \pi_n(X, Y) \to H_n(X, Y)$ (that generalize the notion of *degree*), and then show that they are isomorphisms in favourable circumstances.

Definition 8.3.1 The *Hurewicz homomorphism* $h_n: \pi_n(X) \to H_n(X)$ $(n \geqslant 1)$ is defined as follows. Let $\sigma_n \in H_n(S^n)$ be the standard generator; then if $[f] \in \pi_n(X)$ is represented by a map $f: S^n \to X$, define $h_n[f] = f_*(\sigma_n)$. Clearly this is independent of the representative map f chosen.

Similarly, an element of $\pi_n(X, Y)$ is represented by a map $f: (I^n, \partial I^n) \to (X, Y)$, and by using the standard homeomorphism this may be regarded as a map $f: (E^n, S^{n-1}) \to (X, Y)$. Thus we may define $h_n: \pi_n(X, Y) \to H_n(X, Y)$ by $h_n[f] = f_*(\bar{\sigma}_n)$.

Observe that if Y is the base point x_0 of X, and the relative homotopy group $\pi_n(X, x_0)$ is identified with $\pi_n(X)$ via the standard homeomorphism $\theta: E^n/S^{n-1} \to S^n$, then the two definitions of h_n coincide, since by Example 8.2.8 we have $\theta_*(\bar{\sigma}_n) = \sigma_n$.

Proposition 8.3.2 *h_n is a homomorphism if $n \geqslant 1$ ($n \geqslant 2$ in the relative case).*

Proof. The proof of Proposition 7.4.2 easily extends to show that $h_n: \pi_n(X) \to H_n(X)$ is a homomorphism for $n \geqslant 1$, since if $f, g: S^n \to X$ and $x \oplus y \in H_n(S^n) \oplus H_n(S^n) \cong H_n(S^n \vee S^n)$, we have $[\nabla(f \vee g)]_*(x \oplus y) = f_*(x) + g_*(y)$. As for $h_n: \pi_n(X, Y) \to H_n(X, Y)$, the diagram

$$\begin{array}{ccccc} \pi_n(X, Y) & \xrightarrow{i_*} & \pi_n(C_i, cY) & \xleftarrow{j_*} & \pi_n(C_i) \\ \downarrow h_n & & \downarrow h_n & & \downarrow h_n \\ H_n(X, Y) & \xrightarrow[i_*]{} & H_n(C_i, cY) & \xleftarrow[j_*]{} & H_n(C_i) \end{array}$$

is easily seen to be commutative, where C_i is the mapping cone of $i: Y \to X$. Since cY is contractible and $n \geqslant 2$, both j_*'s are isomorphisms; but the lower row is just the isomorphism α of Theorem 8.2.6. Hence $h_n: \pi_n(X, Y) \to H_n(X, Y)$ is a composite of homomorphisms, and thus is itself a homomorphism. ∎

Theorem 8.3.3 *If X is a path-connected space, $h_1: \pi_1(X) \to H_1(X)$ is onto, with kernel the commutator subgroup $[\pi, \pi]$ of $\pi_1(X)$.*

Proof. It is easy to see that the standard map $\theta l: I \to S^1$ is a singular 1-simplex that represents the generator σ_1 of $H_1(S^1)$. Hence h_1 may be regarded as the homomorphism induced by sending the loop u in X (based at x_0) to the singular 1-simplex u in X.

Now let $\sum n_i \lambda_i$ be an element of $Z_1(X)$, where the n_i are integers and the λ_i are singular 1-simplexes. For each point $x \in X$, choose a path $v(x)$ from x_0 to x, and let u_i be the product path

$$v(\lambda_i(0)) . \lambda_i . v(\lambda_i(1))^{-1}.$$

Since $\partial(\sum n_i \lambda_i) = 0$, we have $\sum n_i \lambda_i = \sum n_i\{v(\lambda_i(0)) + \lambda_i - v(\lambda_i(1))\}$, and as in the proof of Theorem 8.2.1 this represents the same element of $H_1(X)$ as does $\sum n_i u_i$. But the coset $[\sum n_i u_i] = \sum n_i[u_i]$ is plainly in the image of h_1, so that h_1 is onto.

Since $H_1(X)$ is abelian, h_1 induces a homomorphism $h\colon \pi_1(X)/[\pi, \pi] \to H_1(X)$, and to show that $\operatorname{Ker} h_1 = [\pi, \pi]$ it is sufficient to prove that h is (1-1). Suppose then that u is a loop based at x_0, and $u = \partial(\sum n_i \lambda_i)$ in $S(X)$, where the λ_i are singular 2-simplexes: thus $u = F^r \lambda_i$ for some r, i, and all other terms in $\partial(\sum n_i \lambda_i)$ cancel. Now write

$$a_i = v(\lambda_i(0)).F^2\lambda_i.v(\lambda_i(1))^{-1},$$

$$b_i = v(\lambda_i(1)).F^0\lambda_i.v(\lambda_i(2))^{-1},$$

$$c_i = v(\lambda_i(2)).(F^1\lambda_i)^{-1}.v(\lambda_i(0))^{-1},$$

and let $u_i = a_i.b_i.c_i$, $w = (u_1)^n \cdots (u_m)^n$. It is easy to see that $u_i \simeq e_{x_0}$ rel 0, 1 (since Δ_2 is contractible), and hence that w represents the identity element of $\pi_1(X)$. On the other hand, by 'abelianizing' u_i and w, we have $[u] = [w]$ in $\pi_1(X)/[\pi, \pi]$: hence $[u] = 0$ and h is (1-1). ∎

In proving the general Hurewicz theorem for CW-complexes, we shall make extensive use of the fact that the homomorphisms h_n connect the homotopy and homology exact sequences of pairs and triples in diagrams that commute up to sign.

Proposition 8.3.4 *Let (X, Y, Z) be a triple of spaces. Then in the diagram*

$$\begin{array}{ccccccccc}
\cdots \longrightarrow & \pi_n(Y, Z) & \xrightarrow{i_*} & \pi_n(X, Z) & \xrightarrow{j_*} & \pi_n(X, Y) & \xrightarrow{\partial_*} & \pi_{n-1}(Y, Z) & \longrightarrow \cdots \\
& \downarrow h_n & & \downarrow h_n & & \downarrow h_n & & \downarrow h_{n-1} & \\
\cdots \longrightarrow & H_n(Y, Z) & \xrightarrow[i_*]{} & H_n(X, Z) & \xrightarrow[j_*]{} & H_n(X, Y) & \xrightarrow[\partial]{} & H_{n-1}(Y, Z) & \longrightarrow \cdots,
\end{array}$$

the first two squares are commutative and the third commutes up to a sign $(-1)^n$. There is a similar result involving the exact sequences of a pair (X, Y).

Proof. The first two squares are easily seen to be commutative. As for the third, it is sufficient to prove that the square

$$\begin{array}{ccc}
\pi_n(X, Y) & \xrightarrow{\partial_*} & \pi_{n-1}(Y) \\
\downarrow h_n & & \downarrow h_{n-1} \\
H_n(X, Y) & \xrightarrow[\partial_*]{} & H_{n-1}(Y)
\end{array}$$

is commutative up to a sign $(-1)^n$. But if $f: (E^n, S^{n-1}) \to (X, Y)$ represents an element of $\pi_n(X, Y)$, Proposition 7.4.11 shows that $\partial_*[f]$ is $(-1)^n$ times the homotopy class of $f|S^{n-1}$; hence

$$\begin{aligned}\partial_* h_n[f] &= \partial_* f_*(\bar{\sigma}_n)\\ &= f_*\partial_*(\bar{\sigma}_n)\\ &= f_*(\sigma_{n-1})\\ &= (-1)^n h_{n-1}\partial_*[f]. \quad \blacksquare\end{aligned}$$

We shall prove the Hurewicz theorem by examining diagrams such as that in Proposition 8.3.4, based on exact sequences of pairs such as (K^n, K^{n-1}); and in order to make this approach work, two lemmas are necessary.

Lemma 8.3.5 *Let (K, L) be a CW-pair, where L is connected. If $\pi_r(K, L) = 0$ for $1 \leqslant r < n$, $K^{n-1} \cup L$ can be deformed in $K^n \cup L$,* rel L, *into L; more precisely, there exists a homotopy $F: (K^{n-1} \cup L) \times I \to K^n \cup L$, such that F is constant on L, F starts with the inclusion map, and F ends with a map into L.*

Proof. Write $M^n = K^n \cup L$, and consider the exact sequence

$$\cdots \to \pi_{r+1}(K, M^n) \to \pi_r(M^n, L) \to \pi_r(K, L) \to \cdots.$$

Now for $1 \leqslant r < n$, $\pi_r(K, L) = 0$, and also $\pi_{r+1}(K, M^n) = 0$ by Theorem 7.4.17; hence $\pi_r(M^n, L) = 0$ for $1 \leqslant r < n$. The argument used in the proof of Theorem 7.5.2 now shows that we can construct F by induction on the skeletons of K. $\blacksquare$

Lemma 8.3.6 *If L is connected, and contains all the cells of K except for some of dimension n, then $h_n: \pi_n(K, L) \to H_n(K, L)$ is onto $(n \geqslant 2)$. If moreover (K, L) is relatively n-simple, then h_n is an isomorphism.*

Proof. Let the indexing sets for the n cells of K, L be A_n, B_n respectively. Now, as in Proposition 8.2.9, $H_n(K, L)$ is a free abelian group with generators in (1-1)-correspondence with the elements of $A_n - B_n$. If $\alpha \in A_n - B_n$, $\phi_\alpha^n: (E^n, S^{n-1}) \to (K, L)$ may be regarded as a map of $(I^n, \partial I^n)$ to (K, L), and if l_0 is the base point of L, this is homotopic to a map $\psi_\alpha^n: (I^n, \partial I^n) \to (K, L)$ that sends D^{n-1} to l_0: this is proved as in Proposition 7.2.15. But then

$$\begin{aligned}h_n[\psi_\alpha^n] &= (\psi_\alpha^n)_*(\bar{\sigma}_n)\\ &= (\phi_\alpha^n)_*(\bar{\sigma}_n),\end{aligned}$$

which is the generator of $H_n(K, L)$ corresponding to α. Hence h_n is onto.

If (K, L) is n-simple, consider the composite

$$\bigoplus_{A_n - B_n} \pi_n(E^n_\alpha, S^{n-1}_\alpha) \xrightarrow{\phi^n_*} \pi_n(K, L) \xrightarrow{h_n} H_n(K, L),$$

where ϕ^n_* is as in Theorem 7.4.15. Now the homology homomorphism corresponding to ϕ^n_* is an isomorphism; and $h_n: \pi_n(E^n, S^{n-1}) \to H_n(E^n, S^{n-1})$ is an isomorphism by Proposition 7.4.13. Hence $h_n\phi^n_*$ is an isomorphism; but by Theorem 7.4.15 ϕ^n_* is onto ($i_*: \pi_n(K^n, L^n) \to \pi_n(K, L)$ is onto, as in Theorem 7.4.17). Hence ϕ^n_* and therefore h_n are also isomorphic. ∎

Observe that we may now amend the statement of Theorem 7.4.15(b) to read: ϕ^n_* is an isomorphism.

We are now in a position to prove the Hurewicz theorem.

Theorem 8.3.7 *Let K be a connected CW-complex, and let L be a connected subcomplex.*

(a) *If* $\pi_r(K) = 0$ *for* $1 \leqslant r < n$ $(n \geqslant 2)$, *then* $h_n: \pi_n(K) \to H_n(K)$ *is an isomorphism.*

(b) *If* $\pi_1(K) = \pi_1(L) = 0$, *and* $\pi_r(K, L) = 0$ *for* $1 \leqslant r < n$ $(n \geqslant 2)$, *then* $h_n: \pi_n(K, L) \to H_n(K, L)$ *is an isomorphism.*

Proof. Write $M^n = K^n \cup L$. We first remark that it is sufficient to prove the theorem with K replaced by K^{n+1}, M^{n+1} in (a), (b), respectively, since for example if $i: (M^{n+1}, L) \to (K, L)$ is the inclusion map, the diagram

$$\begin{array}{ccc} \pi_n(M^{n+1}, L) & \xrightarrow{i_*} & \pi_n(K, L) \\ \downarrow{\scriptstyle h_n} & & \downarrow{\scriptstyle h_n} \\ H_n(M^{n+1}, L) & \xrightarrow[i_*]{} & H_n(K, L) \end{array}$$

is commutative, and both maps i_* are isomorphic: the lower one by Theorem 8.2.10, and the upper one by Theorem 7.4.17.

Now consider the diagram

$$\begin{array}{ccccccc} \cdots\to \pi_{n+1}(M^{n+1}, M^n) & \to & \pi_n(M^n, L) & \to & \pi_n(M^{n+1}, L) & \to & \pi_n(M^{n+1}, M^n) \\ \downarrow{\scriptstyle h_{n+1}} & & \downarrow{\scriptstyle h_n} & & \downarrow{\scriptstyle h_n} & & \downarrow{\scriptstyle h_n} \\ \cdots\to H_{n+1}(M^{n+1}, M^n) & \to & H_n(M^n, L) & \to & H_n(M^{n+1}, L) & \to & H_n(M^{n+1}, M^n), \end{array}$$

which is commutative, except for a sign $(-1)^{n+1}$ in the first square, by Proposition 8.3.4. Here, $\pi_n(M^{n+1}, M^n) = H_n(M^{n+1}, M^n) = 0$, using Theorem 7.4.17, and h_{n+1} is onto by Lemma 8.3.6, so that the second h_n will be isomorphic if the first is, by Proposition 1.3.35. That is to say, we may even replace M^{n+1} by M^n in (b). A similar argument shows that we may replace K^{n+1} by K^n in (a).

To complete the proof of (a), note first that, by Lemma 8.3.5 (with the base point as L), K^{n-1} can be deformed in K^n, relative to the base point, to the base point. By Theorem 7.3.19 this homotopy can be extended to a (based) homotopy between the identity map of K^n and a map $f\colon K^n \to K^n$ that sends K^{n-1} to the base point. The map f induces $g\colon K^n/K^{n-1} \to K^n$, such that if $p\colon K^n \to K^n/K^{n-1}$ is the identification map, $gp \simeq 1$. This means that there is a commutative diagram

$$\begin{array}{ccccc} \pi_n(K^n) & \xrightarrow{p_*} & \pi_n(K^n/K^{n-1}) & \xrightarrow{g_*} & \pi_n(K^n) \\ \downarrow{\scriptstyle h_n} & & \downarrow{\scriptstyle h_n} & & \downarrow{\scriptstyle h_n} \\ H_n(K^n) & \xrightarrow[p_*]{} & H_n(K^n/K^{n-1}) & \xrightarrow[g_*]{} & H_n(K^n), \end{array}$$

where $g_* p_* = 1$, so that p_* is (1-1) and g_* is onto. But by Lemma 8.3.6 (with L = base point), $h_n\colon \pi_n(K^n/K^{n-1}) \to H_n(K^n/K^{n-1})$ is isomorphic: hence $h_n\colon \pi_n(K^n) \to H_n(K^n)$ is both (1-1) and onto.

It remains to prove (b). In the diagram

$$\begin{array}{ccccccc} \cdots \longrightarrow \pi_2(L) & \xrightarrow{i_*} & \pi_2(M^2) & \xrightarrow{j_*} & \pi_2(M^2, L) & \longrightarrow & 0 \\ \downarrow{\scriptstyle h_2} & & \downarrow{\scriptstyle h_2} & & \downarrow{\scriptstyle h_2} & & \\ \cdots \longrightarrow H_2(L) & \xrightarrow[i_*]{} & H_2(M^2) & \xrightarrow[j_*]{} & H_2(M^2, L) & \longrightarrow & 0, \end{array}$$

the first two maps are isomorphic by case (a), since $\pi_1(M^2) = \pi_1(K) = 0$ by Theorem 7.4.17 (and $H_1(L) = 0$ by Theorem 8.3.3). Hence $h_2\colon \pi_2(M^2, L) \to H_2(M^2, L)$ is isomorphic by Proposition 1.3.35, and this proves (b) in the case $n = 2$. More generally, consider the following diagram, which is commutative up to a sign $(-1)^n$ in the third square.

$$\begin{array}{ccccccc} \cdots \to \pi_n(M^{n-1}, L) & \xrightarrow{i_*} & \pi_n(M^n, L) & \to & \pi_n(M^n, M^{n-1}) & \to & \pi_{n-1}(M^{n-1}, L) \to \cdots \\ \downarrow{\scriptstyle h_n} & & \downarrow{\scriptstyle h_n} & & \downarrow{\scriptstyle h_n} & & \downarrow{\scriptstyle h_{n-1}} \\ \cdots \to H_n(M^{n-1}, L) & \xrightarrow[i_*]{} & H_n(M^n, L) & \to & H_n(M^n, M^{n-1}) & \to & H_{n-1}(M^{n-1}, L) \to \cdots \end{array}$$

If $n \geqslant 3$, then $h_n\colon \pi_n(M^n, M^{n-1}) \to H_n(M^n, M^{n-1})$ is isomorphic by Lemma 8.3.6, since $\pi_1(M^{n-1}) = \pi_1(K) = 0$, and so (M^n, M^{n-1}) is relatively n-simple (we cannot use this argument for $n = 2$, since

$\pi_1(M^1)$ may not vanish). But $i_* = 0$ by Lemma 8.3.5, so that at least $h_n: \pi_n(M^n, L) \to H_n(M^n, L)$ is (1-1). Similarly, h_{n-1} is (1-1), and then Proposition 1.3.35 shows that therefore $h_n: \pi_n(M^n, L) \to H_n(M^n, L)$ is isomorphic. And we have already seen that this is sufficient to prove (b). ∎

Example 8.3.8 If $p, q > 1$, then $\pi_1(S^p \times S^q) = \pi_1(S^p \vee S^q) = 0$. Also, $\pi_r(S^p \times S^q, S^p \vee S^q) = 0$ for $r < p + q$, by Theorem 7.4.15. Hence

$$\pi_{p+q}(S^p \times S^q, S^p \vee S^q) \cong H_{p+q}(S^p \times S^q, S^p \vee S^q) \cong Z,$$

and so we can add to Example 7.4.16 the result

$$\pi_{p+q-1}(S^p \vee S^q) \cong \pi_{p+q-1}(S^p) \oplus \pi_{p+q-1}(S^q) \oplus Z. \ ∎$$

The Hurewicz theorem itself can be used to establish the following alternative form of the hypotheses.

Corollary 8.3.9 *Let (K, L) be as in Theorem 8.3.7. Then*

(a) *If $\pi_1(K) = 0$ and $H_r(K) = 0$ for $r < n$ ($n \geqslant 2$), then $h_n: \pi_n(K) \to H_n(K)$ is an isomorphism.*

(b) *If $\pi_1(K) = \pi_1(L) = 0$, and $H_r(K, L) = 0$ for $r < n$ ($n \geqslant 2$), then $h_n: \pi_n(K, L) \to H_n(K, L)$ is an isomorphism.*

Proof. In (a), since $\pi_1(K) = 0$, we have $\pi_2(K) = H_2(K) = 0$. Hence $\pi_3(K) = H_3(K) = 0$, and so on: in fact $\pi_r(K) = 0$ for $r < n$.

Similarly, in (b) $\pi_2(K, L) = H_2(K, L)$ and so on: hence $\pi_r(K, L) = 0$ for $1 \leqslant r < n$ ($\pi_1(K, L) = 0$ anyway, by the exact homotopy sequence). ∎

One of the most useful corollaries of the Hurewicz theorem is a version of the Whitehead theorem involving homology rather than homotopy. Theorem 7.5.4 is all very well, but as a means of proving that two given CW-complexes are homotopy-equivalent, it is only of theoretical interest, since we would at least need to know all the homotopy groups of the two complexes. However, the following version is of much more practical use, since it is often quite possible to calculate all the *homology* groups of a CW-complex.

Theorem 8.3.10 *If K and L are connected CW-complexes, such that $\pi_1(K) = \pi_1(L) = 0$, and $f: K \to L$ is a based map such that $f_*: H_n(K) \to H_n(L)$ is isomorphic for all n, then f is a (based) homotopy equivalence.*

Proof. By Theorem 6.5.5, f is the composite

$$K \xrightarrow{g} M \xrightarrow{h} L,$$

where M is the mapping cylinder of f, g is an inclusion, and h is a homotopy equivalence. Now by Theorem 7.5.7 we may as well assume that f is a cellular map, in which case it is easy to see that M is a CW-complex. For if we write $\overline{M}$ for the space obtained from L and $K \times I$ by identifying $(k, 1)$ with $f(k)$ for all $k \in K$, then $\overline{M}$ is a CW-complex, whose cells are those of $K \times I - K \times 1$ and L: properties (a)–(d) of Definition 7.3.1 are clear, and (e) follows since a (closed) cell of $K \times I$ is contained in a subcomplex with a finite number of cells in $K \times I$ and L, the latter since the part of the boundary of the cell lying in L is the image under f of the part lying in $K \times 0$, and is hence compact. Hence (provided the base points of K and L are 0-cells) M is also a CW-complex, by Theorem 7.3.13.

Since h is a homotopy equivalence it induces isomorphisms in homology, and so $g_*: H_n(K) \to H_n(M)$ is isomorphic for all n. By the exact homology sequence, $H_n(M, K) = 0$ for all n, and so by Corollary 8.3.9 $\pi_n(M, K) = 0$ for all n, since $\pi_1(M) = \pi_1(L) = 0$. So by the exact homotopy sequence, $g_*: \pi_n(K) \to \pi_n(M)$ is isomorphic for all n, and hence by Theorem 7.5.4 g is a homotopy equivalence. Hence so also is $f = hg$. ∎

Of course, if f induces homology isomorphisms in each dimension, but is not a based map, it is easy to construct a homotopic based map by using Theorem 7.3.19, since L is path-connected. Thus the word 'based' can be removed from the hypotheses of Theorem 8.3.10, provided it is also removed from the conclusion.

A particular case of Theorem 8.3.10 is

Corollary 8.3.11 *If K is a connected CW-complex, and $\pi_1(K) = 0$, $\tilde{H}_n(K) = 0$ for all n, then K is contractible.*

Proof. By Theorem 8.3.10, the map that sends K to a single point is a homotopy equivalence. ∎

Example 8.3.12 The reader is warned, however, that $\tilde{H}_n(K)$ may be zero for all n, without K being contractible. For example, consider the space X constructed in Example 3.3.22; this is certainly a connected CW-complex, since it is triangulable. To calculate $\tilde{H}_*(X)$, let T be the maximal tree consisting of the 1-simplexes (a^0, a^1), (a^0, a^2), (a^0, a^3) and (a^0, a^4); since T is contractible, $\tilde{H}_*(X) \cong \tilde{H}_*(X/T)$ by Corollary 6.2.7. Now X/T has a CW-decomposition with one 0-cell a^0, six

1-cells $\alpha, \ldots, \zeta$, and six 2-cells $A, \ldots, F$; and since $\pi_1(X)$ becomes zero on abelianizing, $H_1(X) = H_1(X/T) = 0$. Thus $B_1(C(X/T)) = Z_1(C(X/T)) = C_1(X/T)$, since $\partial\colon C_1 \to C_0$ is clearly zero; and this implies that $\partial\colon C_2 \to C_1$ must be an isomorphism, since ∂ is onto and both C_2 and C_1 are free abelian with six generators. Hence $H_2(X) = H_2(X/T) = 0$, and of course $H_n(X) = 0$ for $n > 2$.

On the other hand, X cannot be contractible since $\pi_1(X)$ is non-trivial. ▮

Example 8.3.13 As an example of the use of Theorem 8.3.10 to show that two non-contractible CW-complexes are homotopy-equivalent, we shall prove that a simply-connected homology 3-manifold X is homotopy-equivalent to S^3.

Now since $\pi_1(X) = 0$, $H_1(X) = 0$ and so $H_1(X; Z_2) = 0$ by Theorem 4.4.15. Thus by Corollary 5.3.18 X is orientable. Moreover $H^1(X) = 0$ by Proposition 5.2.17, so that by Theorem 5.3.17 $H_2(X) = 0$. Finally $H_3(X) \cong Z$ since X is orientable.

It follows that $\pi_2(X) = 0$ and $\pi_3(X) \cong Z$. Let $f\colon S^3 \to X$ be a map representing a generator of $\pi_3(X)$, so that $f_*\colon \pi_3(S^3) \to \pi_3(X)$ is an isomorphism. Hence, since $h_3 f_* = f_* h_3$ and h_3 is an isomorphism for both S^3 and X, $f_*\colon H_3(S^3) \to H_3(X)$ is also isomorphic. All other reduced homology groups of S^3 and X are zero, so that f induces isomorphisms in homology in all dimensions. Hence, by Theorem 8.3.10, f is a homotopy equivalence, since $\pi_1(S^3) = \pi_1(X) = 0$. ▮

The Poincaré conjecture can thus be restated for 3-manifolds in the form: a simply-connected 3-manifold is homeomorphic to S^3.

8.4 Cohomology and Eilenberg–MacLane spaces

In this section we shall see how, for CW-complexes at least, cohomology theory can be fitted into the general scheme of Chapter 6. This will be done by showing that, for any CW-complex K and abelian group G, the group $\tilde{H}^n(K; G)$ can be identified with the group of homotopy classes $[K, K(G, n)]$, where $K(G, n)$ is a CW-complex with the property that

$$\pi_r(K(G, n)) \cong \begin{cases} G, & r = n \\ 0, & \text{otherwise.} \end{cases}$$

Indeed, if (K, L) is a CW-pair the exact (reduced) cohomology sequence of (K, L) can be identified with the corresponding exact sequence of groups of homotopy classes, obtained from the pair (K, L) as in Corollary 6.5.3.

It will be seen that this approach is capable of generalization. For we could replace the spaces $K(G, n)$ by a different set of spaces indexed by the integers, and thus obtain a 'cohomology theory', defined for CW-complexes, with the same formal properties as ordinary cohomology; indeed, the only virtue of the spaces $K(G, n)$ is that they are the particular set of spaces that happens to give cohomology groups that coincide with the ordinary (singular) cohomology groups. Since these more general cohomology theories have been much used in recent years, and are no more difficult to describe, we shall start by considering them, and will specialize to ordinary cohomology afterwards.

In fact if we wish to retain the 'exact sequence property' for these general cohomology theories, the spaces $K(G, n)$ must be replaced, not by any set of spaces, but by what is called an Ω-spectrum.

Definition 8.4.1 An *Ω-spectrum* $\mathbf{E}$ is a sequence of based spaces E_n, one for each integer n, together with based weak homotopy equivalences $\epsilon_n\colon \Omega E_n \to E_{n+1}$.

Given an Ω-spectrum, it is very easy to define the associated cohomology theory. What we should like to do is to define the cohomology groups of K associated with the Ω-spectrum $\mathbf{E}$ by the rule $H^n(K; \mathbf{E}) = [K, E_n]$; but since E_n may not be an AHI this may not be a group. However, if K is a CW-complex, then by Corollary 7.5.3 ϵ_n induces a (1-1)-correspondence $(\epsilon_n)_*\colon [K, E_n] \to [K, \Omega E_{n+1}]$. The latter set is a group, and hence the former set can be made into a group by requiring that $(\epsilon_n)_*$ should be an isomorphism, not merely a (1-1)-correspondence; we shall call this the multiplication in $[K, E_n]$ induced by ϵ_n.

Definition 8.4.2 Given an *Ω-spectrum* $\mathbf{E}$ and a CW-pair (K, L), the *cohomology groups of* (K, L) *associated with* $\mathbf{E}$ are defined by $H^n(K, L; \mathbf{E}) = [K/L, E_n]$, with multiplication induced by ϵ_n. We write $H^*(K, L; \mathbf{E}) = \bigoplus H^r(K, L; \mathbf{E})$.

The corresponding non-relative groups are defined by

$$H^n(K; \mathbf{E}) = H^n(K, \varnothing; \mathbf{E}), \qquad \tilde{H}^n(K; \mathbf{E}) = H^n(K, k_0; \mathbf{E}),$$

where k_0 is the base point (assumed to be a 0-cell). Observe that $H^n(K; \mathbf{E}) = [K^+, E_n]$, the set of unbased homotopy classes of maps of K into E_n, and $\tilde{H}^n(K; \mathbf{E}) = [K, E_n]$.

Proposition 8.4.3 *The groups* $H^n(K, L; \mathbf{E})$ (*and hence also* $H^n(K; \mathbf{E})$, $\tilde{H}^n(K; \mathbf{E})$) *are all abelian.*

Proof.

$$\begin{aligned} H^n(K, L; \mathbf{E}) &\cong [K/L, \Omega E_{n+1}], \\ &\cong [s(K/L), E_{n+1}], \quad \text{by Corollary 6.3.29}, \\ &\cong [s(K/L), \Omega E_{n+2}], \quad \text{by Corollary 7.5.3}, \\ &\cong [K/L, \Omega(\Omega E_{n+2})], \end{aligned}$$

which by Corollary 6.3.26 is an abelian group. ∎

It is easy to check that the cohomology groups associated with $\mathbf{E}$ have all the expected formal properties.

Theorem 8.4.4 *If (M, N) is another CW-pair, a map $f: (K, L) \to (M, N)$ induces homomorphisms $f^*: H^n(M, N; \mathbf{E}) \to H^n(K, L; \mathbf{E})$, with the following properties.*

(a) *If f is the identity map, f^* is the identity isomorphism.*
(b) *If $g: (M, N) \to (P, Q)$ is another map, $(gf)^* = f^*g^*$.*
(c) *If $f' \simeq f: (K, L) \to (M, N)$ (as a map of pairs), then $(f')^* = f^*$.*

Moreover, there exist homomorphisms $\delta^*: H^{n-1}(L; \mathbf{E}) \to H^n(K, L; \mathbf{E})$, $\delta^*: \tilde{H}^{n-1}(L; \mathbf{E}) \to H^n(K, L; \mathbf{E})$ such that the sequence

$$\cdots \longrightarrow H^{n-1}(L; \mathbf{E}) \xrightarrow{\delta^*} H^n(K, L; \mathbf{E}) \xrightarrow{j^*} H^n(K; \mathbf{E}) \xrightarrow{i^*} H^n(L; \mathbf{E}) \longrightarrow \cdots$$

and the corresponding sequence of reduced cohomology groups, are exact: a map $f: (K, L) \to (M, N)$ gives rise to commutative diagrams of exact sequences. Finally, if L and M are subcomplexes of K, such that $L \cup M = K$, then the 'excision homomorphism' $i^*: H^n(K, L; \mathbf{E}) \to H^n(M, L \cap M; \mathbf{E})$ is an isomorphism for all n.

Proof. Let f^* be the function $f^*: [M/N, E_n] \to [K/L, E_n]$, as defined in Theorem 6.3.4. Since the group structures are defined by replacing E_n by ΩE_{n+1}, f^* is a homomorphism; and properties (a)–(c) follow immediately from Theorem 6.3.4.

Let $\delta^*: \tilde{H}^{n-1}(L; \mathbf{E}) \to H^n(K, L; \mathbf{E})$ be $(-1)^{n-1}$ times the composite

$$[L, E_{n-1}] \xrightarrow{(\varepsilon_{n-1})_*} [L, \Omega E_n] \underset{\cong}{\xleftarrow{\bar{\alpha}}} [sL, E_n] \xrightarrow{(i_2\mu)^*} [K/L, E_n],$$

where $\bar{\alpha}$ is the isomorphism of Corollary 6.3.29, and $i_2\mu$ is as in Corollary 6.5.3 (the sign $(-1)^{n-1}$ is introduced to make the analogue of Corollary 8.2.7 hold). This is a homomorphism, since by Corollary 6.3.26 the two possible group structures in $[L, \Omega(\Omega E_{n+1})]$ coincide. Since $(\epsilon_{n-1})_*$ is therefore an isomorphism, Corollary 6.5.3 shows that, in the reduced cohomology sequence, Im $\delta^* = \operatorname{Ker} j^*$, and, of course, Im $j^* = \operatorname{Ker} i^*$; moreover Im $i^* = \operatorname{Ker} \delta^*$, since the diagram

$$\begin{array}{ccccc} [K, E_{n-1}] & \xrightarrow{(\varepsilon_{n-1})_*} & [K, \Omega E_n] & \xleftarrow{\bar{\alpha}} & [sK, E_n] \\ \downarrow i^* & & \downarrow i^* & & \downarrow i^* \\ [L, E_{n-1}] & \xrightarrow[(\varepsilon_{n-1})_*]{} & [L, \Omega E_n] & \xleftarrow[\bar{\alpha}]{} & [sL, E_n] \end{array}$$

is clearly commutative. That a map $f: (K, L) \to (M, N)$ gives rise to commutative diagrams of exact sequences follows at once from Corollary 6.5.3.

The definition of δ^* and the exactness of the cohomology sequence, in the case of unreduced cohomology, follow immediately on replacing (K, L) by (K^+, L^+) (the 'extra point' being the same for both K and L), since $K^+/L^+ = K/L$. Observe that a similar trick yields the exact cohomology sequence of a triple (K, L, M), since $(K/M)/(K/L) = L/M$.

Finally, the excision homomorphism is isomorphic because K/L and $M/(L \cap M)$ are clearly homeomorphic. ∎

Notice also that, if we define the suspension isomorphism $s^*: \tilde{H}^n(sK; \mathbf{E}) \to \tilde{H}^{n-1}(K; \mathbf{E})$ to be the composite

$$[sK, E_n] \xrightarrow{\bar{\alpha}} [K, \Omega E_n] \xleftarrow{(\varepsilon_{n-1})_*} [K, E_{n-1}],$$

then

(a) if $f: K \to L$ is a based map, $f^* s^* = s^*(f \wedge 1)^*$;
(b) if (K, L) is a CW-pair, $(i_2\mu)^* = (-1)^{n-1}\delta^* s^*: \tilde{H}^n(sL; \mathbf{E}) \to H^n(K, L; \mathbf{E})$.

The exact cohomology sequence allows us to prove the following generalization of a result in Example 4.2.12.

Corollary 8.4.5 $H^n(K; \mathbf{E}) \cong \tilde{H}^n(K; \mathbf{E}) \oplus \tilde{H}^n(S^0; \mathbf{E})$.

Proof. Let k_0 be the base point of K, and consider the exact cohomology sequence of the pair (K, k_0):

$$\cdots \longrightarrow H^n(K, k_0; \mathbf{E}) \xrightarrow{j^*} H^n(K; \mathbf{E}) \xrightarrow{i^*} H^n(k_0; \mathbf{E}) \longrightarrow \cdots.$$

If $p: K \to k_0$ is the constant map, $pi = 1: k_0 \to k_0$, so that $i^*p^* = 1$. Thus by Proposition 1.3.36,

$$H^n(K; \mathbf{E}) \cong H^n(K, k_0; \mathbf{E}) \oplus H^n(k_0; \mathbf{E});$$

but $H^n(K, k_0; \mathbf{E}) = \tilde{H}^n(K; \mathbf{E})$ and $H^n(k_0; \mathbf{E}) = \tilde{H}^n(S^0; \mathbf{E})$. ∎

For a general Ω-spectrum $\mathbf{E}$, there is no reason why the groups $\tilde{H}^n(S^0; \mathbf{E})$ should vanish if $n \neq 0$. In fact it is easy to see that a necessary and sufficient condition for this to happen is precisely that the homotopy groups of E_n should vanish in dimensions other than n. This brings us back to the particular case in which we are most interested, where each E_n is an Eilenberg–MacLane space. We now give the precise definition and existence theorem.

Definition 8.4.6 Given an integer $n \geqslant 0$ and an abelian group G, a CW-complex K is called an *Eilenberg–MacLane space* $K(G, n)$ if

$$\pi_r(K) \cong \begin{cases} G, & r = n \\ 0, & \text{otherwise.} \end{cases}$$

If $n = 0$, we require only that $\pi_0(K)$ should be in (1-1)-correspondence with G, and we may take $K(G, 0) = G$, with the discrete topology.

Theorem 8.4.7 *For any $n \geqslant 1$ and any abelian group G, $K(G, n)$ exists.*

Proof. We first construct a CW-complex B such that

$$\pi_r(B) \cong \begin{cases} G, & r = n \\ 0, & r < n, \end{cases}$$

and then use Theorem 7.5.9 to 'kill' the higher homotopy groups.

Write $G = F/R$, where F is a free group and R is a subgroup (for example, F may be the free group with the elements of G as generators). Let $A = \bigvee S^n_\alpha$, one for each generator α of F, and define $\theta: F \to \pi_n(A)$ by $\theta(\alpha) = [i_\alpha]$, where $i_\alpha: S^n \to A$ is the inclusion map onto S^n_α. For each element x of R, let $\phi_x: S^n \to A$ be a map representing $\theta(x) \in \pi_n(A)$; let B be the space obtained from A by attaching $(n+1)$-cells E^{n+1} by the maps ϕ_x, one for each element $x \in R$. Then certainly B is a CW-complex, and it is easy to see that $\pi_r(B) = 0$ for $r < n$. Moreover, there is a commutative diagram

$$\begin{array}{ccccccc} & & \pi_n(A) & \xrightarrow{i_*} & \pi_n(B) & \longrightarrow & 0 \\ & & \downarrow{\scriptstyle h_n} & & \downarrow{\scriptstyle h_n} & & \\ H_{n+1}(B, A) & \xrightarrow[\partial_*]{} & H_n(A) & \xrightarrow[i_*]{} & H_n(B) & \longrightarrow & 0, \end{array}$$

where h_n is the Hurewicz homomorphism. Now if $n = 1$, $\pi_1(A) \cong F$ and h_1 is the quotient homomorphism onto $F/[F, F]$; otherwise, if $n > 1$, $h_n: \pi_n(A) \to H_n(A)$ is isomorphic and $\pi_n(A) \cong F/[F, F]$. Further, $H_{n+1}(B, A) \cong \tilde{H}_{n+1}(B/A)$ is the free abelian group on the elements of R as generators, and ∂_* sends each generator to its coset in $F/[F, F]$, so that $H_n(B) \cong F/R = G$. If $n > 1$, $h_n: \pi_n(B) \to H_n(B)$ is an isomorphism; so that $\pi_n(B) \cong G$ as well; if $n = 1$, we at least know that $h_1 i_* = i_* h_1: \pi_1(A) \to H_1(B)$ is the quotient homomorphism $F \to F/R$, so that Ker $[i_*: \pi_1(A) \to \pi_1(B)]$ is contained in R: however, $i_*\theta(x) = i_*[\phi_x] = 0$ for all $x \in R$, since $\phi_x: S^1 \to B$ extends to a map of E^2. Thus in all cases $\pi_n(B) \cong G$.

The proof is now completed by using Theorem 7.5.9 to 'kill' the homotopy groups of B in dimensions greater than n. ▌

Corollary 8.4.8 *Given an abelian group G, there exists an Ω-spectrum* **E** *with*

$$E_n = \begin{cases} K(G, n), & n \geqslant 0 \\ \textit{point}, & \textit{otherwise.} \end{cases}$$

Proof. Define E_n to be a point or $K(G, n)$ as the case may be; we have only to construct the weak homotopy equivalences $\epsilon_n: E_n \to \Omega E_{n+1}$. Since $\Omega(\text{point}) = \Omega K(G, 0) = \text{point}$, the only possible map $\epsilon_n: E_n \to \Omega E_{n+1}$ is obviously a weak homotopy equivalence if $n < 0$, so that in fact it is sufficient to consider only the case $n \geqslant 0$.

If $n = 0$, note that

$$\pi_r(\Omega K(G, 1)) \cong \begin{cases} 0, & r > 0 \\ G, & r = 0. \end{cases}$$

Thus the map ϵ_0 that sends each element of $G = K(G, 0)$ into the corresponding path component of $\Omega K(G, 1)$ is a weak homotopy equivalence.

If $n \geqslant 1$, construct $K(G, n)$ as in Theorem 8.4.7, and define $f: A \to \Omega K(G, n + 1)$ by mapping each S^n_α by a representative map for the coset of α in $F/R \cong \pi_n(\Omega K(G, n + 1))$; thus $f_*: \pi_n(A) \to \pi_n(\Omega K(G, n + 1))$ is just the quotient map $F \to F/R$ ($F/[F, F] \to F/R$ if $n > 1$). Each $(n + 1)$-cell E^{n+1} of B is attached by a map ϕ_x that represents an element x of R; thus each map $f\phi_x$ is homotopic to a constant map and we can extend f to a map $g: B \to \Omega K(G, n + 1)$. Since $f^* = g_* i_*: \pi_n(A) \to \pi_n(\Omega K(G, n + 1))$, $g_*: \pi_n(B) \to \pi_n(\Omega K(G, n + 1))$ is an isomorphism. Finally, $K(G, n)$ is obtained from B by attaching cells of dimension at least $(n + 2)$, so that, since $\pi_r(\Omega K(G, n + 1)) = 0$ for $r > n$, g can be extended to a map ϵ_n of the whole of $K(G, n)$, that still induces isomorphisms in π_n, and so is a weak homotopy equivalence. ▌

It remains now to prove that, if (K, L) is a CW-pair, then the groups $H^n(K, L; \mathbf{E})$ are isomorphic to $H^n(K, L; G)$, where **E** is the

Ω-spectrum of Corollary 8.4.8. In fact rather more than this is true: the definitions of induced homomorphisms and the exact sequence of a pair, given in Theorem 8.4.4, coincide with those of Chapter 5.

The proof of these results is similar to that of Theorem 8.2.15, and depends on the following proposition.

Proposition 8.4.9 *Let (K, L) be a CW-pair, G be an abelian group, and $\mathbf{E}$ be the Ω-spectrum of Corollary 8.4.8. There is a homomorphism $\beta: H^n(K, L; \mathbf{E}) \to H_n(K, L) \pitchfork G$, with the following properties.*

(a) *Given a map $f: (K, L) \to (P, Q)$, the diagram*

$$\begin{array}{ccc} H^n(P, Q; \mathbf{E}) & \xrightarrow{f^*} & H^n(K, L; \mathbf{E}) \\ \downarrow \beta & & \downarrow \beta \\ H_n(P, Q) \pitchfork G & \xrightarrow[f_* \pitchfork 1]{} & H_n(K, L) \pitchfork G \end{array}$$

is commutative.

(b) *The diagram*

$$\begin{array}{ccc} \tilde{H}^n(sK; \mathbf{E}) & \xrightarrow{s^*} & \tilde{H}^{n-1}(K; \mathbf{E}) \\ \downarrow \beta & & \downarrow \beta \\ \tilde{H}_n(sK) \pitchfork G & \xrightarrow[s_* \pitchfork 1]{} & \tilde{H}_{n-1}(K) \pitchfork G \end{array}$$

is commutative.

Proof. Represent an element $x \in H^n(K, L; \mathbf{E})$ by a map $\xi: K/L \to K(G, n)$, and let $\bar{\xi}: s(K/L) \to K(G, n+1)$ be the map that corresponds under the association map to $\epsilon_n \xi: K/L \to \Omega K(G, n+1)$. Now define

$$\beta(x) = \bar{\xi}_* s_*: \tilde{H}_n(K/L) \to G,$$

where $\tilde{H}_{n+1}(K(G, n+1))$ is identified with G via the Hurewicz isomorphism

$$h_{n+1}: G \cong \pi_{n+1}(K(G, n+1)) \to H_{n+1}(K(G, n+1)).$$

The proof that β is a homomorphism is like that of Proposition 7.4.1: if $y \in H^n(K, L; \mathbf{E})$ is represented by $\bar{\eta}: s(K/L) \to K(G, n+1)$ $x + y$ is represented by the composite

$$s(K/L) \xrightarrow{\mu} s(K/L) \vee s(K/L) \xrightarrow{\nabla(\bar{\xi} \vee \bar{\eta})} K(G, n+1),$$

so that

$$\begin{aligned}\beta(x + y) &= (\nabla(\bar{\xi} \vee \bar{\eta}))_* \mu_* s_* \\ &= (\bar{\xi}_* + \bar{\eta}_*) s_* \\ &= \beta(x) + \beta(y).\end{aligned}$$

Property (a) is easy, since f^* corresponds to

$$(f \wedge 1)^*: [s(P/Q), K(G, n + 1)] \to [s(K/L), K(G, n + 1)];$$

hence

$$\begin{aligned}(f_* \wedge 1)\beta(x) &= \bar{\xi}_* s_* f_* \\ &= \bar{\xi}_* (f \wedge 1)_* s_* \\ &= [\bar{\xi}(f \wedge 1)]_* s_* \\ &= \beta(f^* x).\end{aligned}$$

To prove property (b), we first observe that, if $x \in \tilde{H}^n(sK; \mathbf{E})$ is represented by $\xi: sK \to K(G, n)$, then $\beta(x) = \xi_*: \tilde{H}_n(sK) \to \tilde{H}_n(K(G, n)) \cong G$ $(n \geqslant 1)$. For $\bar{\xi}: s(sK) \to K(G, n + 1)$ is the same as the composite

$$s(sK) \xrightarrow{\xi \wedge 1} sK(G, n) \xrightarrow{\bar{\epsilon}_n} K(G, n + 1),$$

where $\bar{\epsilon}_n$ corresponds to ϵ_n under the association map. Hence

$$\begin{aligned}\beta(x) &= (\bar{\epsilon}_n)_* (\xi \wedge 1)_* s_* \\ &= (\bar{\epsilon}_n)_* s_* \xi_*.\end{aligned}$$

But if $\theta: \pi_n(K(G, n)) \to \pi_{n+1}(K(G, n + 1))$ is the homomorphism defined by sending the homotopy class of $g: S^n \to K(G, n)$ to $[\bar{\epsilon}_n(g \wedge 1)]$, there is a commutative diagram

$$\begin{array}{ccc}\pi_n(K(G, n)) & \xrightarrow{\theta} & \pi_{n+1}(K(G, n + 1)) \\ \downarrow h_n & & \downarrow h_{n+1} \\ H_n(K(G, n)) & \xrightarrow[(\bar{\epsilon}_n)_* s_*]{} & H_{n+1}(K(G, n + 1)).\end{array}$$

However, θ is the same as the homomorphism that sends $[g]$ to the class of the map corresponding under the association map to $\epsilon_n g$, and this is the identity isomorphism of G. Hence $(\bar{\epsilon}_n)_* s_* = 1$ and $\beta(x) = \xi_*$.

Now if $s^*x \in \tilde{H}^{n-1}(K; \mathbf{E})$ is represented by $\eta: K \to K(G, n-1)$, we have $\bar{\eta} = \xi$, so that

$$\begin{aligned} \beta s^* x &= \bar{\eta}_* s_* \\ &= \xi_* s_* \\ &= (s_* \pitchfork 1)\beta(x). \ \blacksquare \end{aligned}$$

Theorem 8.4.10 *With the notation of Proposition 8.4.9, there is an isomorphism* γ: $H^n(K, L; \mathbf{E}) \to H^n(K, L; G)$, *with the following properties:*

(a) *Given* $f: (K, L) \to (P, Q)$, *the diagram*

$$\begin{array}{ccc} H^n(P, Q; \mathbf{E}) & \xrightarrow{f^*} & H^n(K, L; \mathbf{E}) \\ \gamma\downarrow & & \downarrow\gamma \\ H^n(P, Q; G) & \xrightarrow[f^*]{} & H^n(K, L; G) \end{array}$$

is commutative.

(b) *The diagram*

$$\begin{array}{ccccccccc} \cdots \to \tilde{H}^{n-1}(L; \mathbf{E}) & \xrightarrow{\delta^*} & H^n(K, L; \mathbf{E}) & \xrightarrow{j^*} & \tilde{H}^n(K; \mathbf{E}) & \xrightarrow{i^*} & \tilde{H}^n(L; \mathbf{E}) \to \cdots \\ \gamma\downarrow & & \gamma\downarrow & & \downarrow\gamma & & \downarrow\gamma \\ \cdots \to \tilde{H}^{n-1}(L; G) & \xrightarrow[\delta^*]{} & H^n(K, L; G) & \xrightarrow[j^*]{} & \tilde{H}^n(K; G) & \xrightarrow[i^*]{} & \tilde{H}^n(L; G) \to \cdots \end{array}$$

is commutative.

Proof. We show first that $H^*(K, L; \mathbf{E})$ can be calculated from the chain complex $D(K, L)$, defined by $D_{-n}(K, L) = H^n(M^n, M^{n-1}; \mathbf{E})$ (and $M^n = K^n \cup L$), and then show that the homomorphism β of Proposition 8.4.9 yields a chain isomorphism

$$D(K, L) \to C(K, L) \pitchfork G.$$

Now the argument of Theorem 8.2.10 will certainly show that $H^n(M^p, L; \mathbf{E}) \cong H_{-n}(D(K, L))$ for all $p > n$, but once again a special argument is needed to show that $H^n(M^{n+1}, L; \mathbf{E}) \cong H^n(K, L; \mathbf{E})$. This time, however, it is sufficient to remark that, for any CW-complex K, a map $f: K^{n+1} \to K(G, n)$ can be extended over the remaining cells of K, since $\pi_r(K(G, n)) = 0$ for $r > n$, and similarly the homotopy class of such an extension depends only on that of f.

By properties (a) and (b) of Proposition 8.4.9, and Corollary 8.2.7,

$$\beta: D(K, L) \to C(K, L) \pitchfork G$$

is certainly a chain map. But in fact β is a chain isomorphism, since

$$\begin{aligned} H^n(M^n, M^{n-1}; \mathbf{E}) &\cong [s(M^n/M^{n-1}), K(G, n+1)] \\ &\cong \pi_{n+1}(s(M^n/M^{n-1})) \pitchfork G \\ &\cong \tilde{H}_{n+1}(s(M^n/M^{n-1})) \pitchfork G \\ &\cong \tilde{H}_n(M^n/M^{n-1}) \pitchfork G. \end{aligned}$$

Hence β induces an isomorphism $\gamma: H^n(K, L; \mathbf{E}) \to H^n(K, L; G)$.

Property (a) follows from Proposition 8.4.9(a) and the argument used to prove Corollary 8.2.11, at least if $f: (K, L) \to (P, Q)$ is a cellular map. But since by Theorem 7.5.7 any continuous map is homotopic to a cellular map, property (a) immediately extends to any continuous map. As for property (b), we need only consider δ^*, and by the cohomology analogue of Corollary 8.2.7 it is sufficient to show that the diagram

$$\begin{array}{ccc} \tilde{H}^n(sK; \mathbf{E}) & \xrightarrow{s^*} & \tilde{H}^{n-1}(K; \mathbf{E}) \\ \downarrow{\scriptstyle \gamma} & & \downarrow{\scriptstyle \gamma} \\ \tilde{H}^n(sK; G) & \xrightarrow[s^*]{} & \tilde{H}^{n-1}(K; G) \end{array}$$

is commutative. But this follows from Proposition 8.4.9(b) in the same way that (a) follows from Proposition 8.4.9(a). ∎

8.5 Products

It has already been hinted, at the beginning of Chapter 5 and elsewhere, that cohomology theory has a real advantage over homology theory, in that it is possible to introduce products, so as to make the direct sum of the cohomology groups into a ring. This makes cohomology a more delicate algebraic invariant, which will often distinguish between spaces that have isomorphic homology groups.

It is possible to set up the general theory for the (singular) cohomology of arbitrary topological spaces (see the notes at the end of the chapter). However, it appears more illuminating—and it is certainly a good deal easier—to confine attention to CW-complexes and make use of the 'cellular chain groups' of Section 8.2. The product will be defined by a set of axioms; and since these axioms are just as easy to state for a general cohomology theory, we shall give the definition in terms of the cohomology theory associated with an arbitrary Ω-spectrum $\mathbf{E}$.

Since the axioms involve the cohomology groups of a product of two CW-complexes, it is convenient to use the products $\bar{\times}$ and $\bar{\wedge}$, in order to ensure that all products are again CW-complexes.

Definition 8.5.1 Let $\mathbf{E}$ be an Ω-spectrum. In the cohomology theory associated with $\mathbf{E}$, a *product* is a set of homomorphisms

$$\wedge : \tilde{H}^r(K; \mathbf{E}) \otimes \tilde{H}^s(L; \mathbf{E}) \to \tilde{H}^{r+s}(K \barwedge L; \mathbf{E}),$$

for all integers r, s and all (based) CW-complexes K, L (we shall write $x \wedge y$ for $\wedge(x \otimes y)$). These homomorphisms are required to satisfy the following four axioms.

Axiom 1. Given based maps $f: K \to M$, $g: L \to N$, the following diagram is commutative:

$$\begin{array}{ccc} \tilde{H}^r(M; \mathbf{E}) \otimes \tilde{H}^s(N; \mathbf{E}) & \xrightarrow{\wedge} & \tilde{H}^{r+s}(M \barwedge N; \mathbf{E}) \\ {\scriptstyle f^* \otimes g^*} \downarrow & & \downarrow {\scriptstyle (f \barwedge g)^*} \\ \tilde{H}^r(K; \mathbf{E}) \otimes \tilde{H}^s(L; \mathbf{E}) & \xrightarrow[\wedge]{} & \tilde{H}^{r+s}(K \barwedge L; \mathbf{E}). \end{array}$$

Axiom 2. The product is *associative*, that is,

$$\wedge(\wedge \otimes 1) = \wedge(1 \otimes \wedge): \tilde{H}^r(K; \mathbf{E}) \otimes \tilde{H}^s(L; \mathbf{E}) \otimes \tilde{H}^t(M; \mathbf{E}) \to \tilde{H}^{r+s+t}(K \barwedge L \barwedge M; \mathbf{E}).$$

Axiom 3. The product is *anti-commutative*, that is, if $x \in \tilde{H}^r(K; \mathbf{E})$ and $y \in \tilde{H}^s(L; \mathbf{E})$, then $x \wedge y = (-1)^{rs}\tau^*(y \wedge x)$, where $\tau: K \barwedge L \to L \barwedge K$ is the map that exchanges the two factors.

Axiom 4. There exists an element $z \in \tilde{H}^1(S^1; \mathbf{E})$ such that, for each $x \in \tilde{H}^r(K; \mathbf{E})$, $s^*(x \wedge z) = x$.

Of course, by replacing K, L by K^+, L^+, we obtain a product in unreduced cohomology, of the form $\times : H^r(K; \mathbf{E}) \otimes H^s(L; \mathbf{E}) \to H^{r+s}(K \mathbin{\bar{\times}} L; \mathbf{E})$; again we write $x \times y$ for $\times(x \otimes y)$. This product satisfies axioms similar to Axioms 1–4 above; in particular, the analogue of Axiom 1 holds for *unbased* maps f and g.

Moreover, by taking $K = L$ and using the diagonal map $\Delta: K \to K \barwedge K$, $\tilde{H}^*(K; \mathbf{E})$ can be made into a ring. As has already been suggested, this is the real object in introducing products.

Theorem 8.5.2 *If K is a CW-complex, a product $\wedge$ induces a product between elements $x, y \in \tilde{H}^*(K; \mathbf{E})$, written $x \cup y$, in such a way as to make $\tilde{H}^*(K; \mathbf{E})$ into a ring. Moreover, the following properties hold:*

(a) *If $x \in \tilde{H}^r(K; \mathbf{E})$ and $y \in \tilde{H}^s(K; \mathbf{E})$, then $x \cup y \in \tilde{H}^{r+s}(K; \mathbf{E})$ and $x \cup y = (-1)^{rs} y \cup x$.*

(b) *If $f: K \to L$ is a based map, then f^* is a ring homomorphism.*

(c) *All products are zero in $\tilde{H}^*(sK; \mathbf{E})$.*

Proof. If $x \in \tilde{H}^r(K; \mathbf{E})$ and $y \in \tilde{H}^s(K; \mathbf{E})$, define

$$x \cup y = \Delta^*(x \wedge y) \in \tilde{H}^{r+s}(K; \mathbf{E}).$$

By definition of $\wedge$ this is distributive and associative, and remains so when the product $\wedge$ is extended in the obvious way to more general elements of $\tilde{H}^*(K; \mathbf{E})$: hence $\tilde{H}^*(K; \mathbf{E})$ is a ring. Properties (a) and (b) are immediate from Axioms 1 and 3, since $\tau\Delta = \Delta$ and $(f \barwedge f)\Delta = \Delta f$.

To prove (c), we have only to remark that $\Delta: sK \to sK \barwedge sK$ is the same (up to rearrangement of the factors) as $\Delta \wedge \Delta: K \wedge S^1 \to (K \barwedge K) \wedge (S^1 \wedge S^1)$; but this is homotopic to the constant map, since $\pi_1(S^1 \wedge S^1) = \pi_1(S^2) = 0$. ∎

Naturally we can replace K by K^+, so as to obtain a similar product in $H^*(K; \mathbf{E})$, and then any unbased map $f: K \to L$ induces a ring homomorphism.

Theorem 8.5.2 shows why it is cohomology, rather than homology, that can be made into a ring. The point is that the diagonal map induces a homomorphism $\Delta^*: \tilde{H}^*(K \barwedge K; \mathbf{E}) \to \tilde{H}^*(K; \mathbf{E})$ in cohomology, but goes in the opposite direction in homology, and so cannot be used to form a product.

The next step is to justify Definition 8.5.1 by showing that there exists a product in ordinary cohomology. We use the results of Section 8.2 (in particular Theorem 8.2.15), and the first step is to construct a homomorphism $C(K) \otimes C(L) \to C(K \times L)$ for any two CW-complexes K and L. Let the indexing sets and characteristic maps for K, L be A_n, B_m, ϕ^n_α, ψ^m_β respectively. By Theorem 7.3.16, the indexing sets for $K \bar{\times} L$ are $C_n = \bigcup_{r+s=n} A_r \times B_s$, and the characteristic maps are $(\phi^r_\alpha \times \psi^s_\beta)h_{r,s}$, where $h_{r,s}: E^{r+s} \to E^r \times E^s$ is the standard homeomorphism. A homomorphism $\times : C_r(K) \otimes C_s(L) \to C_{r+s}(K \bar{\times} L)$ may therefore be defined by setting $\times(\alpha \otimes \beta) = \alpha \times \beta$, where the generators of, for example, $C_r(K)$ are identified with the elements of A_r as in Proposition 8.2.12.

Proposition 8.5.3 $\partial(\alpha \times \beta) = \partial\alpha \times \beta + (-1)^r \alpha \times \partial\beta.$

Proof. For each $\gamma \times \delta$ in C_{r+s-1}, consider the composite map $d_{\alpha\times\beta,\gamma\times\delta}$, as in Proposition 8.2.12:

$$S^{r+s-1} \xrightarrow{h_{r,s}} S^{r-1} \times E^s \cup E^r \times S^{s-1}$$

$$\xrightarrow{p(\phi^r_\alpha \bar{\times} \psi^s_\beta)} \frac{K^{r-1} \bar{\times} L^s \cup K^r \bar{\times} L^{s-1}}{K^{r-2} \bar{\times} L^s \cup K^{r-1} \bar{\times} L^{s-1} \cup K^r \bar{\times} L^{s-2}}$$

$$\longleftarrow \bigvee (E^{r+s-1}/S^{r+s-2}) \xrightarrow{\theta q_{\gamma\times\beta}} S^{r+s-1}.$$

This is clearly the constant map unless $\gamma \times \delta \in A_{r-1} \times B_s \cup A_r \times B_{s-1}$, and is homotopic to the constant map (since it is not onto) unless $\gamma \times \delta$ is of form $\gamma \times \beta$ or $\alpha \times \delta$. On the other hand, by Proposition 7.3.21, $d_{\alpha \times \beta, \gamma \times \beta}$ is the composite

$$S^{r+s-1} \xrightarrow{h_{r,s}} S^{r-1} \wedge (E^s/S^{s-1}) \xrightarrow{\phi_\alpha^r \bar{\wedge} \psi_\beta^s} (K^{r-1}/K^{r-2}) \bar{\wedge} (L^s/L^{s-1})$$
$$\xleftarrow{\phi^{r-1} \bar{\wedge} \psi^s} \bigvee (E^{r-1}/S^{r-2}) \wedge (E^s/S^{s-1})$$
$$\xleftarrow{h_{r-1,s}} \bigvee (E^{r+s-1}/S^{r+s-2}) \xrightarrow{\theta q_{\gamma \times \beta}} S^{r+s-1}.$$

($\phi_\alpha^r: S^{r-1} \to K^{r-1}/K^{r-2}$ may be assumed to be a based map if $r > 1$, by choosing a suitable base point in $(\phi_\alpha^r)^{-1}K^{r-2}$. If $\phi_\alpha^r(S^{r-1}) \cap K^{r-2} = \varnothing$, both $d_{\alpha \times \beta, \gamma \times \beta}$ and $d_{\alpha,\gamma}$ are homotopic to the constant map, since they are not onto. The case $r = 1$ is discussed below.)

Now this composite is the same as

$$S^{r+s-1} \xrightarrow{h_{r,s}} S^{r-1} \wedge (E^s/S^{s-1}) \xrightarrow{d_{\alpha\gamma} \wedge 1} S^{r-1} \wedge (E^s/S^{s-1})$$
$$\xleftarrow{\theta \wedge 1} (E^{r-1}/S^{r-2}) \wedge (E^s/S^{s-1})$$
$$\xleftarrow{h_{r-1,s}} E^{r+s-1}/S^{r+s-2} \xrightarrow{\theta} S^{r+s-1}.$$

But by Proposition 6.2.16 this is homotopic to

$$S^{r+s-1} \xrightarrow{h_{r,s}} S^{r-1} \wedge (E^s/S^{s-1}) \xrightarrow{d_{\alpha\gamma} \wedge 1}$$
$$S^{r-1} \wedge (E^s/S^{s-1}) \xrightarrow{\bar{h}_{r,s}} S^{r+s-1},$$

where $\bar{h}_{r,s}$ is a homotopy inverse to $h_{r,s}$; thus, as in Example 7.4.10, $\partial_{\alpha \times \beta, \gamma \times \beta} = \partial_{\alpha\gamma}$.

If $r = 1$, the above argument does not work, since it is not possible to make $\phi_\alpha^1: S^0 \to K^0/K^{-1} = (K^0)^+$ into a based map. However, the reader should have no difficulty in making the necessary modifications to deal with this case: the space $S^0 \wedge (E^s/S^{s-1})$ should be replaced by $(S^0 \times E^s)/(S^0 \times S^{s-1})$.

A similar argument shows that $\partial_{\alpha \times \beta, \alpha \times \delta} = (-1)^r \partial_{\beta\delta}$, so that in $C(K \bar{\times} L)$ we have $\partial(\alpha \times \beta) = \partial\alpha \times \beta + (-1)^r \alpha \times \partial\beta$. ∎

Notice that, since the obvious homomorphism

$$\bigoplus_{r+s=n} C_r(K) \otimes C_s(L) \to C_n(K \bar{\times} L)$$

is clearly an isomorphism, Proposition 8.5.3 can be used to give a formula for the homology groups of $K \bar{\times} L$ in terms of those of K and L: see Exercise 9.

Proposition 8.5.4 *Given cellular maps $f: K \to M$, $g: L \to N$,*

$$(f \bar{\times} g).(\alpha \times \beta) = (f.\alpha) \times (g.\beta).$$

Proof. By similar methods to those in the proof of Proposition 8.5.3, it is easy to see that $(f \bar{\times} g)_{\alpha\times\beta,\gamma\times\delta}$ may be identified with $f_{\alpha\gamma} \wedge g_{\beta\delta}: S^r \wedge S^s \to S^r \wedge S^s$. But

$$\begin{aligned} d(f_{\alpha\gamma} \wedge g_{\beta\delta}) &= d(f_{\alpha\gamma} \wedge 1)d(1 \wedge g_{\beta\delta}) \\ &= d(f_{\alpha\gamma})d(g_{\beta\delta}), \end{aligned}$$

by Example 7.4.10. Hence

$$\begin{aligned} (f \bar{\times} g).(\alpha \times \beta) &= \sum_{\gamma,\delta} d((f \bar{\times} g)_{\alpha\times\beta,\gamma\times\delta})(\gamma \times \delta) \\ &= \sum_{\gamma,\delta} d(f_{\alpha\gamma})d(g_{\beta\delta})(\gamma \times \delta) \\ &= \Big(\sum_{\gamma} d(f_{\alpha\gamma})\gamma\Big) \times \Big(\sum_{\delta} d(g_{\beta\delta})\delta\Big) \\ &= (f.\beta) \times (g.\beta). \ \blacksquare \end{aligned}$$

By composing with the identification map $p: K \bar{\times} L \to K \bar{\wedge} L$, the homomorphism $\bar{\times}$ can be turned into a homomorphism into $C(K \bar{\wedge} L)$ (at least if the base points are 0-cells); more precisely, define

$$\wedge : C(K) \otimes C(L) \to C(K \bar{\wedge} L)$$

by $\alpha \wedge \beta \to p.(\alpha \times \beta)$. Naturally, the analogues of Propositions 8.5.3 and 8.5.4 remain true, since $p.$ is a chain map.

Theorem 8.5.5 *If G is a commutative ring with a* 1, *there exists a product in the cohomology with coefficients in G of CW-complexes.*

Proof. By interpreting $\tilde{C}(K)$ as $C(K, k_0)$, where k_0 is the base point, the homomorphism $\wedge$ may be regarded as an isomorphism

$$\wedge : \bigoplus_{r+s=n} \tilde{C}_r(K) \otimes \tilde{C}_s(L) \to \tilde{C}_n(K \bar{\wedge} L),$$

where $\partial(\alpha \wedge \beta) = (\partial\alpha) \wedge \beta + (-1)^r\alpha \wedge (\partial\beta)$ if $\alpha \in \tilde{C}_r(K)$. If G is a commutative ring with a 1, this gives rise to a homomorphism

$$\wedge : (\tilde{C}_r(K) \pitchfork G) \otimes (\tilde{C}_s(L) \pitchfork G) \to \tilde{C}_{r+s}(K \bar{\wedge} L) \pitchfork G$$

as follows. Let A_r, B_s be the indexing sets for the r-cells of K and the

s-cells of L, respectively. If $x\colon \tilde{C}_r(K) \to G$ and $y\colon \tilde{C}_s(L) \to G$, define $x \wedge y$ on the generators of $\tilde{C}_{r+s}(K \bar{\wedge} L)$ by the rule

$$(x \wedge y)(\alpha \wedge \beta) = \begin{cases} x(\alpha).y(\beta), & \alpha \in A_r, \beta \in B_s \\ 0, & \text{otherwise.} \end{cases}$$

Let $\alpha \in \tilde{C}_p(K)$, $\beta \in \tilde{C}_q(L)$, where $p + q = r + s + 1$. If we write δ for $\partial ⋔ 1$, then

$$\begin{aligned} [\delta(x \wedge y)](\alpha \wedge \beta) &= (x \wedge y)[\partial(\alpha \wedge \beta)] \\ &= (x \wedge y)[(\partial\alpha) \wedge \beta + (-1)^p\alpha \wedge (\partial\beta)] \\ &= \begin{cases} x(\partial\alpha).y(\beta), & p = r + 1, q = s \\ (-1)^r x(\alpha).y(\partial\beta), & p = r, q = s + 1 \\ 0, \quad \text{otherwise,} \end{cases} \end{aligned}$$

so that $\delta(x \wedge y) = \delta x \wedge y + (-1)^r x \wedge \delta y$. It follows that if $\delta x = \delta y = 0$, then $\delta(x \wedge y) = 0$; if $x = \delta x'$, $\delta y = 0$, then $x \wedge y = \delta(x' \wedge y)$; and if $\delta x = 0$, $y = \delta y'$, then $x \wedge y = (-1)^r \delta(x \wedge y')$. Thus if $[x]$ denotes the homology class of the cycle x, we can define

$$\wedge : \tilde{H}^r(K; G) \otimes \tilde{H}^s(L; G) \to \tilde{H}^{r+s}(K \bar{\wedge} L; G)$$

unambiguously by $[x] \wedge [y] = [x \wedge y]$.

It remains to check Axioms 1–4.

Axiom 1. Without loss of generality we may assume that f and g are cellular. In this case $(f \bar{\wedge} g)_.(\alpha \wedge \beta) = (f_.\alpha) \wedge (g_.\beta)$ by Proposition 8.5.4, so that Axiom 1 follows since f^*, g^* and $(f \bar{\wedge} g)^*$ are induced by $f_. ⋔ 1$, $g_. ⋔ 1$ and $(f \bar{\wedge} g)_. ⋔ 1$ respectively.

Axiom 2. This is trivial, since $(\alpha \wedge \beta) \wedge \gamma = \alpha \wedge (\beta \wedge \gamma)$ in $\tilde{C}_{r+s+t}(K \bar{\wedge} L \bar{\wedge} M)$.

Axiom 3. By Example 7.4.9, the map $\tau\colon S^s \wedge S^r \to S^r \wedge S^s$ that exchanges the two factors has degree $(-1)^{rs}$. Thus if we write τ also for the 'exchange map' $\tau\colon L \bar{\wedge} K \to K \bar{\wedge} L$, it is easy to see that

$$\tau_.\colon \tilde{C}_{r+s}(L \bar{\wedge} K) \to \tilde{C}_{r+s}(K \bar{\wedge} L)$$

is given by $\tau_.(\beta \wedge \alpha) = (-1)^{rs}(\alpha \wedge \beta)$, if $\alpha \in \tilde{C}_r(K)$ and $\beta \in \tilde{C}_s(L)$. Hence in cohomology $\tau^*(y \wedge x) = (-1)^{rs}(x \wedge y)$, since G is commutative.

Axiom 4. It is easy to see that $s^*\colon \tilde{H}^{r+1}(sK; G) \to \tilde{H}^r(K; G)$ is induced by $s_*\colon \tilde{C}_r(K) = \tilde{H}_r(M^r/M^{r-1}) \to \tilde{H}_{r+1}(s(M^r/M^{r-1})) = \tilde{C}_{r+1}(sK)$, where $M^r = K^r \cup k_0$. Now for a CW-complex X, $s_*\colon \tilde{H}_r(X) \to \tilde{H}_{r+1}(sX)$ is $(-1)^{r+1}$ times the composite $\tilde{H}_r(X) \xrightarrow{\partial^*} H_{r+1}(cX, X) \xrightarrow{p_*}$

$\tilde{H}_{r+1}(sX)$, and if X is a one-point union of S^r's (for example if $X = M^r/M^{r-1}$), then the r-skeleton of cX is X, so that this composite may be identified with

$$\tilde{C}_r(X) = \tilde{C}_r(cX) \xleftarrow{\partial} \tilde{C}_{r+1}(cX) \xrightarrow{p.} \tilde{C}_{r+1}(sX).$$

Now if I denotes the obvious 1-cell of I, then $\partial(I) = 1 - 0$ in $C(I)$; hence if $\alpha \in \tilde{C}_r(X)$ we have $\partial(\alpha \wedge I) = (-1)^{r+1}\alpha$ in $\tilde{C}(cX)$. That is, $s_*: \tilde{H}_r(X) \to \tilde{H}_{r+1}(sX)$ is given by sending α to $\alpha \wedge s_1$, where $s_1 = p_.(I) \in \tilde{C}_1(S^1)$. It follows that, if $z \in \tilde{H}^1(S^1; G)$ is the class of the homomorphism $\tilde{C}_1(S^1) \to G$ that sends s_1 to 1 in G, then for $x \in \tilde{H}^r(K; G)$ we have $s^*(x \wedge z) = x$. ▮

Of course, the corresponding product in unreduced cohomology is induced by the homomorphism

$$\times: C(K) \otimes C(L) \to C(K \mathbin{\bar{\times}} L).$$

It can be shown that Axioms 1–4 characterize the product uniquely, if G is one of the groups Z or Z_n: see Exercise 12.

We have now set up the theory of products in cohomology, and certainly two homotopy-equivalent CW-complexes will have ring-isomorphic cohomology rings. In order to apply this theory in practice, however, we clearly need an effective method of calculation. This reduces to a calculation of the cohomology homomorphism induced by the diagonal map $\Delta: K \to K \mathbin{\bar{\times}} K$ (in the case of unreduced cohomology). It is not very easy to do this for arbitrary CW-complexes, but if K is a polyhedron there is a simple formula for a chain map

$$d: C(K) \to C(K \times K)$$

that induces the same homology and cohomology homomorphisms as Δ (we may now write $\times$ rather than $\bar{\times}$, since obviously a polyhedron is a countable CW-complex).

Let K be a simplicial complex, which by Proposition 7.3.2 is also a CW-complex. Moreover Proposition 4.4.21 shows that we may identify the simplicial chain complex $C(K)$ with the cellular chain complex $C(|K|)$. Suppose that the vertices of K are totally ordered, as $a^0 < a^1 < a^2 < \cdots$, say.

Theorem 8.5.6 *For each $n \geqslant 0$, define $d: C_n(K) \to C_n(|K| \times |K|)$ by*

$$d[a^{i_0}, \ldots, a^{i_n}] = \sum_r [a^{i_0}, \ldots, a^{i_r}] \times [a^{i_r}, \ldots, a^{i_n}],$$

where $i_0 < \cdots < i_n$. Then d is a chain map, and induces the same homology and cohomology homomorphisms as Δ.

Proof. To show that d is a chain map, observe that

$$\begin{aligned}\partial d[a^0,\ldots,a^n] &= \sum_{r=0}^{n} (\partial[a^0,\ldots,a^r]\times[a^r,\ldots,a^n] + (-1)^r[a^0,\ldots,a^r] \\ &\qquad \times \partial[a^r,\ldots,a^n]) \\ &= \sum_{r=0}^{n}\Bigg(\sum_{s=0}^{r-1}(-1)^s[a^0,\ldots,\hat{a}^s,\ldots,a^r]\times[a^r,\ldots,a^n] \\ &\qquad + (-1)^r[a^0,\ldots,a^{r-1}]\times[a^r,\ldots,a^n] \\ &\qquad + \sum_{s=r+1}^{n}(-1)^s[a^0,\ldots,a^n]\times[a^r,\ldots,\hat{a}^s,\ldots,a^n] \\ &\qquad + (-1)^r[a^0,\ldots,a^r]\times[a^{r+1},\ldots,a^n]\Bigg) \\ &= d\partial[a^0,\ldots,a^n].\end{aligned}$$

The rest of the proof consists in constructing a chain homotopy h between d and $\bar{\Delta}_.$, where $\bar{\Delta}$ is a cellular approximation to Δ. This will show that $d_* = \Delta_*$, and dually $h \pitchfork 1$ will be a chain homotopy between $d \pitchfork 1$ and $\bar{\Delta}_. \pitchfork 1$, so that $d^* = \Delta^*$. To construct h, note first that we may assume that, for each simplex σ of K, $\bar{\Delta}(\sigma) \subset \sigma \times \sigma$. This follows by induction on the dimension of σ: in constructing $\bar{\Delta}$ by the method of Theorem 7.5.7, assume that Δ has been constructed on $|K^{n-1}|$ with the homotopy F between Δ and $\bar{\Delta}$ sending $\sigma \times I$ into $\sigma \times \sigma$ for each $\sigma \in K^{n-1}$ (this is certainly possible if $n = 1$). Then F can be extended to $|K^n| \times I$, with the same property, since

$$\pi_n(\sigma \times \sigma, (\sigma \times \sigma)^n) = 0.$$

We can now construct h on $C_n(K)$ by induction on n. Suppose that $h\colon C_{n-1}(K) \to C_n(|K| \times |K|)$ has been defined, such that

$$\partial h(\sigma) + h\partial(\sigma) = \bar{\Delta}_.(\sigma) - d(\sigma),$$

and $h(\sigma) \in C_n(\sigma \times \sigma)$, for all $(n-1)$-simplexes σ. This can certainly be done for $n = 1$, since $\bar{\Delta}_. = d$ on $C_0(K)$ and we may define $h = 0\colon C_0(K) \to C_1(|K| \times |K|)$. Now if σ is an n-simplex, h is already defined on $\partial(\sigma)$, and

$$\begin{aligned}\partial[-h\partial(\sigma) + \bar{\Delta}_.(\sigma) - d(\sigma)] &= -(\partial h)(\partial\sigma) + \bar{\Delta}_.\partial(\sigma) - d\partial(\sigma) \\ &= (h\partial)(\partial\sigma) - \bar{\Delta}_.\partial(\sigma) + d\partial(\sigma) \\ &\qquad + \bar{\Delta}_.\partial(\sigma) - d\partial(\sigma) \\ &= 0.\end{aligned}$$

But $-h\partial(\sigma) + \bar{\Delta}_{.}(\sigma) - d(\sigma)$ is an element of $C_n(\sigma \times \sigma)$, and $\sigma \times \sigma$ is contractible, so that $H_n(\sigma \times \sigma) = 0$ $(n \geqslant 1)$. Hence there exists an element $h(\sigma) \in C_{n+1}(\sigma \times \sigma)$, such that

$$\partial h(\sigma) + h\partial(\sigma) = \bar{\Delta}_{.}(\sigma) - d(\sigma),$$

as required. ∎

Corollary 8.5.7 *The ring structure in $H^*(K; G)$ is induced by a product*

$$\cup : (C_r(K) \pitchfork G) \otimes (C_s(K) \pitchfork G) \to C_{r+s}(K) \pitchfork G,$$

where if $x \in C_r(K) \pitchfork G$, $y \in C_s(K) \pitchfork G$, and $(a^0, \ldots, a^{r+s})$ is an $(r + s)$-simplex with its vertices in the correct order, we have

$$(x \cup y)[a^0, \ldots, a^{r+s}] = x[a^0, \ldots, a^r].y[a^r, \ldots, a^{r+s}].$$ ∎

Corollary 8.5.8 *If $|K|$ is connected, $1 \in H^0(K; G) \cong G$ acts as an identity element for $H^*(K; G)$.* ∎

Theoretically, Corollary 8.5.7 gives all the information necessary to calculate the product in $H^*(K; G)$. However, as we saw in Chapters 4 and 5, it is much too laborious in practice to use the individual simplexes in making calculations with homology or cohomology. Since it appears difficult to give an analogue of Theorem 8.5.6 involving a block dissection or CW-decomposition of K, we cannot hope to improve on Corollary 8.5.7 in general; but if $|K|$ is a homology manifold, the ring structure in $H^*(K)$ can easily be computed by relating it to the Poincaré duality isomorphism.

Theorem 8.5.9 *Let K be a triangulation of an orientable homology n-manifold, and let $\bar{D}: H^r(K') \to H_{n-r}(K')$ be the Poincaré duality isomorphism. Then if $x \in H^r(K')$, $y \in H^{n-r}(K')$, we have*

$$\bar{D}(x \cup y) = \langle x, \bar{D}y \rangle h_0,$$

where h_0 is the homology class of any vertex of K'.

Proof. As remarked after Theorem 5.3.17, $\bar{D}$ is given at chain level by $\bar{D}(x) = x \cap \phi(z)$, where $x \in C(K') \pitchfork Z$ and $z \in C_n(K)$ is the sum of all the n-simplexes of K. Here, the definition of $x \cap \phi(z)$ involves an ordering of the vertices of K', which we assume done in such a way that $\hat{\sigma} < \hat{\tau}$ if dim $\sigma >$ dim τ. Using the same ordering to give the ring structure in $H^*(K)$, if $x \in C_r(K') \pitchfork Z$, $y \in C_{n-r}(K') \pitchfork Z$, and

$(a^0, \ldots, a^n)$ is an n-simplex of K' with its vertices in the correct order, we have

$$\begin{aligned}(x \cup y) \cap [a^0, \ldots, a^n] &= [a^0].(x \cup y)[a^0, \ldots, a^n] \\ &= [a^0].x[a^0, \ldots, a^r].y[a^r, \ldots, a^n] \\ &= \langle x, y \cap [a^0, \ldots, a^n]\rangle [a^0].\end{aligned}$$

It follows that $(x \cup y) \cap \phi(z) = \langle x, y \cap \phi(z)\rangle [a^0]$, so that, passing to homology classes, we have $\bar{D}(x \cup y) = \langle x, \bar{D}(y)\rangle h_0$. ∎

The same argument shows that a similar formula holds using Z_m or Q coefficients instead of Z, and (with Z_2 coefficients only) if K is non-orientable.

Examples 8.5.10

(a) We know that RP^n is a homology n-manifold, and that $H_r(RP^n; Z_2) \cong H^r(RP^n; Z_2) \cong Z_2$ if $0 \leqslant r \leqslant n$. Moreover the CW-decomposition of RP^n given in Examples 7.3.3(c) shows that, if $i: RP^{n-1} \to RP^n$ is the inclusion map, $i_*: H_r(RP^{n-1}; Z_2) \to H_r(RP^n; Z_2)$ is isomorphic for $0 \leqslant r \leqslant n - 1$; similar remarks apply to cohomology.

Let x be the generator of $H^1(RP^n; Z_2)$ corresponding under the inclusion map to a choice of generator of $H^1(RP^1; Z_2)$; we shall show that x^r, the r-fold product of x with itself, is a generator of $H^r(RP^n; Z_2)$, $0 \leqslant r \leqslant n$. For suppose this is true in RP^{n-1} (it is certainly true in RP^1). Then by Axiom 1 of Definition 8.5.1 x^r is a generator of $H^r(RP^n; Z_2)$ for $0 \leqslant r \leqslant n - 1$. Moreover by Theorem 8.5.9

$$\begin{aligned}\bar{D}(x^n) &= \bar{D}(x \cup x^{n-1}) \\ &= \langle x, \bar{D}(x^{n-1})\rangle h_0.\end{aligned}$$

But $\langle x, \bar{D}(x^{n-1})\rangle = 1$ by Proposition 5.2.11, since $\bar{D}$ is an isomorphism, so that $x^n \neq 0$ and therefore x^n is a generator of $H^n(RP^n; Z_2)$.

In other words, $H^*(RP^n; Z_2)$ is isomorphic to the polynomial algebra $Z_2[x]$, subject to the relation $x^{n+1} = 0$.

(b) To deduce the ring structure of $H^*(CP^n; Z_2)$ and $H^*(HP^n; Z_2)$, we make use of a cellular map $c: RP^{2n} \to CP^n$, defined by

$$c[x_1, \ldots, x_{2n+1}] = [x_1 + ix_2, \ldots, x_{2n-1} + ix_{2n}, x_{2n+1}]$$

(this is easily seen to be well-defined and continuous). If $i_R: RP^{2n-2} \to RP^{2n}$ and $i_C: CP^{n-1} \to CP^n$ are the inclusion maps, then $ci_R = i_C c$; and if $c': RP^{2n-1} \to CP^{n-1}$ is defined by

$$c'[x_1, \ldots, x_{2n}] = [x_1 + ix_2, \ldots, x_{2n-1} + ix_{2n}],$$

then there is a commutative diagram

$$\begin{array}{ccccc} & & RP^{2n-1} & \xrightarrow{i_R} & RP^{2n} \\ & \overset{p_R}{\nearrow} & & & \\ S^{2n-1} & & \downarrow{\scriptstyle c'} & & \downarrow{\scriptstyle c} \\ & \underset{p_C}{\searrow} & & & \\ & & CP^{n-1} & \xrightarrow[i_C]{} & CP^n \end{array}$$

where p_R and p_C are the local product maps as in Chapter 7, Exercise 19. It follows easily that $c_{.}: C_{2n}(RP^{2n}) \to C_{2n}(CP^n)$ is an isomorphism, and hence that $c_{.}: C_{2r}(RP^{2n}) \to C_{2r}(CP^n)$ is isomorphic for $0 \leqslant r \leqslant n$.

Thus $c^*: H^{2r}(CP^n; Z_2) \to H^{2r}(RP^{2n}; Z_2)$ is isomorphic for $0 \leqslant r \leqslant n$; and if $y \in H^2(CP^n; Z_2)$ is the generator such that $c^*(y) = x^2$, then $c^*(y^r) = x^{2r}$. Hence y^r generates $H^{2r}(CP^n; Z_2)$, and so $H^*(CP^n; Z_2)$ is isomorphic to $Z_2[y]$, subject to the relation $y^{n+1} = 0$.

A similar argument shows that $H^*(HP^n; Z_2)$ is isomorphic to $Z_2[z]$, subject to $z^{n+1} = 0$, where $z \in H^4(HP^n; Z_2)$. ∎

Note. It is possible to prove (b) directly from Theorem 8.5.9, by showing that CP^n and HP^n are homology manifolds. Since they are in fact orientable, similar results will hold with Z rather than Z_2 coefficients.

As a corollary of (b), we can easily prove that CP^2 and $S^2 \vee S^4$ are not homotopy-equivalent. For if $p, q: S^2 \vee S^4 \to S^2, S^4$ are the projection maps, then $H^r(S^2 \vee S^4; Z_2) \cong Z_2$ for $r = 2, 4$, the generators being $p^*(s_2)$ and $q^*(s_4)$, where s_2 and s_4 are the generators of $H^2(S^2; Z_2)$, $H^4(S^4; Z_2)$ respectively. Now

$$p^*(s_2) \cup p^*(s_2) = p^*(s_2 \cup s_2) = 0,$$

so that although $H^*(S^2 \vee S^4; Z_2)$ and $H^*(CP^2; Z_2)$ are isomorphic as groups, there is no ring isomorphism between them. Thus $S^2 \vee S^4$ and CP^2 cannot be homotopy-equivalent.

In order to calculate the cohomology ring of a product of two CW-complexes, the following proposition is useful.

Proposition 8.5.11 *Given CW-complexes K and L, let $p, q: K \bar{\times} L \to K, L$ be the projection maps (which are continuous by Proposition 7.3.23). Then if G is any commutative ring with* 1, *the following diagram is commutative:*

$$\begin{array}{ccc} H^*(K;G)\otimes H^*(L;G) & \xrightarrow{\times} & H^*(K\bar{\times} L;G) \\ \downarrow{\scriptstyle p^*\otimes q^*} & \nearrow{\scriptstyle \cup} & \\ H^*(K\bar{\times} L;G)\otimes H^*(K\bar{\times} L;G) & & \end{array}$$

Proof. By Axiom 1 of Definition 8.5.1, the diagram

$$\begin{array}{ccc} H^*(K;G)\otimes H^*(L;G) & \xrightarrow{\times} & H^*(K\bar{\times} L;G) \\ \downarrow{\scriptstyle p^*\otimes q^*} & & \downarrow{\scriptstyle (p\bar{\times} q)^*} \\ H^*(K\bar{\times} L;G)\otimes H^*(K\bar{\times} L;G) & \xrightarrow[\times]{} & H^*(K\bar{\times} L\bar{\times} K\bar{\times} L;G) \end{array}$$

is commutative. But clearly $(p \bar{\times} q)\Delta = 1: K \bar{\times} L \to K \bar{\times} L$. ∎

Example 8.5.12 It follows that $H^*(S^n \times S^m)$ has generators $s_n \in H^n$, $s_m \in H^m$ and $z \in H^{n+m}$, where $z = s_n \cup s_m$. All other products of s_n, s_m and z are zero, since s_n and s_m are in the images of p^* and q^* respectively, and $H^r(S^n \times S^m) = 0$ for $r > n + m$. A similar result holds using Z_2 coefficients.

We can now, at last, prove that $S^2 \times S^4$ and CP^3 are not homotopy-equivalent. For if y is the generator of $H^2(CP^3; Z_2)$, then $y^3 \neq 0$; but $(s_2)^3 = 0$ in $H^4(S^2 \times S^4; Z_2)$. ∎

We end this chapter with an important geometrical application of the theory of products, on the non-existence of antipodal maps of S^n to S^m, $m < n$.

Definition 8.5.13 A map $f: S^n \to S^m$ $(n, m \geqslant 0)$ is called *antipodal* if, for all points $x \in S^n$, $f(-x) = -f(x)$.

Theorem 8.5.14 *There is no antipodal map $f: S^n \to S^m$, if $n > m \geqslant 0$.*

Proof. Suppose that there were such a map f. Then f would induce a map $g: RP^n \to RP^m$, such that the diagram

$$\begin{array}{ccc} S^n & \xrightarrow{f} & S^m \\ \downarrow{\scriptstyle p_n} & & \downarrow{\scriptstyle p_m} \\ RP^n & \xrightarrow[g]{} & RP^m \end{array}$$

is commutative, where p_n and p_m are the local product maps of Chapter 7, Exercise 19. Now by Chapter 6, Exercise 19, p_n and p_m are Serre fibre maps; and if we choose base points in all four spaces so as to make all the maps based, f induces a homeomorphism on the fibres F_n and F_m, which are both homeomorphic to S^0. So by Chapter 6, Exercise 18, there is a commutative diagram of exact sequences

$$\begin{array}{ccccccccc} 0 & \longrightarrow & \pi_1(S^n) & \xrightarrow{(p_n)_*} & \pi_1(RP^n) & \longrightarrow & \pi_0(F_n) & \longrightarrow & 0 \\ & & \downarrow f_* & & \downarrow g_* & & \downarrow f_* & & \downarrow \\ 0 & \longrightarrow & \pi_1(S^m) & \xrightarrow[(p_m)_*]{} & \pi_1(RP^m) & \longrightarrow & \pi_0(F_m) & \longrightarrow & \pi_0(S^m) \longrightarrow 0. \end{array}$$

Here, $f_*: \pi_0(F_n) \to \pi_0(F_m)$ is a (1-1)-correspondence. This is an immediate contradiction if $m = 0$ (the third square cannot be commutative), and if $m = 1$ (the second square cannot be commutative, since $\pi_1(RP^n) \to \pi_0(F_n)$ must be a (1-1)-correspondence and $g_* = 0$ because $\pi_1(RP^n) \cong Z_2$ and $\pi_1(RP^m) \cong Z$). On the other hand, if $n > m \geqslant 2$, then $\pi_1(S^n) = \pi_1(S^m) = 0$, and so $g_*: \pi_1(RP^n) \to \pi_1(RP^m)$ is an isomorphism. Thus by Theorem 8.3.7 $g_*: H_1(RP^n) \to H_1(RP^m)$ is also an isomorphism, both groups being isomorphic to Z_2. Using the exact coefficient sequence associated with Z_2, this shows that $g_*: H_1(RP^n; Z_2) \to H_1(RP^m; Z_2)$ is an isomorphism, and so $g^*: H^1(RP^m; Z_2) \to H^1(RP^n; Z_2)$ is isomorphic by Proposition 5.2.11. Thus if x is the generator of $H^1(RP^m; Z_2)$, $g^*(x)$ generates $H^1(RP^n; Z_2)$, and hence $g^*(x^n) = [g^*(x)]^n \neq 0$. But this is a contradiction, since $x^n = 0$ because $m < n$. ∎

An interesting corollary of Theorem 8.5.14 is the Fixed-Point Theorem of Borsuk.

Theorem 8.5.15 *Given any continuous map $f: S^n \to R^n$, there exists a point $x \in S^n$ such that $f(x) = f(-x)$.*

Proof. Suppose no such point x exists. Then a continuous map $g: S^n \to S^{n-1}$ can be defined by setting

$$g(x) = [f(x) - f(-x)]/\|f(x) - f(-x)\|.$$

But g is clearly antipodal. ∎

Corollary 8.5.16 *Let $X_1, \ldots, X_n$ be bounded measurable subsets of R^n. Then there exists an $(n - 1)$-dimensional hyperplane Y in R^n that bisects each of $X_1, \ldots, X_n$.*

Proof. Given a point $x \in S^n \subset R^{n+1}$, let Z_x be the n-dimensional hyperplane of R^{n+1} through $(0, \ldots, 0, 1)$, perpendicular to the vector x. For $1 \leqslant r \leqslant n$, let $f_r(x)$ be the measure of that part of X_r that lies on the same side of Z_x as $x + (0, \ldots, 0, 1)$. It is easy to see that f_r is a continuous function from S^n to R^1, and hence that

$$f(x) = (f_1(x), \ldots, f_n(x))$$

is a continuous function from S^n to R^n. By Theorem 8.5.15 there is a point $x \in S^n$ such that $f_r(x) = f_r(-x)$ for all r; but since $Z_x = Z_{-x}$ and $x + (0, \ldots, 0, 1)$ and $-x + (0, \ldots, 0, 1)$ are on opposite sides of Z_x, this means that Z_x bisects each X_r. Hence $Y = R^n \cap Z_x$ is an $(n - 1)$-dimensional hyperplane in R^n that bisects each X_r. ∎

If $n = 3$, Corollary 8.5.16 says that three bounded measurable sets in R^3 can be simultaneously bisected with one plane. This result is popularly known as the Ham Sandwich Theorem: no matter how the slices of bread and the slice of ham are arranged, it is always possible to cut the sandwich in half with a single knife cut.

EXERCISES

1. Let $\Omega_n(X)$ and $\Omega_n(X, Y)$ be as in Chapter 7, Exercises 2 and 4. Show that, for any pair (X, Y), $h_n\Omega_n(X) = h_n\Omega_n(X, Y) = 0$, so that the Hurewicz homomorphisms may be regarded as homomorphisms

$$h_n\colon \pi_n^*(X) \to H_n(X), \qquad h_n\colon \pi_n^*(X, Y) \to H_n(X, Y).$$

Prove also that, if K is a connected CW-complex and $n \geqslant 2$, $h_n\colon \pi_n^*(K^n, K^{n-1}) \to H_n(K^n, K^{n-1})$ is always an isomorphism. (*Hint:* use Chapter 7, Exercise 16.)

2. The *dunce hat* D is the space obtained from an equilateral triangle by identifying edges as shown in Fig. 8.1.

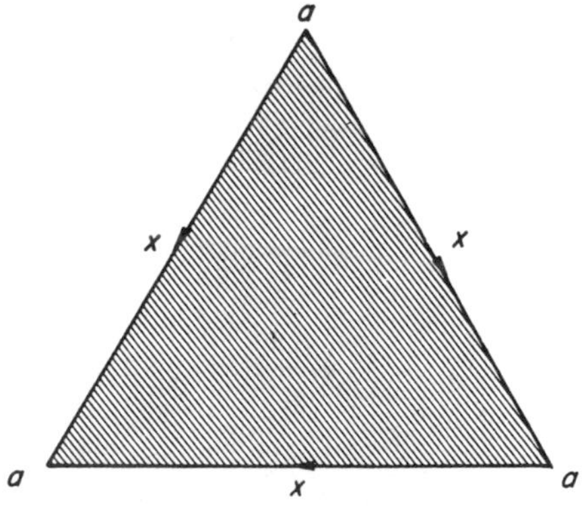

Fig. 8.1

Show that D is triangulable, and use Theorem 8.3.10 to show that D is contractible. However, D is not collapsible.

3. Show that a simply connected homology n-manifold X is homotopy-equivalent to S^n if $\tilde{H}_r(X) = 0$ for $r \leqslant [n/2]$, where $[n/2]$ is the integer part of $n/2$. Deduce that if X is the 3-manifold of Chapter 5, Exercise 9, then SX is a homology 4-manifold that is homotopy-equivalent to S^4, but not homeomorphic to S^4 (thus the Poincaré conjecture is false for *homology* manifolds).

4. Let $f\colon K \to L$ be a based map between connected CW-complexes, and let $\tilde{f}\colon \tilde{K} \to \tilde{L}$ be the corresponding map of their universal covers, so that there is a commutative diagram

$$\begin{array}{ccc} \tilde{K} & \xrightarrow{\tilde{f}} & \tilde{L} \\ {\scriptstyle g}\downarrow & & \downarrow{\scriptstyle h} \\ K & \xrightarrow[f]{} & L, \end{array}$$

where g and h are the covering maps (see Chapter 6, Exercises 23 and 25). Show that, if

$$f_*\colon \pi_1(K) \to \pi_1(L)$$

and

$$\tilde{f}_*\colon H_r(\tilde{K}) \to H_r(\tilde{L})$$

are isomorphic for all r, then f is a (based) homotopy equivalence.

5. Let L, M be subcomplexes of a CW-complex K, where $K = L \cup M$, and suppose that a 0-cell k_0 of $L \cap M$ is taken as the base point. By considering the inclusion map of $L \vee M$ in

$$K' = (L \times 0) \cup \frac{(L \cap M) \times I}{k_0 \times I} \cup (M \times 1),$$

show that for any Ω-spectrum $\mathbf{E}$ there is an exact Mayer–Vietoris sequence

$$\cdots \longrightarrow \tilde{H}^n(K; \mathbf{E}) \xrightarrow{\zeta^*} \tilde{H}^n(L; \mathbf{E}) \oplus \tilde{H}^n(M; \mathbf{E}) \xrightarrow{\eta^*}$$
$$\tilde{H}^n(L \cap M; \mathbf{E}) \xrightarrow{\partial^*} \tilde{H}^{n+1}(K; \mathbf{E}) \longrightarrow \cdots,$$

where ζ^* is induced by the inclusion maps of L, M in K, and $\eta^*(x \oplus y) = i_1^*(x) - i_2^*(y)$, where $i_1, i_2\colon L \cap M \to L, M$ are the inclusion maps.

6. Let $i\colon L \to K$ be the inclusion of a subcomplex in a CW-complex, and let M be another CW-complex. Show that $(C_i) \barwedge M$ and $C_{(i \barwedge 1)}$ are homeomorphic, where $i \barwedge 1\colon L \barwedge M \to K \barwedge M$.

Given any integer $r > 1$, let L_r be the space $S^1 \cup_f E^2$, where $f\colon S^1 \to S^1$ is a map of degree r. For any Ω-spectrum $\mathbf{E}$, define

$$H^n(K; \mathbf{E}; Z_r) = H^{n+2}(K \wedge L_r; \mathbf{E}).$$

Prove that there is an exact sequence

$$0 \to H^n(K; \mathbf{E}) \otimes Z_r \to H^n(K; \mathbf{E}; Z_r) \to \operatorname{Tor}(H^{n+1}(K; \mathbf{E}), Z_r) \to 0,$$

and that if $E_n = K(Z, n)$ then $H^n(K; \mathbf{E}; Z_r) = H^n(K; Z_r)$.

7. Show that, if G is a finitely generated abelian group, then $K(G, n)$ can be constructed as a countable CW-complex. (*Hint:* use Chapter 7, Exercise 12.) Deduce from Corollary 7.5.3 that $K(G, n)$ is an AHI.

8. Let K be a CW-complex, and let Y be a path-connected $(n-1)$-simple space. Given a map $f: K^{n-1} \to Y$, define $c(f) \in C_n(K) \pitchfork \pi_{n-1}(Y)$ by

$$c(f)(\alpha) = [f\phi_\alpha^n],$$

where the characteristic map ϕ_α^n is regarded as a map of S^{n-1} to K^{n-1}. Similarly, given two maps $f, g: K^{n-1} \to Y$, such that $f \simeq g$ on K^{n-2} by a homotopy $F: K^{n-2} \times I \to Y$, define $d(f, g) \in C_{n-1}(K) \pitchfork \pi_{n-1}(Y)$ by regarding f, g, F as a map $\bar{F}: (K \times I)^{n-1} \to Y$, and setting

$$d(f, g) = (\theta \pitchfork 1)c(\bar{F}),$$

where $\theta: C(K) \to C(K \times I)$ is defined by $\theta(\alpha) = \alpha \times I$. Prove the following results ($n \geqslant 2$).

(a) $c(f)$ depends only on the homotopy class of f, and is zero if and only if f has an extension to a map $f: K^n \to Y$.
(b) $\delta c(f) = 0$. (*Hint:* use Exercise 1, and show that $c(f)$ may alternatively be defined to be the composite

$$H_n(K^n, K^{n-1}) \xleftarrow{h_n} \pi_n^*(K^n, K^{n-1}) \xrightarrow{\partial_*} \pi_{n-1}^*(K^{n-1}) \xrightarrow{f_*} \pi_{n-1}(Y).)$$

(c) $\delta d(f, g) = (-1)^n(c(f) - c(g))$.
(d) Given $d \in C_{n-1}(K) \pitchfork \pi_{n-1}(Y)$, and $f: K^{n-1} \to Y$, there exists $g: K^{n-1} \to Y$, such that $g = f$ on K^{n-2}, and $d(f, g) = d$.
(e) If $\gamma(f)$ denotes the homology class of $c(f)$ in $H^n(K; \pi_{n-1}(Y))$, then $\gamma(f) = 0$ if and only if there exists a map $g: K^n \to Y$ such that $f = g$ on K^{n-2}.

9. Let K and L be CW-complexes. Show that there is an exact sequence

$$0 \longrightarrow Z_r(C(K)) \otimes C_s(L) \longrightarrow C_r(K) \otimes C_s(L) \xrightarrow{\partial \otimes 1} B_{r-1}(C(K)) \otimes C_s(L) \longrightarrow 0,$$

and deduce from Theorem 4.4.2 and Proposition 8.5.3 that there is an exact sequence

$$0 \to \bigoplus_{r+s=n} H_r(K) \otimes H_s(L) \to H_n(K \mathbin{\bar{\times}} L) \to \bigoplus_{r+s=n-1} \operatorname{Tor}(H_r(K), H_s(L)) \to 0.$$

By constructing a suitable homomorphism

$$\theta: H_n(K \mathbin{\bar{\times}} L) \to \bigoplus_{r+s=n} H_r(K) \otimes H_s(L),$$

prove that this sequence splits, so that

$$H_n(K \mathbin{\bar{\times}} L) \cong \bigoplus_{r+s=n} H_r(K) \otimes H_s(L) \oplus \bigoplus_{r+s=n-1} \operatorname{Tor}(H_r(K), H_s(L)).$$

Show similarly that

$$\tilde{H}_n(K \mathbin{\bar{\wedge}} L) \cong \bigoplus_{r+s=n} \tilde{H}_r(K) \otimes \tilde{H}_s(L) \oplus \bigoplus_{r+s=n-1} \operatorname{Tor}(\tilde{H}_r(K), \tilde{H}_s(L)).$$

10. Let (K, L) be a CW-pair, such that L is a retract of K. Show that, if G is a commutative ring with a 1,

$$H^*(K; G) \cong \operatorname{Im} r^* \oplus \operatorname{Ker} i^*$$

as a direct sum of groups, and that $\operatorname{Im} r^*$ is a subring and $\operatorname{Ker} i^*$ is an ideal ($i: L \to K$ and $r: K \to L$ are the inclusion and retraction maps respectively).

By taking $K = CP^2$, $L = CP^1$ and $G = Z_2$, show that CP^1 is not a retract of CP^2.

11. Given an element $x \in \pi_{2n-1}(S^n)$ $(n \geqslant 1)$, the *Hopf invariant* of x is defined as follows. Represent x by a map $f: S^{2n-1} \to S^n$, and let $Y = C_f$. Then $H^n(Y) \cong Z$, $H^{2n}(Y) \cong Z$, the generators being y_n, y_{2n}, where $f_1^*(y_n) = s_n$, $y_{2n} = f_2^*(s_{2n})$, and s_n, s_{2n} are the generators of $H^n(S^n)$, $H^{2n}(S^{2n})$ respectively. The Hopf invariant of x, $\theta(x)$, is then defined by

$$(y_n)^2 = \theta(x) \cdot y_{2n}.$$

Prove the following results.

(a) $\theta(x)$ depends only on x, and not on the choice of f (use Chapter 6, Exercise 11).

(b) $\theta: \pi_{2n-1}(S^n) \to Z$ is a homomorphism. (*Hint:* consider $C_{\nabla(f \vee g)}$, where $\nabla(f \vee g): S^{2n-1} \vee S^{2n-1} \to S^n$.)

(c) $\theta = 0$ if n is odd.

(d) If n is even, $\theta[\iota_n, \iota_n] = \pm 2$, where ι_n is the generator of $\pi_n(S^n)$ represented by the identity map. Deduce that, if n is even, $\pi_{2n-1}(S^n)$ has an element of infinite order, and that S^n cannot be an H-space.

12. Show that

$$\tilde{H}^{r+s}(K(Z, r) \mathbin{\bar{\wedge}} K(Z, s)) \cong Z$$

and

$$\tilde{H}^{r+s}(K(Z_n, r) \mathbin{\bar{\wedge}} K(Z_n, s); Z_n) \cong Z_n,$$

for any integer $n > 1$. Deduce that Axioms 1–4 of Definition 8.5.1 determine the product $\wedge$ uniquely for cohomology with coefficients Z or Z_n.

13. Given CW-complexes K, L and a commutative ring with a 1, G, define homomorphisms $\backslash : (C_r(L) \bar{\times} G) \otimes (C_n(K \bar{\times} L) \otimes G) \to C_{n-r}(K) \otimes G$ by the rule

$$x\backslash[(\alpha \times \beta) \otimes g] = \begin{cases} \alpha \otimes (x(\beta).g), & \alpha \in C_{n-r}(K), \beta \in C_r(L) \\ 0, & \text{otherwise.} \end{cases}$$

Show that $(\partial \otimes 1)(x\backslash y) = x\backslash[(\partial \otimes 1)y] + (-1)^{n-r}\delta x\backslash y$, and deduce that $\backslash$ induces 'products'

$$\backslash : H^r(L; G) \otimes H_n(K \bar{\times} L; G) \to H_{n-r}(K; G).$$

Establish the following properties of $\backslash$.

(a) Given maps $f: K \to M$, $g: L \to N$, then $f_*(g^* x\backslash y) = x\backslash[(f \bar{\times} g)_* y]$.
(b) Given $x \in H^r(L; G)$, $y \in H_s(M; G)$ and $z \in H_n(K \bar{\times} L \bar{\times} M; G)$, then

$$(x \times y)\backslash z = x\backslash(y\backslash z).$$

(c) Given $x \in H^n(K; G)$ and $y \in H_n(K; G)$, then $x\backslash y = \langle x, y \rangle$ in $H_0(P; G) = G$, where P is a single point, and y is regarded as an element of $H_n(P \times K; G)$.
(d) If K is a polyhedron, $x \in H^r(K; G)$ and $y \in H_n(K; G)$, then $x \cap y = x\backslash\Delta_*(y)$, where $\Delta: |K| \to |K| \times |K|$ is the diagonal map.

Establish similar results for reduced cohomology, involving $\bar{\wedge}$ instead of $\bar{\times}$.

14. Let K be a triangulation of an orientable homology n-manifold, and let L be a subcomplex of K. Let $\bar{L}$ be the supplement of L in K, so that a typical simplex of K' (with the usual ordering) is $(a^0, \ldots, a^n)$ where $(a^0, \ldots, a^{r-1}) \in \bar{L}$ and $(a^r, \ldots, a^n) \in L'$. Let $N(\bar{L})$ be the set of points in such simplexes of the form $\sum \lambda_i a^i$, where $\lambda_0 + \cdots + \lambda_{r-1} \geqslant \frac{1}{2}$, and let $N(L)$ be the set of points where $\lambda_r + \cdots + \lambda_n \geqslant \frac{1}{2}$; show that $|\bar{L}|$, $|L|$ are strong deformation retracts of $N(\bar{L})$, $N(L)$ respectively.

Now let M be a subcomplex of L, so that $N(M) \subset N(L)$, $N(\bar{L}) \subset N(\bar{M})$, and $|K| = N(\bar{L}) \cup N(M) \cup (N(\bar{M}) \cap N(L))$. Show that there are homotopy equivalences $p: (N(\bar{M}), N(\bar{L})) \to (|\bar{M}|, |\bar{L}|)$, $q: (N(L), N(M)) \to (|L|, |M|)$. Hence define a map

$$\Delta: |K| \to (|\bar{M}|/|\bar{L}|) \wedge (|L|/|M|)$$

by the rule

$$\Delta(x) = \begin{cases} p(x) \wedge q(x), & x \in N(\bar{M}) \cap N(L) \\ \text{base point}, & \text{otherwise.} \end{cases}$$

Also define a homomorphism $d: C(K') \to C((|\bar{M}|/|\bar{L}|) \wedge (|L'|/|M'|))$ by the rule $d[a^0, \ldots, a^n] = \sum [a^0, \ldots, a^r] \wedge [a^r, \ldots, a^n]$, the sum being taken over those values of r for which $(a^0, \ldots, a^r)$ is a simplex of $\bar{M} - \bar{L}$ and $(a^r, \dots, a^n)$ is a simplex of $L' - M'$. Show that d is a chain map, and induces the same homology and cohomology homomorphisms as Δ. (*Hint:* given a simplex $\sigma = (a^0, \dots, a^n) \in K'$, where $a^0, \dots, a^{r-1} \in \bar{L}$, $a^r, \dots, a^{s-1} \in \bar{M} \cap L'$ and $a^s, \dots, a^n \in M'$, use the method of Theorem 8.5.6, with $\sigma \times \sigma$ replaced by $(a^0, \dots, a^{s-1}) \wedge (a^r, \dots, a^n)$ in the definition of h.)

Let $\bar{D}\colon H^r(L', M') \to H_{n-r}(\bar{M}, \bar{L})$ be the duality isomorphism of Theorem 5.3.13, composed with $(h^*)^{-1}$, where $h\colon (|L'|, |M'|) \to (|L|, |M|)$ is a simplicial approximation to the identity. Deduce that

$$\bar{D}(x) = x \backslash \Delta_*(z),$$

where $z \in H_n(K) = \tilde{H}_n(K)$ is the standard generator, and hence show that, if $y \in H^{n-r}(\bar{M}, \bar{L})$, then

$$\bar{D}\Delta^*(y \wedge x) = \langle y, \bar{D}(x) \rangle h_0,$$

where $\bar{D}$ also denotes the Poincaré duality isomorphism $H^n(K') \cong H_0(K')$, and h_0 is the homology class of any vertex of K'.

Establish similar results for non-orientable homology manifolds, using Z_2 coefficients.

15. Let K be a simplicial complex. Show that, for each integer $r \geqslant 0$, there exists a homomorphism

$$d_r\colon C(K) \otimes Z_2 \to C(|K| \times |K|) \otimes Z_2,$$

such that the following properties hold.

(a) $d_0^* = \Delta^*$, where Δ is the diagonal map.
(b) $d_r[C_n(K) \otimes Z_2] \subset C_{n+r}(|K| \times |K|) \otimes Z_2$.
(c) $\partial d_r + d_r \partial = (1 + \tau) d_{r-1}$, where $\tau\colon C(|K| \times |K|) \otimes Z_2 \to C(|K| \times |K|) \otimes Z_2$ is the chain map exchanging the two factors.
(d) For each simplex σ, $d_r(\sigma \otimes 1) \in C(\sigma \times \sigma) \otimes Z_2$.
(e) If $d_0', d_1', d_2', \dots$ are another set of such homomorphisms, then there exist homomorphisms $h_r\colon C(K) \otimes Z_2 \to C(|K| \times |K|) \otimes Z_2$ $(r \geqslant 0)$, such that $d_r + d_r' = \partial h_r + h_r \partial + (1 + \tau) h_{r-1}$.

Now suppose that $x \in C_n(K) \pitchfork Z_2$, $y \in C_m(K) \pitchfork Z_2$, and define

$$x \cup_r y = (d_r \pitchfork 1)(x \times y) \in C_{n+m-r}(K) \pitchfork Z_2.$$

Show that $\delta(x \cup_r y) = \delta(x) \cup_r y + x \cup_r \delta(y) + x \cup_{r-1} y + y \cup_{r-1} x$, and hence define $Sq^r\colon H^n(K; Z_2) \to H^{n+r}(K; Z_2)$ by the rule

$$Sq^r[x] = [x \cup_{n-r} x].$$

Show that Sq^r is well defined (that is, is independent of the choice of d_{n-r} and x), and has the following properties.

(a) Sq^r is a homomorphism.
(b) If $f\colon |K| \to |L|$ is a continuous map, then $f^* Sq^r = Sq^r f^*$. (*Hint:*

use the Simplicial Approximation Theorem and a suitable modification of (e) above.)

(c) Sq^0 is the identity isomorphism.

(d) $Sq^r(x) = x^2$ if $x \in H^r(K; Z_2)$.

(e) $Sq^r(x) = 0$ if $x \in H^n(K; Z_2)$, $n < r$.

(f) If $x \in H^n(K; Z_2)$ and $y \in H^m(L; Z_2)$, then

$$Sq^r(x \times y) = \sum_{i=0}^{r} (Sq^{r-i}x) \times (Sq^i y) \quad \text{in} \quad H^{n+m+r}(|K| \times |L|; Z_2).$$

(*Hint:* show that Sq^r may be calculated in (the polyhedron) $|K| \times |L|$ by using the homomorphisms

$$D_r\colon C(|K| \times |L|) \otimes Z_2 \to C(|K| \times |K| \times |L| \times |L|) \otimes Z_2$$

defined by $D_r((\sigma \times \mu) \otimes 1) = \sum_{i=0}^{r} d_{r-i}(\sigma \otimes 1) \times \tau d_i(\mu \otimes 1)$.)

Extend the definition of Sq^r to $\tilde{H}^n(K; Z_2)$ and $H^n(K, L; Z_2)$, and prove:

(g) If δ^* is the homomorphism in the exact cohomology sequence of (K, L), then $\delta^* Sq^r = Sq^r \delta^*$.

(h) If $s^*\colon H^{n+1}(SK; Z_2) \to H^n(K; Z_2)$ is the suspension isomorphism, then $s^* Sq^r = Sq^r s^*$.

16. Show that Sq^r can be calculated in $H^*(RP^n; Z_2)$ by the rule

$$Sq^r(x^s) = \binom{s}{r} x^{s+r},$$

where x generates $H^1(RP^n; Z_2)$, $\binom{s}{r}$ is the binomial coefficient reduced mod 2, and $\binom{s}{r}$, x^s are interpreted as zero if $r > s$, $s > n$ respectively. Deduce that RP^4/RP^2 is not a retract of RP^5/RP^2 (both spaces have the homotopy type of polyhedra).

17. Let L be a subcomplex of some triangulation of S^n, and let $\bar{L}$ be the supplement of L; let

$$D\colon \tilde{H}^r(L; Z_2) \to \tilde{H}_{n-r-1}(\bar{L})$$

be the Alexander duality isomorphism, as in Theorem 5.3.19. Define a homomorphism $c_r(Sq^t)\colon \tilde{H}^{r-t}(L; Z_2) \to \tilde{H}^r(L; Z_2)$ by the rule

$$\langle y, Dc_r(Sq^t)x\rangle = \langle Sq^t y, Dx\rangle,$$

where $x \in \tilde{H}^{r-t}(L; Z_2)$ and $y \in \tilde{H}^{n-r-1}(\bar{L}; Z_2)$. Use Exercise 14 and Exercise 15(f) to show that

$$\sum_{i=0}^{t} c_r(Sq^{t-i}) Sq^i = 0 \qquad (t > 0),$$

and deduce that $c_r(Sq^t)$ depends only on Sq^t: we therefore drop the suffix r and write $c(Sq^t)$.

Now suppose, if possible, that some triangulation of RP^{2^n} can be regarded as a subcomplex of some triangulation of $S^{2^{n+1}-1}$. Use Exercise 15(e) to show that

$$c(Sq^{2^n-1}) = 0: H^1(RP^{2^n}; Z_2) \to H^{2^n}(RP^{2^n}; Z_2).$$

On the other hand, show by using Exercise 16 that $Sq^1, Sq^2, Sq^3, \ldots, Sq^{2^r-1}$ are all zero on x^{2^r}, and deduce that $c(Sq^{2^n-1})x = x^{2^n}$. This contradiction shows that RP^{2^n} cannot be embedded as a subcomplex of a triangulation of $S^{2^{n+1}-1}$.

NOTES ON CHAPTER 8

The Hurewicz theorem. Hurewicz first stated Theorem 8.3.7(a), and gave a sketch of the proof, in [74]. Theorem 8.3.7(b) can be refined slightly to read: if $\pi_r(K, L) = 0$ for $1 \leqslant r < n$ $(n \geqslant 2)$, then $h_n: \pi_n^*(K, L) \to H_n(K, L)$ is an isomorphism.

For an 'elementary' proof of the Hurewicz theorem for arbitrary spaces, see Spanier [131], Chapter 7; however, a much easier proof can be given, based on the work of Serre [125] on the homology of fibre spaces: see for example Hu [73], Chapter 10.

General cohomology theories. The cohomology theory associated with an Ω-spectrum **E** was first defined by G. W. Whitehead [155]. It is interesting to observe that, under mild restrictions, any 'cohomology theory' $h^*(K, L)$, having the properties in the statement of Theorem 8.4.4, is the cohomology theory associated with some Ω-spectrum: this is a theorem of E. H. Brown [30, 31]. It is possible to give a definition of the *homology* groups associated with an Ω-spectrum, although this is more complicated than the corresponding cohomology theory: for details, see G. W. Whitehead [155].

For the original definition of Eilenberg–MacLane spaces, see Eilenberg and MacLane [52, 54].

Important cohomology theories associated with other Ω-spectra include the groups $K^*(X)$ of Atiyah and Hirzebruch (see for example Atiyah and Hirzebruch [18], Adams [3]), and the groups $MU^*(X)$ of Conner and Floyd [42] and Atiyah [16] (see also Novikov [110]).

G. W. Whitehead has proved that the duality theorems of Section 5.3 extend to the homology and cohomology theories associated with an arbitrary Ω-spectrum **E**, provided that the manifolds involved are 'orientable with respect to $H_*(\ ; \mathbf{E})$'. The method is that of Exercise 14.

Products. Instead of defining products axiomatically, it is possible to work directly with the Ω-spectra; this is the approach of G. W. Whitehead [155]. The explanation why it is cohomology rather than homology that admits a ring structure is due originally to Lefschetz [91].

For the theory of products in the homology and cohomology of arbitrary spaces, see for example Spanier [131], Chapter 5.

The Borsuk–Ulam Theorem. Theorem 8.5.15 was conjectured by Ulam, and first proved by Borsuk [21].

The Dunce Hat. For more details, see Zeeman [168].

Exercise 6. This proof of the Universal Coefficient Theorem for general cohomology theories is based on a result of Puppe [119]. For details of the proof, see Araki and Toda [14] or Maunder [99].

Obstruction theory (Exercise 8). Most of this is due to Steenrod [137], Part III, who also deals with the problem of obstructions to cross-sections of fibre bundles. The theory can be extended to arbitrary topological spaces: see Olum [111].

The Hopf invariant. The definition in Exercise 11 is that of Steenrod [136], and is somewhat different from Hopf's original definition [70]. The Hopf invariant has been generalized by G. W. Whitehead [153] to a homomorphism $H: \pi_m(S^n) \to \pi_m(S^{2n-1})$ $(m \leqslant 4n - 4)$, and by Hilton [62] to a homomorphism $H^*: \pi_m(S^n) \to \pi_{m+1}(S^{2n})$ $(m > 0)$. It is the homomorphism H that occurs in the EHP sequence.

Adams has proved in [2] that there exist elements of Hopf invariant one in $\pi_{2n-1}(S^n)$ only if $n = 2$, 4 or 8.

Cohomology operations (Exercise 15). The operations Sq^r were first defined by Steenrod [135], who also constructed similar operations in cohomology with coefficients Z_p, for odd primes p [138]; for an elegant account of the theory, see Steenrod [139]. The operations can be extended to the cohomology of arbitrary spaces: see for example Spanier [131], Chapter 5.

There are many other applications of cohomology operations besides Exercise 17. For applications to obstruction theory, see Steenrod [135, 139]; and for applications to the calculation of the homotopy groups of spheres, via the Adams spectral sequence, see Adams [1, 5]. A general account of the operations and their uses will be found in Steenrod and Epstein [140].

More complicated cohomology operations, known as higher-order operations, have been studied by Adams [2] and Maunder [96].

Exercise 17. This proof that RP^{2^n} cannot be embedded in $S^{2^{n+1}-1}$ is due to Peterson [114], although the formula for $c(Sq^r)$ was first established by Thom [142] (see also Maunder [100]).

REFERENCES

[1] ADAMS, J. F. 'On the structure and applications of the Steenrod algebra' *Comment. math. helvet.* **32**, 180–214 (1958).

[2] ADAMS, J. F. 'On the non-existence of elements of Hopf invariant one' *Ann. Math.* **72**, 20–104 (1960).

[3] ADAMS, J. F. 'Vector fields on spheres' *Ann. Math.* **75**, 603–632 (1962).

[4] ADAMS, J. F. 'On the groups $J(X)$' *Topology* **2**, 181–195 (1963); **3** 137–171; 193–222 (1965); **5**, 21–71 (1966).

[5] ADAMS, J. F. *Stable Homotopy Theory*, Springer Verlag, Berlin (1964).

[6] AHLFORS, L. and SARIO, L. *Riemann Surfaces*, Princeton (1960).

[7] ALEXANDER, J. W. 'A proof of the invariance of certain constants of analysis situs' *Trans. Am. math. Soc.* **16**, 148–154 (1915).

[8] ALEXANDER, J. W. 'A proof and extension of the Jordan–Brouwer separation theorem' *Trans. Am. math. Soc.* **23**, 333–349 (1922).

[9] ALEXANDER, J. W. 'Combinatorial analysis situs' *Trans. Am. math. Soc.* **28**, 301–329 (1926).

[10] ALEXANDER, J. W. 'On the chains of a complex and their duals' *Proc. natn. Acad. Sci. U.S.A.* **21**, 509–511 (1935).

[11] ALEXANDER, J. W. 'On the connectivity ring of an abstract space' *Ann. Math.* **37**, 698–708 (1936).

[12] ALEXANDER, J. W. and VEBLEN, O. 'Manifolds of n dimensions' *Ann. Math.* **14**, 163–178 (1913).

[13] ALEXANDROFF, P. 'Untersuchungen über Gestalt und Lage abgeschlossener Mengen beliebiger Dimension' *Ann. Math.* **30**, 101–187 (1928).

[14] ARAKI, S. and TODA, H. 'Multiplicative structures in mod q cohomology theories' *Osaka J. Math.* **2**, 71–115 (1965).

[15] ARMSTRONG, M. A. and ZEEMAN, E. C. 'Transversality for piecewise-linear manifolds' *Topology* **8**, 433–466 (1967).

[16] ATIYAH, M. F. 'Bordism and cobordism' *Proc. Camb. phil. Soc.* **57**, 200–208 (1961).

[17] ATIYAH, M. F. *K-Theory*, Benjamin, New York (1967).

[18] ATIYAH, M. F. and HIRZEBRUCH, F. 'Vector bundles and homogeneous spaces' *Proc. Symposia pure Math.* **3**, 7–38 (American Mathematical Society, Providence, R.I., 1961).

[19] BARRATT, M. G. 'Track groups: I' *Proc. Lond. math. Soc.* (*3*) **5**, 71–106 (1955).

[20] BARRATT, M. G. and MAHOWALD, M. E. 'The metastable homotopy of $O(n)$' *Bull. Am. math. Soc.* **70**, 758–760 (1964).

[21] BORSUK, K. 'Drei Sätze über die *n*-dimensionale euklidische Sphäre' *Fundam. Math.* **20**, 177–190 (1933).

[22] BOTT, R. 'The stable homotopy of the classical groups' *Ann. Math.* **70**, 313–337 (1959).

[23] BOUSFIELD, A., CURTIS, E. B., KAN, D. M., QUILLEN, D. G., RECTOR, D. L. and SCHLESINGER, J. W. 'The mod *p* lower central series and the Adams spectral sequence' *Topology* **5**, 331–342 (1966).

[24] BRAHANA, H. R. 'Systems of circuits on two-dimensional manifolds' *Ann. Math.* **23**, 144–168 (1921).

[25] BROUWER, L. E. J. 'Über Abbildungen von Mannigfaltigkeiten' *Math. Annln* **71**, 97–115 (1912).

[26] BROWDER, W. 'Torsion in *H*-spaces' *Ann. Math.* **74**, 24–51 (1961).

[27] BROWDER, W. 'Homotopy-commutative *H*-spaces' *Ann. Math.* **75** 283–311 (1962).

[28] BROWDER, W., LIULEVICIUS, A. L. and PETERSON, F. P. 'Cobordism theories' *Ann. Math.* **84**, 91–101 (1966).

[29] BROWDER, W. and THOMAS, E. 'On the projective plane of an *H*-space' *Illinois J. Math.* **7**, 492–502 (1963).

[30] BROWN, E. H. 'Cohomology theories' *Ann. Math.* **75**, 467–484 (1962).

[31] BROWN, E. H. 'Abstract homotopy theory' *Trans. Am. math. Soc.* **119**, 79–85 (1965).

[32] BROWN, R. 'Ten topologies for $X \times Y$' *Q. Jl Math. Oxford* (*2*) **14**, 303–319 (1963).

[33] BROWN, R. 'Function spaces and product topologies' *Q. Jl Math. Oxford* (*2*) **15**, 238–250 (1964).

[34] BROWN, R. 'Two examples in homotopy theory' *Proc. Camb. phil. Soc.* **62**, 575–576 (1966).

[35] BROWN, R. 'Groupoids and van Kampen's theorem' *Proc. Lond. math. Soc.* (*3*) **17**, 385–401 (1967).

[36] CAUCHY, A.-L. 'Récherches sur les polyèdres: II' *J. Ec. polytech.* **9**, 76–86 (1813).

[37] ČECH, E. 'Théorie générale de l'homologie dans un espace quelconque' *Fundam. Math.* **19**, 149–183 (1932).

[38] ČECH, E. 'Höherdimensionale Homotopiegruppen' *Proc. Int. Congr. Mathematicians* **3**, 203 (Zürich, 1932).

[39] ČECH, E. 'Les groupes de Betti d'un complexe infini' *Fundam. Math.* **25**, 33–44 (1935).

[40] CHEVALLEY, C. *Theory of Lie Groups: I*, Princeton (1946).

[41] COHEN, D. E. 'Products and carrier theory' *Proc. Lond. math. Soc.* (*3*) **7**, 219–248 (1957).

[42] CONNER, P. E. and FLOYD, E. E. 'Differentiable periodic maps' *Bull. Am. math. Soc.* **68**, 76–86 (1962).

[43] DEHN, M. and HEEGAARD, P. 'Analysis situs' *Encyklopädie der mathematischen Wissenschaften*, III, AB 3, pp. 153–220, Leipzig (1907).

[44] Dold, A. 'Partitions of unity in the theory of fibrations' *Ann. Math.* **78**, 223–255 (1963).

[45] Dold, A. and Lashof, R. 'Principal quasifibrations and fibre homotopy equivalence of bundles' *Illinois J. Math.* **3**, 285–305 (1959).

[46] Dold, A. and Thom, R. 'Quasifaserungen und unendliche symmetrische Produkte' *Ann. Math.* **67**, 239–281 (1958).

[47] Dowker, C. H. 'Homology groups of relations' *Ann. Math.* **56**, 84–95 (1952).

[48] Dowker, C. H. 'Topology of metric complexes' *Am. J. Math.* **74**, 555–577 (1952).

[49] Eckmann, B. and Hilton, P. J. 'Groupes d'homotopie et dualité' *C. r. Acad. Sci. Paris Sér. A.-B.* **246**, 2444–2447 (1958).

[50] Eilenberg, S. 'Singular homology theory' *Ann. Math.* **45**, 407–447 (1944).

[51] Eilenberg, S. and MacLane, S. 'Group extensions and homology' *Ann. Math.* **43**, 757–831 (1942).

[52] Eilenberg, S. and MacLane, S. 'Relations between homology and homotopy groups' *Proc. natn. Acad. Sci. U.S.A.* **29**, 155–158 (1943).

[53] Eilenberg, S. and MacLane, S. 'General theory of natural equivalences' *Trans. Am. math. Soc.* **58**, 231–294 (1945).

[54] Eilenberg, S. and MacLane, S. 'On the groups $H(\pi, n)$' *Ann. Math.* **58**, 55–106 (1953); **60**, 49–139; 513–557 (1954).

[55] Eilenberg, S. and Steenrod, N. E. 'Axiomatic approach to homology theory' *Proc. natn. Acad. Sci. U.S.A.* **31**, 117–120 (1945).

[56] Eilenberg, S. and Steenrod, N. E. *Foundations of Algebraic Topology*, Princeton (1952).

[57] Eilenberg, S. and Zilber, J. A. 'Semi-simplicial complexes and singular homology' *Ann. Math.* **51**, 499–513 (1950).

[58] Fox, R. H. 'On topologies for function spaces' *Bull. Am. math. Soc.* **51**, 429–432 (1945).

[59] Freudenthal, H. 'Über die Klassen von Sphärenabbildungen' *Compositio math.* **5**, 299–314 (1937).

[60] Greenberg, M. J. *Lectures on Algebraic Topology*, Benjamin, New York (1967).

[61] Griffiths, H. B. 'The fundamental group of two spaces with a common point' *Q. Jl Math. Oxford* (*2*) **5**, 175–190 (1954); correction, (*2*) **6**, 154–155 (1955).

[62] Hilton, P. J. 'Suspension theorems and the generalized Hopf invariant' *Proc. Lond. math. Soc.* (*3*) **1**, 462–492 (1951).

[63] Hilton, P. J. *Homotopy Theory and Duality*, Nelson, London (1965).

[64] Hilton, P. J. and Wylie, S. *Homology Theory*, Cambridge (1960).

[65] Hirsch, M. W. 'A proof of the nonretractability of a cell onto its boundary' *Proc. Am. math. Soc.* **14**, 364–365 (1963).

[66] Hopf, H. 'Abbildungsklassen n-dimensionaler Mannigfaltigkeiten' *Math. Annln* **92**, 209–224 (1926).

[67] Hopf, H. 'A new proof of the Lefschetz formula on invariant points' *Proc. natn. Acad. Sci. U.S.A.* **14**, 149–153 (1928).

[68] Hopf, H. 'Über die algebraische Anzahl von Fixpunkten' *Math. Z.* **29**, 493–524 (1929).

[69] Hopf, H. 'Über die Abbildungen der 3-Sphäre auf die Kugelflache' *Math. Annln* **104**, 637–665 (1931).

[70] Hopf, H. 'Über die Abbildungen von Sphären auf Sphären niedrigerer Dimension' *Fundam. Math.* **25**, 427–440 (1935).

[71] Hopf, H. 'Über die Topologie der Gruppen-Mannigfaltigkeiten und ihre Verallgemeinerungen' *Ann. Math.* **42**, 22–52 (1941).

[72] Hu, S.-T. 'A theorem on homotopy extension' *Dokl. Akad. Nauk SSSR* **57**, 231–234 (1947).

[73] Hu, S.-T. *Homotopy Theory*, Academic Press, New York (1959).

[74] Hurewicz, W. 'Beiträge der Topologie der Deformationen' *Proc. K. Akad. Wet., Ser. A* **38**, 112–119; 521–528 (1935).

[75] Hurewicz, W. 'On duality theorems' *Bull. Am. math. Soc.* **47**, 562–563 (1941).

[76] Hurewicz, W. 'On the concept of fiber space' *Proc. natn. Acad. Sci. U.S.A.* **41**, 956–961 (1955).

[77] Jacobson, N. *Lectures in Abstract Algebra*, Vol. I, Van Nostrand, Princeton, N.J. (1951).

[78] James, I. M. 'The suspension triad of a sphere' *Ann. Math.* **63**, 407–429 (1956).

[79] James, I. M. 'On *H*-spaces and their homotopy groups' *Q. Jl Math. Oxford (2)* **11**, 161–179 (1960).

[80] James, I. M. 'On homotopy-commutativity' *Topology* **6**, 405–410 (1967).

[81] Jordan, C. 'Des contours tracés sur les surfaces' *J. Math. pures appl. (2)* **11**, 110–130 (1866).

[82] Jordan, C. *Cours d'analyse*, Paris (1893).

[83] Kampen, E. H. van. 'On the connection between the fundamental groups of some related spaces' *Am. J. Math.* **55**, 261–267 (1933).

[84] Kan, D. M. 'Abstract homotopy' *Proc. natn. Acad. Sci. U.S.A.* **41**, 1092–1096 (1955); **42**, 255–258; 419–421; 542–544 (1956).

[85] Kelley, J. L. *General Topology*, Van Nostrand, Princeton, N.J. (1955).

[86] Lefschetz, S. 'Intersections and transformations of complexes and manifolds' *Trans. Am. math. Soc.* **28**, 1–49 (1926).

[87] Lefschetz, S. 'Manifolds with a boundary and their transformations' *Trans. Am. math. Soc.* **29**, 429–462 (1927).

[88] Lefschetz, S. 'The residual set of a complex on a manifold and related questions' *Proc. natn. Acad. Sci. U.S.A.* **13**, 614–622 (1927).

[89] Lefschetz, S. *Topology* (American Mathematical Society Colloquium Publications No. 12), New York (1930).

[90] Lefschetz, S. 'On singular chains and cycles' *Bull. Am. math. Soc.* **39**, 124–129 (1933).

[91] LEFSCHETZ, S. *Algebraic Topology* (American Mathematical Society Colloquium Publications No. 27), New York (1942).

[92] LISTING, J. B. *Der Census räumliche Complexe*, Göttingen (1862).

[93] MAHOWALD, M. E. *On the Metastable Homotopy of S^n* (Memoirs of the American Mathematical Society No. 72), Providence, R.I. (1967).

[94] MAHOWALD, M. E. 'On the order of the image of *J*' *Topology* **6**, 371–378 (1967).

[95] MAHOWALD, M. E. and TANGORA, M. 'Some differentials in the Adams spectral sequence' *Topology* **6**, 349–369 (1967).

[96] MAUNDER, C. R. F. 'Cohomology operations of the *N*th kind' *Proc. Lond. math. Soc.* (*3*) **13**, 125–154 (1963).

[97] MAUNDER, C. R. F. 'On the differentials in the Adams spectral sequence' *Proc. Camb. phil. Soc.* **60**, 409–420 (1964).

[98] MAUNDER, C. R. F. 'On the differentials in the Adams spectral sequence for the stable homotopy groups of spheres' *Proc. Camb. phil. Soc.* **61**, 53–60; 855–868 (1965).

[99] MAUNDER, C. R. F. 'Mod *p* cohomology theories and the Bockstein spectral sequence' *Proc. Camb. phil. Soc.* **63**, 23–43 (1967).

[100] MAUNDER, C. R. F. 'Cohomology operations and duality' *Proc. Camb. phil. Soc.* **64**, 15–30 (1968).

[101] MAY, J. P. 'The cohomology of restricted Lie algebras and of Hopf algebras' *J. Algebra* **3**, 123–146 (1966).

[102] MAY, J. P. 'The cohomology of the Steenrod algebra; stable homotopy groups of spheres' *Bull. Am. math. Soc.* **71**, 377–380 (1965).

[103] MAYER, W. 'Über abstrakte Topologie' *Mh. Math.* **36**, 1–42 (1929).

[104] MILNOR, J. W. 'Spaces having the homotopy type of a CW-complex' *Trans. Am. math. Soc.* **90**, 272–280 (1959).

[105] MILNOR, J. W. 'On the cobordism ring Ω_* and a complex analogue' *Am. J. Math.* **82**, 505–521 (1960).

[106] MILNOR, J. W. 'A procedure for killing homotopy groups of differentiable manifolds' *Proc. Symposia pure Math.* **3**, 39–55 (American Mathematical Society, Providence, R.I., 1961).

[107] MILNOR, J. W. 'Microbundles: I' *Topology* **3**, Supplement, 53–80 (1964).

[108] MOSTERT, P. S. 'Local cross-sections in locally compact groups' *Proc. Am. math. Soc.* **4**, 645–649 (1953).

[109] NEWMAN, M. H. A. 'On the foundations of combinatory analysis situs' *Proc. K. ned. Akad. Wet. Ser. A* **29**, 611–626 (1926).

[110] NOVIKOV, S. P. 'The methods of algebraic topology from the viewpoint of cobordism theories' *Izv. Akad. Nauk SSSR Ser. mat.* **31**, 855–951 (1967).

[111] OLUM, P. 'Obstructions to extensions and homotopies' *Ann. Math.* **25**, 1–50 (1950).

[112] OLUM, P. 'Non-abelian cohomology and van Kampen's theorem' *Ann. Math.* **68**, 658–668 (1958).

[113] Paechter, G. F. 'On the groups $\pi_r(V_{n,m})$' *Q. Jl Math. Oxford* (*2*) **7**, 249–268 (1956); **9**, 8–27 (1958); **10**, 17–37; 241–260 (1959); **11**, 1–16 (1960).

[114] Peterson, F. P. 'Some non-embedding problems' *Boln Soc. mat. mex.* (*2*) **2**, 9–15 (1957).

[115] Peterson, F. P. 'Functional cohomology operations' *Trans. Am. math. Soc.* **86**, 197–211 (1957).

[116] Poincaré, H. 'Analysis situs' *J. Ecole polytech.* (*2*) **1**, 1–121 (1895).

[117] Poincaré, H. 'Second complément à l'analysis situs' *Proc. Lond. math. Soc.* **32**, 277–308 (1900).

[118] Poincaré, H. 'Cinquième complément à l'analysis situs' *Rc. Circ. mat. Palermo* **18**, 45–110 (1904).

[119] Puppe, D. 'Homotopiemengen und ihre induzierten Abbildungen: I' *Math. Z.* **69**, 299–344 (1958).

[120] Radó, T. 'Über den Begriff der Riemannschen Flache' *Acta Sci. Math.* (*Szeged*) **2**, 101–121 (1925).

[121] Rector, D. L. 'An unstable Adams spectral sequence' *Topology* **5** 343–346 (1966).

[122] Reidemeister, K. 'Homotopieringe und Linsenräume' *Abh. math. Semin. Univ. Hamburg* **11**, 102–109 (1935).

[123] Rourke, C. P. and Sanderson, B. J. 'Block bundles' *Ann. Math.* **87** 1–28; 256–278; 431–483 (1968).

[124] Seifert, H. and Threlfall, W. *Lehrbuch der Topologie*, Teubner, Leipzig (1934).

[125] Serre, J.-P. 'Homologie singulière des espaces fibrés' *Ann. Math.* **54**, 425–505 (1951).

[126] Sibson, R. 'Existence theorems for *H*-space inverses' *Proc. Camb. phil. Soc.* **65**, 19–21 (1969).

[127] Smale, S. 'Generalized Poincaré's conjecture in dimensions greater than 4' *Ann. Math.* **74**, 391–406 (1961).

[128] Smale, S. 'A survey of some recent developments in differential topology' *Bull. Am. math. Soc.* **69**, 131–145 (1963).

[129] Spanier, E. H. 'Infinite symmetric products, function spaces and duality' *Ann. Math.* **69**, 142–198 (1959).

[130] Spanier, E. H. 'Quasi-topologies' *Duke math. J.* **30**, 1–14 (1964).

[131] Spanier, E. H. *Algebraic Topology*, McGraw-Hill, New York (1966).

[132] Stallings, J. R. 'Polyhedral homotopy spheres' *Bull. Am. math.. Soc.* **66**, 485–488 (1960).

[133] Stasheff, J. D. 'A classification theorem for fibre spaces' *Topology* **2**, 239–246 (1963).

[134] Stasheff, J. D. 'Homotopy associativity of *H*-spaces' *Trans. Am. math. Soc.* **108**, 275–292; 293–312 (1963).

[135] Steenrod, N. E. 'Products of cocycles and extensions of mappings' *Ann. Math.* **48**, 290–320 (1947).

[136] Steenrod, N. E. 'Cohomology invariants of mappings' *Ann. Math.* **50**, 954–988 (1949).

[137] STEENROD, N. E. *The Topology of Fibre Bundles*, Princeton (1951).
[138] STEENROD, N. E. 'Homology groups of symmetric groups and reduced power operations' *Proc. natn. Acad. Sci. U.S.A.* **39**, 213–223 (1953).
[139] STEENROD, N.E. 'Cohomology operations and obstructions to extending continuous functions' (mimeographed notes), Princeton (1957).
[140] STEENROD, N. E. and EPSTEIN, D. B. A. *Cohomology Operations* (Annals of Mathematics Studies No. 50), Princeton (1962).
[141] Swan, R. *Theory of Sheaves*, Chicago (1964).
[142] THOM, R. 'Espaces fibrés en sphères et carrés de Steenrod' *Annls scient. Éc. norm. sup., Paris* (*3*) **69**, 109–182 (1952).
[143] THOM, R. 'Quelques propriétés globales des variétés differentiables' *Comment. math. helvet.* **28**, 17–86 (1954).
[144] TIETZE, H. 'Über die topologischen Invarianten mehrdimensionaler Mannigfaltigkeiten' *Mh. Math. Phys.* **19**, 1–118 (1908).
[145] TODA, H. *Composition Methods in Homotopy Groups of Spheres* (Annals of Mathematics Studies No. 49), Princeton (1963).
[146] VEBLEN, O. 'Theory of plane curves in non-metrical analysis situs' *Trans. Am. math. Soc.* **6**, 83–98 (1905).
[147] VEBLEN, O. *Analysis situs* (American Mathematical Society Colloquium Publications No. 5, Part II), New York (1922).
[148] VIETORIS, L. 'Über die höheren Zusammenhang kompakter Räume und eine Klasse von zusammenhangstreuen Abbildungen' *Math. Annln* **97**, 454–472 (1927).
[149] VIETORIS, L. 'Über die Homologiegruppen der Vereinigung zweier Komplexe' *Mh. Math.* **37**, 159–162 (1930).
[150] WALL, C. T. C. 'Determination of the cobordism ring' *Ann. Math.* **72**, 292–311 (1960).
[151] WALL, C. T. C. 'Cobordism of combinatorial *n*-manifolds for $n \leq 8$' *Proc. Camb. phil. Soc.* **60**, 807–811 (1964).
[152] WHITEHEAD, G. W. 'On the homotopy groups of spheres and rotation groups' *Ann. Math.* **43**, 634–640 (1942).
[153] WHITEHEAD, G. W. 'A generalization of the Hopf invariant' *Ann. Math.* **51**, 192–237 (1950).
[154] WHITEHEAD, G. W. 'On the Freudenthal theorem' *Ann. Math.* **57**, 209–228 (1953).
[155] WHITEHEAD, G. W. 'Generalized homology theories' *Trans. Am. math. Soc.* **102**, 227–283 (1962).
[156] WHITEHEAD, J. H. C. 'Simplicial spaces, nuclei and *m*-groups' *Proc. Lond. math. Soc.* (*2*) **45**, 243–327 (1939).
[157] WHITEHEAD, J. H. C. 'On adding relations to homotopy groups' *Ann. Math.* **42**, 409–428 (1941).
[158] WHITEHEAD, J. H. C. 'On incidence matrices, nuclei and homotopy types' *Ann. Math.* **42**, 1197–1239 (1941).
[159] WHITEHEAD, J. H. C. 'On the homotopy type of ANR's' *Bull. Am. math. Soc.* **54**, 1133–1145 (1948).

[160] WHITEHEAD, J. H. C. 'Combinatorial homotopy: I' *Bull. Am. math. Soc.* **55**, 213–245 (1949).

[161] WHITEHEAD, J. H. C. 'A certain exact sequence' *Ann. Math.* **52**, 51–110 (1950).

[162] WHITNEY, H. 'Sphere spaces' *Proc. natn. Acad. Sci. U.S.A.* **21**, 462–468 (1935).

[163] WHITNEY, H. 'On products in a complex' *Ann. Math.* **39**, 397–432 (1938).

[164] WILLIAMSON, R. E. 'Cobordism of combinatorial manifolds' *Ann. Math.* **83**, 1–33 (1966).

[165] ZEEMAN, E. C. 'The generalized Poincaré conjecture' *Bull. Am. math. Soc.* **67**, 270 (1961).

[166] ZEEMAN, E. C. 'The Poincaré conjecture for $n \geq 5$' *Topology of 3-Manifolds and Related Topics*, pp. 198–204, Prentice-Hall, Englewood Cliffs, N.J. (1961).

[167] ZEEMAN, E. C. *Seminar on Combinatorial Topology*, Institut des Hautes Études Scientifiques, Paris (1963).

[168] ZEEMAN, E. C. 'On the dunce hat' *Topology* **2**, 341–358 (1963).

[169] ZEEMAN, E. C. 'Relative simplicial approximation' *Proc. Camb. phil. Soc.* **60**, 39–43 (1964).

INDEX

abelian group 6
absolute homotopy extension property 40, 245, 255, 284
abstract
  simplex 37
  simplicial complex 37
abstraction 37
Adams spectral sequence 310, 360
adjunction space 19
Alexander Duality Theorem 182
Alexander–Poincaré Duality Theorem 179
annihilator 12
antipodal map 350
approximation, simplicial 45, 46
association map 220, 251

Barratt–Puppe sequence 238
barycentre 32
barycentric subdivision 49
base
  of open sets 16
  of vector space 11
based
  homotopy 31, 210
  map 31, 210
  set 234
  sets, exact sequence of 234
block 134
  boundary of 134
  dissection 134
  interior of 134
  skeleton 134
  subcomplex 134
Bockstein boundary 148, 154
Borsuk Fixed-Point Theorem 351
boundaries, group of 108
boundary
  of a block 134
  homomorphism 105, 106
  of a manifold 102, 187
  of a simplex 33
Brouwer Fixed-Point Theorem 57

cap product 165, 195, 356
category 61
Cayley numbers 307
Čech
  cohomology 198
  homology 156
n-cell 21
cell (in a CW-complex) 274
cellular
  Approximation Theorem 302
  map 302
  space 278
chain
  complex 107
    simplicial 119
    singular 107
  equivalence 151
  group, singular 105
  homotopy 110
  isomorphism 109
  map 109
    subdivision 115
characteristic map 274
closed
  (homology) manifold 187
  set 15
closure 15
closure-finite 274
cobordism 194, 198, 199
cofibre map 244, 285
cohomology
  associated with an Ω spectrum 331, 359
  Čech 198
  group, 158, 162
    reduced 162
  homomorphism 163
  operation 357, 360
  suspension isomorphism 164, 340
collapse 72
  elementary 72
commutative
  diagram 8
  ring 10

commutator subgroup 6
compact-open topology 214
compact space 16
complex
  abstract simplicial 37
  chain 107
  CW- 61, 257, 273
  derived 49
  geometric simplicial 33
  projective space 305
    cohomology ring of 349
    homology and cohomology groups of 321
    homotopy groups of 307
  semi-simplicial 62
composite (of functions) 2
cone
  mapping 234
  (on a simplicial complex) 152
  reduced 211
conjugate elements 4
connected
  space 15
  sum 101, 197
constant homotopy 25
continuous map 16
contractible 30
  locally 306
convex 11
correspondence, (1-1) 2
coset 4
countable set 2
covering
  map 254
  space 256
    universal 254, 306
cross-section, local 308
cup product 340
CW-complex 61, 257, 273
cycles, group of 108
cyclic group 6

deformation retract(ion) 30
  strong 30
degree 288, 293
derived complex 49
determinant 14
diagonal map 201, 345
diameter 20
dimension
  of a CW-complex 274
  of a simplex 32
  of a simplicial complex 33
  of a vector space 12
direct sum
  of groups 4
  of rings 10
  of vector spaces 11
disconnected space 15
discrete
  space 15
  topology 15
disjoint union 17
double
  of a homology manifold with boundary 190
  unit interval 21
dual
  linear map 12
  space 12
Duality Theorem
  Alexander 182
  Alexander–Poincaré 179
  Lefschetz 192
  Poincaré 181
dunce hat 352

edge-group 74
edge-loop 73
edge-path 73
EHP sequence 256, 310
Eilenberg–MacLane space 304, 334
enumerable set 2
equivalence
  class 2
  relation 2
Euclidean space 15
Euler–Poincaré characteristic 150, 156, 183, 194
evaluation map 217, 251
even permutation 3
exact
  coefficient sequence 148, 154
  cohomology sequence 163
  homology sequence
    of pair 124, 126
    of triple 127
  homotopy sequence
    of fibre map 248

of pair 270
of triple 270
sequence 7
of based sets 234
of chain complexes 125
split 9
Excision Theorem 118, 312
extension product (Ext) 196
extraordinary product 255

face
map 106
of simplex 32
proper 32
fibre 247
bundle 255
map 247
exact homotopy sequence of 248
Hurewicz 255
Serre 253
field 10
finitely-generated group 6
Fixed-Point Theorem
Borsuk 351
Brouwer 57
Lefschetz 150
folding map 205
free
abelian group 6
group 5
product 79
full subcomplex 52
function 2
identity 2
functor 61
fundamental group 63
change of base point in 69

geometric
simplex 31
simplicial complex 33
group 3
abelian 6
cyclic 6
finitely-generated 6
free 5
free abelian 6
of homomorphisms 159
order of a 3
quotient 4
topological 100, 307
trivial 3

Hairy Ball Theorem 131
Ham Sandwich Theorem 352
Hausdorff space 15
homeomorphism 16
of pairs 17
homology
associated with an Ω-spectrum 359
Čech 156
group 108
reduced 108
relative 108
simplicial 119
singular 108
homomorphism 110
manifold 171
orientable 173, 182, 190
with boundary 187
special 190, 193
with coefficients 144
homomorphism 3
group 159
ring 10
homotopy 25
based 31, 201
classes of maps 30, 222
constant 25
equivalence 27
of pairs 27
weak 298
extension property absolute 40, 245, 255, 284
group 228, 258
change of base point in 260, 265
of a one-point union 272
of a product 272
of a retract 305
relative 264
inverse 27
linear 26
of pairs 27
relative 27
semi-constant 56
type 27
Hopf
fibrings 307, 310
invariant 355, 360

H-space 224, 251, 252, 255, 262, 355
associative 224
commutative 231
with inverse 224
H'-space 225
associative 225
commutative 229
with inverse 226
Hurewicz
fibre map 255
homomorphism 322
Theorem 326, 359

ideal 10
identification
map 18, 202
space 18
identity function 2
image (of a function) 2
independent points (of R^n) 31
inner product 14
interior
of a block 134
of a simplex 31
intersection (of sets) 1
inverse
function 2
path 64
isomorphism 3
chain 109
ring 10

J-homomorphism 310
join
of simplicial complexes 39
of spaces 251
reduced 251
Jordan Curve Theorem 184, 199

van **Kampen's** Theorem 80
kernel (of a homomorphism) 3
Klein bottle 59
Kronecker product 164, 356
Künneth formula 355

Lebesgue number 20
Lefschetz
Duality Theorem 192
Fixed-Point Theorem 150
number 149
length (of a vector) 14
Lens space 153, 157
Lie group 310
linear
homotopy 26
map 11
dual 12
linearly dependent (vectors) 11
linearly independent (vectors) 11
link 42
local
cross-section 308
product 253, 307, 308
locally
compact 16
contractible 306
path-connected 254
simply-connected, weakly 254
loop 63
space 216

manifold 88
homology 171
orientable (homology) 173, 182, 190
triangulable 89
with boundary 102, 185
homology 187
map 16
antipodal 350
based, 31, 201
cellular 302
cofibre 244
continuous 16
covering 254
fibre 247
Hurewicz fibre 255
identification 18
linear 11
of pairs 17
simplicial 35
Serre fibre 253
mapping
cone 234
cylinder 246
path-space 240
space 213

matrix 13
  non-singular 13
  trace of 13
Mayer–Vietoris sequence 128, 353
  relative 129
mesh 53
metric 19
  space 19
metrizable 20

neighbourhood 15
  simplicial 42
norm (of vector) 14
normal
  space 15
  subgroup 4

obstruction theory 354
odd permutation 3
one-point union 205
  homotopy groups of 272
  of CW-complexes 281
open set 15
order 3
ordering
  partial 1
  total 1
orientable (homology) manifold 173, 182
  with boundary 190
orthogonal group 308

pair 17
  CW 280
  of spaces 17
  simplicial 33
pairs
  homeomorphism of 17
  map of 17
partial ordering 1
path 63
  component 67, 277
  inverse 64
  product 64
  -space 216
    mapping 240
path-connected space 67
permutation 3
  even 3
  odd 3
Poincaré
  conjecture 98, 103, 330
  duality isomorphism 347
  Duality Theorem 181
pointwise convergence, topology of 254
polyhedron 33
polynomial ring 10
product
  cap 165, 195, 356
  cup 340
  extraordinary 255
  in cohomology 340
  Kronecker 164, 356
  of CW-complexes 282
  of sets 1
  of spaces 17
  path 64
  reduced 207
  slant 356
  smash 207
  Whitehead 252, 309
projective
  space 21, 305
    complex 305
    quaternionic 305
    real 21
pseudocycles 198
Puppe sequence 238

quasifibration 255
quaternionic projective space 305
  cohomology ring of 349
  homology and cohomology 322
  homotopy groups of 307
quotient
  group 4
  of CW-complexes 280
  space
    (topological spaces) 18
    (vector spaces) 11

real projective space 21
  cohomology ring of 348
  homology groups of 140
  homotopy groups of 307

realization (of an abstract simplicial complex) 37, 38
reduced
  cohomology group 162
  cone 211
  homology group 108
  join 251
  product 207
    of CW-complexes 283
  suspension 209
regular space 15
relation
  between sets 1
  equivalence 2
  in a group 6, 7
relative
  homology group 108
  homotopy group 264
  Simplicial Approximation Theorem 55
relatively n-simple 266
restriction (of a function) 2
retract(ion) 30
  deformation 30
  homotopy groups of a 305
  strong deformation 30
ring 10
  commutative 10
  homomorphism 10
  isomorphism 10
  polynomial 10
  with an identity element (with a 1) 10

semi-constant homotopy 56
semi-simplicial complex 62
Serre fibre map 253
n-simple 262
  relatively 266
simplex
  abstract 37
  geometric 31
  interior of 31
  singular 105
  standard 105
simplicial
  approximation 45, 46
    Theorem 53
      relative 55
  chain complex 119
  complex
    abstract 37
    geometric 33
  homology groups 119
  map 35
  neighbourhood 42
  pair 33
simply-connected 71
  weakly locally 254
singular
  chain complex 107
    reduced 107
    relative 107
  chain group 105
  homology group 108
  simplex 105
skeleton
  block 134
  CW 274
  simplicial 33
slant product 356
smash product 207
space
  adjunction 19
  compact 16
  connected 15
  disconnected 15
  discrete 15
  Euclidean 15
  Hausdorff 15
  identification 18
  locally compact 16
  loop 216
  mapping 213
  metric 19
  normal 15
  path 216
  quotient 18
  regular 15
  topological 15
  triangulated 36
  vector 11
spaces, pair of 17
special homology manifold with boundary 190, 193
spectral sequence, Adams 310, 360
Ω-spectrum 331
  cohomology associated with 331, 359
  homology associated with 359
n-sphere 21

split exact sequence 9
standard simplex 105
star 42
  covering 52
Steenrod square 357
Stiefel manifold 308
sub-base 16, 218
subcomplex
  block 134
  CW 276
  finite 276
  full 52
  simplicial 33
subdivision
  barycentric 49
  chain map 115
subgroup 3
  commutator 6
  normal 4
subpolyhedron 33
subring 10
subspace
  topology 15
  (vector spaces) 11
supplement 54, 174
surgery 198
suspension
  isomorphism 130, 314, 315
    cohomology, 164, 333, 340
  of simplicial complex 130
  reduced 209
symplectic group 309

tensor product 141
topological
  group 100, 307
  space 15
topology 15
  compact-open 214
  discrete 15
  of pointwise convergence 254
  subspace 15
  weak 274
torsion product (Tor) 147, 155
torus 59
total ordering 1
trace (of a matrix) 13
transposition 3
tree 76
  maximal 76
triangulated space 36, 306
triangulation 36
triple of spaces 17
trivial group 3

union
  disjoint 17
  of sets 1
  one-point 205
unit interval 21
  double 21
unitary group 309
universal
  Coefficient Theorems, 156, 196, 354
  cover 254
    of a CW-complex 306

vector space 11
  finite-dimensional 11
vertices (of a simplex) 31, 37

weak
  homotopy equivalence 298
  topology 274
weakly locally simply-connected 254
wedge 205
Whitehead
  product 252, 309
  Theorem 300

A CATALOG OF SELECTED

DOVER BOOKS

IN SCIENCE AND MATHEMATICS

A CATALOG OF SELECTED

DOVER BOOKS

IN SCIENCE AND MATHEMATICS

QUALITATIVE THEORY OF DIFFERENTIAL EQUATIONS, V.V. Nemytskii and V.V. Stepanov. Classic graduate-level text by two prominent Soviet mathematicians covers classical differential equations as well as topological dynamics and ergodic theory. Bibliographies. 523pp. 5⅜ × 8½. 65954-2 Pa. $14.95

MATRICES AND LINEAR ALGEBRA, Hans Schneider and George Phillip Barker. Basic textbook covers theory of matrices and its applications to systems of linear equations and related topics such as determinants, eigenvalues and differential equations. Numerous exercises. 432pp. 5⅜ × 8½. 66014-1 Pa. $10.95

QUANTUM THEORY, David Bohm. This advanced undergraduate-level text presents the quantum theory in terms of qualitative and imaginative concepts, followed by specific applications worked out in mathematical detail. Preface. Index. 655pp. 5⅜ × 8½. 65969-0 Pa. $14.95

ATOMIC PHYSICS (8th edition), Max Born. Nobel laureate's lucid treatment of kinetic theory of gases, elementary particles, nuclear atom, wave-corpuscles, atomic structure and spectral lines, much more. Over 40 appendices, bibliography. 495pp. 5⅜ × 8½. 65984-4 Pa. $12.95

ELECTRONIC STRUCTURE AND THE PROPERTIES OF SOLIDS: The Physics of the Chemical Bond, Walter A. Harrison. Innovative text offers basic understanding of the electronic structure of covalent and ionic solids, simple metals, transition metals and their compounds. Problems. 1980 edition. 582pp. 6⅛ × 9¼. 66021-4 Pa. $16.95

BOUNDARY VALUE PROBLEMS OF HEAT CONDUCTION, M. Necati Özisik. Systematic, comprehensive treatment of modern mathematical methods of solving problems in heat conduction and diffusion. Numerous examples and problems. Selected references. Appendices. 505pp. 5⅜ × 8½. 65990-9 Pa. $12.95

A SHORT HISTORY OF CHEMISTRY (3rd edition), J.R. Partington. Classic exposition explores origins of chemistry, alchemy, early medical chemistry, nature of atmosphere, theory of valency, laws and structure of atomic theory, much more. 428pp. 5⅜ × 8½. (Available in U.S. only) 65977-1 Pa. $11.95

A HISTORY OF ASTRONOMY, A. Pannekoek. Well-balanced, carefully reasoned study covers such topics as Ptolemaic theory, work of Copernicus, Kepler, Newton, Eddington's work on stars, much more. Illustrated. References. 521pp. 5⅜ × 8½. 65994-1 Pa. $12.95

PRINCIPLES OF METEOROLOGICAL ANALYSIS, Walter J. Saucier. Highly respected, abundantly illustrated classic reviews atmospheric variables, hydrostatics, static stability, various analyses (scalar, cross-section, isobaric, isentropic, more). For intermediate meteorology students. 454pp. 6⅛ × 9¼. 65979-8 Pa. $14.95

CATALOG OF DOVER BOOKS

THE ELECTROMAGNETIC FIELD, Albert Shadowitz. Comprehensive undergraduate text covers basics of electric and magnetic fields, builds up to electromagnetic theory. Also related topics, including relativity. Over 900 problems. 768pp. 5⅜ × 8¼. 65660-8 Pa. $18.95

FOURIER SERIES, Georgi P. Tolstov. Translated by Richard A. Silverman. A valuable addition to the literature on the subject, moving clearly from subject to subject and theorem to theorem. 107 problems, answers. 336pp. 5⅜ × 8½. 63317-9 Pa. $9.95

THEORY OF ELECTROMAGNETIC WAVE PROPAGATION, Charles Herach Papas. Graduate-level study discusses the Maxwell field equations, radiation from wire antennas, the Doppler effect and more. xiii + 244pp. 5⅜ × 8½. 65678-0 Pa. $6.95

DISTRIBUTION THEORY AND TRANSFORM ANALYSIS: An Introduction to Generalized Functions, with Applications, A.H. Zemanian. Provides basics of distribution theory, describes generalized Fourier and Laplace transformations. Numerous problems. 384pp. 5⅜ × 8½. 65479-6 Pa. $11.95

THE PHYSICS OF WAVES, William C. Elmore and Mark A. Heald. Unique overview of classical wave theory. Acoustics, optics, electromagnetic radiation, more. Ideal as classroom text or for self-study. Problems. 477pp. 5⅜ × 8½. 64926-1 Pa. $12.95

CALCULUS OF VARIATIONS WITH APPLICATIONS, George M. Ewing. Applications-oriented introduction to variational theory develops insight and promotes understanding of specialized books, research papers. Suitable for advanced undergraduate/graduate students as primary, supplementary text. 352pp. 5⅜ × 8½. 64856-7 Pa. $9.95

A TREATISE ON ELECTRICITY AND MAGNETISM, James Clerk Maxwell. Important foundation work of modern physics. Brings to final form Maxwell's theory of electromagnetism and rigorously derives his general equations of field theory. 1,084pp. 5⅜ × 8½. 60636-8, 60637-6 Pa., Two-vol. set $23.90

AN INTRODUCTION TO THE CALCULUS OF VARIATIONS, Charles Fox. Graduate-level text covers variations of an integral, isoperimetrical problems, least action, special relativity, approximations, more. References. 279pp. 5⅜ × 8½. 65499-0 Pa. $8.95

HYDRODYNAMIC AND HYDROMAGNETIC STABILITY, S. Chandrasekhar. Lucid examination of the Rayleigh-Benard problem; clear coverage of the theory of instabilities causing convection. 704pp. 5⅜ × 8¼. 64071-X Pa. $14.95

CALCULUS OF VARIATIONS, Robert Weinstock. Basic introduction covering isoperimetric problems, theory of elasticity, quantum mechanics, electrostatics, etc. Exercises throughout. 326pp. 5⅜ × 8½. 63069-2 Pa. $8.95

DYNAMICS OF FLUIDS IN POROUS MEDIA, Jacob Bear. For advanced students of ground water hydrology, soil mechanics and physics, drainage and irrigation engineering and more. 335 illustrations. Exercises, with answers. 784pp. 6⅛ × 9¼. 65675-6 Pa. $19.95

DE RE METALLICA, Georgius Agricola. The famous Hoover translation of greatest treatise on technological chemistry, engineering, geology, mining of early modern times (1556). All 289 original woodcuts. 638pp. 6¾ × 11.
60006-8 Pa. $18.95

SOME THEORY OF SAMPLING, William Edwards Deming. Analysis of the problems, theory and design of sampling techniques for social scientists, industrial managers and others who find statistics increasingly important in their work. 61 tables. 90 figures. xvii + 602pp. 5⅜ × 8½. 64684-X Pa. $15.95

THE VARIOUS AND INGENIOUS MACHINES OF AGOSTINO RAMELLI: A Classic Sixteenth-Century Illustrated Treatise on Technology, Agostino Ramelli. One of the most widely known and copied works on machinery in the 16th century. 194 detailed plates of water pumps, grain mills, cranes, more. 608pp. 9 × 12.
28180-9 Pa. $24.95

LINEAR PROGRAMMING AND ECONOMIC ANALYSIS, Robert Dorfman, Paul A. Samuelson and Robert M. Solow. First comprehensive treatment of linear programming in standard economic analysis. Game theory, modern welfare economics, Leontief input-output, more. 525pp. 5⅜ × 8½. 65491-5 Pa. $14.95

ELEMENTARY DECISION THEORY, Herman Chernoff and Lincoln E. Moses. Clear introduction to statistics and statistical theory covers data processing, probability and random variables, testing hypotheses, much more. Exercises. 364pp. 5⅜ × 8½. 65218-1 Pa. $10.95

THE COMPLEAT STRATEGYST: Being a Primer on the Theory of Games of Strategy, J.D. Williams. Highly entertaining classic describes, with many illustrated examples, how to select best strategies in conflict situations. Prefaces. Appendices. 268pp. 5⅜ × 8½. 25101-2 Pa. $7.95

CONSTRUCTIONS AND COMBINATORIAL PROBLEMS IN DESIGN OF EXPERIMENTS, Damaraju Raghavarao. In-depth reference work examines orthogonal Latin squares, incomplete block designs, tactical configuration, partial geometry, much more. Abundant explanations, examples. 416pp. 5⅜ × 8¼.
65685-3 Pa. $10.95

THE ABSOLUTE DIFFERENTIAL CALCULUS (CALCULUS OF TENSORS), Tullio Levi-Civita. Great 20th-century mathematician's classic work on material necessary for mathematical grasp of theory of relativity. 452pp. 5⅜ × 8½.
63401-9 Pa. $11.95

VECTOR AND TENSOR ANALYSIS WITH APPLICATIONS, A.I. Borisenko and I.E. Tarapov. Concise introduction. Worked-out problems, solutions, exercises. 257pp. 5⅜ × 8¼. 63833-2 Pa. $8.95

CATALOG OF DOVER BOOKS

TENSOR CALCULUS, J.L. Synge and A. Schild. Widely used introductory text covers spaces and tensors, basic operations in Riemannian space, non-Riemannian spaces, etc. 324pp. 5⅜ × 8¼. 63612-7 Pa. $9.95

A CONCISE HISTORY OF MATHEMATICS, Dirk J. Struik. The best brief history of mathematics. Stresses origins and covers every major figure from ancient Near East to 19th century. 41 illustrations. 195pp. 5⅜ × 8½. 60255-9 Pa. $7.95

A SHORT ACCOUNT OF THE HISTORY OF MATHEMATICS, W.W. Rouse Ball. One of clearest, most authoritative surveys from the Egyptians and Phoenicians through 19th-century figures such as Grassman, Galois, Riemann. Fourth edition. 522pp. 5⅜ × 8½. 20630-0 Pa. $11.95

HISTORY OF MATHEMATICS, David E. Smith. Nontechnical survey from ancient Greece and Orient to late 19th century; evolution of arithmetic, geometry, trigonometry, calculating devices, algebra, the calculus. 362 illustrations. 1,355pp. 5⅜ × 8½. 20429-4, 20430-8 Pa., Two-vol. set $26.90

THE GEOMETRY OF RENÉ DESCARTES, René Descartes. The great work founded analytical geometry. Original French text, Descartes' own diagrams, together with definitive Smith-Latham translation. 244pp. 5⅜ × 8½. 60068-8 Pa. $7.95

THE ORIGINS OF THE INFINITESIMAL CALCULUS, Margaret E. Baron. Only fully detailed and documented account of crucial discipline: origins; development by Galileo, Kepler, Cavalieri; contributions of Newton, Leibniz, more. 304pp. 5⅜ × 8½. (Available in U.S. and Canada only) 65371-4 Pa. $9.95

THE HISTORY OF THE CALCULUS AND ITS CONCEPTUAL DEVELOPMENT, Carl B. Boyer. Origins in antiquity, medieval contributions, work of Newton, Leibniz, rigorous formulation. Treatment is verbal. 346pp. 5⅜ × 8½. 60509-4 Pa. $9.95

THE THIRTEEN BOOKS OF EUCLID'S ELEMENTS, translated with introduction and commentary by Sir Thomas L. Heath. Definitive edition. Textual and linguistic notes, mathematical analysis. 2,500 years of critical commentary. Not abridged. 1,414pp. 5⅜ × 8½. 60088-2, 60089-0, 60090-4 Pa., Three-vol. set $31.85

GAMES AND DECISIONS: Introduction and Critical Survey, R. Duncan Luce and Howard Raiffa. Superb nontechnical introduction to game theory, primarily applied to social sciences. Utility theory, zero-sum games, n-person games, decision-making, much more. Bibliography. 509pp. 5⅜ × 8½. 65943-7 Pa. $12.95

THE HISTORICAL ROOTS OF ELEMENTARY MATHEMATICS, Lucas N.H. Bunt, Phillip S. Jones, and Jack D. Bedient. Fundamental underpinnings of modern arithmetic, algebra, geometry and number systems derived from ancient civilizations. 320pp. 5⅜ × 8½. 25563-8 Pa. $8.95

CALCULUS REFRESHER FOR TECHNICAL PEOPLE, A. Albert Klaf. Covers important aspects of integral and differential calculus via 756 questions. 566 problems, most answered. 431pp. 5⅜ × 8½. 20370-0 Pa. $8.95